Designing Embedded Systems with PIC Microcontrollers

Principles and applications

Tim Wilmshurst

ELSEVIER

AMSTERDAM • BOSTON • HEIDELBERG • LONDON • NEW YORK • OXFORD
PARIS • SAN DIEGO • SAN FRANCISCO • SINGAPORE • SYDNEY • TOKYO
Newnes is an imprint of Elsevier

Newnes

Newnes is an imprint of Elsevier
Linacre House, Jordan Hill, Oxford OX2 8DP, UK
30 Corporate Drive, Suite 400, Burlington, MA 01803, USA

First edition 2007
Reprinted 2007

British Library Cataloguing in Publication Data
Wilmshurst, Tim
  Designing embedded systems with PIC microcontrollers:
  principles and applications
  1. Embedded computer systems – Design and Construction
  2. Microprocessors – Design and Construction
  I. Title
  004.1'6

Library of Congress Control Number: 2006933361

ISBN–13: 978-0-7506-6755-5
ISBN–10: 0-7506-6755-9

For information on all Newnes publications
visit our website at www.newnespress.com

Printed and bound in *Great Britain*

07 08 09 10 10 9 8 7 6 5 4 3 2

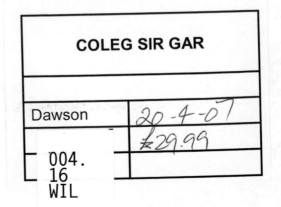

Contents

Introduction   xxi
Acknowledgements   xxv

Section 1   Getting Started with Embedded Systems   1

1   Tiny computers, hidden control   3

1.1   The main idea – embedded systems in today's world   3
    1.1.1   What is an embedded system?   3
1.2   Some example embedded systems   4
    1.2.1   The domestic refrigerator   4
    1.2.2   A car door mechanism   5
    1.2.3   The electronic 'ping-pong'   6
    1.2.4   The Derbot Autonomous Guided Vehicle   7
1.3   Some computer essentials   8
    1.3.1   Elements of a computer   9
    1.3.2   Instruction sets – CISC and RISC   9
    1.3.3   Memory types   10
    1.3.4   Organising memory   10
1.4   Microprocessors and microcontrollers   11
    1.4.1   Microprocessors   11
    1.4.2   Microcontrollers   12
    1.4.3   Microcontroller families   13
    1.4.4   Microcontroller packaging and appearance   14
1.5   Microchip and the PIC microcontroller   15
    1.5.1   Background   15
    1.5.2   PIC microcontrollers today   15
1.6   An introduction to PIC microcontrollers using the 12 Series   17
    1.6.1   The 12F508 architecture   18
1.7   What others do – a Freescale microcontroller   20
Summary   22
References   22

Section 2   Minimum Systems and the PIC® 16F84A   23

2   Introducing the PIC® 16 Series and the 16F84A   25

2.1   The main idea – the PIC 16 Series family   25
    2.1.1   A family overview   25

	2.1.2	The 16F84A	27
	2.1.3	A caution on upgrades	27
2.2		An architecture overview of the 16F84A	27
	2.2.1	The Status register	29
2.3		A review of memory technologies	29
	2.3.1	Static RAM (SRAM)	30
	2.3.2	EPROM (Erasable Programmable Read-Only Memory)	31
	2.3.3	EEPROM (Electrically Erasable Programmable Read-Only Memory)	31
	2.3.4	Flash	31
2.4		The 16F84A memory	32
	2.4.1	The 16F84A program memory	32
	2.4.2	The 16F84A data and Special Function Register memory ('RAM')	33
	2.4.3	The Configuration Word	35
	2.4.4	EEPROM	35
2.5		Some issues of timing	37
	2.5.1	Clock oscillator and instruction cycle	37
	2.5.2	Pipelining	38
2.6		Power-up and Reset	38
2.7		What others do – the Atmel AT89C2051	40
2.8		Taking things further – the 16F84A on-chip reset circuit	41
	Summary		44
	References		44

3	**Parallel ports, power supply and the clock oscillator**		**45**
3.1		The main idea – parallel input/output	46
3.2		The technical challenge of parallel input/output	46
	3.2.1	Building a parallel interface	46
	3.2.2	Port electrical characteristics	49
	3.2.3	Some special cases	49
3.3		Connecting to the parallel port	52
	3.3.1	Switches	52
	3.3.2	Light-emitting diodes	53
3.4		The PIC 16F84A parallel ports	55
	3.4.1	The 16F84A Port B	55
	3.4.2	The 16F84A Port A	55
	3.4.3	Port output characteristics	56
3.5		The clock oscillator	59
	3.5.1	Clock oscillator types	59
	3.5.2	Practical oscillator considerations	60
	3.5.3	The 16F84A clock oscillator	60
3.6		Power supply	61
	3.6.1	The need for power, and its sources	61
	3.6.2	16F84A operating conditions	62
3.7		The hardware design of the electronic ping-pong	63

Summary 64
References 64

4 Starting to program – an introduction to Assembler **65**

4.1 The main idea – what programs do and how we develop them 66
   4.1.1 The problem of programming and the Assembler compromise 66
   4.1.2 The process of writing in Assembler 67
   4.1.3 The program development process 68
4.2 The PIC 16 Series instruction set, with a little more on the ALU 69
   4.2.1 More on the PIC 16 Series ALU 69
   4.2.2 The PIC 16 Series instruction set – an introduction 70
4.3 Assemblers and Assembler format 71
   4.3.1 Introducing Assemblers and the Microchip MPASM^{TM} Assembler 71
   4.3.2 Assembler format 71
   4.3.3 Assembler directives 72
   4.3.4 Number representation 72
4.4 Creating simple programs 73
   4.4.1 A simple data transfer program 73
4.5 Adopting a development environment 76
   4.5.1 Introducing MPLAB 76
   4.5.2 The elements of MPLAB 76
   4.5.3 The MPLAB file structure 77
4.6 An introductory MPLAB tutorial 77
   4.6.1 Creating a project 77
   4.6.2 Entering source code 79
   4.6.3 Assembling the project 80
4.7 An introduction to simulation 81
   4.7.1 Getting started 81
   4.7.2 Generating port inputs 81
   4.7.3 Viewing microcontroller features 82
   4.7.4 Resetting and running the program 82
4.8 Downloading the program to a microcontroller 83
4.9 What others do – a brief comparison of CISC and RISC instruction sets 86
4.10 Taking things further – the 16 Series instruction set format 87
Summary 88
References 88

5 Building Assembler programs **89**

5.1 The main idea – building structured programs 89
   5.1.1 Flow diagrams 89
   5.1.2 State diagrams 91
5.2 Flow control – branching and subroutines 92
   5.2.1 Conditional branching and working with bits 92
   5.2.2 Subroutines and the Stack 94

5.3	Generating time delays and intervals	95
5.4	Dealing with data	97
	5.4.1 Indirect addressing and the File Select Register	97
	5.4.2 Look-up tables	98
	5.4.3 Example program with delays and look-up table	99
5.5	Introducing logical instructions	101
5.6	Introducing arithmetic instructions and the Carry flag	102
	5.6.1 Using add instructions	102
	5.6.2 Using subtract instructions	102
	5.6.3 An arithmetic program example	102
	5.6.4 Using indirect addressing to save the Fibonacci series	104
5.7	Taming Assembler complexity	106
	5.7.1 Include Files	106
	5.7.2 Macros	107
	5.7.3 MPLAB special instructions	108
5.8	More use of the MPLAB simulator	109
	5.8.1 Breakpoints	109
	5.8.2 Stopwatch	110
	5.8.3 Trace	110
5.9	The ping-pong program	112
	5.9.1 A structure for the ping-pong program	112
	5.9.2 Exploring the ping-pong program code	115
5.10	Simulating the ping-pong program – tutorial	116
	5.10.1 Setting up input stimulus	116
	5.10.2 Setting up the Watch window	116
	5.10.3 Single stepping	116
	5.10.4 Animate	117
	5.10.5 Run	117
	5.10.6 Breakpoints	117
	5.10.7 Stopwatch	117
	5.10.8 Trace	117
	5.10.9 Debugging the full program	118
5.11	What others do – graphical simulators	118
	Summary	119
	References	119
6	**Working with time: interrupts, counters and timers**	**120**
6.1	The main idea – interrupts	121
	6.1.1 Interrupt structures	121
	6.1.2 The 16F84A interrupt structure	122
	6.1.3 The CPU response to an interrupt	124
6.2	Working with interrupts	125
	6.2.1 Programming with a single interrupt	125
	6.2.2 Moving to multiple interrupts – identifying the source	126

	6.2.3	Stopping interrupts from wrecking your program 1 – context saving	127
	6.2.4	Stopping interrupts from wrecking your program 2 – critical regions and masking	130
6.3	The main idea – counters and timers		131
	6.3.1	The digital counter reviewed	131
	6.3.2	The counter as timer	132
	6.3.3	The 16F84A Timer 0 module	134
6.4	Applying the 16F84A Timer 0, with examples using the electronic ping-pong		136
	6.4.1	Object or event counting	136
	6.4.2	Hardware-generated delays	137
6.5	The Watchdog Timer		138
6.6	Sleep mode		139
6.7	What others do		140
6.8	Taking things further – interrupt latency		141
	Summary		142

Section 3   Larger Systems and the PIC® 16F873A — **143**

7   Larger systems and the PIC® 16F873A — **145**

7.1	The main idea – the PIC 16F87XA		146
7.2	The 16F873A block diagram and CPU		146
	7.2.1	Overview of CPU and core	146
	7.2.2	Overview of memory	147
	7.2.3	Overview of peripherals	150
7.3	16F873A memory and memory maps		150
	7.3.1	The 16F873A program memory	150
	7.3.2	The 16F873A data memory and Special Function Registers	152
	7.3.3	The Configuration Word	154
7.4	'Special' memory operations		155
	7.4.1	Accessing EEPROM and program memory	155
	7.4.2	In-Circuit Serial Programming (ICSP™)	156
7.5	The 16F873A interrupts		158
	7.5.1	The interrupt structure	158
	7.5.2	The interrupt registers	159
	7.5.3	Interrupt identification and context saving	161
7.6	The 16F873A oscillator, reset and power supply		161
	7.6.1	The clock oscillator	161
	7.6.2	Reset and power supply	161
7.7	The 16F873A parallel ports		161
	7.7.1	The 16F873A Port A	163
	7.7.2	The 16F873A Port B	164
	7.7.3	The 16F873A Port C	164

7.8	Test, commission and diagnostic tools	165
	7.8.1 The challenge of testing an embedded system	165
	7.8.2 Oscilloscopes and logic analysers	167
	7.8.3 In-circuit emulators	170
	7.8.4 On-chip debuggers	170
7.9	The Microchip in-circuit debugger (ICD 2)	171
7.10	Applying the 16F873A: the Derbot AGV	172
	7.10.1 Power supply, oscillator and reset	172
	7.10.2 Use of the parallel ports	173
	7.10.3 Assembling the hardware	174
7.11	Downloading, testing and running a simple program with ICD 2	176
	7.11.1 A first Derbot program	176
	7.11.2 Applying the ICD 2	178
	7.11.3 Setting the configuration bits within the program	179
7.12	Taking things further – the 16F874A/16F877A Ports D and E	180
Summary		182
References		183
8	**The human and physical interfaces**	**184**
8.1	The main idea – the human interface	184
8.2	From switches to keypads	187
	8.2.1 The keypad	187
	8.2.2 Design example: use of keypad in Derbot hand controller	188
8.3	LED displays	193
	8.3.1 LED arrays: seven-segment displays	193
	8.3.2 Design example: the Derbot hand controller seven-segment display	194
8.4	Liquid crystal displays	199
	8.4.1 The HD44780 LCD driver and its derivatives	199
	8.4.2 Design example: use of LCD display in Derbot hand controller	200
8.5	The main idea – interfacing to the physical world	203
8.6	Some simple sensors	203
	8.6.1 The microswitch	204
	8.6.2 Light-dependent resistors	204
	8.6.3 Optical object sensing	205
	8.6.4 The opto-sensor applied as a shaft encoder	205
	8.6.5 Ultrasonic object sensor	207
8.7	More on digital input	207
	8.7.1 16F873A input characteristics	207
	8.7.2 Ensuring legal logic levels, and input protection	208
	8.7.3 Switch debouncing	212

8.8	Actuators: motors and servos	212
	8.8.1 DC and stepper motors	212
	8.8.2 Angular positioning: the 'servo'	214
8.9	Interfacing to actuators	215
	8.9.1 Simple DC switching	215
	8.9.2 Simple switching on the Derbot	217
	8.9.3 Reversible switching: the H-bridge	218
	8.9.4 Motor switching on the Derbot	220
8.10	Building up the Derbot	220
8.11	Applying sensors and actuators – a 'blind' navigation Derbot program	222
	Summary	223
	References	223

9	**Taking timing further**	**225**
9.1	The main ideas – taking counting and timing further	225
9.2	The 16F87XA Timer 0 and Timer 1	226
	9.2.1 Timer 0	226
	9.2.2 Timer 1	226
	9.2.3 Application of Timer 0 and Timer 1 as counters for Derbot odometry	228
	9.2.4 Using Timer 0 and Timer 1 to generate repetitive interrupts	231
9.3	The 16F87XA Timer 2, comparator and PR2 register	232
	9.3.1 Timer 2	232
	9.3.2 The PR2 register, comparator and postscaler	234
9.4	The capture/compare/PWM (CCP) modules	235
	9.4.1 A capture/compare/PWM overview	235
	9.4.2 Capture mode	235
	9.4.3 Compare mode	237
9.5	Pulse width modulation	237
	9.5.1 The principle of PWM	237
	9.5.2 Generating PWM signals in hardware – the 16F87XA PWM	239
	9.5.3 PWM applied in the Derbot for motor control	241
9.6	Generating PWM in software	244
	9.6.1 An example of software-generated PWM	245
	9.6.2 Further Assembler directives for memory definition and branching	248
9.7	PWM used for digital-to-analog conversion	249
	9.7.1 An example of PWM used for digital-to-analog conversion	249
9.8	Frequency measurement	252
	9.8.1 The principle of frequency measurement	252
	9.8.2 Frequency (speed) measurement in the Derbot	252
9.9	Speed control applied to the Derbot	255
9.10	Where there is no timer	258
9.11	Sleep mode	260
9.12	Where do we go from here?	261
9.13	Building up the Derbot	262

| | | Summary | 262 |
| | | References | 262 |

10   Starting with serial   **263**

10.1	The main idea – introducing serial	263
10.2	Simple serial links – synchronous data communication	265
	10.2.1 Synchronous basics	265
	10.2.2 Implementing synchronous serial I/O in the microcontroller	266
	10.2.3 Microwire and SPI (Serial Peripheral Interface)	266
	10.2.4 Introducing multiple nodes	267
10.3	The 16F87XA Master Synchronous Serial Port (MSSP) module in SPI mode	267
	10.3.1 Port overview	268
	10.3.2 Port configuration	268
	10.3.3 Setting the clock	270
	10.3.4 Managing data transfer	271
10.4	A simple SPI example	273
10.5	The limitations of Microwire and SPI, and of simple synchronous serial transfer	275
10.6	Enhancing synchronous serial, and the Inter-Integrated Circuit bus	275
	10.6.1 Main I^2C features and physical interconnection	275
	10.6.2 The pull-up resistor	275
	10.6.3 I^2C signal characteristics	276
10.7	The MSSP configured for I^2C	277
	10.7.1 The MSSP I^2C registers and their preliminary use	277
	10.7.2 The MSSP in I^2C Slave mode	281
	10.7.3 The MSSP in I^2C Master mode	283
10.8	I^2C applied in the Derbot AGV	286
	10.8.1 The Derbot hand controller as a serial node	286
	10.8.2 The AGV as an I^2C master	286
	10.8.3 The hand controller as an I^2C slave	290
	10.8.4 Evaluation of the Derbot I^2C programs	292
10.9	Evaluation of synchronous serial data communication and an introduction to asynchronous	293
	10.9.1 Asynchronous principles	293
	10.9.2 Synchronising serial data – without an incoming clock	293
10.10	The 16F87XA Addressable Universal Synchronous Asynchronous Receiver Transmitter (USART)	295
	10.10.1 Port overview	295
	10.10.2 The USART asynchronous transmitter	295
	10.10.3 The USART baud rate generator	298
	10.10.4 The USART asynchronous receiver	299
	10.10.5 An asynchronous example	300
	10.10.6 Using address detection with the USART receive mode	302
	10.10.7 The USART in synchronous mode	302

10.11 Implementing serial without a serial port – 'bit banging'   303
10.12 Building up the Derbot   303
Summary   303
References   303

11   Data acquisition and manipulation   304

11.1   The main idea – analog and digital quantities, their
       acquisition and use   304
11.2   The data acquisition system   305
       11.2.1   The analog-to-digital converter   306
       11.2.2   Signal conditioning – amplification and filtering   308
       11.2.3   The analog multiplexer   308
       11.2.4   Sample and hold, and acquisition time   309
       11.2.5   Timing and microprocessor control   310
       11.2.6   Data acquisition in the microcontroller environment 311
11.3   The PIC® 16F87XA ADC module   312
       11.3.1   Overview and block diagram   312
       11.3.2   Controlling the ADC   313
       11.3.3   The analog input model   317
       11.3.4   Calculating acquisition time   318
       11.3.5   Repeated conversions   319
       11.3.6   Trading off conversion speed and resolution   319
11.4   Applying the ADC in the Derbot light meter program   319
       11.4.1   Configuration of the ADC   319
       11.4.2   Acquisition time   320
       11.4.3   Data conversion   321
11.5   Some simple data manipulation techniques   321
       11.5.1   Fixed- and floating-point arithmetic   322
       11.5.2   Binary to Binary Coded Decimal conversion   323
       11.5.3   Multiplication   324
       11.5.4   Scaling and the Derbot light meter example   324
       11.5.5   Using the voltage reference for scaling   326
11.6   The Derbot light-seeking program   326
11.7   The comparator module   327
       11.7.1   Review of comparator action   327
       11.7.2   The 16F87XA comparators and voltage reference   329
11.8   Applying the Derbot circuit for measurement purposes   329
       11.8.1   The electronic tape measure   329
       11.8.2   The light meter   331
       11.8.3   The voltmeter   331
       11.8.4   Other measurement systems   331
11.9   Configuring the Derbot AGV as a light-seeking robot   332
Summary   332
References   332

Section 4   Smarter Systems and the PIC® 18FXX2   333

12   Smarter systems and the PIC® 18FXX2   **335**

12.1   The main idea – the PIC 18 Series and the 18FXX2   336
12.2   The 18F2X2 block diagram and Status register   337
12.3   The 18 Series instruction set   340
    12.3.1   Instructions which are unchanged   344
    12.3.2   Instructions which have been upgraded   344
    12.3.3   New, variant, instructions   345
    12.3.4   New instructions   345
12.4   Data memory and Special Function Registers   345
    12.4.1   The data memory map   345
    12.4.2   Access RAM   347
    12.4.3   Indirect addressing and accessing tables in data memory   347
12.5   Program memory   347
    12.5.1   The program memory map   349
    12.5.2   The Program Counter   349
    12.5.3   Upgrading from the 16 Series and computed *goto* instructions   349
    12.5.4   The Configuration registers   350
12.6   The Stacks   352
    12.6.1   Automatic Stack operations   352
    12.6.2   Programmer access to the Stack   352
    12.6.3   The Fast Register Stack   352
12.7   The interrupts   353
    12.7.1   An interrupt structure overview   353
    12.7.2   The interrupt sources, their enabling and prioritisation   353
    12.7.3   Overall interrupt prioritisation enabling   354
    12.7.4   Global enabling   354
    12.7.5   Other aspects of the interrupt logic   355
    12.7.6   The Interrupt registers   355
    12.7.7   Context saving with interrupts   356
12.8   Power supply and reset   358
    12.8.1   Power supply   358
    12.8.2   Power-up and Reset   358
12.9   The oscillator sources   360
    12.9.1   LP, XT, HS and RC oscillator modes   362
    12.9.2   EC, ECIO and RCIO oscillator modes   363
    12.9.3   **HS + PLL** oscillator mode   363
    12.9.4   Clock source switching   363
12.10   Introductory programming with the 18F242   364
    12.10.1   Using the MPLAB IDE for the 18 Series   364
    12.10.2   The Fibonacci program   365
Summary   367
References   367

13 The PIC® 18FXX2 peripherals   **368**

13.1   The main idea – the 18FXX2 peripherals   368
13.2   The parallel ports   369
       13.2.1 The 18FXX2 Port A   369
       13.2.2 The 18FXX2 Port B   369
       13.2.3 The 18FXX2 Port C   371
       13.2.4 The parallel slave port   371
13.3   The timers   371
       13.3.1 Timer 0   371
       13.3.2 Timer 1   373
       13.3.3 Timer 2   373
       13.3.4 Timer 3   373
       13.3.5 The Watchdog Timer   376
13.4   The capture/compare/PWM (CCP) modules   376
       13.4.1 The control registers   376
       13.4.2 Capture mode   376
       13.4.3 Compare mode   377
       13.4.4 Pulse width modulation   378
13.5   The serial ports   378
       13.5.1 The MSSP in SPI mode   379
       13.5.2 The MSSP in I²C mode   379
       13.5.3 The USART   380
13.6   The analog-to-digital converter (ADC)   380
13.7   Low-voltage detect   380
13.8   Applying the 18 Series in the Derbot-18   382
13.9   The 18F2420 and the extended instruction set   383
       13.9.1 Nanowatt technology   383
       13.9.2 The extended instruction set   384
       13.9.3 Enhanced peripherals   384
Summary   385
Reference   385

14 Introducing C   **386**

14.1   The main idea – why C?   387
14.2   An introduction to C   387
       14.2.1 A little history   387
       14.2.2 A first program   388
       14.2.3 Laying out the program – declarations, statements, comments and space   388
       14.2.4 C keywords   390
       14.2.5 The C function   391
       14.2.6 Data type and storage   392
       14.2.7 C operators   392
       14.2.8 Control of program flow, and the **while** keyword   393

14.2.9	The C preprocessor and its directives	394
14.2.10	Use of libraries, and the Standard Library	394
14.3	Compiling the C program	394
14.4	The MPLAB C18 compiler	395
14.4.1	Specification of radix	396
14.4.2	Arithmetic operations	396
14.5	A C18 tutorial	396
14.5.1	The Linker and Linker Scripts	396
14.5.2	Linking header and library files	397
14.5.3	Building the project	397
14.5.4	Project files	398
14.6	Simulating a C program	400
14.7	A second C example – the Fibonacci program	401
14.7.1	Program preliminaries – more on declaring variables	402
14.7.2	The **do–while** construct	403
14.7.3	Labels and the **goto** keyword	403
14.7.4	Simulating the Fibonacci program	403
14.8	The MPLAB C18 libraries	403
14.8.1	Hardware peripheral functions	404
14.8.2	The software peripheral library	404
14.8.3	The general software library	405
14.8.4	The maths library	406
14.9	Further reading	406
Summary		407
References		407

15	**C and the embedded environment**	**409**
15.1	The main idea – adapting C to the embedded environment	409
15.2	Controlling and branching on bit values	409
15.2.1	Controlling individual bits	411
15.2.2	The **if** and **if–else** conditional branch structures	411
15.2.3	Setting the configuration bits	412
15.2.4	Simulating and running the example program	412
15.3	More on functions	413
15.3.1	The function prototype	413
15.3.2	The function definition	414
15.3.3	Function calls and data passing	414
15.3.4	Library delay functions, and **Delay10KTCYx( )**	415
15.4	More branching and looping	415
15.4.1	Using the **break** keyword	415
15.4.2	Using the **for** keyword	416
15.5	Using the timer and PWM peripherals	417
15.5.1	Using the timer peripherals	420

	15.5.2 Using PWM	421
	15.5.3 The main program loop	421
	Summary	422

16   Acquiring and using data with C   **423**

16.1	The main idea – using C for data manipulation	423
16.2	Using the 18FXX2 ADC	423
	16.2.1 The light-seeking program structure	427
	16.2.2 Use of the ADC	428
	16.2.3 Further use of **if–else**	429
	16.2.4 Simulating the light-seeking program	429
16.3	Pointers, arrays and strings	431
	16.3.1 Pointers	431
	16.3.2 Arrays	432
	16.3.3 Using pointers with arrays	432
	16.3.4 Strings	433
	16.3.5 An example program: using pointers, arrays and strings	433
	16.3.6 A word on evaluating the **while** condition	434
	16.3.7 Simulating the program example	435
16.4	Using the I^2C peripheral	437
	16.4.1 An example I^2C program	437
	16.4.2 Use of $++$ and $--$ operators	439
16.5	Formatting data for display	440
	16.5.1 Overview of example program	440
	16.5.2 Using library functions for data formatting	442
	16.5.3 Program evaluation	442
	Summary	443

17   More C and the wider C environment   **444**

17.1	The main idea – more C and the wider C environment	444
17.2	Assembler inserts	445
17.3	Controlling memory allocation	446
	17.3.1 Memory allocation pragmas	447
	17.3.2 Setting the Configuration Words	447
17.4	Interrupts	448
	17.4.1 The Interrupt Service Routine	448
	17.4.2 Locating and identifying the ISR	449
17.5	Example with interrupt on overflow – flashing LEDs on the Derbot	449
	17.5.1 Using Timer 0	450
	17.5.2 Using interrupts, and the ISR action	451
	17.5.3 Simulating the flashing LEDs program	452
17.6	Storage classes and their application	453
	17.6.1 Storage classes	453

17.6.2 Scope   454
17.6.3 Duration   454
17.6.4 Linkage   455
17.6.5 Working with 18 Series memory   455
17.6.6 Storage class examples   455
17.7 Start-up code: **c018i.c**   456
17.7.1 The C18 start-up files   456
17.7.2 The **c018i.c** structure   457
17.7.3 Simulating **c018i.c**   457
17.8 Structures, unions and bit-fields   459
17.9 Processor-specific header files   460
17.9.1 SFR definitions   460
17.9.2 Assembler utilities in the header file   461
17.10 Taking things further – the MPLAB Linker and the .map file 462
17.10.1 What the Linker does   462
17.10.2 The Linker Script   462
17.10.3 The **.map** file   464
Summary   465
References   465

18 Multi-tasking and the Real Time Operating System   466

18.1 The main ideas – the challenge of multi-tasking
      and real time   466
18.1.1 Multi-tasking – tasks, priorities and deadlines   467
18.1.2 So what is 'real time'?   468
18.2 Achieving multi-tasking with sequential programming   469
18.2.1 Evaluating the super loop   469
18.2.2 Time-triggered and event-triggered tasks   469
18.2.3 Using interrupts for prioritisation – the foreground/background structure 469
18.2.4 Introducing a 'clock tick' to synchronise program activity 470
18.2.5 A general-purpose 'operating system'   471
18.2.6 The limits of sequential programming when multi-tasking   471
18.3 The Real Time Operating System (RTOS)   472
18.4 Scheduling and the scheduler   473
18.4.1 Cyclic scheduling   473
18.4.2 Round robin scheduling and context switching   473
18.4.3 Task states   474
18.4.4 Prioritised pre-emptive scheduling   475
18.4.5 Cooperative scheduling   476
18.4.6 The role of interrupts in scheduling   477
18.5 Developing tasks   477
18.5.1 Defining tasks   477
18.5.2 Writing tasks and setting priority   478

18.6 Data and resource protection – the semaphore   478
18.7 Where do we go from here?   479
Summary   479
References   479

19   The Salvo™ Real Time Operating System   **480**

19.1 The main idea – Salvo, an example RTOS   480
      19.1.1 Basic Salvo features   480
      19.1.2 Salvo versions and references   481
19.2 Configuring the Salvo application   482
      19.2.1 Building Salvo applications – the library build   482
      19.2.2 Salvo libraries   482
      19.2.3 Using Salvo with C18   483
19.3 Writing Salvo programs   483
      19.3.1 Initialisation and scheduling   484
      19.3.2 Writing Salvo tasks   485
19.4 A first Salvo example   485
      19.4.1 Program overview and the **main** function   487
      19.4.2 Tasks and scheduling   488
      19.4.3 Creating a Salvo/C18 project   488
      19.4.4 Setting the configuration file   489
      19.4.5 Building the Salvo example   489
      19.4.6 Simulating the Salvo program   490
19.5 Using interrupts, delays and semaphores with Salvo   491
      19.5.1 An example program using an interrupt based clock tick   492
      19.5.2 Selecting the library and configuration   494
      19.5.3 Using interrupts and establishing the clock tick   494
      19.5.4 Using delays   496
      19.5.5 Using a binary semaphore   496
      19.5.6 Simulating the program   497
      19.5.7 Running the program   499
19.6 Using Salvo messages and increasing RTOS complexity   499
19.7 A program example with messages   500
      19.7.1 Selecting the library and configuration   505
      19.7.2 The task: **USnd_ Task**   505
      19.7.3 The task: **Motor_ Task**   505
      19.7.4 The use of messages   506
      19.7.5 The use of interrupts, and the ISRs   507
      19.7.6 Simulating or running the program   509
19.8 The RTOS overhead   509
Summary   510
References   510

Section 5   Techniques of Connectivity and Networking   **511**

20   Connectivity and networks   **513**

20.1   The main idea – networking and connectivity   513
    20.1.1 A word on protocols   514
20.2   Infrared connectivity   515
    20.2.1 The IrDA and the PIC microcontroller   515
20.3   Radio connectivity   516
    20.3.1 Bluetooth   516
    20.3.2 Zigbee   517
    20.3.3 Zigbee and the PIC microcontroller   517
20.4   Controller Area Network (CAN) and Local Interconnect Network (LIN) 518
    20.4.1 Controller Area Network (CAN)   518
    20.4.2 CAN and the PIC microcontroller   520
    20.4.3 Local Interconnect Network (LIN)   520
    20.4.4 LIN and the PIC microcontroller   521
20.5   Embedded systems and the Internet   522
    20.5.1 Connecting to the Internet with the PIC microcontroller   523
20.6   Conclusion   523
Summary   524
References   524

Appendix 1   The PIC® 16 Series instruction set   527
Appendix 2   The electronic ping-pong   528
Appendix 3   The Derbot AGV – hardware design details   533
Appendix 4   Some basics of Autonomous Guided Vehicles   537
Appendix 5   PIC® 18 Series instruction set (non-extended)   541
Appendix 6   Essentials of C   544

Index   549

Introduction

This is a book about embedded systems, introduced primarily through the application of three PIC® micro-controllers. Starting from an introductory level, the book aims to make the reader into a competent and independent practitioner in the field of embedded systems, to a level whereby he or she has the skills necessary to gain entry to professional practice in the embedded world.

The book achieves its aims by developing the underlying knowledge and skills appropriate to today's embedded systems, in both hardware and software development. On the hardware side, it includes in-depth study both of microcontroller design, and of the circuits and transducers to which the microcontroller must interface. On the software side, programming in both Assembler and C is covered. This culminates in the study and application of a Real Time Operating System, representing the most elegant way that an embedded system can be programmed.

The book is divided into introductory and concluding sections and three main parts, and develops its themes primarily around three example PIC microcontrollers, which form the basis of each part. These are the 16F84A, the 16F873A and the 18F242. It works through these in turn, using each to develop the sophistication of the ideas introduced. Nevertheless, the book should not be viewed just as a manual on PIC microcontrollers. Using these as the medium of study, the main issues of embedded design are explored. The skills and knowledge acquired through the study of this set of microcontrollers can readily be transferred to others.

A distinctive feature of the book is its combination of practical and theoretical. The vast majority of topics are directly illustrated by practical application, in hardware or in program simulation. Thus, at no point is there abstract theory presented without application. The main project in the book is the Derbot AGV (Autonomous Guided Vehicle). This is a customisable design, which can be used as a self-contained development platform. As an AGV it can be developed into many different forms. It can also be adapted into plenty of other things as well, for example a waveform generator, an electronic tape measure or a light meter. Before the Derbot is introduced, use is made of a very simple project, the electronic ping-pong game. The example projects can be built by the reader, with design information being given on the book CD. Alternatively, projects can simply be used as theoretical case studies.

This book is aimed primarily at second- or third-year undergraduate engineering or technology students. It will also be of interest to the informed hobbyist, and parts to the practising professional. Readers are expected to have a reasonable knowledge of electronics, equivalent to, say, a first-year undergraduate course. This will include an understanding of the operation of transistors and diodes, and simple analog and digital electronic subsystems. It is also beneficial to have some knowledge of computer architecture, for example gained by an introductory course on microprocessors.

Because the book moves in three distinct stages from the introductory to the advanced, it will in general provide material for more than one course or module. The first six chapters can be used for a short and self-contained one-semester course, covering an introduction to microcontrollers and their programming in Assembler. The 16F84A is chosen as the example for these chapters. It is an excellent introductory microcontroller, due to its simplicity. Chapters 7–11 can form an intermediate course, using Assembler to program more complex systems. This leads to a detailed knowledge of microcontroller peripherals and their use, as exemplified by the 16F873A. Chapters 12–20 can then be used to form an advanced course, working with C and the 18F242, and leading up to use of the RTOS. Alternatively, lecturers may wish to 'pick and choose' in Chapters 7–20, depending on their preference for C or Assembler, and their preference for the microcontroller used. Having worked through Chapters 1–6, it is just possible to go directly to Chapter 12, thereby apparently skipping Chapters 7–11. The detail of the middle chapters is missed, but this approach can also work. Using C demands less detailed knowledge of the peripherals than is required if using Assembler, and cross-reference is made to the middle chapters where it is needed.

Whatever sequence of reading is chosen, the reader is expected as a minimum to have ready access to the Microchip MPLAB® Integrated Development Environment, which is available on the book CD. This allows the example programs in the book to be simulated and then modified and developed. Almost inevitably the book starts with some study of hardware, so that the reader has a basic knowledge of the system that the software will run on. To some extent the first few chapters, on PIC microcontroller architecture, represent a steep learning curve for the beginner. The fun then starts in Chapter 4, when programming and simulation can begin. From here, with the foundations laid, hardware and software run more or less in parallel, each gaining in sophistication and complementing the other. For the final third of the book, the Microchip C18 C compiler should be used. The student version of this is also available on the book CD. For Chapter 19, the 'Lite' version of the Salvo™ RTOS can be installed, again from the book CD.

Beyond program simulation, it is hoped that the reader has access to electronic build and test facilities, whether at home, college, university or workplace. With these, it is possible to build up some of the example project material or work on equivalent systems. By so doing, the satisfaction of actually implementing real embedded systems will be achieved. When working through the middle or later chapters, the best thing a lecturer or instructor can do is to get a Derbot printed circuit board into the hands of every student on the course, along with a basic set of components. Guide them through initial development and then give them suggestions for further customisation. It is wonderful what ideas they then come up with. Design details are on the book CD.

An essential skill of any professional designer in this field is the ability to work with the manufacturer's data sheets. These are the main source of information when designing with microcontrollers and the ultimate point of reference in the professional world. It is in general *not* desirable to work from intermediate drawings by a third party, even if these are meant to simplify the information. Therefore, this book unashamedly uses (with permission) a large number of diagrams straight from the Microchip data sheets. Many are made more accessible by the inclusion of supplementary labelling. The reader is encouraged to download the full version of the data sheet in use and to refer directly to it.

A complete knowledge of the field of embedded systems requires both breadth and depth. This is particularly true of embedded systems, which combine elements of hardware and software, semiconductor technology, analog and digital electronics, computer architecture, sensors and actuators, and more. With its focus on the PIC microcontroller, this book cannot cover all these areas. For the wider contextual background, the author's earlier book, *An Introduction to the Design of Small-Scale Embedded Systems*, is recommended. With whole chapters on memory technology, power supply, numerical algorithms, interfacing to transducers and the design process, it provides a ready complement to this book.

I hope that you enjoy working through this book. In particular I hope you go on to enjoy the challenge and pleasure of designing and building embedded systems.

Tim Wilmshurst
University of Derby, UK

Acknowledgements

Certain materials contained herein are reprinted with permission of the copyright holder, Microchip Technology Incorporated. All rights reserved. No further reprints or reproductions may be made without Microchip Technology Inc.'s prior written consent.

PIC®, PICSTART® and MPLAB® are all registered trademarks of Microchip Technology Inc. PICBASIC™, PICBASIC PRO™, ECAN™, In-Circuit Serial Programming™, ICSP™, MPASM™, MPLIB™, MPLINK™, MPSIM™ and PICDEM.net™ are all trademarks of Microchip Technology Inc.

Figures 1.11, 1.13, 2.2–2.10, 3.8, 3.10–3.12, 3.14–3.16, 4.4, 4.13, 6.2, 6.3, 6.8–6.10, 7.1–7.4, 7.6, 7.7, 7.9–7.11, 7.14–7.16, 7.25, 7.26, 8.7, 9.1, 9.2, 9.4, 9.5, 9.7 9.9, 9.11, 10.7 10.9, 10.14 10.21, 10.25 10.28, 11.6–11.10, 11.15, 12.1–12.10, 12.13, 12.14 and 13.1–13.11 are taken from Microchip Data Sheets: PIC12F508/509/16F505 (DS41236A); PIC16F84A (DS35007B); PIC16F87XA (DS39582B); PIC18FXX2 (DS39564B) and the PIC micro™ Mid-Range MCU Family Reference Manual (DS31004A, DS31005A) and are reproduced by kind permission of Microchip Technology Inc.

Grateful acknowledgement is made to David Manley and Mike Vernon, for allowing me the sabbatical leave which made this project possible. Thanks to many students who have designed their own AGVs and contributed ideas to the Derbot project, including Jonathan Guinet, David Coterill-Drew, Grigorios Dedes and Kelvin Brammer. Thanks also to Naoko Evans and Nick Roberts for reading and commenting on sections of the script. Thanks to Trevor Noble, who has, over the years, with skill and enthusiasm, made many an embedded system, including a series of Derbot prototypes. Thanks to staff at Microchip Technology, who have answered numerous questions, both technical and on copyright and related issues, and who freely gave permission for much copyright material to be reproduced. Similar thanks is due to Pumpkin Inc., the authors of the Salvo™ Operating System, for their technical support and permission to place Salvo 'Lite' on the book CD. Salvo™ is a trademark of Pumpkin Inc.

Thanks especially to my family, Beate, Imogen, Jez and Naomi, for supporting this project for the 15 months in which it has been a significant part of our lives. They are the joy of my life, and this book is dedicated to them.

Section 1
Getting Started with
Embedded Systems

This introductory chapter introduces embedded systems and the microcontroller, leading to a survey of the Microchip range of PIC® microcontrollers.

1
Tiny computers, hidden control

We are living in the age of information revolution, with computers of astonishing power available for our use. Computers find their way into every realm of activity. Some are developed to be as powerful as possible, without concern for price, for high-powered applications in industry and research. Others are designed for the home and office, less powerful but also less costly. Another category of computer is little recognised, partly because it is little seen. This is the type of computer that is designed into a product, in order to provide its control. The computer is hidden from view, such that the user often doesn't know it's even there. This sort of product is called an embedded system, and it is what this book is about. Those little computers we generally call microcontrollers; it is one extended family of these that the book studies.

In this chapter you will learn about:

- The meaning of the term 'embedded system'
- The microcontroller which lies at the heart of the embedded system
- The Microchip PIC® family
- A first PIC microcontroller, the 12F508
- An alternative microcontroller structure from Freescale.

1.1 The main idea – embedded systems in today's world

1.1.1 What is an embedded system?

The basic idea of an embedded system is a simple one. If we take any engineering product that needs control, and if a computer is incorporated within that product to undertake the control, then we have an embedded system. An embedded system can be defined as [Ref. 1.1]:

> A system whose principal function is not computational, but which is controlled by a computer embedded within it.

These days embedded systems are everywhere, appearing in the home, office, factory, car or hospital. Table 1.1 lists some example products that are likely to be embedded systems, all chosen for their familiarity. While many of these examples seem very different from each other, they all draw on the same principles as far as their characteristics as an embedded system are concerned.

The vast majority of users will not recognise that what they are using is controlled by one or more embedded computers. Indeed, if they ever saw the controlling computer they would barely recognise it as such. Most people, after all, recognise computers by their screen, keyboard, disc drives and so on. This embedded computer would have none of those.

Table 1.1 Some familiar examples of embedded systems

Home	Office and commerce	Motor car
Washing machine	Photocopier	Door mechanism
Fridge	Checkout machine	Climate control
Burglar alarm	Printer	Brakes
Microwave	Scanner	Engine control
Central heating controller		In-car entertainment
Toys and games		

1.2   Some example embedded systems

Let's take a look at some example embedded systems, first from everyday life and then from the projects used to illustrate this book.

1.2.1   The domestic refrigerator

A simple domestic refrigerator is shown in Figure 1.1. It needs to maintain a moderately stable, low temperature within it. It does this by sensing its internal temperature and comparing that with the temperature required. It lowers the temperature by switching on a compressor. The temperature measurement requires one or more sensors, and then whatever signal conditioning and data acquisition circuitry that is needed. Some sort of data processing is required to compare the signal representing the measured temperature to that representing the required temperature and deduce an output. Controlling the compressor requires some form of electronic interface, which accepts a low-level input control signal and then converts this to the electrical drive necessary to switch the compressor power.

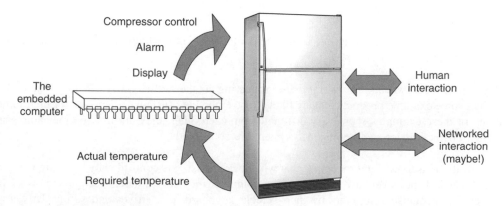

Figure 1.1 Embedded system example 1: the refrigerator

This process of control can be done by a conventional electronic circuit or it can be done by a small embedded computer. If used, the embedded computer could be designed simply to replicate the minimalist control process described above. Once a little computer is in place, however, there is tremendous opportunity for *added value*. With the signal in digital form and processing power now readily available, it is an easy step to add features like intelligent displays, more advanced control features, a better user control mechanism and so on.

Taking the idea of added value one step further, once an embedded computer is in place it is possible to network it to other computers, embedded or otherwise. This opens up big new horizons, allowing a small system to become a subset of a much larger system and to share information with that system. This is now happening with domestic products, like the refrigerator, as well as much more complex items.

The diagram of Figure 1.1, while specific for a fridge, actually represents very well the overall concept of an embedded system. There is an embedded computer, engaged in reading internal variables, and outputting signals to control the performance of the system. It *may* have human interaction (but in general terms does not have to) and it *may* have networked interaction. Generally, the user has no idea that there's a computer inside the fridge!

1.2.2 A car door mechanism

A very different example of an embedded system is the car door, as shown in Figure 1.2. Once again there are some sensors, some human interaction and a set of actuators that must respond to the requirements of the system. One set of sensors relates to the door lock and another to the window. There are two actuators, the window motor and the lock actuator.

It might appear that a car door could be designed as a self-contained embedded system, in a similar way to the fridge. Initially, one might even question whether it is worthy of any form of computer control

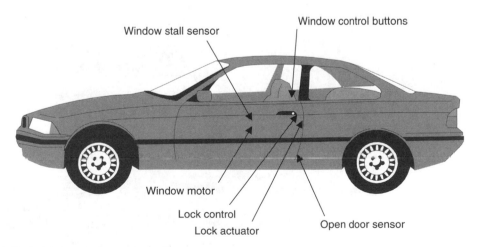

Figure 1.2 Embedded system example 2: the car door

whatsoever, as the functions seem so simple. Once again, by creating it as an embedded system, we see the opportunity to enhance functionality. Now we have the door status and actuators under electronic control, they can be integrated with the rest of the car. Central locking can be introduced or an alarm sounded if the door is not locked when the driver tries to pull away. There is therefore considerable advantage in having a network which links the humble actions of the door control to other functions of the car. We will see in later chapters that networked interaction is an important feature of the embedded system.

1.2.3   The electronic 'ping-pong'

This little game, shown in Figure 1.3, is one of several projects used to illustrate the material of this book. It is a game for two players, who each have a push button 'paddle'. Either player can start the game by pressing his/her paddle. The ball, represented by the row of eight LEDs (light-emitting diodes), then flies through the air to the opposing player, who must press his paddle only when the ball is at the end LED and at none other. The ball continues in play until either player violates this rule of play. Once this happens, the non-violating player scores and the associated LED is briefly lit up. When the ball is out of play, an 'out-of-play' LED is lit.

All the above action is controlled by a tiny embedded computer, a microcontroller, made by a company called Microchip [Ref. 1.2]. It takes the form of an 18-pin integrated circuit (IC), and has none of the visible features that one would normally associate with a computer. Nevertheless, electronic technology is now so advanced that inside that little IC there is a Central Processing Unit (CPU), a complex array of memories, and a set of timing and interface circuits. One of its memories contains a stored program, which it executes to run the game. It is able to read in as inputs the position of the switches (the player paddles) and calculate the required LED positions. It then has the output capability to actually power the LEDs to which it is connected. All of this computing action is powered from only two AAA cells!

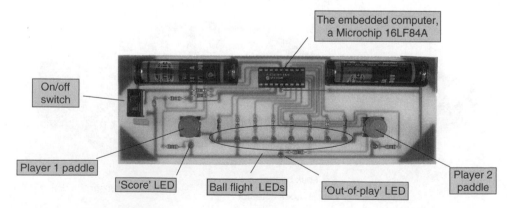

Figure 1.3 The electronic 'ping-pong'

1.2.4 The Derbot Autonomous Guided Vehicle

Another project used later in this book is the Derbot Autonomous Guided Vehicle (AGV), pictured in Figure 1.4. How do its features compare with the examples seen to date? Looking at the photograph, we can see from the front that it bristles with sensors and actuators. Two microswitch bump detectors sense if the Derbot hits an obstacle. An ultrasound detector, mounted on a servo actuator, is there with the aim of ensuring that the Derbot never needs to have an unexpected collision! Two light sensors on either side of the servo are used for light tracking applications – a third, not seen in the photo, is mounted at the rear. A further navigational option is a compass, so that direction can be determined from the earth's magnetic field. Locomotion is provided by two geared DC motors, while a sensor on each (again not seen in this picture) counts wheel revolutions to calculate actual distance moved. Steering is achieved by driving the wheels at different speeds. A piezo-electric sounder is included for the AGV to alert its human user. The Derbot is powered from six AA Alkaline cells, which it carries on a power pack almost directly above its wheels. Its block diagram is shown in Figure 1.5.

As with earlier examples, the Derbot operates as an embedded system, reading in values from its diverse sensors and computing outputs to its actuators. It is controlled by another Microchip microcontroller, hidden from view in the picture by the battery pack. This microcontroller is seemingly more powerful than the one in the ping-pong game, as it needs to interface with far more inputs and drive its outputs in a more complex way.

Figure 1.4 A Derbot AGV

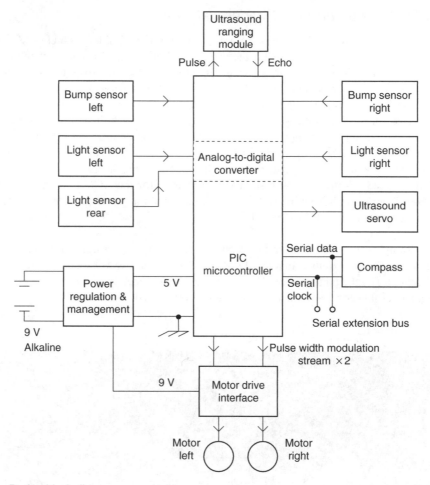

Figure 1.5 The Derbot block diagram

Interestingly, as we shall see, the CPU of each microcontroller is the same. They are differentiated primarily by their interface capabilities. It is this difference that gives the Derbot microcontroller its aura of far greater power.

1.3   Some computer essentials

When designing embedded systems we usually need to understand in some detail the features of the embedded computer that we are working with. This is quite unlike working with a desktop computer used for word processing or computer-aided design, where the internal workings are skilfully hidden from the user. As a preliminary to developing our knowledge, let us undertake a rapid survey of some important computer features.

1.3.1 Elements of a computer

Figure 1.6 shows the essential elements of any computer system. Fundamentally, it must be able to perform arithmetic or logical calculations. This function is provided by the Central Processing Unit (CPU). It operates by working through a series of instructions, called a program, which is held in memory. Any one of these instructions performs a very simple function. However, because the typical computer runs so incredibly fast, the overall effect is one of very great computational power. Many instructions cause mathematical and logical operations to occur. These take place in a part of the CPU called the ALU, the Arithmetic Logic Unit.

To be of any use the computer must be able to communicate with the outside world, and it does this through its input/output. On a personal computer this implies human interaction, through keyboard, VDU (Visual Display Unit) and printer. In an embedded system the communication is likely to be primarily with the physical world around it, through sensors and actuators.

The computer revolution that is taking place is due not only to the incredible processing power now at our disposal, but also to the equally incredible ability that we now have to store and access data. Broadly speaking there are two main applications for memory in a computer, as shown in Figure 1.6. One memory holds the program that the computer will execute. This memory needs to be permanent. If it is, then the program is retained indefinitely, whether power is applied or not, and it is ready to run as soon as power is applied. The other memory is used for holding temporary data, which the program works on as it runs. This memory type need not be permanent, although there is no harm if it is.

Finally, there must be data paths between each of these main blocks, as shown by the block arrows in the diagram.

1.3.2 Instruction sets – CISC and RISC

Any CPU has a set of instructions that it recognises and responds to; all programs are built up in one way or another from this instruction set. We want computers to execute code as fast as possible, but how to achieve this aim is not always an obvious matter. One approach is to build sophisticated CPUs with exotic instruction sets, with an instruction ready for every foreseeable operation. This leads to the CISC, the *Complex Instruction Set Computer*. A CISC has many instructions and considerable sophistication. Yet the complexity of the design needed to achieve this tends to lead to slow operation. One characteristic of the CISC approach is that instructions have different levels of complexity. Simple ones can be expressed

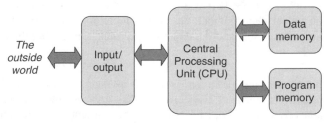

Figure 1.6 Essentials of a computer

in a short instruction code, say one byte of data, and execute quickly. Complex ones may need several bytes of code to define them and take a long time to execute.

Another approach is to keep the CPU very simple and have a limited instruction set. This leads to the RISC approach – the *Reduced Instruction Set Computer*. The instruction set, and hence overall design, is kept simple. This leads to fast operation. One characteristic of the RISC approach is that each instruction is contained within a single binary word. That word must hold all information necessary, including the instruction code itself, as well as any address or data information also needed. A further characteristic, an outcome of the simplicity of the approach, is that every instruction normally takes the same amount of time to execute.

1.3.3   Memory types

Traditionally, memory technology has been divided into two categories:

(1) *Volatile*. This is memory that only works as long as it is powered. It loses its stored value when power is removed, but can be used as memory for temporary data storage. Generally, this type of memory uses simple semiconductor technology and is easier to write to from an electrical point of view. For historical reasons it has commonly been called RAM (Random Access Memory). A slightly more descriptive name is simply 'data memory'.
(2) *Non-volatile*. This is memory that retains its stored value even when power is removed. On a desktop computer this function is achieved primarily by the hard disk, a huge non-volatile store of data. In an embedded system it is achieved using non-volatile semiconductor memory. It is a greater challenge to make non-volatile memory and sophisticated semiconductor technology is applied. Generally, this type of memory has been more difficult to write to electrically, for example in terms of time or power taken, or complexity of the writing process. Non-volatile memory is used for holding the computer program and for historical reasons has commonly been called ROM (Read-Only Memory). A more descriptive name is 'program memory'.

With the very sophisticated memory technology that is now available, we will see that the division of function between these two memory categories is becoming increasingly blurred. We return to the issue of memory technology and its applications in Chapter 2.

1.3.4   Organising memory

To interact with memory, there must be two types of number moved around, the address of the memory location required and the actual data that belongs in the location. These are connected on two sets of interconnections, called the address bus and the data bus. We must ensure that the data bus and address bus (or a subset of it) reach every memory area.

A simple way of meeting the need just described is shown in Figure 1.7(a). It is called the *Von Neumann* structure or architecture, after its inventor. The computer has just one address bus and one data bus, and the same address and data buses serve both program and data memories. The input/output may also be interconnected in this way and made to behave like memory as far as the CPU is concerned.

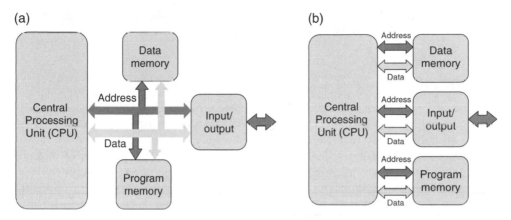

Figure 1.7 Organising memory access. (a) The Von Neumann way. (b) The Harvard way

An alternative to the Von Neumann structure is seen in Figure 1.7(b). Every memory area gets its *own* address bus and its *own* data bus. Because this structure was invented in the university of the same name, this is called a *Harvard* structure.

The Von Neumann structure is simple and logical, and gives a certain type of flexibility. The addressable memory area can be divided up in any way between program memory and data memory. However, it suffers from two disadvantages. One is that it is a 'one size fits all' approach. It's the same data bus for all areas of memory, even if one area wants to deal with large words and another wants to deal with small. It also has the problem of all things that are shared. If one person is using it, another can't. Therefore, if the CPU is accessing program memory, then data memory must be idle and vice versa.

In the Harvard approach we get greater flexibility in bus size, but pay for it with a little more complexity. With program memory and data memory each having their own address and data buses, each can be a different size, appropriate to its need, *and* data and program can be accessed simultaneously. On the minus side, Harvard reinforces the distinction between program and data memory, even when this distinction is not wanted. This disadvantage may be experienced, for example, when data is stored in program memory as a table, but is actually needed in the data domain.

1.4   Microprocessors and microcontrollers

1.4.1   *Microprocessors*

The first microprocessors appeared in the 1970s. These were amazing devices, which for the first time put a computer CPU onto a single IC. For the first time, significant processing power was available at rather low cost, in comparatively small space. At first, all other functions, like memory and input/output interfacing, were outside the microprocessor, and a working system still had to be made of a good number of ICs. Gradually, the microprocessor became more self-contained, with the possibility, for example, of including different memory types on the same chip as the CPU. At the same time, the CPU was becoming more powerful and faster, and moved rapidly from 8-bit to 16- and 32-bit devices. The development of the microprocessor led very directly to applications like the personal computer.

1.4.2   Microcontrollers

While people quickly recognised and exploited the computing power of the microprocessor, they also saw another use for them, and that was in control. Designers started putting microprocessors into all sorts of products that had nothing to do with computing, like the fridge or the car door that we have just seen. Here the need was not necessarily for high computational power, or huge quantities of memory, or very high speed. A special category of microprocessor emerged that was intended for control activities, *not* for crunching big numbers. After a while this type of microprocessor gained an identity of its own, and became called a *microcontroller*. The microcontroller took over the role of the embedded computer in embedded systems.

So what distinguishes a microcontroller from a microprocessor? Like a microprocessor, a microcontroller needs to be able to compute, although not necessarily with big numbers. But it has other needs as well. Primarily, it must have excellent input/output capability, for example so that it can interface directly with the ins and outs of the fridge or the car door. Because many embedded systems are both size and cost conscious, it must be small, self-contained and low cost. Nor will it sit in the nice controlled environment that a conventional computer might expect. No, the microcontroller may need to put up with the harsh conditions of the industrial or motor car environment, and be able to operate in extremes of temperature.

A generic view of a microcontroller is shown in Figure 1.8. Essentially, it contains a simple microprocessor core, along with all necessary data and program memory. To this it adds all the peripherals that allow it

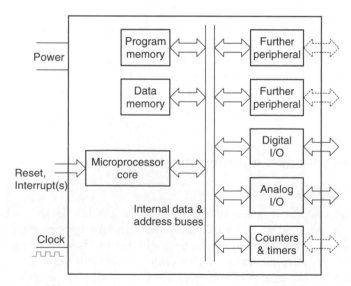

Figure 1.8 A generic microcontroller

to do the interfacing it needs to do. These may include digital and analog input and output, or counting and timing elements. Other more sophisticated functions are also available, which you will encounter later in the book. Like any electronic circuit the microcontroller needs to be powered, and needs a clock signal (which in some controllers is generated internally) to drive the internal logic circuits.

1.4.3 Microcontroller families

There are thousands of different microcontroller types in the world today, made by numerous different manufacturers. All reflect in one way or another the block diagram of Figure 1.8. A manufacturer builds a microcontroller *family* around a fixed microprocessor core. Different family members are created by using the *same* core, combining with it *different* combinations of peripherals and different memory sizes. This is shown symbolically in Figure 1.9. This manufacturer has three microcontroller families, each with its own core. One core might be 8-bit with limited power, another 16-bit and another a sophisticated 32-bit machine. To each core is added different combinations of peripheral and memory size, to make a number of family members. Because the core is fixed for all members of one family, the instruction set is fixed and users have little difficulty in moving from one family member to another.

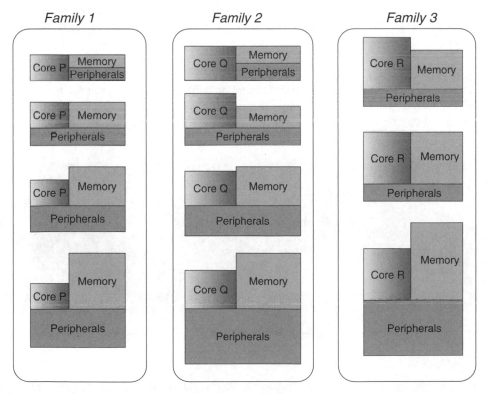

Figure 1.9 A manufacturer's microcontroller portfolio

While Figure 1.9 suggests only a few members of each family, in practice this is not the case; there can be more than 100 microcontrollers in any one family, each one with slightly different capabilities and some targeted at very specific applications.

1.4.4   Microcontroller packaging and appearance

Integrated circuits are made in a number of different forms, usually using plastic or ceramic as the packaging material. Interconnection with the outside world is provided by the pins on the package. Where possible microcontrollers should be made as physically small as possible, so it is worth asking: what determines the size? Interestingly, it is not usually the size of the integrated circuit chip itself, in a conventional microcontroller, which determines the overall size. Instead, this is set by the number of interconnection pins provided on the IC and their spacing.

It is worth, therefore, pausing to consider what these pins carry in a microcontroller. The point has been made that a microcontroller is usually input/output intensive. It is reasonable then to assume that a good number of pins will be used for input/output. Power must also be supplied and an earth connection made. It is reasonable to assume for the sort of systems we will be looking at that the microcontroller has all the memory it needs on-chip. Therefore, it will not require the huge number of pins that earlier microprocessors needed, simply for connecting external data and address buses. It will, however, be necessary to provide pin interconnection to transfer program information into the memory and possibly provide extra power for the programming process. There is then usually a need to connect a clock signal, a reset and possibly some interrupt inputs.

Figure 1.10, which shows a selection of microprocessors and microcontrollers, demonstrates the stunning diversity of package and size that is available. On the far right, the massive (and far from recent) 64-pin Motorola 68000 dwarfs almost everything else. The package is a dual-in-line package (DIP), with its pins arranged in two rows along the longer sides of the IC, the pin spacing being 0.1 inches. Because the 68000

Figure 1.10 A gathering of microprocessors and microcontrollers – old and new. From left to right: PIC 12F508, PIC 16F84A, PIC 16C72, Motorola 68HC05B16, PIC 16F877, Motorola 68000

depends on external memory, many of its pins are committed to data and address bus functions, which forces the large size. Second from right is the comparatively recent 40-pin PIC 16F877. While this looks similar to the 68000, it actually makes very different use of its pins. With its on-chip program and data memory it has no need for external data or address buses. Its high pin count is now put to good use, allowing a high number of digital input/output and other lines. In the middle is the 52-pin Motorola 68HC705. This is in a square ceramic package, windowed to allow the on-chip EPROM (Erasable Programmable Read-Only Memory) to be erased. The pin spacing here is 0.05 inches, so the overall IC size is considerably more compact than the 68000, even though the pin count is still high. To the left of this is a 28-pin PIC 16C72. Again, this has EPROM program memory and thus is also in a ceramic DIP package. On the far left is the tiny 8-pin surface-mounted PIC 12F508 and to the right of this is an 18-pin PIC 16F84A.

1.5   Microchip and the PIC microcontroller

1.5.1   Background

The PIC was originally a design of the company General Instruments. It was intended for simple control applications, hence the name – Peripheral Interface Controller. In the late 1970s General Instruments produced the PIC® 1650 and 1655 processors. Although the design was comparatively crude and unorthodox, it was completely stand-alone, and contained some important and forward-looking features. The simple CPU was a RISC structure, with a single Working register and just 30 instructions. The output pins could source or sink much more current than most other microprocessors of the time. Already the trademarks of the PIC were emerging – simplicity, stand-alone, high speed and low cost.

General Instruments sold off its semiconductor division to a group of venture capitalists, who must have realised the immense potential of these odd little devices. Throughout the 1990s the range of available PIC microcontrollers grew, and as they did so they gradually overtook many of their better-established competitors. In many cases PIC microcontrollers could run faster, needed a simpler chip-set and were quicker to prototype with than their competitors. Unlike many competitors, Microchip made their development tools simple and low cost or free. Moreover, Microchip stayed firmly entrenched in the 8-bit world. Despite the huge advances that have been made, we can still see features of the old General Instruments microcontroller, even in the most recent designs.

1.5.2   PIC microcontrollers today

Looking at the range of PIC microcontrollers today, anyone can be forgiven for a sense of complete bewilderment. There are literally hundreds of different devices, offered in different packages, for different applications. Let us therefore try to identify the characteristics that all of these have in common. At the time of writing, all PIC microcontrollers are low-cost, self-contained, 8-bit, Harvard structure, pipelined, RISC, single accumulator (the Working or W register), with fixed reset and interrupt vectors.

Today, Microchip offers five main families of microcontrollers, whose features are summarised in Table 1.2. It is possible to see clear evolution from one family to the other, so knowledge of one readily leads to knowledge of another. Every member of any one family shares the same core architecture and instruction set. The families are identified primarily by the first two digits of the device code. The alphabetic character

Table 1.2 Comparison of PIC families

PIC family	Stack size (words)	Instruction word size	Number of instructions	Interrupt vectors
12CXXX/12FXXX	2	12- or 14-bit	33	None
16C5XX/16F5XX	2	12-bit	33	None
16CXXX/16FXXX	8	14-bit	35	1
17CXXX	16	16-bit	58, including hardware multiply	4
18CXXX/18FXXX	32	16-bit	75, including hardware multiply	2 (prioritised)

that follows gives some indication of the technology used. The 'C' insert implies CMOS technology, where CMOS stands for Complementary Metal Oxide Semiconductor, the leading semiconductor technology for implementing low-power logic systems. The 'F' insert indicates incorporation of Flash memory technology (still using CMOS as the core technology). An 'A' after the number indicates a technological upgrade on the first issue device. An 'X' indicates that a certain digit can take a number of values, the one taken being unimportant to the overall number quoted.

For example, the 16C84 was the first of its kind. It was later reissued as the 16F84, incorporating Flash memory technology. It was then reissued as the 16F84A, with certain further technological upgrades.

Microchip also used to give each family a name. Thus, their first family, the 16C5XX, was called the 'baseline' family. The development of this, with device numbers starting '16C' or '16F' (and a fourth digit that was *not* 5), was called the 'mid-range' family. The powerful evolution of this, with codes starting '17C', was called the 'high-end' family. As the further families developed, with both very simple and advanced architectures, this naming convention has lost prominence, although the terminology is still found. For simplicity, to identify a PIC family, this book will refer to '12 Series', '16 Series', '18 Series' and so on. Let us survey each family in turn.

The 16C5X Series family

This PIC microcontroller family represents the most direct descendant of its General Instrument ancestors and displays all the core features of the original PIC design. With only a two-level stack and no interrupts, there is significant limitation on the program and hardware complexity that can be developed. Particularly without interrupts there is restriction on the type of on-chip peripheral that can be included, as most peripherals use interrupts to enhance their interface with the CPU. The 16C5X family has also been issued with Flash memory, with 16F5X codes. While this family is well established, it has a limited number of members and is not being given much prominence by Microchip.

The PIC 16 Series family

This, the 'mid-range' family, represents an improved version of the 16C5XX Series, in which interrupts (albeit with a single interrupt vector) are introduced and the stack size increased. The instruction set is a slight extension of that of the 16C5X. A very wide range of family members exists, with many different peripherals and technical enhancements. The larger devices, with many peripherals and significant on-chip memory, are both powerful and versatile.

The 12 Series family

The 12 series microcontrollers are designed for really tiny applications, being packaged in small ICs (for example, 8- or 14-pin). They have a simple architecture and can be viewed as 'stripped-down' versions of the 16C5XX series, with the same instruction set. Despite their small size, 12 Series microcontrollers carry some interesting peripherals, including analog-to-digital converters and EEPROM (Electrically Erasable Programmable Read-Only Memory) data memory. Although a small family, there is strong interest at this end of the size range, and further interesting additions to the family can be expected.

The 17 Series family

This family was introduced to give a real step-up in CPU performance compared with any of the 16 Series devices. While retaining the RISC strategy, the instruction set size is nearly doubled and the instruction word size increased to 16-bit. Thus, some programming activities that are awkward in the mid-range family, like table reads or data moves, are here much simpler. A hardware multiplier is also available. The single, often overloaded, interrupt vector of the mid-range family becomes four. Although much more powerful than the 16 Series, this family is limited in number, and Microchip appear to be focusing on the 18 Series family to move forward developments at the more powerful end of their range.

The 18 Series family

In this family Microchip comes to grips with some of the issues of sophisticated processors. The instruction set has increased again, now to 75 instructions, and is designed to facilitate use of the C programming language. In certain versions there is also an 'extended' instruction set, with a further small set of instructions. There are two interrupt vectors, which can be prioritised. This is an extremely powerful family of microcontrollers and a number of new members can be expected in the future.

1.6   An introduction to PIC microcontrollers using the 12 Series

As the simplest of the PIC microcontrollers, this is a very useful series to introduce the range. Features identified here will be recognisable in the more advanced PIC microcontrollers, where they appear alongside the more advanced features that are added.

We will look at the PIC 12F508/509, whose pin connection diagram is shown in Figure 1.11. The only difference between the 508 and 509 is that the latter has slightly larger program and data memories. Most (if not all) labels on the pins in the diagram may initially make no sense – don't worry, their meanings will emerge.

The staggeringly small size of this microcontroller is reinforced in Figure 1.12. While the 12F508 has been chosen as a simple microcontroller for introductory purposes, we also need to recognise that we are

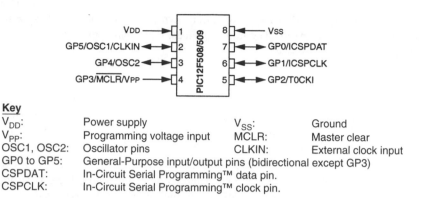

Key

V_{DD}:	Power supply	V_{SS}:	Ground
V_{PP}:	Programming voltage input	MCLR:	Master clear
OSC1, OSC2:	Oscillator pins	CLKIN:	External clock input
GP0 to GP5:	General-Purpose input/output pins (bidirectional except GP3)		
CSPDAT:	In-Circuit Serial Programming™ data pin.		
CSPCLK:	In-Circuit Serial Programming™ clock pin.		

Figure 1.11 PIC 12F508/509 pin connection diagram

Figure 1.12 *How small* is a 12F508?!

looking at almost a conjuring trick as well. Remember that it has been said earlier that a microcontroller should be input/output intensive. Then consider: how can a microcontroller be useful if it has only eight pins interconnecting with the outside world? We will attempt to answer this question as we look at the microcontroller's architecture.

1.6.1   The 12F508 architecture

The annotated block diagram of the 12F508 appears in Figure 1.13. This may be the first Microchip diagram that you have ever looked at. Don't worry if it initially appears complex – we will aim to break it into digestible pieces.

Let's start by finding the microcontroller essentials identified in Figure 1.8: the core (containing the CPU), program memory, data memory (or RAM), data paths and any peripherals. We should be able to relate some of these features to the microcontroller pins of Figure 1.11.

The CPU, enclosed in a dotted line bottom right, is made up essentially of the ALU (the Arithmetic Logic Unit), the Working register ('W Reg') and the Status register. This register carries a number of bits that

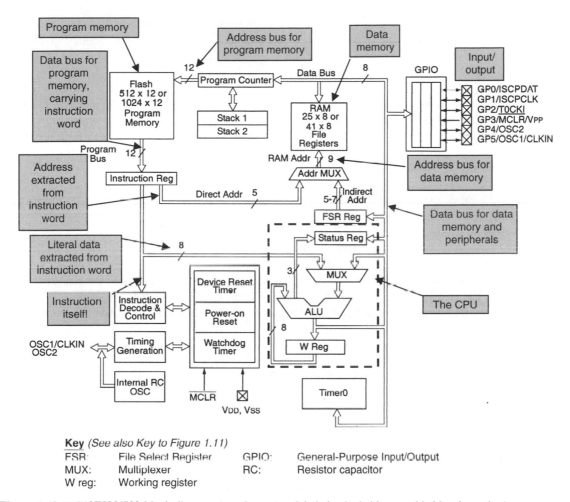

Figure 1.13 PIC12F508/509 block diagram (supplementary labels in shaded boxes added by the author)

give information on the outcome of the instruction most recently carried out. A multiplexer ('MUX') selects from two sources which data is presented to the ALU.

The data memory is just 25 bytes for the 508 or 41 for the 509. Notice that Microchip call the RAM memory locations 'file registers' or elsewhere just 'registers'. Program memory appears top left, with 512 12-bit words for the 12F508 or 1024 for the 509.

A distinctive feature of the PIC architecture is that it is Harvard structure, as discussed above. We should therefore be able to find *two* address buses (one for program memory, and the other for data memory and all peripherals) and *two* data buses (again, one for program memory, and one for data memory and peripherals). Easiest to find is the data bus for data memory and peripherals. This is simply labelled 'data bus' and is seen to the right of the diagram. It is 8-bit, and serves primarily the data memory, the General-Purpose Input/Output (GPIO) and the 'Timer 0' peripheral. The address bus for data memory is labelled 'RAM addr' and feeds into the RAM data memory. It is derived from the address multiplexer ('addr MUX'), which selects the address from one of two sources.

The program address bus arises from the Program Counter and goes only to the program memory, as shown. It is 12-bit, and hence can address 2^{12} memory locations, or 4096 locations. As the program memory itself is given as only 512 or 1024 words, we recognise that the address bus is larger than necessary for this memory size. Coming from the program memory we see the 12-bit 'program bus'. This carries the instruction words from the memory to the 'Instruction register'.

It is interesting to track the way the instruction word from program memory is divided up. As the PIC is a RISC computer, each instruction word must carry not only the instruction code itself, but also any address or data information needed as well. In the diagram the 'Instruction Reg' receives the instruction word and then starts the process of dividing this up into its component parts. Depending on the instruction itself, 5 bits of the instruction word *may* carry address information and hence be sent down the 'Direct Addr' bus to the address multiplexer ('Addr MUX'). Eight bits of the instruction word *may* carry a data byte that is to be used as literal data for the execution of that instruction. This goes to the multiplexer ('MUX'), which feeds into the ALU. Finally, there is the instruction data itself, which feeds into the 'Instruction Decode and Control' unit.

This microcontroller has only two on-chip peripheral devices, a Timer ('Timer 0') and the General-Purpose Input/Output port, with pins GP0 to GP5. The IC pins themselves appear in the block diagram as squares with crosses inside. Each of these pins is dual or triple function, so each has a second function identified in the diagram. We do not need at this moment to understand what each of these is, although that understanding is soon coming.

Towards the bottom left of the diagram are a number of functions relating to the clock oscillator, power supply and reset. Power supply and ground are connected via pins V_{DD} and V_{SS} respectively. A 'Power-on Reset' function detects when power is applied and holds the microcontroller in a Reset condition while the power supply stabilises. The $\overline{\text{MCLR}}$ input can be used to place the CPU in a Reset condition and to force the program to start again. An internal clock oscillator ('Internal RC OSC') is provided so that no external pins whatsoever need be committed to this function. External oscillator connections *can*, however, be made, using input/output pins GP4 and GP5. The oscillator signal is conditioned for use through the microcontroller in the 'Timing Generation' unit. The Watchdog Timer is a safety feature, used to force a reset in the processor if it crashes.

Having worked through this section, it should be possible for the reader to appreciate that the diagram of Figure 1.13 is a direct embodiment of the generic microcontroller shown in Figure 1.8. While the detail at this stage is incomplete, it will fall into place in the coming chapters.

1.7   What others do – a Freescale microcontroller

At around the same time that General Instruments were producing their PIC 1650, Motorola were at work on their first 8-bit microprocessor, the 6800. Both devices have become the ancestor of a truly impressive succession of descendants. While keeping a presence in the 8-bit world, Motorola also moved on to 16- and 32-bit devices. They continued to recognise the importance of the small 8-bit device, however, and in the mid-1980s produced the first of their 6805 8-bit microcontrollers. This family evolved over the next 20 years, and from it was developed the 68HC08 family. Motorola's Semiconductor Division was relaunched under the name *Freescale* in 2004.

While Microchip have become known for their very small microcontrollers, Freescale have also been active in this area. Let us take an overview look at the Freescale 8-pin microcontroller shown in Figure 1.14, with its simplified block diagram in Figure 1.15.

Although it is not clear from the figure, this microcontroller has a Von Neumann structure. The same data interconnection goes to both data and program memory, and to the peripherals. The data bus is 8-bit,

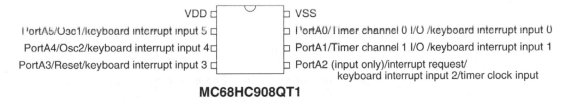

VDD — VSS
PortA5/Osc1/keyboard interrupt input 5 — PortA0/Timer channel 0 I/O /keyboard interrupt input 0
PortA4/Osc2/keyboard interrupt input 4 — PortA1/Timer channel 1 I/O /keyboard interrupt input 1
PortA3/Reset/keyboard interrupt input 3 — PortA2 (input only)/interrupt request/
keyboard interrupt input 2/timer clock input

MC68HC908QT1

Figure 1.14 A Freescale 68HC908 microcontroller

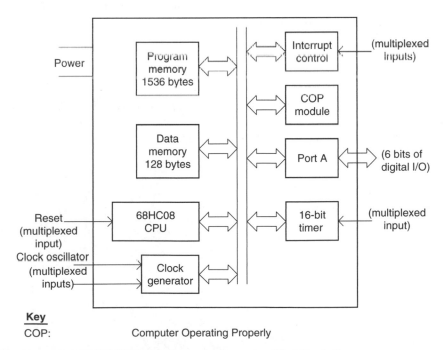

Figure 1.15 The Freescale MC68HC908QT1 microcontroller – simplified block diagram

which means that both the program and data word size are 8-bit. Like the 12F508, the Freescale device has on-chip program and data memory, with somewhat larger sizes. Remember, however, that the F509 memory is quoted for 12-bit words, not bytes. The Freescale device has two peripherals with direct F508/9 equivalents, the input/output port and a timer. The Freescale port, Port A, is 6-bit, almost directly equivalent to the 12F508 GPIO. Both devices have equivalent reliability features, Watchdog Timer for Microchip, 'Computer Operating Properly' for Freescale. A significant difference between the two is that Freescale have chosen, even for so small a microcontroller, to include interrupts in the structure.

Summary

- An embedded system is a product that has one or more computers embedded within it, which exercise primarily a control function.
- The embedded computer is usually a microcontroller: a microprocessor adapted for embedded control applications.
- Microcontrollers are designed according to accepted electronic and computer principles, and are fundamentally made up of microprocessor core, memory and peripherals; it is important to be able to recognise their principal features.
- Microchip offers a wide range of microcontrollers, divided into a number of different families. Each family has identical (or very similar) central architecture and instruction set. However, common features also appear across all their microcontrollers, and knowledge of one family can lead with ease to knowledge of another.
- The Microchip 12F508 is a good microcontroller to introduce a range of features of microcontrollers in general and of PIC microcontrollers in particular.

References

1.1. Wilmshurst, T. (2001). *An Introduction to the Design of Small-Scale Embedded Systems*. Palgrave. ISBN 0-333-92994-2.
1.2. Website of Microchip Technology Inc.: www.microchip.com

Section 2
Minimum Systems and the
PIC® 16F84A

This section, of five chapters, introduces the main concepts of a microcontroller, using a 'small' 16 Series PIC microcontroller. Emphasis is placed on understanding the core architecture, and using simple peripherals. Programming is in Assembler, as this allows the closest possible contact with the underlying hardware.

2
Introducing the PIC® 16 Series and the 16F84A

In Chapter 1 we introduced embedded systems and surveyed the different PIC® microcontroller families that are available, using the 12F508 as an introductory device. We are now going to step up a gear and begin to look at the detail of the PIC 16 Series 'mid-range' family. As an example device we will use the 16F84A, as it is a comparatively small member of the family. Six chapters later the focus of study will change to the 16F873A, a larger member of the same family. Note carefully that the 'F84A is an almost direct subset of the 'F873A. Therefore, don't worry if you are more interested in the latter device. Everything you learn about the smaller PIC is directly applicable to the bigger, and forms part of it. Indeed, just about everything we meet in the following chapters applies to all of the PIC 16 Series microcontrollers.

We will explore the overall architecture of the device and take time to get into some detail over its memory – both the technology and the memory maps.

In this chapter you will therefore learn about:

- The PIC 16 Series family, in overview
- The overall architecture of the 16F84A
- The 16F84A memory system, along with a review of memory technologies
- Other hardware features of the 16F84A, including the reset system.

If you wish, you will also learn about:

- Alternative approaches to microcontroller structure, through an example from another microcontroller family.

2.1 The main idea – the PIC 16 Series family

2.1.1 A family overview

The PIC 16 Series family is growing rapidly, with a huge and almost bewildering diversity of members. Therefore, when we talk of 'family' here, we are applying the concept of 'extended family', and a very large one at that. Nevertheless, the 16 Series stays true to the concept that all family members have identical core and instruction set, with the difference arising from different peripherals and other features being implemented, and different package sizes. Hence the pattern of Figure 1.9 is followed.

Table 2.1 gives summary details of some members of the 16 Series family, choosing the ones we meet in this book. Even with this limitation, there is considerable diversity. Within the 16 Series extended family, we find groupings of very closely related controllers, of which two are represented here, the 16F84A and the 16F87XA. The 16F84A is listed first, with features we are about to explore in detail. It has a very close

Table 2.1 Some members of the PIC 16 Series family

Device number	No. of pins*	Clock speed	Memory (K = Kbytes, i.e. 1024 bytes)	Peripherals/special features
16F84A	18	DC to 20 MHz	1K program memory, 68 bytes RAM, 64 bytes EEPROM	1 8-bit timer 1 5-bit parallel port 1 8-bit parallel port
16LF84A	As above	As above	As above	As above, with extended supply voltage range
16F84A-04	As above	DC to 4 MHz	As above	As above
16F873A	28	DC to 20 MHz	4K program memory 192 bytes RAM, 128 bytes EEPROM	3 parallel ports, 3 counter/timers, 2 capture/compare/PWM modules, 2 serial communication modules, 5 10-bit ADC channels, 2 analog comparators
16F874A	40	DC to 20 MHz	4K program memory 192 bytes RAM, 128 bytes EEPROM	5 parallel ports, 3 counter/timers, 2 capture/compare/PWM modules, 2 serial communication modules, 8 10-bit ADC channels, 2 analog comparators
16F876A	28	DC to 20 MHz	8K program memory 368 bytes RAM, 256 bytes EEPROM	3 parallel ports, 3 counter/timers, 2 capture/compare/PWM modules, 2 serial communication modules, 5 10-bit ADC channels, 2 analog comparators
16F877A	40	DC to 20 MHz	8K program memory 368 bytes RAM, 256 bytes EEPROM	5 parallel ports, 3 counter/timers, 2 capture/compare/PWM modules, 2 serial communication modules, 8 10-bit ADC channels, 2 analog comparators

*For DIP package only.

ADC, analog-to-digital converter; PWM, pulse width modulation.

relative, the 16LF84A, whose extended supply voltage range allows operation at lower voltages. Either of these controllers is available in different packages, different operating temperature ranges, and different clock speed ranges. For example, the 16F84A is available in 4 and 20 MHz versions.

The 16F87XA is a diverse grouping, as can be seen. There are two package sizes and two memory sizes. It is easy to see that package size is driven by the number of input/outputs that are available. The 40-pin versions have five parallel ports (which translates to 33 lines of parallel digital input/output), as well as more analog input, compared with their 28-pin relatives. There is otherwise not much difference. Each package size, however, comes with two different memory sizes. The bigger memory of course gives the opportunity for longer programs and more data storage, but also costs a little more.

As is normal Microchip practice, each member of the 16 Series family has its own comprehensive data sheet, available from Microchip's website. Reference 2.1 is the data sheet for the 16F84A. As well as this, there is a manual covering all the features that are common to all members of the family [Ref. 2.2]. While it is not necessary to refer to these while reading this chapter, it is worth knowing they are there, and extremely useful for looking up the finer details of a microcontroller's design and use.

2.1.2 The 16F84A

The 16F84A, along with its direct predecessors, has been one of many PIC success stories. It first appeared as the 16C84. At a time when most microcontroller manufacturers were trying to make their products bigger, more sophisticated and more complex, Microchip took the bold decision to stay small, simple and easy to use. While many microcontrollers of the day did have on-chip program memory, it was usually EPROM (Erasable Programmable Read-Only Memory), with the attendant time-consuming EPROM erase cycle. With the 16C84, Microchip chose to use EEPROM (Electrically Erasable Programmable Read-Only Memory) for program memory. Thus, it could be programmed rapidly, and repeatedly changed. Then, as Flash memory technology became more accessible, the 'C84 was reissued as the 16F84, with the new memory technology. With further upgrading it became the 16F84A. At the time of writing, this is the current version. A 16LF84A, intended for low-power applications, is also available.

2.1.3 A caution on upgrades

As technological expertise develops, any microcontroller design is inevitably upgraded. These are normally spelled out in documentation published by the manufacturer (e.g. Ref. 2.3). While each upgrade is generally to be welcomed, the changes introduced need to be watched with care. Some are of obvious benefit. For example, the 'A' version of the 16F84 can run at a higher speed than before (20 MHz maximum instead of 10 MHz). However, the technical upgrade sometimes has side-effects. These are of no direct advantage and sometimes make it difficult to replace a microcontroller in an existing product with its upgraded version. For example, operating power supply voltages and logic input thresholds are different between the 'F84 and the 'F84A.

2.2 An architecture overview of the 16F84A

The pin connection diagram of the 16F84A is shown in Figure 2.1 and its block diagram in Figure 2.2. A comparison of these figures with the equivalent ones for the PIC 12F508 in Chapter 1 shows some

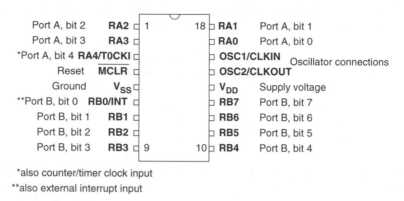

Figure 2.1 The PIC 16F84A pin connection diagram

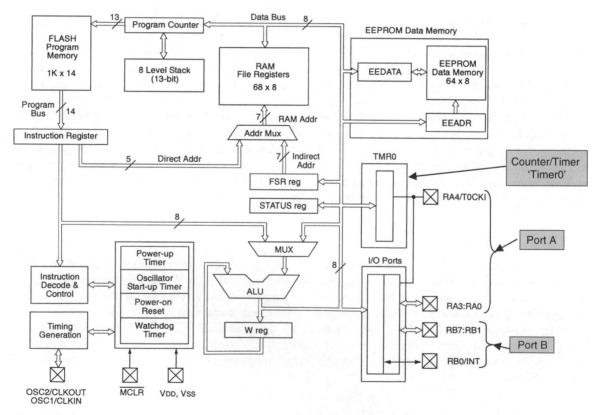

Figure 2.2 Block diagram of the 16F84A (supplementary labels in shaded boxes added by the author)

interesting similarities and differences. With 18 pins in play, there isn't the intense pressure to squeeze several functions onto each pin. Separate and dedicated pins are now provided, for example, for clock oscillator (pins 15 and 16) and Reset (pin 4 – $\overline{\text{MCLR}}$). Nevertheless, compared to most, the 'F84A remains a small microcontroller.

Architecturally there is clear similarity between the 12F508 and the 16F84A. In fact, the former is a direct subset of the 'F84A, with near identical CPU, memory, bus structure and counter/timer (TMR0) peripheral. Notice first, however, that the address bus sizes have been increased, to meet the needs of the whole PIC 16 Series family. As a smaller member of that family, the 'F84A doesn't fully exploit all these developments. The program address bus is now 13-bit and the instruction word size is 14-bit. Therefore, 2^{13} (i.e. 8192) memory locations *could* be addressed. Program memory size, at 1 K, is however only one-eighth of this. The larger bus size will, however, be useful in the larger 16 Series devices, as can be seen in the program memory size of the 16F876A and 16F877A (Table 2.1). RAM size has crept up cautiously to 68 locations and the Stack to eight locations.

A number of important new additions have appeared. The addition of an EEPROM memory gives the valuable capability of being able to store data values even when the chip is powered down. There are now two digital input/output ports. These are Port A, with five pins, and Port B, with eight. Importantly, there is the addition of an interrupt capability (which we explore in detail in Chapter 6). This can be seen externally on pin 6, where bit 0 of Port B is shared with the external interrupt input. We will also see that there are three further internal interrupt sources, generated by the peripherals.

Overall, we have a microcontroller that, while only modestly more complex than the 12F508, has proved incredibly diverse and useful in small applications.

2.2.1 The Status register

The result of any CPU operation is held in the Working register, but this does not necessarily tell everything about the operation which has just occurred. What if, for example, the 8-bit range has been exceeded by an addition instruction? The Working register has no way of indicating this and would simply hold an incorrect result. Therefore, a set of logic bits, sometimes called *condition code* flags, is built in to any computer CPU. These are used to carry extra information about the result of the instruction most recently executed, for example whether the result is zero, negative or positive. For the 16F84A, these flags are held in the Status register, shown in Figure 2.3. Only three of these Status register bits genuinely fall into the category of condition codes. These are bits 0 to 2, i.e. bits **C**, **DC** and **Z**. As the key to the figure shows, these indicate respectively whether a Carry or Digit Carry has been generated, or if the result is Zero. Their use is explained in Chapters 4 and 5.

2.3 A review of memory technologies

In order to examine the memory capabilities of the 16F84A, and to work with embedded systems in general, it is important to have some knowledge of the characteristics of the memory technologies in use. A detailed survey can be found in Chapter 4 of Ref. 1.1. The following section gives just a brief overview of the different memory technologies currently used by Microchip.

R/W-0	R/W-0	R/W-0	R-1	R-1	R/W-x	R/W-x	R/W-x
IRP	RP1	RP0	$\overline{\text{TO}}$	$\overline{\text{PD}}$	Z	DC	C

bit 7   bit 0

bit 7-6   **Unimplemented:** Maintain as '0'

bit 5   **RP0:** Register Bank Select bits (used for direct addressing)
01 = Bank 1 (80h - FFh)
00 = Bank 0 (00h - 7Fh)

bit 4 $\overline{\text{TO}}$: Time-out bit
1 = After power-up, CLRWDT instruction, or SLEEP instruction
0 = A WDT time-out occurred

bit 3 $\overline{\text{PD}}$: Power-down bit
1 = After power-up or by the CLRWDT instruction
0 = By execution of the SLEEP instruction

bit 2   **Z:** Zero bit
1 = The result of an arithmetic or logic operation is zero
0 = The result of an arithmetic or logic operation is not zero

bit 1   **DC:** Digit Carry/borrow bit (ADDWF, ADDLW, SUBLW, SUBWF instructions) (for borrow, the polarity is reversed)
1 = A carry-out from the 4th low order bit of the result occurred
0 = No carry-out from the 4th low order bit of the result

bit 0   **C:** Carry/borrow bit (ADDWF, ADDLW, SUBLW, SUBWF instructions) (for borrow, the polarity is reversed)
1 = A carry-out from the Most Significant Bit of the result occurred
0 = No carry-out from the Most Significant Bit of the result occurred

Note: A subtraction is executed by adding the twos complement of the second operand. For rotate (RRF, RLF) instructions, this bit is loaded with either the high or low order bit of the source register.

Figure 2.3 The 16F84A Status register

An ideal memory reads and writes in negligible time, retains its stored value indefinitely, occupies negligible space and consumes negligible power. In practice no memory technology meets all these happy ideals! In general, different technologies are strong in one or more of these characteristics and weaker in others. There is not one best memory technology, and different technologies are therefore applied for different applications, according to the need.

Any memory is made up of an *array* of memory *cells*, where each cell holds one bit of data. The characteristics of the single cell reflect the characteristics of the overall array; therefore, each technology is described here simply in terms of its cell design.

2.3.1   Static RAM (SRAM)

Here each memory cell is designed as a simple flip-flop, using two back-to-back transistor pairs. Two further transistors allow the cell to connect into the main array. Data is held only as long as power is supplied. Hence the SRAM technology is volatile. With each cell taking six transistors, SRAM is not dense. However, if made from CMOS (Complementary Metal Oxide Semiconductor) it can be made to consume *very* little power and can retain its data down to a low voltage (around 2 V). It has thus been a popular technology in battery-powered systems. SRAM is mainly used for data memory (RAM) in a microcontroller.

2.3.2   EPROM (Erasable Programmable Read-Only Memory)

In this technology each memory cell is made of a single MOS transistor – but with a difference. Within the transistor there is embedded a 'floating gate'. Using a technique known as *hot electron injection* (HEI), the floating gate can be charged. When it is *not* charged, the transistor behaves normally and the cell output takes one logic state when activated. When it *is* charged, the transistor no longer works properly and it no longer responds when it is activated. The charge placed on the floating gate is totally trapped by the surrounding insulator. Hence EPROM technology is non-volatile. EPROM can, however, be erased by exposing it to intense ultraviolet light. This gives the trapped electrons the energy to leave the floating gate.

A special version of EPROM is 'OTP' – One Time Programmable. Here the EPROM is packaged in plastic, without a window. Therefore, OTP can be programmed only once and never erased.

With a single transistor for a cell, EPROM is very high density and robust. Its requirement for a quartz window and ceramic packaging, to enable erasing, raises its price and reduces its flexibility. EPROM used to be integrated onto many microcontrollers for program memory, forcing the whole microcontroller to be ceramic-packaged, with a quartz window (as seen in Figure 1.10). As a technology, EPROM is now rapidly giving way to Flash, which follows shortly.

2.3.3   EEPROM (Electrically Erasable Programmable Read-Only Memory)

EEPROM also uses floating gate technology. Its dimensions are finer, so that it can exploit another means of charging its floating gate. This is known as Nordheim Fowler tunnelling (NFT). With NFT, it is possible to electrically erase the memory cell, as well as write to it. To allow this to happen, a number of switching transistors need to be included around the memory element itself, so the high density of EPROM is lost.

Generally, EEPROM can be written to and erased on a byte-by-byte basis. This makes it especially useful for storing single items of data, like television settings or mobile phone numbers. Both writing and erasing take finite time, up to several milliseconds, although a read can be accomplished at normal semiconductor memory access times, i.e. within microseconds or less. Again, like EPROM, because the charge on the floating gate is totally trapped by the surrounding insulator, EEPROM is non-volatile. Because the EEPROM structure is now so fine, it suffers from certain wear-out mechanisms. Manufacturers usually therefore define a guaranteed minimum number of erase/write cycles that their memory can successfully undergo.

2.3.4   Flash

Flash represents a further evolution of floating gate technology. With a single transistor per memory cell, it uses both HEI and NFT to allow electrical writing and erase. It does not include the extra switch transistors that EEPROM has, so can only erase in blocks. It therefore returns to the exceptionally high density of EPROM. Like EEPROM, it has wear-out mechanisms, so cannot be written and erased indefinitely.

Apart from its inability to erase byte by byte, Flash is an incredibly powerful technology. It is now a central feature of a huge range of products, including digital cameras, 'memory sticks', laptop computers and microcontroller program memory.

2.4 The 16F84A memory

As Figure 2.2 shows, there are no less than *four* areas of memory in the 16F84A, as summarised in Table 2.2. Each memory has its own distinct function and means of access.

2.4.1 The 16F84A program memory

The 16F84A program memory map is shown in Figure 2.4. Looking at this diagram, we can see that it actually shows three things: the Program Counter, the Stack and the actual program memory. The three work inextricably together. The program memory is loaded with the program code that the microcontroller executes. The program is in the form of a list of instructions, and the Program Counter holds the address of the next instruction that is to be executed by the microcontroller. Therefore, it acts as a pointer to program memory, as indicated in the diagram. The value of the Program Counter can also be moved onto the Stack. This occurs when either a subroutine or an interrupt occurs. The instructions indicated in the diagram, **CALL**, **RETURN**, **RETFIE** and **RETLW**, all relate to subroutines and interrupts. We will meet them in the coming chapters – don't worry if they have no meaning to you at present!

We can see that the address range of program memory is from 0000 to 03FF$_H$. With its 13-bit Program Counter, the microcontroller can theoretically address a range from 0000 to 1FFF$_H$. The extra address space is shown (in grey), although it is of no use here.

The very first location in the program memory is labelled the *reset vector*. When the program starts running for the first time, for example on power-up, the Program Counter is set to 0000. Therefore, the first memory location that it points to is the reset vector. The programmer must therefore place his/her first instruction

Table 2.2 16F84A memory features

Memory function	Technology	Size	Volatile/non-volatile	Special characteristics*
Program	Flash	1K × 14 bits	Non-volatile	10 000 erase/write cycles, typically
Data memory (file registers)	SRAM	68 bytes	Volatile	Retains data down to supply voltage of 1.5 V
Data memory (EEPROM)	EEPROM	64 bytes	Non-volatile	10 000 000 erase/write cycles, typically
Stack	SRAM	8 × 13 bits	Volatile	

*Information obtained from full 16F84A data sheet [Ref. 2.1].

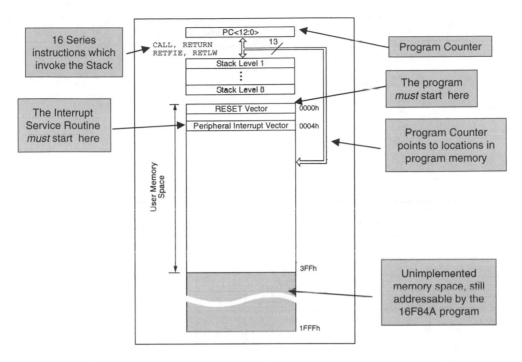

Figure 2.4 The 16F84A – program memory and Stack (supplementary labels in shaded boxes added by the author)

at this location. The *peripheral interrupt vector* acts in a similar way for interrupt service routines, as we shall see in Chapter 6.

2.4.2 The 16F84A data and Special Function Register memory ('RAM')

The RAM memory map is shown in Figure 2.5. The memory area is banked and is divided into two important areas. The first is the general-purpose data memory, which occupies locations $0C_H$ to $4F_H$. Above that are the Special Function Registers (SFRs). Let us explore the two concepts just mentioned, as they are likely to be unfamiliar.

'Banked' addressing

A problem with any memory space is that the larger the memory is, the larger the address bus must be. One way of avoiding big address buses is to divide the memory into a number of smaller blocks – called banks – each identical in size. Now a smaller address bus can be used. It can access all banks in an identical way, with just one of the banks being identified at any one time as the target of the address specified.

PIC microcontrollers adopt a banked structure for their RAM, with the 16F84A having just two banks. The address of either bank is the 7-bit RAM address ('RAM addr') seen in Figure 2.2. The active bank is

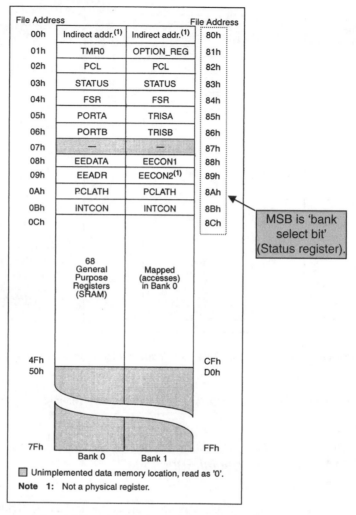

Figure 2.5 Data memory and Special Function Register map of the 16F84A (supplementary labels in shaded boxes added by the author)

selected by bit 5 in the Status register (Figure 2.3). The programmer must ensure that the bank bit in the Status register is correctly set before making any access to memory.

Special Function Registers

The SFRs are the gateway to interaction between the CPU and the peripherals, and we will get to know them very well. To the CPU, an SFR acts more or less like a normal memory location – you can usually write to it or read from it. What makes it 'special' is that the bits of that memory location have a dual purpose. Each bit is wired across to one or other of the microcontroller peripherals. Each is then used either to set up the operating mode of the peripheral or to transfer data between the peripheral and the microcontroller core. As we get to know the peripheral of the 16F84A, we will get to know each of the SFRs shown in Figure 2.5. Note that four SFRs appear in Figure 2.2. Can you identify what they are?

RAM addressing

Figure 2.2 shows that there are two possible sources of the RAM address, selected through the address multiplexer ('Addr Mux'). Either the address forms part of the instruction, and is routed across to the address multiplexer from the Instruction register. This is called *direct* addressing. Alternatively, the address is taken from the File Select Register, or FSR, which can be found as one of the SFRs in Figure 2.5. If the user loads an address into the FSR, that can then be used as an address to data memory, a technique know as *indirect* addressing. This will be described in Chapter 5, Section 5.4.1.

The actual memory addresses are shown in Figure 2.5, labelled as 'file address'. These addresses, at least in the right-hand column, appear to be 8-bit. We know, however, from Figure 2.2 that the RAM address bus is only 7-bit, or only 5 valid bits if direct addressing is used. It is important to understand that the addresses shown are made up of this 7-bit RAM address, *with* the bank select bit from the Status register inserted as the eighth, most significant, bit. When programming it is necessary to separate the two, ensuring that the MSB in Figure 2.5 is used for the bank select bit. This will become clear as we start to program.

2.4.3 The Configuration Word

A special part of the 16F84A program memory is its *Configuration Word* (Figure 2.6). This allows the user to define certain configurable features of the microcontroller, at the time of program download. These are fixed until the next time the microcontroller is programmed. This is distinct from those many selectable features, like the setting of SFRs, which are under normal program control. While the Configuration Word is part of program memory, it is not accessible within the program, or in any way while the program is running. The actual features it controls, which can be read on the diagram, are explained in this and later chapters.

2.4.4 EEPROM

The EEPROM is non-volatile and is particularly useful for holding data variables that can be changed, but are likely to be needed for the medium to long term. Examples include TV tuner settings, phone numbers stored in a cell phone or calibration settings on a measuring instrument.

In the 16F84A (and indeed any PIC microcontroller), the EEPROM is not placed in the main data memory map. Instead (as the top right of Figure 2.2 neatly shows) it is addressed through the **EEADR** register and data is transferred through the **EEDATA** register. These are both SFRs, seen in Figure 2.5.

As the review of memory technology earlier suggests, reading from EEPROM is a simple process, but writing to it is not. This latter takes significant time in electronic terms (i.e. milliseconds) and care must be taken to avoid accidental writes. A set of controls is therefore required, to start the process and (for write) to detect when it is ended. These are found in the bits of the **EECON1** register (Figure 2.7). To *read*

R/P-u	R/P-u	R/P-u	R/P-u	R/P-u	R/P-u	R/P-u	R/P-u	R/P-u	R/P-u	R/P-u	R/P-u	R/P-u	R/P-u
CP	CP	CP	CP	CP	CP	CP	CP	CP	CP	$\overline{\text{PWRTE}}$	WDTE	F0SC1	F0SC0

bit13   bit0

bit 13-4   **CP:** Code Protection bit
                            1 = Code protection disabled
                            0 = All program memory is code protected

bit 3 $\overline{\text{PWRTE}}$: Power-up Timer Enable bit
                            1 = Power-up Timer is disabled
                            0 = Power-up Timer is enabled

bit 2   **WDTE:** Watchdog Timer Enable bit
                            1 = WDT enabled
                            0 = WDT disabled

bit 1-0   **FOSC1:FOSC0:** Oscillator Selection bits
                            11 = RC oscillator
                            10 = HS oscillator
                            01 = XT oscillator
                            00 = LP oscillator

Figure 2.6 16F84A Configuration Word

U-0	U-0	U-0	R/W-0	R/W-x	R/W-0	R/S-0	R/S-0
—	—	—	EEIF	WRERR	WREN	WR	RD

bit 7   bit 0

bit 7-5   **Unimplemented:** Read as '0'

bit 4   **EEIF:** EEPROM Write Operation Interrupt Flag bit
                            1 = The write operation completed (must be cleared in software)
                            0 = The write operation is not complete or has not been started

bit 3   **WRERR:** EEPROM Error Flag bit
                            1 = A write operation is prematurely terminated
                                 (any $\overline{\text{MCLR}}$ Reset or any WDT Reset during normal operation)
                            0 = The write operation completed

bit 2   **WREN:** EEPROM Write Enable bit
                            1 = Allows write cycles
                            0 = Inhibits write to the EEPROM

bit 1   **WR:** Write Control bit
                            1 = Initiates a write cycle. The bit is cleared by hardware once write is complete. The WR bit
                                 can only be set (not cleared) in software.
                            0 = Write cycle to the EEPROM is complete

bit 0   **RD:** Read Control bit
                            1 = Initiates an EEPROM read RD is cleared in hardware. The RD bit can only be set (not
                                 cleared) in software.
                            0 = Does not initiate an EEPROM read

Figure 2.7 The **EECON1** Special Function Register (address 88_H)

an EEPROM location, the required address must be placed in **EEADR** and the **RD** bit set in **EECON1**. The data in that memory location is then copied to the **EEDATA** register and can be read immediately. To *write* to an EEPROM location, the required data and address must be placed in **EEDATA** and **EEADR** respectively. The write process is enabled by the **WREN** (Write Enable) bit being set high, followed by the bytes 55_H followed by AA_H being sent to the **EECON2** register. The built-in requirement for these codes helps to ensure that accidental writes do not take place, for example on power-up or down. The **WR** bit is then set high and writing actually commences. The write completion is signalled by the setting of bit **EEIF** in **EECON1**.

2.5   Some issues of timing

2.5.1   Clock oscillator and instruction cycle

Any microprocessor or microcontroller is a complex electronic circuit, made up of sequential and combinational logic. At fantastic speed it steps in turn through a series of complex states, each state being dependent on the instruction sequence it is executing. While the detail of this process is invisible to us, it is still necessary to provide the 'clock' signal, a continuously running fixed frequency logic square wave. The overall speed of the microcontroller operation is entirely dependent on this clock frequency. It is not just the CPU that is dependent on the clock. In most microcontrollers many essential timing functions are also derived from it, ranging from counter/timer functions to serial communications. Furthermore, the overall power consumption of the microcontroller has a strong dependence on clock frequency, with high speed operation being much more power hungry than slow speed.

As Table 2.1 shows, every microcontroller has a specified range for its clock frequency. It is up to the designer to determine the clock frequency needed and to select a means of generating the clock source. With so many things depending on the clock frequency and its stability, these can be challenging decisions. These are taken further in Chapter 3.

Within any microprocessor, the main clock signal is immediately divided down by a fixed value into a lower-frequency signal. Each cycle of this slower signal is called either a *machine cycle* or an *instruction cycle*. Microchip use the latter terminology. The instruction cycle becomes the primary unit of time in the action of the processor, for example being used as a measure for how long an instruction takes to execute. The original clock signal is retained to create phases or time stages within the instruction cycle. In PIC 16 Series microcontrollers the main oscillator signal is divided by 4 to produce the instruction cycle time.

Table 2.3 gives some popular clock frequencies, with their resulting instruction cycle durations. For the fastest clock frequency, 20 MHz, the instruction cycle frequency is 5 MHz, with a period of 200 ns. The slightly cheaper version of the controller, the 16F84-04, with maximum clock frequency of 4 MHz, has at this frequency an instruction cycle time of 1 μs. As we will see, this unsurprisingly is a convenient value for a range of simple timing applications, using software delay loops and the counter/timer. A popular clock frequency for very-low-power applications, including wristwatches, is 32.768 kHz. This has an instruction cycle period of 122.07 μs. The result is very low power, but strictly no high-speed calculations!

Table 2.3 PIC 16 Series instruction cycle durations for various clock frequencies

Clock frequency	Instruction cycle	
	Frequency	Period
20 MHz	5 MHz	200 ns
4 MHz	1 MHz	1 μs
1 MHz	250 kHz	4 μs
32.768 kHz	8.192 kHz	122.07 μs

2.5.2   Pipelining

The combination of the RISC instruction set and the Harvard memory map used by PIC microcontrollers has an added advantage: instructions can be *pipelined*. Every instruction in a computer's program memory has first to be fetched and then executed. In many CPUs these two steps are done one after the other – first the CPU fetches and then it executes. If, however, program memory has its own address and data bus, separate from data memory (i.e. a Harvard structure), then there is no reason why a CPU cannot be designed so that while it is executing one instruction, it is already fetching the next. This is called *pipelining*. Pipelining works best if fetch and execute cycles are always of the same duration, such as a RISC structure gives. This fairly simple design upgrade gives a doubling in execution speed!

All PIC microcontrollers implement pipelining, which is one of the reasons for their comparatively high speed of operation. Each instruction is fetched while the previous one is being executed. Pipelining fails only for instructions that cause the value in the Program Counter to be changed, for example a program branch or jump. In this case, the instruction fetched is no longer the one needed. The pipelining process must then start again, with the consequent loss of an instruction cycle.

A diagram representing the pipelining process in 16 Series microcontrollers is shown in Figure 2.8. Here we can see that while instruction 1 is being executed, instruction 2 is already being fetched, the same happening as instruction 2 is executed, and so on. An example sequence of instructions is shown to the left of the diagram. It is not, however, necessary to understand their meaning to understand the diagram, except to know that the **CALL** instruction causes a program branch. The instruction following it, instruction 4, is fetched while instruction 3 is being executed. Due to the program branch, however, instruction 4 is no longer needed, and a cycle has to be lost while the new instruction is fetched.

2.6   Power-up and Reset

When the microcontroller powers up, it must start running its program from its beginning (i.e. for the 16F84A from its reset vector, seen in Figure 2.4). This will only happen if explicit circuitry is built in to detect power-up and force the Program Counter to zero. Along with this, it is also very useful to set SFRs so that peripherals are initially in a safe and disabled state. This 'ready-to-start' condition is called *Reset*. The CPU starts running its program when it leaves the Reset condition.

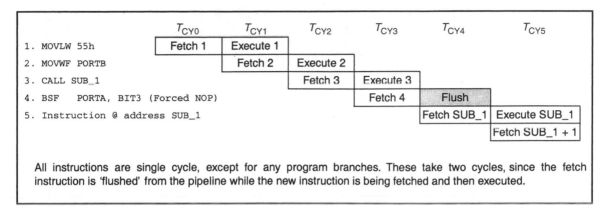

Figure 2.8 contents:

	T_{CY0}	T_{CY1}	T_{CY2}	T_{CY3}	T_{CY4}	T_{CY5}
1. MOVLW 55h	Fetch 1	Execute 1				
2. MOVWF PORTB		Fetch 2	Execute 2			
3. CALL SUB_1			Fetch 3	Execute 3		
4. BSF PORTA, BIT3 (Forced NOP)				Fetch 4	Flush	
5. Instruction @ address SUB_1					Fetch SUB_1	Execute SUB_1
						Fetch SUB_1 + 1

All instructions are single cycle, except for any program branches. These take two cycles, since the fetch instruction is 'flushed' from the pipeline while the new instruction is being fetched and then executed.

Figure 2.8 Instruction pipelining

In the 16F84A there is a Reset input, $\overline{\text{MCLR}}$ ('Master Clear', on pin 4, Figure 2.1). As long as this is held low, the microcontroller is held in Reset. When it is taken high, program execution starts. If the pin is taken low while the program is running, then program execution stops immediately and the microcontroller is forced back into Reset mode.

There remains the question of when program execution should actually be allowed to start. The moment power is applied is a dangerous one for any embedded system. Both the power supply and the clock oscillator take finite time to stabilise, and in a complex system power to different parts of the circuit may become stable at different times. Clearly, this situation takes some careful handling. How can the start of program execution be delayed until power has stabilised?

A simple way to resolve the 'what do we do as power is applied?' question is shown in Figure 2.9, illustrated here for any microcontroller which has an active low Reset input. If a resistor capacitor circuit is connected to the Reset input, then when power is applied the capacitor voltage rises according to the RC time constant, which can be made as big as is wanted. For a certain period of time, because it is rising comparatively slowly, the $\overline{\text{Reset}}$ input is at Logic 0. Thus, the microcontroller can be held in Reset while its power supply stabilises and while the clock oscillator starts up. The diagram for a simple external Reset circuit is shown in Figure 2.9(a).

A small problem arises with this circuit if the power is switched off and then on again quickly (a cruel and challenging thing to do to any electronic device). With the circuit of Figure 2.9(a) the capacitor wouldn't have time to discharge and the Reset condition might not be properly applied when power is applied again. More dangerously, the capacitor voltage might exceed the voltage supplied to the microcontroller and excessive current could then flow from the capacitor into the $\overline{\text{Reset}}$ input. By adding a simple discharge diode, as shown in the circuit in Figure 2.9(b), we can ensure that the capacitor discharges more or less at the same rate as the V_{DD} supply. The resistor R_S is also included, to limit current into the |Reset input if the capacitor voltage does inadvertently exceed the voltage supplied to the microcontroller or another fault condition occurs.

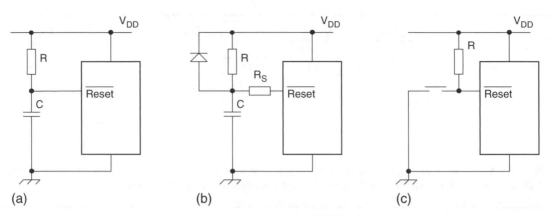

Figure 2.9 External Reset circuits – generic microcontroller with $\overline{\text{Reset}}$ input. (a) Power-on Reset, simplest possible. (b) Power-on Reset, with discharge diode and protective resistor. (c) User Reset button.

If the designer wishes to include a Reset button, then the circuit of Figure 2.9(c) can be applied. This is particularly useful for prototype circuits, where a large amount of testing is expected. Then it is convenient to be able to reset a program that may have crashed. R is a pull-up resistor, whose value can be in the range 10–100 kΩ. In a commercial device it is usually undesirable to have a Reset button; the aim here is design the product so reset by the user is never needed.

One of Microchip's goals is to minimise the number of external components needed for their microcontrollers, and the components of Figure 2.9 fall exactly into this category. Therefore, the 16F84A includes some sophisticated on-chip reset circuitry, which in many situations makes the components of Figure 2.9(a) or (b) unnecessary. A Power-up Timer (PWRT) is included on-chip, which can be enabled by the user with bit 3 of the Configuration Word (Figure 2.6). The 16F84A detects that power has been applied and the Power-up Timer then holds the controller in Reset for a fixed period of time. Once this is over the microcontroller leaves Reset and program execution begins. In practice, the circuit of Figure 2.9(b) need only be applied if the supply voltage rises very slowly. The Power-up Timer, and further details of the internal Reset circuit, are covered in greater detail in Section 2.8.

So what should be done with the 16F84A $\overline{\text{MCLR}}$ input if we don't want to use it? It is essential to recognise that this input must not just be left unconnected. The simplest thing to do is to tie it to the supply rail and then forget about it.

2.7   What others do – the Atmel AT89C2051

Atmel produce a range of small (but not tiny) microcontrollers, comparable in size to the smaller PIC 16 Series devices, though somewhat larger than the 16F84A. They are based on an 8051 core, originally developed by Intel, and since adopted by a number of other manufacturers, including both Atmel and Philips.

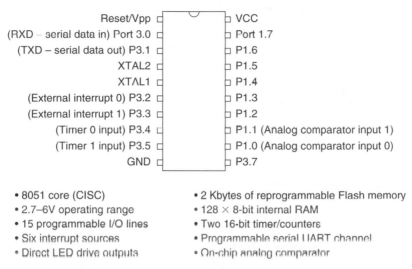

Figure 2.10 The Atmel AT89C2051 pin connection diagram and summary

The Atmel AT89C2051 [Refs 2.4, 2.5] is summarised in Figure 2.10. With 20 pins it is just slightly larger than the 16F84A, and this is reflected in its internal capability. Despite the differences, it is interesting to notice the similarities. Both have on-chip program and data memory, Flash for program and SRAM for data. Crystal inputs, reset and power supply are all extremely similar. The PIC microcontroller has 13 digital input/output lines, the Atmel 15, both with direct LED drive capability. The PIC microcontroller has one external interrupt, the Atmel two. The Atmel has two 16-bit timers, to the one 8-bit timer of the PIC, and the Atmel has a serial capability that the PIC microcontroller just doesn't have. One would not have to look far in the PIC 16 Series family to find relatives of the 'F84A, just slightly larger than it, which have serial capability. An interesting further feature of the '2051 is the on-chip comparator, which allows the controller to host simple analog functions. However, it doesn't all go the way of the Atmel device, as the PIC microcontroller has the EEPROM memory that Atmel does not match.

All this discussion, however, is mainly just comparing peripherals, which manufacturers can add or subtract with ease. What other differences would we find if we put both controllers to use? The answer lies in the core and instruction set. The Atmel is a CISC device, with the 8051 instruction set. We take the instruction set comparison a little further in Chapter 4. Let's just note here the advantages that the PIC microcontroller core enjoys. As a RISC processor, it requires just four oscillator cycles per instruction, as described earlier in this chapter. The 2051 requires 12 for each of its machine cycles, and even then many instructions require more than one machine cycle. The PIC microcontroller also uses pipelining, which the Atmel does not.

2.8 Taking things further – the 16F84A on-chip reset circuit

Let's take a closer look at the 16F84A on-chip reset circuitry, shown in simplified form in Figure 2.11. This takes some understanding, but it is worth doing.

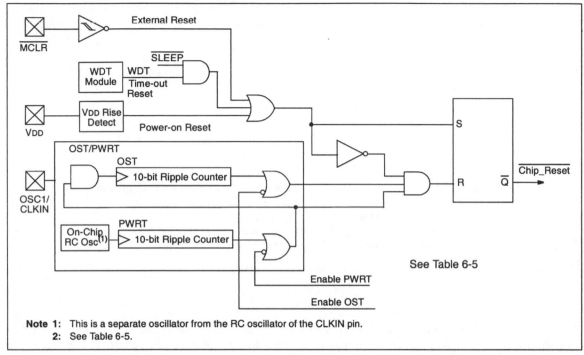

Table 6-5 of Ref. 2.1 gives example reset delay times for different settings of oscillator, and **PWRT** and **OST** enable lines.

Figure 2.11 The 16F84A reset circuitry

The actual reset to the CPU, **Chip_Reset**, is generated by a flip-flop, which appears to the right of the diagram. This has two inputs, **S** (Set) and **R** (Reset). The CPU enters Reset mode when **Chip_Reset** goes low, which is caused by the **S** line going high. It stays there until the flip-flop is cleared, caused by the **R** line going high.

So what causes a reset? The **S** input to the flip-flop goes high, via a three-input OR gate, if any of the following goes high:

- **External Reset**, from the $\overline{\text{MCLR}}$ line we have already seen
- **Time-out Reset**, from the Watchdog Timer (WDT); this is designed to occur if a program crash occurs – the details are given in Chapter 6
- **Power-on Reset**, output of the circuit that detects power being applied ('V$_{DD}$ Rise Detect').

Once any of these occurs, the flip-flop is set, the **Chip_Reset** line goes low and the PIC microcontroller is held in Reset.

The **Chip_Reset** line returns to 1 (and the PIC microcontroller is enabled) if the **R** input to the flip-flop is activated. The three requirements to be satisfied here, determined by the inputs to the associated

AND gate, are that both power supply and oscillator have stabilised, and that any demand for Reset has been cleared. The first two of these requirements are achieved by two interesting timers, the Power-Up Timer (PWRT) and the Oscillator Start-up Timer (OST). The Power-up Timer can be enabled by setting its bit in the Configuration Word (Figure 2.6). The Oscillator Start-up Timer is enabled via the **Enable OST** line. This is set automatically by the user oscillator setting in the Configuration Word, which enables it for all oscillator modes except RC. The Power-Up Timer is clocked by its own on-chip RC oscillator, and when enabled counts 1024 cycles of its oscillator before setting its output to 1. This time duration turns out to be around 72 ms. This is long enough for the average power supply to have stabilised, though is not enough for a slowly rising supply. Once the Power-Up Timer has completed its count, the Oscillator Start-up Timer is then activated, which in turn counts 1024 cycles of the main oscillator signal. This tests for a reliably running clock oscillator − if the oscillator isn't running, then of course it can't count. The outputs of both counters, and the inverse of the **S** input to the flip-flop, are ANDed together, to form the **R** input to the flip-flop. If all lines are high, i.e. both counters have completed their count and there is no demand for a Reset, then the flip-flop is cleared. The CPU accordingly leaves the Reset condition and starts running.

The reset sequence just described is shown in Figure 2.12, for the common situation of $\overline{\text{MCLR}}$ being tied to V_{DD}. The application of power is seen in the rise of the V_{DD} trace, which brings the $\overline{\text{MCLR}}$ line with it. This change is detected, as seen in the change of state of the 'internal POR' line. This in turn triggers the Power-Up Timer, which runs for a period T_{PWRT}. When T_{PWRT} is up, the Oscillator Start-up Timer, whose time delay is T_{OST}, is activated. Notice that T_{OST} depends on the main oscillator running successfully, and is dependent on its frequency. For a 4 MHz oscillator, it will be 1024×250 ns, or 256 µs. When T_{OST} is complete, the **R** line in Figure 2.11 goes high and the microcontroller leaves the Reset state.

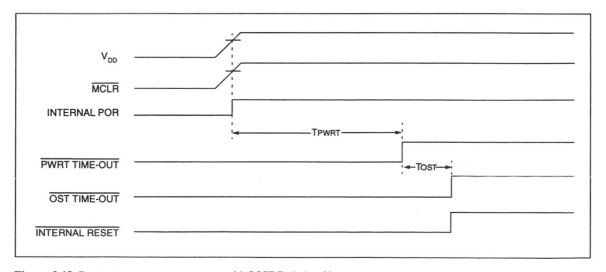

Figure 2.12 Reset sequence on power-up, with **MCLR** tied to V_{DD}

Summary

- The PIC 16 Series is a diverse and effective family of microcontrollers.
- The 16F84A architecture is representative of all 16 Series microcontrollers, with Harvard structure, pipelining and a RISC instruction set.
- The PIC 16F84A has a limited set of peripherals, chosen for small and low-cost applications. It is thus a smaller member of the family, with features that are a subset of any of the larger ones.
- The 16F84A uses three distinct memory technologies for its different memory areas.
- A particular type of memory location is the Special Function Register, which acts as the link between the CPU and the peripherals.
- Reset mechanisms ensure that the CPU starts running when the appropriate operating conditions have been met, and can be used to restart the CPU in case of program failure.

References

2.1. PIC 16F84A Data Sheet (2001). Microchip Technology Inc., Reference no. DS35007B; www.microchip.com

2.2. PICmicro Mid-Range MCU Family Reference Manual (1997). Microchip Technology Inc., Reference no. DS33023A; www.microchip.com

2.3. PIC16F84 to PIC16F84A Migration (2001). Microchip Technology Inc., Reference no. DS30072B; www.microchip.com

2.4. 8-bit Microcontroller with 2K Bytes Flash (2000). Atmel Corporation, AT89C2051, Rev. 0368E-02/00; http://www.atmel.com/

2.5. Atmel 8051 Microcontrollers Hardware Manual (2004). Atmel Corporation, Ref. 4316C-8051-05/04; http://www.atmel.com/

3
Parallel ports, power supply and the clock oscillator

So far we have looked a little at the theory of microcontroller architecture, and its implementation in PIC® microcontrollers. This chapter now begins to move from that theory to the practice of small-scale hardware design.

As we have seen, the microcontroller core has internal data and address buses. In a way these are like motorways, or inter-state freeways, carrying large amounts of traffic in both directions to a variety of different destinations. The microcontroller needs to be provided with a way of allowing that data flow to connect with the outside world, so that it can read in external digital values or output other values. In other words, it needs the equivalent of motorway junctions, where data can leave (or enter) the bus at designated times and locations. In the microcontroller world these junctions have many forms, as there are many different ways that data can be input or output. The most general purpose of these is the parallel input/output port. This is one of the microcontroller's most essential peripherals, and is the opening subject of this chapter.

Given a working car engine, two essentials that it needs to run are fuel and a stream of sparks from the plugs. A microcontroller has similar needs. Its fuel is the low-level electrical power supply that it requires, and instead of a flow of sparks, it needs a regular sequence of clock oscillator pulses. A study of these forms the second half of this chapter.

Putting together our background knowledge already gained, an ability to work with digital input/output, and an ability to design a power supply and clock oscillator, we will be in the happy position of being able to start to design real systems.

In this chapter you will learn about:

- Why we need parallel input/output
- How simple logic circuits can be developed to give a flexible interface between the microcontroller data bus and the outside world – these are the parallel ports
- How external devices can be connected to the parallel port
- The parallel input/output available on the PIC 16F84A
- The essential hardware features of power supply and clock oscillator
- The Microchip approach to power supply and oscillator, with the 16F84A
- The hardware design of the electronic ping-pong game.

3.1   The main idea – parallel input/output

Almost any embedded system needs to transfer digital data between its CPU and the outside world. This transfer falls into a number of categories, which can be summarised as:

- *Direct user interface*, including switches, keypads, light-emitting diodes (LEDs) and displays
- *Input measurement information*, from external sensors, possibly being acquired through an analog-to-digital converter
- *Output control information*, for example to motors or other actuators
- *Bulk data transfer* to or from other systems or subsystems, moving in serial or parallel form, for example sending serial data to an external memory.

With this plethora of data coming and going, it is likely we will need to have a variety of digital inputs and/or outputs. These are divided broadly into serial and parallel. In serial data transfer, the information is transferred one bit at a time. Only a single interconnection is used to carry the data itself, although other lines are usually included for synchronisation and control. In parallel data transfer, a set (for example, eight) of interconnections is used. Each of these can carry 1 bit, and each works in parallel with the others. Data can thus be transferred in groups of bits, for example in bytes. Parallel input/output (I/O) is the workhorse for all the basic data interchange of a microcontroller, including interfacing with switches, LED, displays and so on. A group of parallel I/O interconnections, appearing on the pins of the microcontroller, is called a *parallel port*.

3.2   The technical challenge of parallel input/output

Our immediate challenge is how to provide the required interface between the microcontroller data and address buses and the outside world. As suggested above, we start with the data bus, a multi-purpose data highway. How can we grab the data we want from the bus, and transfer it to the outside world, via the parallel port? Alternatively, how can we take external input data, and introduce it onto the data bus, at the right time and place, so that it gets to the right place within the microcontroller? Finally, given a port that can do these things, how can we make it really flexible, so that it can be used for input, or output, or a mixture of both, and can transfer a combination of data with possibly very different end uses.

3.2.1   Building a parallel interface

It should be simple to create a set of output pins to create an *output* port (Figure 3.1). Let us assign an address in the memory map to the port. Whenever that address is selected by an instruction in the program, it activates a line called *Port Select*. A further line, *Read/$\overline{Write}$*, indicates whether the CPU is undertaking a Read (line is high) or Write (line is low) operation. This is gated with the Port Select line. Each line of the data bus is connected to a bistable, and all of these are clocked by the Port Select line. Then the value of the data bus is latched into the bistable whenever the port memory location is addressed, in Write mode. The outputs of the bistables are made available for connection to the outside world.

It is equally simple to create a set of input pins (Figure 3.2). All that is needed is a tristate buffer gate connected between an external pin and a line of the data bus. When the buffer is enabled, again by a

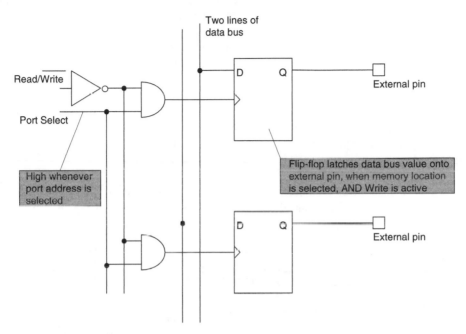

Figure 3.1 Two bits of a possible digital output port

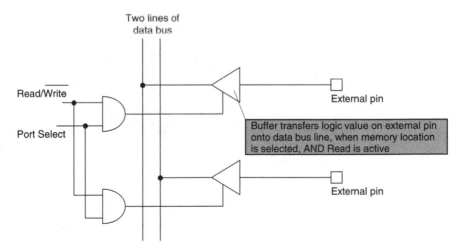

Figure 3.2 Two bits of a possible digital input port

logical combination of Port Select line and Read/Write control, the logic value of the external pin is briefly connected to the data bus line, and can be read by the CPU. Note that in this design the external data is not latched by the port, it must be held at a stable value by the external source.

These ideas are quite attractive, but the reality is that it is inflexible to limit an external pin of an IC to just one function, whether input or output. It would be much neater to combine somehow the two circuits used for input and output, and let the user decide which direction he/she wants the data to move. The diagram of Figure 3.3 does just that. It shows a possible 'pin driver' circuit for one bit of a parallel port. It is easy to pick out in it the circuits of Figures 3.1 and 3.2. What must be added, however, is a further flip-flop ('Direction'), which is set to determine whether this microcontroller pin is to act as an input or output. The state of this flip-flop is set by the program. It controls the 'output buffer', which is enabled when the port bit is in output mode.

This circuit forms the basis for a very useful bi-directional input/output pin driver, and it is easy to find versions of it in many popular microcontrollers. Sets of I/O pins are grouped together to form a parallel

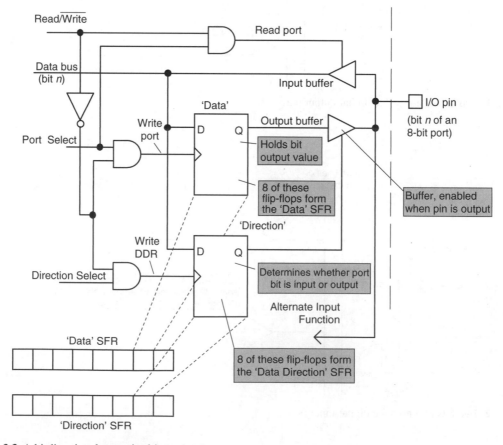

Figure 3.3 A bi-directional port pin driver circuit

I/O port. Each 'Data' flip-flop then forms one bit of a 'Data' SFR (Special Function Register), and each 'Direction' flip-flop forms one bit of a 'Direction' SFR, as seen in Figure 3.3. Each SFR is memory mapped, with its own unique address. Derived from that address is its select line, which goes high when that location is addressed. *Port Select* selects the Data SFR and *Direction Select* selects the Direction SFR.

By writing to the Direction SFR the user can determine which bits are to be input and which are to be output. By writing to the Data SFR he/she can set the value of *all* Data flip-flops, whether that pin is actually set as an output or not. This value is transferred to the I/O pin through the buffer for those pins which are enabled as outputs. By reading from the Data SFR the program can acquire the logical value of the I/O pin. If the pin is set as output, this value is simply the value held by the Data flip-flop and asserted on the I/O pin through the Output Buffer. If the pin is set as an input, then an external signal should be connected to the pin, and the controller will read its value.

Having established this basic design, it is possible to extend it further to add other features. We will see this when we look at some PIC microcontroller examples. One simple extension is already indicated in Figure 3.3, however. This is the 'Alternate Input Function' line, which allows an internal peripheral to share the I/O pin.

3.2.2 Port electrical characteristics

Logic gates are designed to interface easily with each other, and if we connect logic gates from just one family together then we usually don't need to worry about the electrical details of what is going on. If, however, we are connecting logic devices (in this case microcontroller port bits) to *non*-logic elements (like LEDs or switches) then we *do* need to understand the electrical characteristics of the logic. In particular, we need to understand their input and output characteristics.

The output of a logic gate *can* be visualised, or 'modelled', as in Figure 3.4(a). If the output is at Logic high (or '1'), then the internal switch is in the upper position. It is in the lower position for Logic 0. In either case, the output is modelled as a voltage source in series with a resistor (in circuit theory this is called a 'Thevenin equivalent' circuit). V_{LH} is the logic high output voltage, with an output resistance of $R_{S(high)}$. V_{LL} is the logic low output voltage, with an output resistance of $R_{S(low)}$.

In the case of CMOS (Complementary Metal Oxide Semiconductor) the situation is quite simple, as V_{LH} is equal to the supply voltage and V_{LL} is equal to 0 V. This is illustrated in Figure 3.4(b). Thus, if the supply voltage is 5 V, then Logic 0 and 1 will be 0 and 5 V respectively, if no current is being drawn from the gate output.

In practice, $R_{S(high)}$ and $R_{S(low)}$ are not constant, but depend to some extent on the current being sourced or sunk from the gate output. Therefore, manufacturers frequently publish graphical information on the output characteristics. We will see this shortly for the 16F84A.

3.2.3 Some special cases

We review now two special types of I/O characteristic, which will be important as we explore the 16F84A parallel ports.

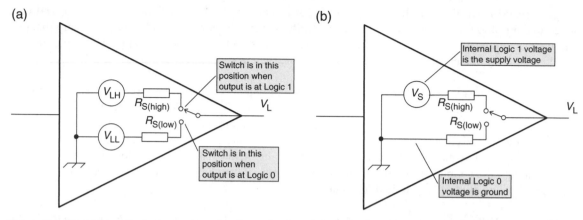

Figure 3.4 Modelling a logic gate output. (a) Generalised model. (b) Model of CMOS logic gate output

Schmitt trigger inputs

A Schmitt trigger (Figure 3.5) is a certain type of logic gate input which is designed to 'clean up' a corrupted logic signal. It has two input thresholds, with the 'positive-going' higher than the 'negative-going'. A signal starting from a low value has to pass the negative-going threshold (at which point nothing happens) and then cross the 'positive-going' threshold, at which point the output changes state. The output will not reverse until the input (now negative-going) has got right back down to the negative-going threshold. Thus, small fluctuations which recross a threshold just crossed do not cause any change in output.

The 'Open Drain' output

The Open Drain output is a flexible style of output that can be adapted either as a standard logic output, *or* as a direct drive for small loads, *or* used for a special logic function known as 'Wired-OR'. The output itself is as illustrated in Figure 3.6(a). A logic gate drives the gate of a MOSFET (Metal Oxide Semiconductor Field Effect Transistor), whose unconnected Drain terminal forms the output. When the MOSFET gate drive is high, the FET conducts and a logic zero is asserted at the terminal. When the gate is low the FET will not conduct and (with no other connection) the terminal will be at an undefined voltage. If a pull-up resistor is

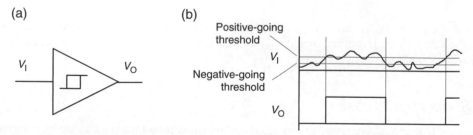

Figure 3.5 Schmitt trigger characteristics. (a) Buffer with Schmitt trigger input. (b) Input/output characteristic

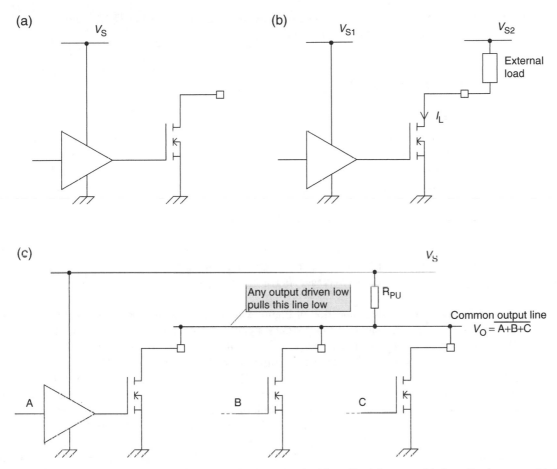

Figure 3.6 The Open Drain output and some applications. (a) An 'Open Drain' output. (b) Open Drain output driving load resistor. (c) The 'Wired-OR' connection

connected from the Drain to the supply voltage, then the output acts more or less as a normal logic output. Without the active pull-up of a normal logic output, however, its rise time will be a little sluggish and the amount of current it can source will be limited by the resistor value.

The Open Drain output can also be used to drive a simple load, acting as illustrated in Figure 3.6(b). Usefully, the load does not have to be supplied from the same voltage as the logic supply, although it would have to be of the same polarity. Therefore, for example, a microcontroller supplied from 5 V (V_{S1} in the diagram) could drive a load supplied from 12 V (V_{S2} in the diagram), if all operating requirements are met.

Another important application of the Open Drain output is the 'Wired-OR' connection, shown in Figure 3.6(c). Here several Open Drain outputs are connected together and tied high through a single pull-up resistor, R$_{PU}$. If all outputs are off, then the common line (V_O) is high. If *any* one output goes low, then the common line is pulled low. This is a possible way of achieving the OR or NOR logic function, and important for certain types of serial link, as we shall see later.

3.3   Connecting to the parallel port

3.3.1   Switches

Switches are extensively used in embedded systems. Our main initial interest is not to switch directly a voltage or current, but to convert the switch position to a logic level that can be read by a microcontroller port bit. Switches are used as direct user interface in the form of push-buttons, toggle switches, slide switches, or as thumbwheel or rotary switches. They are also used, in the form of microswitches, to detect certain types of mechanical movement.

The simplest way of deriving a logic level from a switch is shown in Figure 3.7(a). This shows a single-pole, double-throw (SPDT) switch, with one terminal connected to ground and the other to the supply. The switch wiper simply selects one of these two as the logic input. Some logic families advise against direct connection of a logic input to the supply voltage, so a series resistor (shown dotted) might be in order.

There is a slight disadvantage to the connection of Figure 3.7(a), as it requires the SPDT switch. A simpler option, using just a single-pole, single-throw (SPST) switch, for example a push-button, is shown in Figure 3.7(b). Here a pull-up resistor is connected to one terminal of the switch, with the other terminal connected to ground. If the switch is closed, then the input to the logic gate, V_I, is 0 V and a current V_S/R flows to ground. If the switch is open then V_O is equal to V_S. To reduce wasted current when the switch is closed, the value of R should be high. If it is too high, however, then the Logic 1 level that it is meant to define may not be properly sustained. To evaluate the upper limit of the pull-up resistor, the input leakage current and logic thresholds need to be applied (as demonstrated in Chapter 2 of Ref. 1.1). For PIC microcontrollers, pull-up values in the range 10–100 kΩ are usually appropriate. The circuit of Figure 3.7(b) is very useful and widely applied, as many simple switches (e.g. PCB-mounting slide switches and push-buttons) are only available as SPST.

The switch circuit of Figure 3.7(b) *can* be reconnected as in Figure 3.7(c). The characteristics of some logic families (for example, TTL) do, however, place restrictions on the use of this circuit, as the current sourced

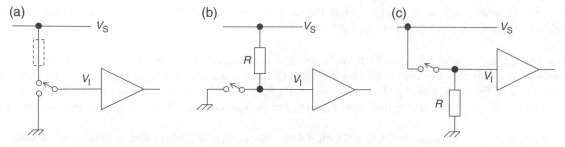

Figure 3.7 Connecting switches to logic inputs. (a) SPDT connection. (b) SPST with pull-up resistor. (c) SPST with pull-down resistor

from the gate input significantly affects the action of the pull-down resistor. The circuit can be applied with PIC microcontrollers.

3.3.2 Light-emitting diodes

In certain semiconductor materials light is emitted as current flows across a forward-biased p-n junction. LEDs exploit this phenomenon. LEDs made of gallium arsenide (GaAs) emit light in the infrared, and if phosphorus is added in increasing proportions, the light moves to visible red and ultimately to green. LEDs are widely available in red, green and yellow, as single devices, and as arrays, bargraphs and alphanumeric displays.

Because they are diodes, LEDs display the normal voltage – current relationship of a forward-biased diode. This means that, to a reasonable approximation, the voltage across an LED is constant, if it is conducting. Note, however, that this forward voltage is considerably higher for GaAs than it is for silicon. Example LED characteristics, for red and green Kingbright LEDs, are shown in Figure 3.8. From these graphs it can be seen that the voltage across the red LED changes from 1.90 to 2.00 V if the current increases from 5 to 20 mA. For the green it changes from 1.95 to 2.20 V for the same current range. These voltage values are typical for all LEDs of similar type, with red having a slightly lower forward voltage, compared to green or yellow.

The different colours also do not give equal intensities for equal drive currents, as shown in the data in Figure 3.8. Red is the most efficient, which may account for its greater popularity. For a single LED to be

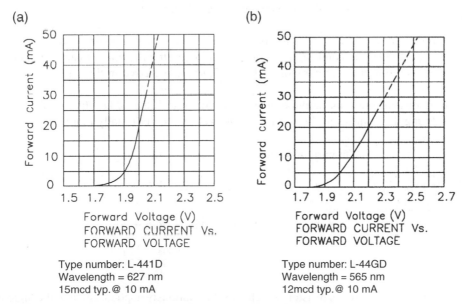

Figure 3.8 Example Kingbright LED characteristics [Ref. 3.1]. (a) High-efficiency red. (b) Green. Reproduced with permission of Kingbright Elec. Co. Ltd

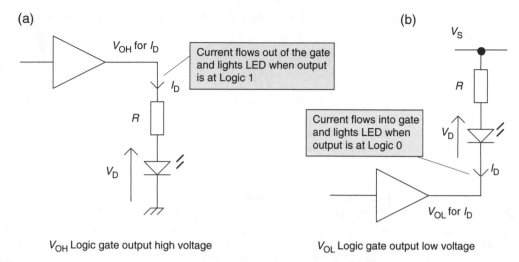

Figure 3.9 Driving LEDs from logic gates. (a) Gate output sourcing current to LED. (b) Gate output sinking current from LED

comfortably visible, it typically requires around 10 mA of current. Brighter ones may require up to 20 mA, but special low-power devices (such as the high-efficiency red) need as little as 1 or 2 mA to be seen.

An LED can be driven from a logic output, for example a microcontroller port, as long as its current requirements can be met. Depending on the capabilities of the port output they can be connected so that the output is sourcing current (Figure 3.9(a)) or sinking current (Figure 3.9(b)).

CMOS logic families have symmetrical outputs and can source or sink almost equally well, so either of these circuits can be applied. In contrast, TTL logic can source little current but can sink a comparatively large amount, and therefore the configuration of Figure 3.9(b) is preferred in this case.

A current-limiting resistor must normally be included in series with the LED. This is calculated as shown below by considering the voltages in the circuit. Precise values are not usually required.

$$\text{For current source:}\quad V_{\text{OH}} = R I_{\text{D}} + V_{\text{D}}$$

$$R = \frac{V_{\text{OH}} - V_{\text{D}}}{I_{\text{D}}} \tag{3.1}$$

$$\text{For current sink:}\quad V_{\text{S}} = V_{\text{OL}} + R I_{\text{D}} + V_{\text{D}}$$

$$R = \frac{V_{\text{S}} - V_{\text{D}} - V_{\text{OL}}}{I_{\text{D}}} \tag{3.2}$$

An exception to the need for a series resistor, which must be cautiously applied, is when the logic is powered from a comparatively low voltage, and its internal output resistance itself forms an appropriate value for the current-limiting resistor.

3.4 The PIC 16F84A parallel ports

We saw in Chapter 2 that the 16F84A has two ports, A and B. A is 5-bit, while B is 8-bit. Notice from Figure 2.1 that some port bits have more than one function. We will see that the 16F84A adapts the generic pin driver circuit of Figure 3.3 and cleverly weaves in these extra functions.

The SFRs that relate to the ports are seen in Figure 2.5. In each case the Port data itself appears in the **PORTA** or **PORTB** register (i.e. these act as the 'Data' SFR of Figure 3.3), while the data direction is determined by the bit values set in the **TRISA** or **TRISB** registers (i.e. these act as the 'Direction' SFR of Figure 3.3).

We will now explore the ports in some further detail. Perhaps the most straightforward is Port B, with which we accordingly start.

3.4.1   The 16F84A Port B

This is a general-purpose 8-bit bi-directional port, with pin driver circuit similar to that in Figure 3.3. The simplest bits, 0 to 3, are illustrated in Figure 3.10(a). The data latch can be seen in each circuit, while the 'TRIS latch' in Figure 3.10 replaces the 'Direction' latch of the earlier diagram. It can be seen that if the 'TRIS latch' output is set to 0, then the buffer that it drives is enabled and the port bit is in output mode.

There are four enhancements to the simple pin driver circuit we saw earlier:

* The incoming data is latched, through the lowest latch in the diagram, rather than just its instantaneous value being read.
* The state of the 'TRIS latch' can be read, via the buffer controlled by the **RD TRIS** line. It follows that the **TRIS** register acts as a normal read/write memory location, and the program can check if necessary the values previously stored there.
* Bit 0 is also the external interrupt input and has a Schmitt trigger interface.
* 'Weak pull-up' resistors can be switched on, for all port bits used as inputs. These can be applied to replace the resistor in circuits like in Figure 3.7(b). The pull-up is implemented with a p-channel MOSFET, seen at the top of the diagram. They are enabled for all port bits set as input by clearing the bit $\overline{\text{RBPU}}$ in the **OPTION** register. (This is seen memory mapped in Figure 2.5 or in full in Figure 6.9.)

Bits 4 to 7 of Port B are seen in Figure 3.10(b). They have a useful 'interrupt on change' facility. As with the lower numbered bits, the data value is latched as input data is read. On these bits, however, the previous input value, from the last time the port was read, is retained on another latch. Its stored value is compared with the current input value. Any difference is detected by an Exclusive OR gate, whose output can generate an interrupt. This capability will be considered in detail in Chapter 6.

3.4.2   The 16F84A Port A

Like Port B, this can be used as a general-purpose bi-directional digital port. The basic port pin driver (Figure 3.11(a)) is very similar to the Port B pin. The diagram this time draws out in full the output

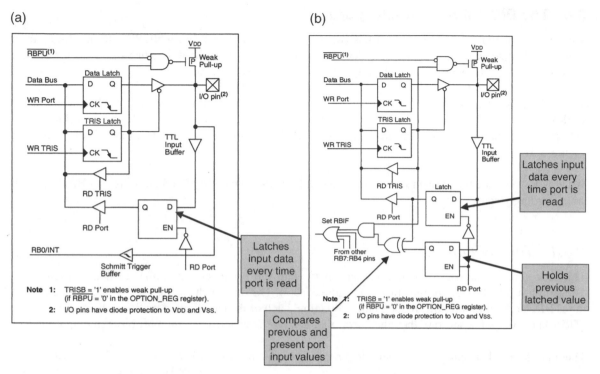

Figure 3.10 Block diagram of Port B pin driver circuits. (a) Pins RB3 to RB0. (b) Pins RB7 to RB4 (supplementary labels in shaded boxes added by the author)

tristate buffer. Bit 4 (Figure 3.11(b)) doubles as the Timer 0 clock input. It also has a Schmitt trigger input characteristic and an Open Drain output, as described in Section 3.2.3. The full device data indicates that the absolute maximum permissible voltage applied to this Open Drain pin is 8.5 V. Therefore, the ability to drive an external load from a supply higher than the microcontroller itself can only be applied in a limited way.

3.4.3 Port output characteristics

The 16F84A port output characteristics are shown in Figure 3.12, for a supply voltage of 3.0 V. In Figure 3.12(a) we see (at 25°C) how the output voltage for Logic 1 is 3 V when the output current is 0, but falls to around 1.7 V when the output current is 10 mA, flowing out of the gate. Similarly, in Figure 3.12(b) we see (at 25°C) how the output voltage for Logic 0 is 0 V when the output current is 0, but *rises* to around 0.8 V when the output current is 22.5 mA (flowing into the gate). It is curves like these that can be used to find the V_{OL} and V_{OH} values used in equations (3.1) and (3.2), once a value for I_D is known. Graphs are also given in the full data for characteristics with a 5 V supply.

Another way of applying these curves is to deduce from them an approximate output resistance. This can be done by measuring the gradient of the curve at a particular point. A simple construction to do

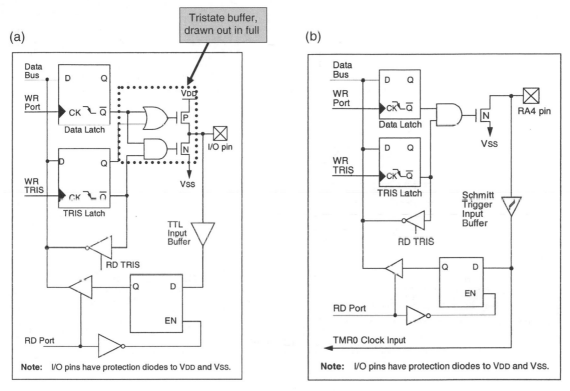

Figure 3.11 Block diagram of Port A pin driver circuits. (a) Pins RA0, 1, 2 and 3. (b) Pin RA4/T0CKI (supplementary labels in shaded boxes added by the author)

this has been added to each plot. By dividing vertical (voltage) by horizontal (current) for each of these, output resistances of *approximately* 130 Ω when at Logic high and 36 Ω when at Logic 0 can be deduced. If we call these two values R_{OH} and R_{OL} respectively, equations (3.1) and (3.2) can be written in a different form:

For current source: $V_S = (R + R_{OH})I_D + V_D$

$$R - \frac{V_S - V_D}{I_D} - R_{UH} \qquad (3.3)$$

For current sink: $V_S = (R + R_{OH})I_D + V_D$

$$R = \frac{V_S - V_D}{I_D} - R_{OH} \qquad (3.4)$$

(a)

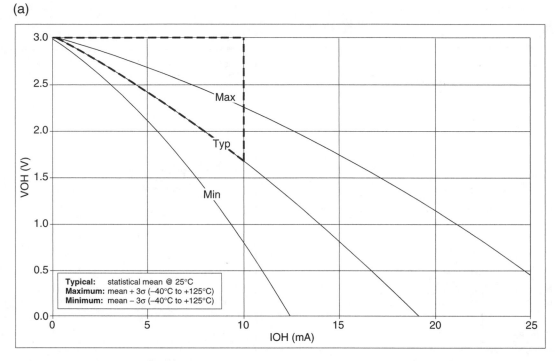

(b)

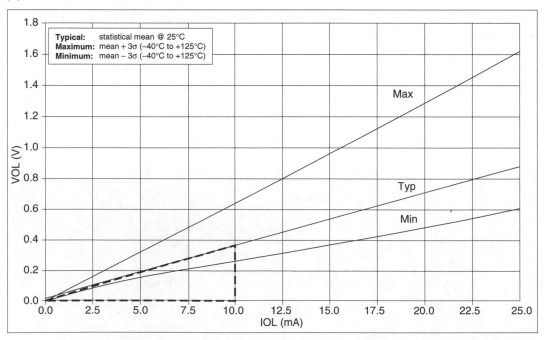

Figure 3.12 16F84A port output characteristics. (a) V_{OH} vs. I_{OH} ($V_{DD} = 3\,V$, -40 to $125°C$). (b) V_{OL} vs. I_{OL} ($V_{DD} = 3\,V$, -40 to $125°C$) (Dashed lines added by the author.)

3.5   The clock oscillator

The choice of microcontroller clock source determines some of its fundamental operating characteristics. While 'faster is better' in terms of operating speed and programming execution, faster is definitely worse in terms of power consumption, and also possibly in terms of electromagnetic interference. All timed elements within the microcontroller almost invariably depend on the clock characteristics. If stable and accurate timing is required, then the clock oscillator must be stable and accurate. With these points in mind, the clock source must be chosen with care and understanding. This section starts with a review of the clock technologies available, before moving on to looking at the options offered with the 16F84A.

3.5.1   Clock oscillator types

Broadly, there are two types of oscillator circuit in common use in microcontrollers, as illustrated in Figure 3.13. In the resistor–capacitor (RC) type (Figure 3.13(a)), a capacitor is charged through a resistor from the supply rail. The capacitor voltage drives the input of a Schmitt trigger buffer. When the Schmitt trigger threshold is exceeded, its output goes high, switching on the MOSFET transistor to which it is connected. The capacitor is quickly discharged, the Schmitt output goes low, the MOSFET is switched off and the charging process starts again. This continues for as long as power is maintained. The clock signal is taken from the rectangular waveform generated at the Schmitt output. This simple circuit is integrated onto many larger ICs requiring a clock signal. Users are then usually required to connect resistor and capacitor externally, choosing these to set the desired frequency. It is important to note, however, that RC oscillators can be implemented entirely on-chip. RC oscillators are very low cost and produce a clock signal very reliably. As resistor, capacitor, power supply and Schmitt trigger threshold values all vary with temperature, their frequency is not very stable. They cannot therefore be used where precise timing is required.

The crystal oscillator (Figure 3.13(b)) depends on the piezo-electric properties of quartz crystal. Any mechanical distortion of the material causes a voltage to be produced across opposite sides of it; similarly, if a voltage is applied to the material, a mechanical distortion results. Crystals are carefully cut into very thin

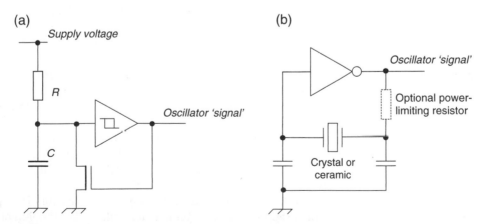

Figure 3.13 Microcontroller oscillator generator circuits. (a) Resistor–capacitor (RC). (b) Crystal or ceramic

slices (usually discs), have tiny electrodes attached and are mounted so that they can vibrate. When connected in the feedback path across a logic inverter, as the figure shows, the crystal can be forced through piezo-electric action into mechanical vibration. This translates into electrical oscillation, an oscillation that is sustained by the action of the logic gate. Small value capacitors connected from either side of the capacitor to ground optimise the electrical conditions needed for this oscillation.

Crystal vibration occurs at a fixed and remarkably stable frequency – this is the great advantage of the crystal oscillator. The crystals themselves tend to be on the expensive side (although cost continues to fall) and mechanically fragile. An alternative is the ceramic resonator. This has similar piezo-electric properties to the crystal and is connected in an identical way. It is, however, both lower in cost and rather less stable in frequency. Crystals are the only option when precise timing functions, derived from the clock oscillator, are required.

3.5.2   Practical oscillator considerations

All microcontroller manufacturers go a long way to making it easy to create a clock waveform for their microcontrollers. Usually, this is done by including the circuits of Figure 3.13, possibly in merged form, on-chip. One may be forgiven, therefore, for thinking that setting up the oscillator on a microcontroller is a straightforward thing – in fact it isn't, and unreliable or non-functioning oscillators are a cause of real frustration with novice builders. Oscillator frequency shows greater or lesser dependence on supply voltage, temperature, humidity, printed circuit board (PCB) layout and possibly other factors. Crystals in particular are sensitive to poor PCB layout. It is important to exclude parasitic resistance, capacitance or inductance by having very short PCB tracks, therefore locating the crystal close to the body of the microcontroller.

3.5.3   The 16F84A clock oscillator

The 16F84A can be configured to operate in four different oscillator modes, allowing implementation of RC, crystal or ceramic oscillators. These are detailed below. It can also accept an external clock source. The user selects which mode is to be used by setting bits in the Configuration Word (Figure 2.6).

- *XT – crystal*. This is the standard crystal configuration. It is intended for crystals or resonators in the range 1–4 MHz.
- *HS – high speed*. This is a higher drive version of the XT configuration. It recognises that higher frequency crystals, and ceramic resonators in general, require a higher drive current. It is intended for crystal frequencies in the region of 4 MHz or greater, and/or ceramic resonators. It leads to the highest current consumption of all the oscillator modes.
- *LP – low power*. This mode is intended for low-frequency crystal applications and gives the lowest power consumption possible. In many cases this will be 32.768 kHz (i.e. 2^{15}), which is the most popular frequency for low-power, time-sensitive applications, for example wristwatches. It will, however, operate at any frequency below around 200 kHz.
- *RC – resistor–capacitor*. For this an external resistor and capacitor must be connected to pin 16, replicating the circuit of Figure 3.13(a). This is the lowest cost way of getting an oscillator, but should not be used when any timing accuracy is required. The nominal frequency of oscillation can be predicted with limited

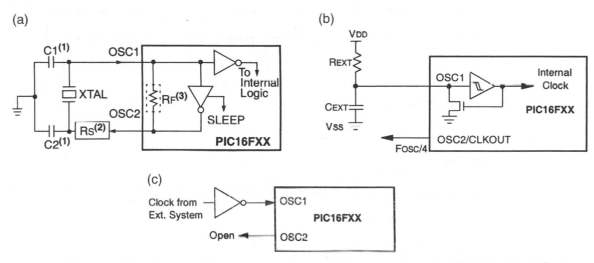

Figure 3.14 Ways of supplying a clock waveform to the 16F84A. (a) Crystal or ceramic, HS, XT or LP. (b) Resistor–capacitor. (c) Externally supplied clock

accuracy only, and even then it will drift with changing temperature, supply voltage and time. An example of use of the RC oscillator appears in the electronic ping-pong case study at the end of this chapter.

As seen in Figure 2.1, the 16F84A has two oscillator pins, OSC1 (pin 16) and OSC2 (pin 15). Between these lies a logic inverter and associated circuitry. Figure 3.14 shows the possible oscillator configurations that can be connected using these pins. Either a crystal or a ceramic can be connected to create the oscillator circuit of Figure 3.14(a). Any of the three speed ranges outlined above can be invoked through the Configuration Word. An RC oscillator can also be used, as shown in Figure 3.14(b). The approximate oscillation frequency can be selected by consulting graphical information given in the Electrical Characteristics section of the data sheet, for example as seen in Figure 3.15. Finally, an external clock source can simply be connected to the OSC1 pin (Figure 3.14(c)). Further guidance on oscillator design for Microchip microcontrollers can be found in Ref. 3.2.

3.6 Power supply

3.6.1 The need for power, and its sources

Like any electronic circuit, a microcontroller and the overall embedded system need to be supplied with electrical power. Traditionally, much logic circuitry is supplied at 5 V, arising from the voltage specified for the TTL (Transistor Transistor Logic) logic family. With the growth in battery-powered equipment and developments in electronic technology, supply voltages have been pushed down, and 3.3 and 3.0 V supplies are now common.

Operating conditions for electronic components are specified in the manufacturer's data sheet. In terms of power supply there are two important issues: the supply voltage required and the current that the device

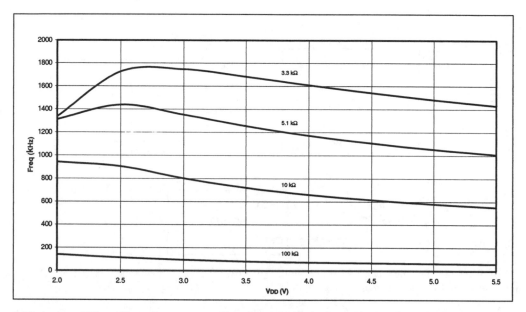

Figure 3.15 Average RC oscillator frequency vs. V_{DD} for variable R, $C = 100\,\text{pF}$, $25°\text{C}$

will then take from the supply. This supply current will be dependent on operating frequency. Also given are *absolute maximum ratings*, which give voltage and power dissipation levels beyond which the device must not be taken.

3.6.2   16F84A operating conditions

The essential operating conditions of the 16F84A are shown in Figure 3.16. From this it can be seen that a supply voltage of between 4.0 and 5.5 V is required, unless the HS oscillator mode is used. In this case the supply voltage must not be below 4.5 V. In Sleep mode (when all program execution is suspended and the oscillator is switched off), the supply voltage can be dropped right down to 1.5 V and the data in RAM is still retained. If operation from lower supply voltages is required, then the 16LF84A should be used.

Looking further down the table, we see how much supply current depends on oscillator frequency. A typical supply current of 1.8 mA can be expected when running at 4 MHz with a supply voltage of 5.5 V. If the oscillator frequency is increased to 20 MHz, then the supply current rises to 10 mA. It's worth mentioning that both these values are actually very good, and compare well with many other, more power-hungry microcontrollers. If we want to operate at really low currents, however, then look what the 16LF84A offers at low frequency – a staggering 15 μA!

You may recognise that, for a battery-powered system, the required supply voltage of the 16F84A makes a three-cell alkaline battery supply a useful option. This gives a supply of around 4.5 V. Suppose you powered with three AA cells, each with a nominal capacity of 800 mAh. Running at 1.8 mA would give a battery life of 444 hours, or 18.5 days. Running at 10 mA would give 80 hours, or 3.3 days, while 15 μA consumption would lead to 53 333 hours, or 2222 days, or just over six years! In this case battery self-discharge would

Param No.	Symbol	Characteristic	Min	Typ†	Max	Units	Conditions
	VDD	**Supply Voltage**					
D001		16LF84A	2.0	—	5.5	V	XT, RC, and LP osc configuration
D001		16F84A	4.0	—	5.5	V	XT, RC and LP osc configuration
D001A			4.5	—	5.5	V	HS osc configuration
D002	VDR	**RAM Data Retention Voltage (Note 1)**	1.5	—	—	V	Device in SLEEP mode
D003	VPOR	**VDD Start Voltage** to ensure internal Power-on Reset signal	—	Vss	—	V	See section on Power-on Reset for details
D004	SVDD	**VDD Rise Rate** to ensure internal Power-on Reset signal	0.05	—	—	V/ms	
	IDD	**Supply Current (Note 2)**					
D010		16LF84A	—	1	4	mA	RC and XT osc configuration (**Note 4**) FOSC = 2.0 MHz, VDD = 5.5V
D010		16F84A	—	1.8	4.5	mA	RC and XT osc configuration (**Note 4**) FOSC = 4.0 MHz, VDD = 5.5V
D010A			—	3	10	mA	RC and XT osc configuration (**Note 4**) FOSC = 4.0 MHz, VDD = 5.5V (During FLASH programming)
D013			—	10	20	mA	HS osc configuration (PIC16F84A-20) FOSC = 20 MHz, VDD = 5.5V
D014		16LF84A	—	15	45	µA	LP osc configuration FOSC = 32 kHz, VDD = 2.0V, WDT disabled

Note 1: This is the limit to which V_{DD} can be lowered without losing RAM data.
Note 2: Gives further information on factors that influence supply current.
Note 3: Gives guidance on how to calculate current consumed by the external RC network, when this is used.

Figure 3.16 The 16F84A basic operating conditions

potentially be significant. The above calculations of course only take account of the consumption of the microcontroller, and not of any other parts of the circuit.

An important opportunity for conserving power is through the 'Sleep' mode. This is introduced in Section 6.6 of Chapter 6.

3.7 The hardware design of the electronic ping-pong

The electronic ping-pong project was introduced in Chapter 1. Its circuit diagram can be seen in Appendix 2, Figure A2.1. We are now in a position to understand every detail of its circuit design. Power is supplied from two AAA cells, which are connected to the V_{SS} and V_{DD} pins of the microcontroller via an on–off switch. Because the power supply is only 3 V, an LF version of the microcontroller is used. A 100 nF decoupling capacitor across the power supply smoothes voltage spikes which may be induced due to the action of the microcontroller internal circuitry. $\overline{MCLR}$ is simply tied to the supply rail, as no Reset function is needed for this simple game.

It can be seen that an RC oscillator is used. This is reasonable, as it is a cost-conscious application, with no time-critical elements. Figure 3.15 shows that for the values used, and with a supply voltage of 3.0 V nominal, the oscillator frequency will be 800 kHz.

Let us now look at the use of the parallel ports. It can be seen that the two player paddles, connected to bits 3 and 4 of Port A, follow the pattern of Figure 3.7(b), with 10k pull-up resistors. The score and out-of-play LEDs take up the remaining bits of Port A, and the 'ball flight' LEDs are all connected to Port B. All LEDs are high-efficiency types and are connected according to Figure 3.9(a). Noting the approximate 130 Ω output resistance derived in Section 3.4.3 of this chapter, the total resistance in series with each LED is $(560 + 130)$ Ω. With a forward voltage across the LED of around 1.8 V, the current is given by applying equation (3.3), i.e.

$$I = (3 - 1.8)/(560 + 130)$$

$$= 1.7\,\text{mA approx.}$$

This current value is just adequate for this type of application and LED, where only close-up viewing is expected. It would in general be viewed as low.

Summary

- The parallel port allows ready exchange of digital data between the outside world and the controller CPU.
- It is important to understand the electrical characteristics of the parallel port and how they interact with external elements.
- While there is considerable diversity in the logic design of ports, they tend to follow similar patterns. The internal circuitry is worth understanding, as it leads to effective use of ports.
- The 16F84A has diverse and flexible parallel ports.
- A microcontroller needs a clock signal in order to operate. The characteristics of the clock oscillator determine speed of operation and timing stability, and strongly influence power consumption. Active elements of the oscillator are usually built in to a microcontroller, but the designer must select the oscillator type, and its frequency and configuration.
- A microcontroller needs a power supply in order to operate. The requirements need to be understood and must be met by a supply of the appropriate type.

References

3.1. Kingbright Elec. Co. Ltd. Taiwan; http://www.kingbright.com.tw
3.2. Overview, Design Tips and Troubleshooting of the PICmicro™ Microcontroller Oscillator (2001). Microchip Technology Inc., Reference no. DS33023A; www.microchip.com

4
Starting to program – an introduction to Assembler

Embedded system design is made up of two main aspects, the hardware and the software. In the early days of microprocessors, systems were built up laboriously using a large number of integrated circuits (ICs). Memory was very limited, so only small programs could be written. Slowly, the available ICs became more and more sophisticated, and the designer had to do less to get a working hardware system. Meanwhile memory was growing, so longer programs could be written. Now we are in a situation where memory is plentiful and cheap, and the hardware is sophisticated and readily available. Complex hardware systems can be built up with comparative ease, and in many projects software development is now the main creative activity. In this chapter we start down the long but exciting road to developing good programs. We start that road using the Assembler programming language, but later in the book continue it using the high-level language C.

We have one problem if we are to start programming. What will the program run on? Ultimately, of course, embedded systems programs are written to run on the target system hardware. You may be working with an educational PIC® hardware system, or you may have the electronic ping-pong. In many cases, however, we don't want to be dependent on hardware to try out a programming idea. What can really cause a study of programming to spring to life is a simulator – a program running on a desktop computer that will run the program we have developed. Therefore, we make it a priority in this chapter to introduce the Microchip MPLAB® Integrated Development Environment, and the simulator in it. Once you have the skill to use this, then most program ideas can be tried out very quickly, and you should be able to make rapid progress in the noble but tricky art of microcontroller programming!

In this chapter you will learn about:

- Some aspects of the underlying issues of computer programming
- The essentials of Assembler programming and how to write simple Assembler programs
- Development environments for programming and the Microchip MPLAB Integrated Development Environment
- The PIC 16 Series instruction set in overview
- The use of certain PIC 16 Series instructions
- Simulating software and the MPLAB software simulator MPSIM™.

You will also, if you wish, be able to learn about:

- How the RISC instruction set of the PIC 16 Series compares with the instruction set of a comparable CISC microcontroller
- The details of how the PIC 16 Series instruction word is constructed.

4.1   The main idea – what programs do and how we develop them

The four main ideas of computer programming, according to this author, are listed here:

(1) A computer has an *instruction set*; it can recognise each instruction and *execute* it.
(2) The program that the computer executes is a list of instructions drawn from its instruction set; it reads these in binary from its program memory. The program in this form is called *machine code*.
(3) To execute, the computer works relentlessly through the instructions of the program, from the beginning, doing exactly what each instruction tells it to do – nothing more, nothing less – except when temporarily diverted by an interrupt.

So far this is simple, but here is the difficult one:

(4) The programmer must find a means of breaking down and translating his/her ideas into steps that the computer can undertake, where each step ultimately must be an instruction from its instruction set.

4.1.1   *The problem of programming and the Assembler compromise*

The problem of programming is summarised in Figure 4.1. We as humans express our ideas in complex and often loosely defined linguistic forms. A computer reads and 'understands' binary, and responds in a precise way to precise instructions. It is ruthlessly logical and does exactly what it is told.

Given this linguistic divide, how can a programmer write programs for a computer? Three ways of bridging the gap present themselves:

(1) *The human learns machine code*. This is what programmers used to do sometimes in the very early days, laboriously writing each instruction in the binary code of the computer, exactly as the computer

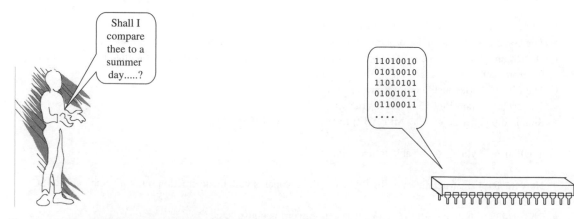

Figure 4.1 The problem of programming

would then read it. This is incredibly slow, tedious and error-prone, but at least the programmer relates directly to the needs and capabilities of the computer.

(2) *Use a high-level language (HLL)*. This is as if we go some way to asking the computer to learn our language. In an HLL, instructions are written in a form that relates in a recognisable way to our own language – in the case of people reading this book, probably English. Another computer program, either a compiler or an interpreter, then converts that program into the machine code that the computer can comprehend. The programmer now has a much easier time and can write very sophisticated programs. He/she is now, however, separated from the resources of the computer, and the program may be comparatively inefficient in terms of its use of memory and in its execution speed.

(3) *Use Assembler*. This is a compromise position. Every one of the computer's instructions set is given a *mnemonic*. This is usually a three- or four-letter word that can be used to represent directly one instruction from the instruction set. The programmer then writes the program using the instruction mnemonics. The programmer has to think at the level of the computer, as he/she is working directly with its instructions, but at least the programmer has the mnemonics to use, rather than actually working with the computer machine code. A special computer program called a *Cross-Assembler*, usually these days running on a PC, converts the code written in mnemonics to the machine code that the computer will see. Because there is a computer doing the conversion from the assembler code to machine code, a number of other benefits can be built into the process. For example, the Cross-Assembler can look after most of the business of allocating memory space in program memory, and it can accept labels for numbers and memory locations, greatly easing the programmer's task.

In the early days of computing, programming in Assembler was used to program almost any type of computer. These days, however, it is pretty much the preserve of embedded designers, particularly when using smaller 8-bit devices. For the embedded designer Assembler offers the huge advantage that it allows him/her to work directly with the resources of the computer, and leads to efficient code, which executes quickly. Because it is so directly linked to the computer structure, working in Assembler helps the user to learn the structure of the computer. Programming in Assembler has the disadvantage that it is rather slow, error-prone and does not always produce well-structured programs. We will aim to resolve this conundrum in later chapters. For now, in order to write simple programs and understand the microcontroller more, we will learn Assembler.

4.1.2   The process of writing in Assembler

The actual process of writing in Assembler is illustrated in Figure 4.2. The programmer writes in the microprocessor or microcontroller Assembly language. This *can* be done using nothing more than a text editor. We will soon recognise the two lines of Assembler program in Figure 4.2 as being from the PIC 16 Series instruction set. The computer he/she is writing on runs the Cross-Assembler. The terminology *Cross-Assembler* implies that one computer is assembling code for one of another type, not for itself. Usually, and somewhat confusingly, Cross-Assembler is shortened simply to Assembler. The Cross-Assembler *assembles* the program, i.e. it converts it from Assembler mnemonics into machine code ready for the microcontroller. In Figure 4.2 the Cross-Assembler is seen converting the two lines of assembler code into the 14-bit machine code words of the PIC 16 Series. For most microcontrollers there are then special programming tools that can download the program in machine code from the main PC and program it into the microcontroller program memory.

Figure 4.2 Programming in Assembler

4.1.3 The program development process

The process of writing in Assembler needs to be placed in the broader context of project development. The possible stages in the development process for the program of a simple embedded system project are shown in Figure 4.3. The programmer writes the program, called the *source code*, in Assembler language.

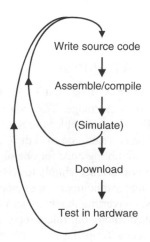

Figure 4.3 Developing a simple project

This is then assembled by the Cross-Assembler running on the host computer. If the programmer has access to a simulator then he/she may choose to test the program by simulation. This is likely to lead to program errors being discovered, which will require alteration to the original source code. When satisfied with the program, the developer will then download it to the program memory of the microcontroller itself, using either a stand-alone 'programmer' linked to the host computer or a programming facility designed into the embedded system itself. He/she will then test the program running in the actual hardware. Again, this may lead to changes being required in the source code.

Clearly, to develop even a simple project, a selection of different software tools is beneficial. These are usually bundled together into what is called an *Integrated Development Environment* (IDE).

4.2 The PIC 16 Series instruction set, with a little more on the ALU

4.2.1 More on the PIC 16 Series ALU

Before looking at the 16 Series instruction set, it is worth taking a more detailed look at the ALU (Figure 4.4). Understanding this will aid in understanding the instruction set. Looking at this, we see that the ALU can operate on data from *two* sources. One is the W (or 'Working') register. The other is *either* a *literal* value *or* a value from a data memory (whose memory locations Microchip call 'register files'). A literal value is a byte of data associated with a particular instruction that the programmer embeds in the program.

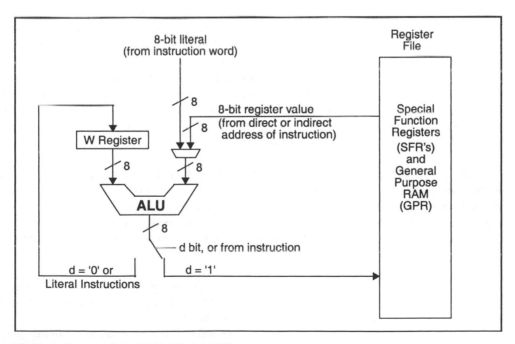

Figure 4.4 Block diagram of the PIC 16 Series ALU

Thus, we can expect to see some instructions that call on data memory and others that require literal data to be specified whenever they are used. Examples of all are coming! The data that the instruction operates on, or uses, is called the *operand*. Operands can be data or addresses. We will see that some types of instructions always need an operand to be specified with them, others do not.

Once an instruction has been executed, where is the result stored? For many instructions Microchip offer a choice, whereby the result can *either* be held in the W register *or* stored back in data memory. Which one is used is fixed by certain instructions; in others it is determined by the state of a special **d** bit, which is specified within the instruction.

4.2.2   The PIC 16 Series instruction set – an introduction

Turn now to the PIC 16 Series instruction set, which can be found in Appendix 1. Take a long hard look at it – we are aiming to get to know it extremely well! You can see that the table is divided into six columns, and each of the 35 instructions gets one line. The first column gives the actual mnemonic, together with the code specifying the type of operand it acts on. There are four such operand codes:

- **f** for file (i.e. memory location in RAM), a 7-bit number
- **b** for bit, to be found within a file also specified, a 3-bit number
- **d** for destination, as described above, a single bit
- **k** for literal, an 8-bit number if data or 11-bit if address.

The second column summarises what the instruction does. In some cases this gives adequate information. A much fuller description of how each instruction works can also be found in the full microcontroller data [Ref. 2.1]. The third column shows how many instruction cycles the instruction takes to execute. As a RISC processor, we expect this to be a single cycle. This turns out to be the case, apart from those instructions that cause a branch in the program. We discuss their use in Chapter 5. The fourth column gives the actual 14-bit opcode of each instruction. This is the code that the Cross-Assembler produces, as it converts the original program in Assembler language to machine code. It is interesting to see here how the operand codes, listed above, become embedded within the opcode. The fifth column shows which bits in the Status register (Figure 2.3) are affected by each instruction.

Let us immediately look at five example instructions, to see how the information is presented. As an aside, let us note now that Assembler programming does not have to be case sensitive, and that all the examples in this book are *not* case sensitive. Therefore, do not worry if you see instruction mnemonics and operands appearing in either upper or lower case in different references. In this book, for stylistic reasons, we choose to write Assembler programs in lower case. Find now each of the instructions below in the Instruction Set Table in the Appendix:

- **clrw** – this clears the value in the W register to zero. There are no operands to specify. Column 5 tells us that the Status register **Z** bit is affected by the instruction. As the result of this instruction is always zero, the bit is always set to 1. No other Status register bits are affected.
- **clrf f** – this clears the value of a memory location, symbolised as **f**. It is up to the programmer to specify a value for **f**. Again, because the result is zero, the Status register **Z** bit is affected.

- **addwf f,d** – this adds the contents of the W register to the contents of a memory location symbolised by **f**. It is up to the programmer to specify a value for **f**. There is a choice of where the result is placed, as discussed above. This is determined by the value of the operand bit **d**. Because of the different values that the result can take, all three condition code bits, i.e. **Z**, the Carry bit **C**, and the Digit Carry bit **DC** are affected by the instruction.

- **bcf f,b** – this instruction clears a single bit in a memory location. Both the bit and the location must be specified by the programmer. The bit number **b** will take a value from 0 to 7, to identify any one of the 8 bits in a memory location. No Status register flags are affected, even though it is possible to imagine that the result of the instruction could be to set a memory location to zero.

- **addlw k** – This instruction adds the value of a literal, whose value **k** must be specified by the programmer, to the value held in the W register. The result is stored in the W register; there is no choice. Like **addwf**, all condition code bits can be affected by this instruction.

4.3   Assemblers and Assembler format

4.3.1   Introducing Assemblers and the Microchip MPASM™ Assembler

For any microprocessor or microcontroller, there are a large number of (Cross-) Assemblers available. Some are distributed free by the makers of the processors to encourage people to buy their products. Others, usually more sophisticated, are written by specialist software houses and sold commercially. Many these days come as part of an IDE, as mentioned in Section 4.1.3. This book uses MPASM, the Assembler offered by Microchip. It is usually used as part of the MPLAB IDE, and both MPASM and MPLAB are introduced in some detail later in this chapter and the next.

While many aspects of Assembler programming are common across all Cross-Assemblers, some are specific to the particular Assembler that is in use.

4.3.2   Assembler format

Having taken a first look at the instruction set, we need now to understand how we can build these instructions into a program. Assembler programs have a simple format, which must be understood and followed. This is shown in Program Example 4.1.

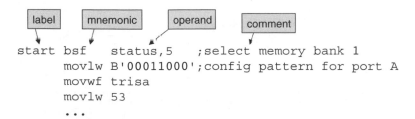

Program Example 4.1 Assembler format

There are four possible elements to an Assembler line of code:

- *Label*. A label for a line is optional. When it is first specified, the label must start in the left-most space of the line. The Assembler will interpret anything starting in this space as a label. Once defined in this way, a label can be used as an operand. Labels must start with an alphabetic character or underscore, but not a number. Labels can stand alone on a line, in which case the label is adopted by the next line that contains an instruction.
- *Instruction mnemonic*. This is drawn from the instruction set. It may be placed anywhere on the line, except starting at the far left. It should be separated from any label by at least one blank space.
- *Operand*. These must conform exactly to the format specified in the instruction set. For better intelligibility, labels are often used rather than numbers. If there is more than one operand they are separated by a comma.
- *Comment*. This is optional, and is used to add information to the program and improve its intelligibility to the human reader. A comment must always start with a semicolon. The Cross-Assembler ignores everything that follows a semicolon in any line. Comments can follow instructions on a line; alternatively, a whole line can be used just for commenting.

A line of the Assembler program can contain an instruction, properly formatted as above, it can be a comment only, or it can be left completely blank (this sometimes helps to improve layout and readability).

4.3.3 Assembler directives

While the Assembler program is written for the target microcontroller, it has to be processed by the Assembler first. To aid this process and make it more powerful and flexible, a way is needed of passing information and instructions to the Assembler, which it recognises as being for its attention only. These instructions are called *Assembler directives*. They are used for very diverse applications, for example defining the target processor or specifying where the program must be placed in memory. A few MPASM examples are shown in Table 4.1. These are written in the code and appear almost like mnemonics from the instruction set. Their very distinct role must, however, be recognised.

4.3.4 Number representation

One of the features of working close to the inner operations of a microcontroller is that sometimes one is thinking in binary, sometimes in decimal and sometimes in hexadecimal, or even octal. Therefore, it is

Table 4.1 Some common MPASM Assembler directives

Assembler directive	Summary of action
list	Implement a listing option*
#include	Include additional source file
org	Set program origin
equ	Define an assembly constant; this allows us to assign a value to a label
end	End program block

*Listing options include setting of radix and of processor type.

Table 4.2 Number representation in MPASM Assembler

Radix	Example representation
Decimal	D'255'
Hexadecimal	H'8d' or 0x8d
Octal	O'574'
Binary	B'01011100'
ASCII	'G' or A'G'

helpful for the Assembler program to be able to recognise and respond to different number bases. MPASM does this first by allowing a default to be set. Thus, for example, if one wants to work only (or mainly) in hexadecimal, then all numbers can be interpreted as such. Any number that the programmer wants to represent in an alternative radix must be prefixed, as shown in Table 4.2. In Program Example 4.1, the programmer is writing for a default radix of hexadecimal. In the second line of the example, however, he wishes to specify a number in binary, so therefore uses the appropriate format from Table 4.2. In the fourth line he is using the hexadecimal number 53 as an operand. As hexadecimal is the default radix, its number base does not need to be specified explicitly.

Note that a hexadecimal number must not start with an alphabetic character, otherwise it might be interpreted as a label. Therefore, any hexadecimal number starting with a, b, c, d, e or f must be preceded with a zero. Thus, for example, the number b2$_H$ must be entered as 0b2.

4.4 Creating simple programs

4.4.1 A simple data transfer program

Let's now look at a simple example program, written for the MPASM Assembler, using the MPLAB IDE. It is shown in Program Example 4.2. The program is written for the electronic ping-pong hardware (Appendix 2), but we will soon use it for simulation.

```
;*********************************************************************
;ELECTRONIC PING-PONG DATA MOVE
;This program moves push button switch values from Port A to the
;leds on Port B
;TJW 21.2.05                                    Tested 22.2.05
;*********************************************************************
;
;Configuration Word: WDT off, power-up timer on,
;                    code protect off, RC oscillator
;
      list p=16F84A
;
;specify SFRs
status     equ     03
porta      equ     05
```

Program Example 4.2 A simple data transfer program

```
trisa    equ    05
portb    equ    06
trisb    equ    06
;
         org    00
;Initialise
start  bsf    status,5        ;select memory bank 1
       movlw  B'00011000'
       movwf  trisa          ;port A according to above pattern
       movlw  00
       movwf  trisb          ;all port B bits output
       bcf    status,5       ;select bank 0
;
;The "main" program starts here
       clrf   porta          ;clear all bits in ports A
loop   movf   porta,0        ;move port A to W register
       movwf  portb          ;move W register to port B
       goto   loop
       end
```

Program Example 4.2 Continued

The program starts with a header made up of five comment lines, each starting with a semicolon. These define the program title, briefly describe what it does, and give the date the program was written and the author. Information on Configuration Word settings (Figure 2.6), fundamental to the running of the program, is then given. We will find that there is more than one way to get this information programmed into the microcontroller. The first active line of the program follows – all lines to here have been comments. This defines the microcontroller to be used, using the **list** directive.

A section follows which uses the **equ** directive to define the memory locations of the SFRs that will be used. It comes as some surprise to many people that it is necessary to do this. Haven't we just 'told' the Assembler what the processor is, so shouldn't it 'know'? The answer is that it doesn't, so we must supply this information. This program just uses the Status register, Ports A and B, and their control registers **TRISA** and **TRISB**. Labels for these are therefore defined, taking memory addresses directly from the memory map of Figure 2.5. Remember (from Section 2.4.2) that the Bank Select bit is held in the Status register. Once this is removed from the SFR addresses shown in Figure 2.5, then the labels **porta** and **trisa**, and **portb** and **trisb** have the same values. In the program it would make some sense to use just one label for each of these pairs, instead of the two. We choose not to do this in this program example, for better clarity when the different locations are used.

Before the actual program starts it is essential to use the **org** directive to define the program start address. We have no choice over the address used – it must be the reset vector address, as seen in Figure 2.4.

The program that follows makes use of seven instructions, all of which manipulate bits and bytes of data, except for one branch instruction. These are:

- **clrf f** – this clears to zero the value in memory location **f**
- **movwf f** – this moves the contents of the W register to the memory location **f**

- **movf f,d** – this instruction moves the contents of the memory location **f** to the W register, *if* the **d** bit is set to 0; if it is set to 1, then the contents of **f** are just returned to **f**
- **movlw k** – this instruction moves the literal value **k**, an 8-bit number which accompanies the instruction, into the W register
- **bcf f,b** – this clears (i.e. sets to Logic 0) the bit **b** in memory location **f**
- **bsf f,b** – this sets to Logic 1 the bit **b** in memory location **f**
- **goto k** – this transfers program execution to the instruction in memory location **k**.

The initialisation section of the program follows. This sets up the direction of each bit of the two ports that are used, and this requires access to the port control registers **TRISA** and **TRISB**. As these are placed in RAM memory bank 1, it is necessary first of all to set bit 5 of the Status register to 1 (as explained in Chapter 2). This is done in the first line of actual program, labelled **start**, using the **bsf** instruction. The label **status** can be used, because it was defined earlier in the program. If this had not been done, then it would have been necessary to write:

```
start      bsf     3,5      ;select memory bank 1
```

which would have been somewhat less intelligible.

The port pin directions needed are derived from the circuit diagram (Figure A2.1). From this we can see that the two push-buttons connect to bits 3 and 4 of Port A, which must accordingly be set up as inputs. The three other bits of Port A are all connected to LEDs, so must be set up as outputs. As described in Section 3.4, to be an output a port pin must have a 0 in its corresponding **TRIS** register bit. It must have a 1 for the bit to be an input. Therefore, we must send the word 00011000 to **TRISA**. Note that **TRISA** is an 8-bit location, even though Port A only has 5 bits. It is therefore necessary to specify a complete 8-bit word to be sent, even for those 3 bits that are not implemented. There is no instruction that allows us to transfer a byte of data directly from the program into a memory location, so two lines of code must be used. First, the required word 00011000 is placed into the W register, using a **movlw** instruction. In doing this, the binary radix is used (Table 4.2), instead of the default hexadecimal. The contents of the W register are then moved to **trisa** using a **movwf** instruction. A similar process is followed for setting up Port B. A quick look at the circuit diagram shows that all Port B pins are connected to LEDs, so all must be set as output. Therefore, the word sent to **trisb** is 00_H. The initialisation section ends with memory bank 0 being selected in the Status register, as from here on the ports themselves will be accessed, whose locations are in bank 0.

Finally, we reach the effective program itself, all five lines of it! The program continuously reads the value of Port A and transfers it to Port B. Thus, if either push-button is pressed, this should be seen on the LEDs connected to bits 3 and 4 of Port B. When Port A is read, all of its 5 bits are read, even though three are set as outputs. For these, the values of the internal data latches (Figure 3.11) are read. All Port A bits are therefore initially cleared to 0 in the program, using a **clrf** instruction.

The actual data transfer part of the program uses a **movf** instruction to shift the value of Port A to the W register, followed by a **movwf** instruction to move the W register value to Port B. A **goto** instruction creates a continuous loop, making use of the earlier defined label **loop**.

It is worth observing here that label values are assigned in two different ways. Some, like **porta** or **portb**, are assigned a specific value by the programmer, using the **equ** directive. Others, like **loop**, are inserted into the program, and the Assembler allocates them a value.

The program ends with an **end** directive.

4.5   Adopting a development environment

4.5.1   Introducing MPLAB

MPLAB is an IDE that can be downloaded free from Microchip's website [Ref. 1.2]. There is also a copy on the book CD. It contains all the software tools necessary to write a program in Assembler, assemble it, simulate it and then download it to a programmer. The latter must be built or bought, or designed in to the target system. Further software tools can be bought and then integrated with MPLAB, both from Microchip and from other suppliers. This includes alternatives to what MPLAB already offers – e.g. Assemblers or simulators, as well as tools which offer much greater development power, like C compilers or emulator drivers.

MPLAB is a continuously evolving package, with its own manuals [Refs 4.1, 4.2] and on-line Help facility. Therefore, this book does not aim to act as a full MPLAB manual. It will, however, aim to give a clear introduction to its use, so that you can begin to apply it with confidence. Screen images from MPLAB Versions 7.00 and 7.22 are used in this chapter and the next.

4.5.2   The elements of MPLAB

MPLAB is made up of a number of distinct elements, which work together to give the overall development environment. These are:

- *Text editor*. This allows entry of the source code. It behaves to some extent like a simple text editor such as Notepad, but it can recognise the main elements of the programming language that is being used. Thus, in Assembler it colour-codes instructions in one colour, labels in another and comments in a third. In this way the programmer can immediately see if there is a misconception in his placing or use of text within the Assembler line.
- *Project manager*. The preferred way of developing programs in MPLAB is by creating a *project*. An MPLAB project groups all the files together that relate to any one project, and ensures that they interact with each other in an appropriate way and are updated as needed.
- *Assembler and Linker*. The function of the Assembler has already been discussed. So far we have assumed that there is a single source file. In advanced projects, however, the code may be created from a number of different files. The role of the Linker is to put these together, give each its correct location in memory, and ensure that branches and calls from one file to the other are correctly established.
- *Software simulator and debugger*. A software simulator allows the program that has been developed to be tested, by running it on a simulated CPU in the host computer. Inputs can also be simulated, and outputs and memory values can be observed. The debugger contains the tools which allow program execution

Table 4.3 Some file extensions used in MPLAB IDE

File extension	Function
.asm	Assembly language source file
.err	Error file
.hex	Machine code in hex format file
.inc	Assembly language include file
.lib	Library file
.lst	Absolute listing file
.o	Object file
.mcp	Project information file
.mcw	Workspace information file

to be fully examined, for example by single stepping through the program, or running at slow speed, or halting at a particular location.

4.5.3 The MPLAB file structure

Even with simple projects, a significant number of files are rapidly generated in MPLAB. Each type is designated by the file extension used, examples of which are given in Table 4.3. Whenever a project is set up, files of type **.mcp** and **.mcw** are created. When using Assembler, the original source code is written in a file with the **.asm** extension. The source code may include an **.inc** file, to be described in Chapter 5. When the source code is assembled successfully, the output appears in **.lst** and **.hex** files. If there is an error, that is placed in an **.err** file.

4.6 An introductory MPLAB tutorial

This tutorial takes you through the stages of creating a project, writing simple source code and assembling that to create output files. To follow the tutorial, you should download and install the current version of MPLAB, if it is not available in your place of work or study.

Open the MPLAB IDE, which should appear as in Figure 4.5. If a blank Output window also opens, close it. The main screen is blank, apart from the Workspace window at top left, which cannot be closed.

4.6.1 Creating a project

Click the Project button on the toolbar to access the pull-down menu, as shown in Figure 4.6.

There are two ways to create a project, both accessible from this menu. One is by using the **Project Wizard** and the other by selecting **New…**. Try following the Project Wizard route, making the following selections

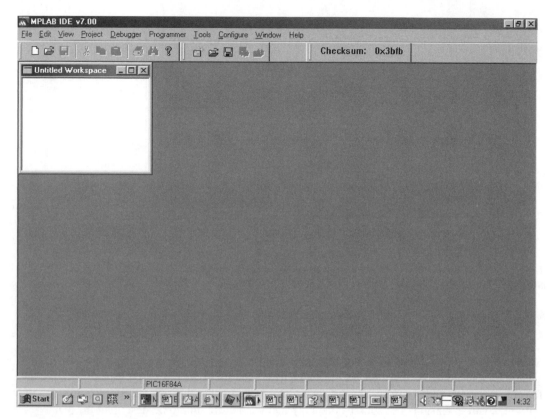

Figure 4.5 The MPLAB IDE screen

as you work through the dialogue boxes:

Device:	**PIC16F84A**
Active Toolsuite:	**Microchip MPASM Toolsuite**
which will display:	

       MPASM Assembler
       MPLINK Object Linker
       MPLIB Librarian

as Toolsuite contents

Project Name	*<your own choice>*
Project Directory	*<your own choice>*
Add existing files…	Make no additions

When you click **Finish**, the Workspace window should be updated to show the filename you have selected, as seen in Figure 4.7(a), for a project called **fred**.

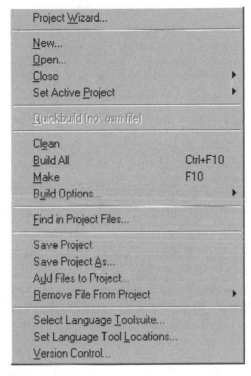

Figure 4.6 Project pull-down menu

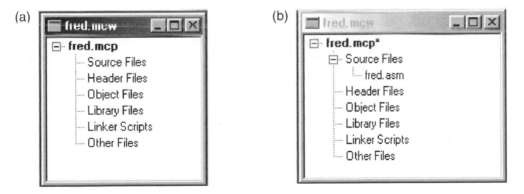

Figure 4.7 Workspace window. (a) For a newly created project. (b) For a project with source file added

4.6.2   Entering source code

Now open a new file by clicking **File > New,** and start to enter into it the program of Program Example 4.2. After a few lines save this using **File > Save As....** Select file type **Assembly Source File,** and save as *<your project name>***.asm**. Continue entering the code and notice now that MPLAB has identified this file as an Assembler Source File. It applies colour-coding to label, instruction mnemonic, numerical data, Assembler directive and comment. When complete, go to the Project menu again, click **Add Files to Project...** and select the one you have just saved. Your Workspace window should now appear as in Figure 4.7(b), with of course your own file name. You will now begin to appreciate how valuable this window is to become, as it shows a complete picture of the files associated with your project.

4.6.3 Assembling the project

Now comes one of the testing moments in any project development. You have entered new source code, and you need to know if it assembles correctly. The Assembler subjects your code to a series of checks. It returns errors if it finds incorrect use of Assembler format, instruction mnemonics, labels or a range of other things. Remember, however, that the Assembler can effectively only check that your program is correct grammatically; it cannot assure you that it is a viable program. Above all else, it has no knowledge of the target hardware, beyond the fact that the microcontroller has been specified. Correct assembly does not guarantee correct program operation!

Check that the default radix is correctly set by clicking **Project > Build Options > Project > MPASM Assembler** and ensuring that **Hexadecimal** is selected in the dialogue box. In the same dialogue box you can enable or disable case sensitivity for all the source code. This is not necessary if you have directly copied Program Example 4.2. You may need to use it in future, however.

Invoke the MPASM Assembler by pressing **Project > Build All**. This also ensures all files are updated as needed. The **Output** window will open, reporting on the progress of the build. In the Output window you will either get a 'Build succeeded' message or a 'Build failed', together with a fleeting box showing a green (for success) or red (for errors) bar.

Whether your build has initially succeeded or failed, open the file *<your project name>*.**lst**. This should be in the directory you specified for the project. Use **File > Open**, and ensure you select **All** Files in the dialogue box against **Files of Type**. The **.lst** file is very informative and gives you the original source code, alongside the assembled machine code, together with any errors and warnings that may have been generated. Part of the list file for Program Example 4.2 is shown in Program Example 4.3. Notice both how the machine code is represented and that, after the main program listing, there is essential underlying program information.

```
                        00029        ;The "main" program starts here
0006    3000            00030            movlw 00     ;clear all bits in ports A and B
0007    0085            00031            movwf porta
0008    0086            00032            movwf portb
0009    0805            00033   loop     movf     porta,0 ;move port A to W register
000A    0086            00034            movwf    portb   ;move W register to port B
000B    2809            00035            goto     loop
                        00036            end
```

Program Example 4.3 Part of the Data_Move list file

```
MPASM 03.90 Released                    DATA MOVE.ASM    3-10-2005
15:55:03              PAGE 2

SYMBOL TABLE
   LABEL                          VALUE

__16F84A                         00000001
loop                             00000009
porta                            00000005
portb                            00000006
start                            00000000
status                           00000003
trisa                            00000005
trisb                            00000006
.....etc
```

Program Example 4.3 Continued

If you have errors, each will be accompanied by an error number and message in the list file. In most cases these are simple typographical errors that can be easily fixed by correcting the source code and building again. If an error proves difficult, then full details on the error should be sought, either in the Help menu or in Ref. 4.1, by looking up its number. At the end of a development session, close the current project using **Project > Close**.

Once you have a source file which builds correctly, you are in a position either to download to microcontroller memory or to simulate. We continue now on the simulation path.

4.7   An introduction to simulation

The following subsection introduces the MPLAB simulator, MPSIM[TM], by means of a tutorial, simulating the program that has just been assembled.

4.7.1   Getting started

Once in MPLAB, select the simulator by invoking **Debugger > Select Tool > MPLAB SIM**. The simulator menu, as seen in Figure 4.8, then appears under **Debugger**.

4.7.2   Generating port inputs

This program applies the ping-pong hardware, so to simulate we will need to create simulated inputs for the two ping-pong paddles, on Port A, pins 3 and 4. There are two ways of generating simulated inputs, which depends on whether the input is to be synchronous with instruction execution or not. We will choose the simpler – asynchronous under user control.

Select **Debugger > Stimulus Controller > New Scenario**. Explore the dialogue box that appears – it allows you to set up different types of inputs at the Port pins, which are initiated by pressing the Fire button

Figure 4.8 Simulator menu

at the appropriate moment. Under **Pin**, select **RA3** and then **RA4**, with Toggle under Action for each. When you close the project your settings are saved as a **.stc** file.

4.7.3   Viewing microcontroller features

You can observe a number of microcontroller features during simulation, including program memory, SFRs, data memory and so on. A window can be opened for each of these, using the **View** menu. If you do this, however, you will find that the screen very quickly becomes cluttered. A Watch window allows you to make selections of only those variables you want to watch, while leaving out the others. Items for the **Watch** window are selected by using the pull-down menus at the top of the window. Open a Watch window, and select **PCL**, **TRISA**, **PORTA**, **TRISB** and **PORTB**. Looking ahead, the Watch window will appear as seen in Figure 4.10.

4.7.4   Resetting and running the program

You can reset the simulated CPU either by the F6 button, or by using **Debugger > Reset**. Using the latter, four Reset categories are offered, reflecting the Reset capabilities of the PIC microcontroller (Section 2.8). Alternatively (and more simply), you can use the Reset button of the Debugger toolbar (Figure 4.9). If this is not displayed, invoke it by **View > Toolbars > Debug**.

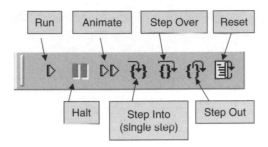

Figure 4.9 MPSIM software simulator debugger toolbar

There are three ways to run the program. Each can be selected under the **Debugger** pull-down menu or by selecting the buttons on the toolbar. These are:

- *Single Step*. This allows you to step through the program one instruction at a time. This version of MPLAB uses the terminology **Step Into** for this mode.
- *Animate*. This is like an automated single step. The program runs slowly but continuously, with the screen being updated after each instruction. The speed it runs at can be set by invoking **Debugger > Settings > Debugger Animation**.
- *Run*. This runs the program, but does not update on-screen windows as it runs. It does, however, accept stimulus input.

It is also possible to **Step Over** a subroutine or **Step Out** of one. Each of these also has a button on the toolbar. These are especially useful for delay routines, which on a simulator may take an unacceptably long time to simulate.

If you have followed the instructions to here, you should have a computer screen similar to that shown in Figure 4.10, although you are likely to have arranged the windows differently. In this image we see the Watch window top left, the Stimulus Controller top right and the source file bottom left. The simulated CPU has been reset, so the arrow representing the Program Counter is pointing to the first instruction. This can be confirmed by checking the **PCL** value in the Watch window.

Now, by repeated pressing of the **Step Into** button, single-step through the program. As the program moves through the initialisation, you will see the SFR values being changed in the Watch window and the **PCL** value being incremented. Program execution then circles around the loop, formed by the last three instructions. Now try 'firing' RA3 or RA4. Display windows are not updated with the value you have forced, until the next instruction execution. Observe Port A and Port B being updated in value, as you continue to execute the program with single-stepping. Try now pressing the **Animate** button. Notice that the program runs continuously, but still responds to stimulus inputs.

4.8   Downloading the program to a microcontroller

Most modern microcontrollers are equipped with on-chip program memory using Flash technology. The process of programming requires data to be transferred into the chip in a precisely timed way and certain

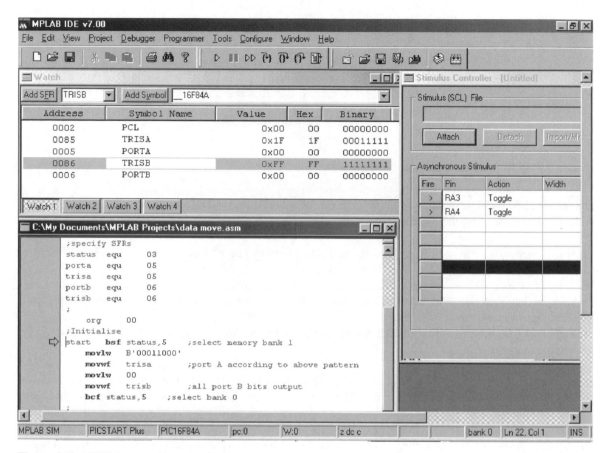

Figure 4.10 MPSIM set-up for a simple simulation

programming voltages to be applied, usually higher than the normal supply voltage. Certain microcontroller pins therefore usually have a secondary function, being used in programming mode to transfer the program data onto the chip and transmit the programming voltages.

In times past, the process of programming always used to require the IC carrying the memory (whether a stand-alone device or memory in a microcontroller) to be placed in a *programmer*. This was linked to a desktop computer for the process to be carried out. As memory technology has improved, however, the process has become simpler and it has become increasingly easy to design the necessary programming circuitry into the target system. This means that many microcontrollers can now be programmed *in situ*, i.e. within the target system. We will see these techniques in later chapters. In this chapter we will stay with traditional programmers, which require the microcontroller to be removed from the target circuit and placed into the programmer.

A popular and low-cost programmer, supplied by Microchip, is the PICSTART® Plus, shown in Figure 4.11. There are many alternatives to this, including many designs intended for home-build, which are available

Figure 4.11 The PICSTART Plus programmer

on the Web. The PICSTART programmer is connected to the host computer by a serial cable and MPLAB has the software to communicate with it. The PICSTART programmer can accept a wide range of dual-in-line microcontroller packages, from 18 to 40 pins. With adaptors it can program other package types.

The following steps take you through actually downloading code to the microcontroller, using the PICSTART Plus programmer. If you have a programmer and the ping-pong hardware, you can immediately download the program you have just created in the preceding tutorial.

You will need to power your PICSTART programmer and connect it to the serial port of your computer. From within MPLAB IDE, select **Programmer > Select Programmer > PICSTART Plus**. Then enable the programmer with **Programmer > Enable Programmer**. A positive response should be given via the Output window. If there is a problem, you may need to check **Programmer > Settings > Communications**. Ensure the programmer toolbar (Figure 4.12) is displayed. If not, find it with **View > Toolbars > Picstart**.

Put the Zero Insertion Force (ZIF) socket on the PICSTART programmer in the Open position. Place a 16F84A into it, ensuring from the legend that the chip is in the right place and is the right way round.

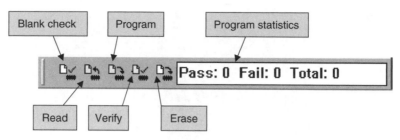

Figure 4.12 The MPLAB programmer toolbar

Then close the ZIF with the lever. With the project you want open on MPLAB, you should now be able to apply the features available to you, as summarised in Figure 4.12.

4.9 What others do – a brief comparison of CISC and RISC instruction sets

As they are CISC CPUs, we expect the Atmel 8051 microcontrollers to have a much larger instruction set, compared to the RISC 16 Series family. Furthermore, we might expect some of those instructions to be more powerful. The price to be paid for these is slower execution time. This expectation is proved correct. The 8051 core has 111 instructions. Most instructions execute in one machine cycle, each of which takes 12 oscillator cycles (compared with the four oscillator cycles for the PIC instruction cycle). Some instructions take two cycles. The two 'advanced' instructions **MUL** (multiply) and **DIV** (divide) take four cycles each.

The versatility offered by a CISC instruction set appears attractive when viewed from a RISC world. Simple actions that require several RISC instructions are reduced to a single instruction. Two simple examples are shown in Program Example 4.4. In the first, a byte of constant data (called literal data in the PIC world as we know, and *immediate data* in the 8051 world) is moved to a memory location called **mem_loc**. This requires two PIC instructions or one 8051 instruction. The programmer gains the advantage with the 8051 of writing less lines of code, but we can see that there is no apparent timing advantage in execution, as each ultimately takes two cycles. Indeed, as the 8051 machine cycle is longer, the timing advantage ultimately lies with the PIC microcontroller.

Interestingly, the advantage is less clear in the next example. The PIC 16 Series only has conditional skip instructions and not branch. Thus, conditional branches have to be built from a skip instruction followed by a **goto**. Here, three cycles are taken if the conditional branch is executed for the PIC microcontroller or two for the 8051, while a single line of code is required for the latter. Now the timing advantage lies with the 8051 CPU, as long as instruction cycles are equal in length.

PIC 16 Series	Atmel 8051

```
movlw   22      ;1 cycle          mov mem_loc,#22   ;2 cycles
movwf mem_loc   ;1 cycle
```

(a)

Program Example 4.4 Comparing RISC and CISC instruction capabilities. (a) Moving immediate/literal data to a memory location. (b) Branching if Carry bit set

```
btfsc status,0  ;1 cycle (no skip)     jc new_place      ;2 cycles
goto new_place  ;2 cycles
```

(b)

Program Example 4.4 Continued

The disadvantage of the extra lines of code required for the RISC processor is not a great one. We will see in the next two chapters that even this slight disadvantage can be ameliorated; in Assembler by using macros and then in a high-level language like C, where the programmer is no longer directly concerned by the number of lines of assembler code produced.

4.10 Taking things further – the 16 Series instruction set format

It is interesting to take a little time here to understand further the way the instruction code is made up of different component parts, as first discussed with the 12F508/9 in Chapter 1. The PIC 16 Series has four possible instruction word formats, as shown in Figure 4.13. The instruction word, which is transferred down the program bus (Figure 2.2), is made up of 14 bits. These appear as bits 0 to 13 in the figure. The opcode, the actual instruction part of the instruction word, always occupies the highest bits of the instruction word.

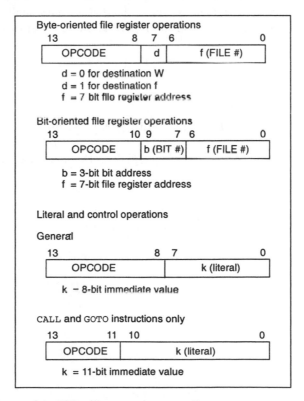

Figure 4.13 Instruction formats of the PIC mid-range microcontrollers

This is the part of the instruction word that ends up in the 'Instruction Decode and Control' unit of Figure 2.2, but it is not always the same length.

If the instruction is the type that contains a file address, then it is of the first format shown. The most significant 6 bits hold the opcode, while the least significant 7 bits are used to hold the address. These bits are transferred onto the 'Direct Addr' bus of Figure 2.2. In fact, because the F84A only has a small memory, only the least significant 5 bits are used, as can be seen from the 'Direct Addr' bus size indicated. Bit 7 holds the **d** bit. Different instruction word patterns are used for the other instruction categories. These can be seen and understood by reading the information in the figure.

Summary

If you have followed the material of this chapter then you have taken an enormous step forward – you are on your way to becoming a programmer of embedded systems! The key points are:

- Assembler is a programming language that is part of the toolset used in embedded systems programming. It comes with its own distinct set of rules and techniques.
- It is essential to adopt and learn an IDE when developing programs. The MPLAB IDE is an excellent tool for PIC microcontrollers, both for learners and professionals. And it can't be beaten on price!
- While some people are eager to get programs into the hardware immediately, it is extremely useful to learn the features of a simulator. The simulator in MPLAB allows the user to test program features with great speed, and is an invaluable learning tool.

References

4.1. MPASM User's Guide, with MPLINK and MPLIB (1999). Microchip Technology Inc., Document No. DS33014G.
4.2. MPLAB User's Guide (2005). Microchip Technology Inc., Document No. DS51519A.

5
Building Assembler programs

In Chapter 4 the basic rules of Assembler programming were introduced, along with some of the instructions from the PIC® 16 Series instruction set. It's as if we have now learned some introductory skills in bricklaying. We need to develop those skills further, but we need to begin to think about the structures that the bricks are going to be built into. Therefore, we now need to develop this introductory knowledge so that we can actually build up programs that have structure, and are functional and reliable.

In this chapter you will learn about:

- How to visualise a program and represent it diagrammatically
- How to use subroutines
- How to implement delays
- How to use look-up tables
- Logical and arithmetic instructions
- How to simplify and optimise Assembler programming
- More advanced features of software simulators.

5.1 The main idea – building structured programs

When we actually design a program that is to do anything more than some minimalist task, it is important to think about and plan its structure *before* starting to write the code. This is especially true in Assembler – Chapter 4 warned that one of the problems of Assembler programming was that it leads to unstructured 'spaghetti' programs. Therefore, we must consider means of representing the program diagrammatically. Let us consider how we might do this, with two examples of commonplace domestic products.

5.1.1 Flow diagrams

A well-established diagramming technique is the flow diagram. While this has many symbols that can be used, we can develop good flow diagrams with just two! – rectangle for process or action and diamond for decision.

Figure 5.1 shows a simple flow diagram example, for a refrigerator controller. The user has a single control, an adjustable potentiometer that allows him/her to set a desired temperature. Within the fridge there is a temperature sensor. Temperature is controlled by switching the compressor on or off – the temperature will fall when it is running. The program reads both the actual and demand temperatures, and determines which is higher. If it is the actual temperature, then the compressor is switched on. If the difference between the two is very great, then an alarm will sound. The flow diagram shows this action, using just the two symbols

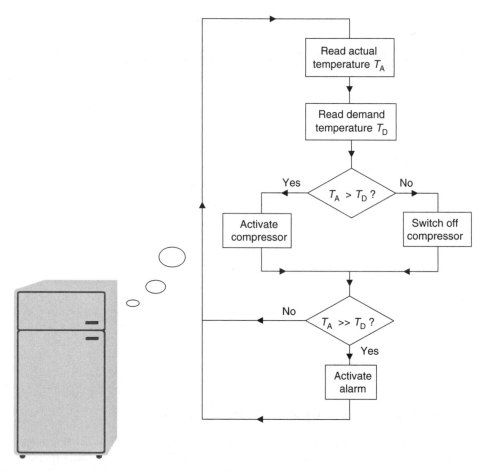

Figure 5.1 Flow diagram of simple refrigerator controller

mentioned above. Notice how each diamond decision symbol contains a question within it, with a yes/no answer. Its two exit points then correspond to the two possible answers. It can be seen that this example program will loop indefinitely. This is a common embedded system program structure and is sometimes called a *super loop*.

It is possible to draw flow diagrams with too much detail or too little. With care and experience, however, it is possible to draw them at a level such that the diagram can be converted to an Assembler program without too much difficulty. In some cases, an overview flow diagram might be appropriate, while different sections within that are then developed as separate diagrams.

Flow diagramming is considered by some to be an old-fashioned technique as, somewhat like Assembler itself, it does not encourage a structured program. Nevertheless, it is easy to learn and use, and it can clearly represent simple program ideas. Therefore, for our purposes we will make use of it.

5.1.2 State diagrams

The flow diagram views life as a series of actions or events which are rapidly passed through. Many products, however, behave in a different sort of way. They tend to move from one state to another, maybe spending a significant period of time in that state and leaving it only when a time period is completed or a specific event occurs. These are best represented by a *state diagram*, which forms an alternative to the flow diagram. As with flow diagrams, there is some sophistication in using state diagrams in their full form. For our purposes, however, all we need to do is to draw each state as a labelled circle and interconnect these with arrows. These show under what condition(s) one state can move to another. Each arrow is labelled with the condition that causes the state to change.

Figure 5.2 shows the function of a domestic washing machine represented as a simple state diagram. When switched on, the machine first enters a Ready state. If the door is closed and the user initiates a wash, then the machine first loads with water. A level sensor detects when this is complete. However, the machine will also measure the time taken to fill. If it does not fill within the allotted time, then a fault is assumed.

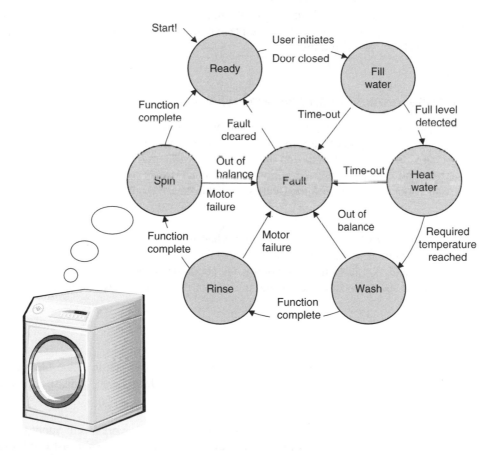

Figure 5.2 A washing machine control program – visualised as a state diagram

This may be due to inadequate water pressure or a faulty valve mechanism. The use of time-out is a low-cost alternative to sensing the water flow or pressure itself. The fill state is followed by a water heating state. Again, a time-out occurs if the water is not heated in an allotted time. The process continues as shown, each state having a 'successful' exit condition, as well as one which leads to the fault state.

From a programming point of view, state diagrams are more abstract than flow diagrams and cannot so easily be translated directly into assembler code. In fact, it is often useful to convert each state into its own flow diagram. To retain clarity of structure and ensure good programming practice, each state should have very clearly defined entry and exit points. The use of both flow diagrams and a state diagram to represent program structure is illustrated later in this chapter, with the electronic ping-pong program.

5.2 Flow control – branching and subroutines

As Figures 5.1 and 5.2 show, programs rarely execute in one continuous and unbroken sequence of instructions. The techniques used to allow program execution to move to different program sections are collectively termed *flow control*, and are the subject of this section. We have already met the PIC instruction **goto**, which unconditionally transfers program execution from one place in the program to another. Now we explore the use of conditional branching and subroutines.

5.2.1 Conditional branching and working with bits

One of the most important features of any microprocessor or microcontroller program is its ability to make 'decisions', i.e. to act differently according to the state of logical variables. Many microprocessors have within their instruction sets a number of instructions which allow them to test a particular bit, and either continue program execution if the condition is not met or branch to another part of the program if it is. This is illustrated in Figure 5.3. Most commonly these variables are bit values in condition code or Status registers.

The PIC 16 Series microcontrollers are a little unusual when it comes to conditional branching, as they do not have branch instructions as such. They have instead four conditional 'skip' instructions. These test for a certain condition, and skip just one instruction if the condition is met and continue normal program execution if it is not. The most versatile and general purpose of these are the instructions:

btfsc   **f,b**
btfss   **f,b**

The first of these tests bit **b** in memory location **f** and skips just one instruction if the bit is set (i.e. at Logic 1). The second does a similar thing, but skips if the tested bit is clear (i.e. at Logic 0). Let us explore this in an example program.

Our first example, the simple data moving program of Program Example 4.2, works well. Suppose, however, we don't want to move the whole of Port B to Port A. Maybe we wanted to transfer just one bit or move a bit from one position in Port B to a different position in Port A. Then we could use the 'bit-oriented'

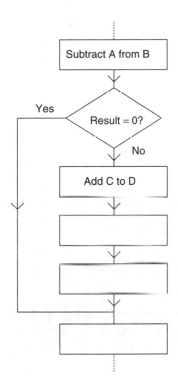

Figure 5.3 Conditional branching

instructions of the 16 Series instruction set, of which there are four: the two we have just seen above, and **bsf** and **bcf**.

The program fragment in Program Example 5.1 performs a near identical function to that of Program Example 4.2, but applies the bit manipulation instructions just identified. As just single bits are manipulated, however, it does not affect any of the other bits in the port to which it is writing. It lights an LED if the associated microswitch is pressed. Even this simple task requires some thought, however. As the port input goes low when the button is pressed, the program needs to *set* the output bit (to light the LED) if the input is low and *clear* it if it is high. This implies a selection process – in a high-level language we might call this an *if...else* structure. The simple skip instruction is not able to do this on its own. One way to do this is to *preset* the output bit with one value and then change it if we find it has been set wrong.

```
;The "main" program starts here
        movlw 00            ;clear all bits in port A and B
        movwf porta
        movwf portb
loop    bcf   portb, 3      ;preclear port B, bit 3
        btfss porta, 3
        bsf   portb, 3      ;but set it if button pressed
;
```

Program Example 5.1 Testing and manipulating single bits

```
          bcf    portb, 4     ;preclear port B, bit 4
          btfss  porta, 4
          bsf    portb, 4     ;but set it if button pressed
          goto   loop
          end
```

Program Example 5.1 Continued

Programming Exercise 5.1

Open a new project under the suggested name Bit_Set, or choose your own name. Copy Program Example 4.2 into it as source file, but replacing the main section of code with Program Example 5.1. Build the project and simulate. With the Stimulus Controller create inputs signals for Port A, pins 3 and 4, selecting Toggle for Action. Open a Watch window with **PCL, PORTA, PORTB** and **W register** as observed variables. Step through the program, 'firing' the inputs at appropriate moments, noting the effect. Change the program so that:

(1) Port B, bits 3 and 4 are *set* if the respective buttons are pressed
(2) Different bits in Port B are set when the buttons are pressed.

5.2.2   Subroutines and the Stack

As we develop bigger programs, we quickly find that there are program sections that are so useful that we would like to use them in different places. Yet it is tedious, and space and memory consuming, to write out the program section whenever it is needed. Enter the subroutine.

The subroutine is a program section structured in such a way that it can be called from anywhere in the program. Once it has been executed the program continues to execute from wherever it was before. The idea is illustrated in Figure 5.4. At some point in the main program there is an instruction 'Call SR1'. Program execution then switches to Subroutine 1, identified by its label. The subroutine must be terminated with a 'return from subroutine' instruction. Program execution then continues from the instruction *after* the Call instruction. A little later in the program another subroutine is called, followed a little later again by another call to the first routine.

The action of the Call instruction is two-fold. It saves the contents of the Program Counter onto the Stack, so that the CPU will know where to come back to after it has finished the subroutine. It then loads the subroutine start address into the Program Counter. Program execution thus continues at the subroutine. The return instruction complements the action of the Call. It loads the Program Counter with the data held at the top of the Stack, which will be the address of the instruction following the Call instruction. Program execution then continues at this address. Subroutine Call and Return instructions *must always* work in pairs.

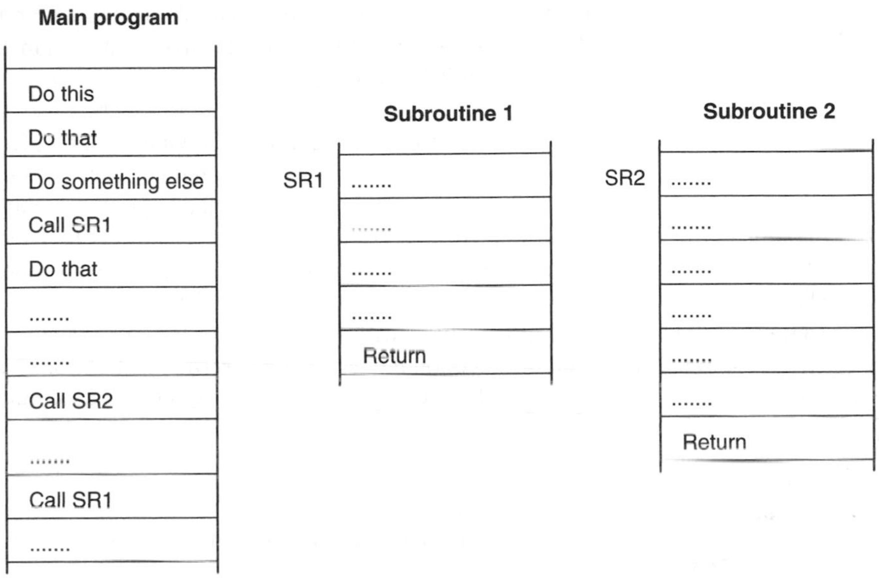

Main program

Figure 5.4 Subroutine calling

The PIC 16 Series subroutine call and return instructions can be seen in Appendix 1, and are simply called **call** and **return**. A special return instruction, **retlw**, is also available. Example subroutines will be explored in the following section.

A subroutine called from *within* another subroutine is called a nested subroutine. In doing this, it must be remembered that every time a subroutine is called one Stack location is taken up, which becomes free again on the subroutine return. If we call a subroutine from within another, then two Stack locations are used up, or three if there is another nested call. As the 16 Series microcontrollers only have an eight-level stack, care must taken that there is not 'stack overflow'.

5.3 Generating time delays and intervals

A recurring theme of embedded systems is how we deal with time – how they respond in a timely way to external events, and how they can measure time and generate time delays. Even with only a limited grasp of programming, we can begin to address the issue of timing by developing program loops that give time delays of known and accurate duration.

The initial concept is simple. A memory location is set up to act as a counter, loaded with a certain value and then decremented repeatedly in a loop until it reaches zero. The time taken will depend on the number first placed in the counter and then the time taken for each decrement loop.

To implement accurate delays, the oscillator frequency needs to be accurate and stable, and we need to know what that frequency is. Here lies one advantage of using a crystal oscillator, as that gives a frequency of excellent accuracy and stability. Approximate delays can of course be implemented with other oscillator sources, as we do with the ping-pong. With the PIC, we need to remind ourselves that each instruction cycle takes four oscillator cycles, as described in Section 2.5.1 of Chapter 2.

A simple example of a delay loop, taken from the ping-pong program, is shown in Program Example 5.2. It takes the form of a subroutine called **delay5**. The subroutine opens by moving a number into the memory location **delcntr1**. In this case the number is 200_D, although this can be varied to produce delays of different lengths (up to a maximum, for an 8-bit location, of 255_D). The actual delay loop is that section of code starting with the label **del1**. Two **nop** instructions, which do nothing at all but take up time, are used to extend the time taken for one loop iteration. The **decfsz** instruction is then implemented. This decrements memory location **delcntr1**, which has been previously loaded. If the result of the decrement is zero, then the subsequent instruction is skipped and program execution moves on to the **return** instruction. For 199 cycles, however, the decrement result will not be zero, there will be no skip and program execution will go back to **del1**.

```
;Delay of 5ms approx. Instruction cycle time is 5us.
delay5  movlw      D'200'        ;200 cycles called,each taking 5x5=25us
        movwf      delcntr1
del1    nop                      ;1 inst. cycle
        nop                      ;1 inst. cycle
        decfsz     delcntr1,1    ;1 inst. cycle, when no skip
        goto del1                ;2 inst. cycles
        return
```

Program Example 5.2 A delay subroutine

The time duration of this delay subroutine can be worked out with ease, by considering the time taken by each instruction in the loop (taken from the instruction set, Appendix 1). This is shown in the program comments. While **delcntr1** is counting down from its initial value, the loop is made up of two **nop** instructions, a **decfsz** and a **goto**. As the **decfsz** instruction does not skip, it takes only one cycle, whereas the **goto** always takes two. Therefore, the total per loop is five. The electronic ping-pong has a clock frequency of approximately 800 kHz, and therefore an instruction cycle frequency of 200 kHz or instruction cycle period of 5 μs. Therefore each loop, with its five instruction cycles, takes 25 μs. Two hundred loops are called; hence the overall duration is 5 ms.

For a precise delay, it is necessary also to take into account the time duration of the final cycle, and the entry and exit to the subroutine. On the final loop iteration, for example, the **decfsz** causes a skip and hence takes two cycles. The **goto** is, however, missed.

The simple delay loop of Program Example 5.1 is useful for comparatively short delays, say up to tens of milliseconds. There are many situations, however, when we want something much longer. A simple way to extend it is to create a second subroutine, similar to the first, but which calls the first from within its loop. An example is shown in Program Example 5.3. This loop makes 100_D calls to the subroutine **delay5**, which we have just seen. The resultant delay is therefore around 500 ms. A further way of writing a longer delay loop, using a 'loop within a loop' within a single subroutine, appears in Program Example 5.4.

```
;500ms delay (approx)      ;100 calls to delay5
delay500 movlw      D'100'
         movwf      delcntr2
del2     call       delay5
         decfsz     delcntr2,1
         goto       del2
         return
```

Program Example 5.3 Nested subroutines for greater delay

Delay routines are very useful things and are widely used. However, they need to be used with care, as when the delay routine is running, the CPU can do nothing else. A delay routine is a bit like asking the scientist Einstein to sit and count beans – it's just not very good use of a powerful resource. In Chapter 6 we meet other ways of creating time delays.

5.4 Dealing with data

We have seen that it is easy to move single bytes of data to or from data memory, whether the memory is SFRs or memory locations dedicated to holding particular variables. We did this in the most recent program example, with the instructions:

```
movlw  D'100'
movwf  delcntr2
```

In using these instructions, we specify the actual address of the memory location required. In this example, **delcntr2** is a label clearly specifying an address. If, however, we want to work with blocks of data, simple data moves like this become restrictive. If working through a list of numbers, for example, it is quite inconvenient to have to specify a fixed address for every memory location. Instead, it is useful to have a pointer that shows where we are in the list, but which can be changed to point to successive locations. It is useful to be able to do this in data memory. In this case the program can develop a block of data and then manipulate it. It is also useful to do it in program memory, in which case the program is able to access a predetermined block of data, but not modify it. This section considers two techniques used for working with blocks of data, in data memory and program memory.

5.4.1 *Indirect addressing and the File Select Register*

In Figures 2.2 and 2.5 we see a register called the FSR, the File Select Register. As Figure 2.2 suggests, instead of embedding an address in the instruction word, the number stored in the FSR can be used as an address to data memory. This is called an *indirect address*. The FSR is invoked whenever the memory location **INDF** is addressed. **INDF** doesn't actually exist as a register, its use simply forces the CPU to implement the indirect addressing mode. The beauty of this mode is that the FSR can be manipulated as a normal memory location, to point to anywhere in data memory that is wanted. When **INDF** is invoked, the instruction used then acts on the memory location pointed to by the FSR.

An example of the use of indirect addressing appears in Program Example 5.7, where we have developed a program that develops a data list. It is discussed in Section 5.6.4.

5.4.2 Look-up tables

The instruction **movlw** allows us to embed within the program a byte of constant data that is then applied in any way we want. We have already seen this in the previous program examples. This is fine for manipulating single bytes, or just a few. But suppose we want to place in the program a whole list of numbers that are needed during program execution, maybe to generate a waveform or to produce output patterns on a display. Suppose also that we wanted to be able to remember where we were in the list with some sort of marker. The **movlw** instruction is then not really up to the job, and we need to apply a way of setting up data known as a *look-up table*.

A look-up table is a block of data that is held in program memory, which can be accessed by the program and used within it. In a Von Neumann structure (Figure 1.7(a)), with its single address and data buses, it is rather easy to set up and use look-up tables, as all memory locations are of equal size and all can be accessed with equal ease. In a Harvard structure (Figure 1.7(b)) it is more difficult, as data must be moved from one distinct memory map to another. The situation is made worse by the difference in memory location size that usually exists between data and program memories. Therefore, in a Harvard structure like the PIC's, a special technique is used to create look-up tables. This introduces several important new ideas.

The PIC 16 Series approach to look-up tables is shown in Figure 5.5. The table is formed as a subroutine. Every byte of data in the table is accompanied by a special instruction, **retlw**. This instruction is another 'return from subroutine', but with a difference – it requires an 8-bit literal operand. As it implements the subroutine return, it picks up its operand and puts it into the W register. The table is essentially a list of **retlw** instructions, each with its byte of data.

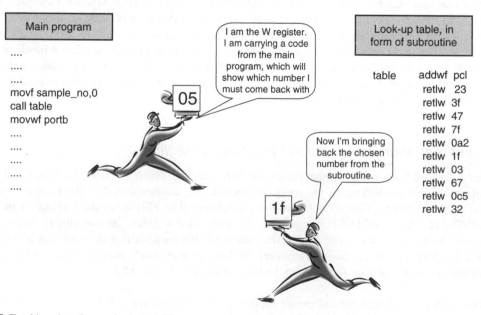

Figure 5.5 Fetching data from a look-up table

What we need now is a technique which allows just one of those **retlw** instructions to be selected from the list – we use something called *computed go to*. Look carefully at the very first instruction in the subroutine, **addwf pcl**. The contents of the W register are added to **pcl**, which is the lower byte of the Program Counter. This sounds like a pretty dangerous sort of thing to do, and for sure it must be done with care. The effect, however, is that once a number has been added to the Program Counter, then program execution jumps forward by whatever that number was. In this example the CPU then executes the **retlw** instruction it lands on and thus goes back to the main program. Note carefully that, however long it appears, on any one iteration *only two instructions in the subroutine are executed*, the **addwf pcl** and the chosen **retlw**. It is obviously up to the programmer to ensure that as the subroutine is called, the W register is already loaded with the offset that is needed.

Let's see this at work in the example of Figure 5.5. Using the **movf** instruction, the main program transfers into the W register the contents of a memory location called **sample_no**. It then calls the subroutine **table**. In this example it is assumed that **sample_no** was holding the number 5, which the W register is holding as the subroutine is entered. As the subroutine starts program execution, the number 5 is added to **pcl**. Program execution therefore jumps forward by 5, to instruction **retlw 1f**. This causes a return from the subroutine, with the number **1f** now placed in the W register. The main program immediately makes use of this number, in this case transferring it to Port B.

In summary, the W register is like a messenger being sent to the subroutine. It goes to the subroutine carrying the code (which acts as a pointer) showing which number is wanted and it comes back carrying that number.

There is one possible problem with this approach – by manipulating only the lower byte of the Program Counter we can only operate within the first 256 words of program memory, or within any page following. If the look-up table is very long, or if it is situated across a page boundary, then problems with the computed go to will occur. In this case it is essential to calculate a fuller version of the computed go to [Ref. 5.1].

5.4.3 Example program with delays and look-up table

Program Example 5.4 is a simple example combining delay loops and use of a look-up table, applied to the ping-pong hardware. It takes 8-bit values from a look-up table and transfers them to the ping-pong LEDs, with a delay between each data transfer. The overall effect is a display of randomly flashing LEDs. The opening sections are very similar to Program Example 4.2, although now we need to specify the address for **pcl** from Figure 2.5. Also specified is an RAM location, called **pointer**, whose address is chosen to lie within the available range of 0C to 4F (also Figure 2.5).

The core of the program starts at the label **loop**. Just as in Figure 5.5, the value of **pointer** is moved to the W register and a subroutine called **table** is called. Upon return from the subroutine, the value held in the W register is transferred to Port B. This lights a certain pattern of LEDs. A delay loop, **delay**, is then called. Note the structure of this, which is similar to the concept implied in Program Example 5.3, but uses a single subroutine and a loop within a loop to extend the delay. The program continues to loop, checking whether **pointer** has reached its maximum value. If so, it is reset to zero before continuing.

```
;****************************************************************
;FLASHING LEDs!
;This program continuously outputs a series of led patterns,
;using simulation or ping-pong hardware.
;TJW 5.3.05.              Tested in simulation 11.3.05.
;****************************************************************
;Clock is 800kHz
;Configuration Word: WDT off, power-up timer on,
;                       code protect off, RC oscillator
;
        list p=16F84A
;
;specify SFRs
pcl      equ  02
status   equ  03
porta    equ  05
trisa    equ  05
portb    equ  06
trisb    equ  06
;
pointer equ  10
delcntr1 equ 11
delcntr2 equ 12
;
        org  00
;Initialise
start   bsf  status,5       ;select memory bank 1
        movlw B'00011000'
        movwf trisa         ;port A according to above pattern
        movlw 00
        movwf trisb         ;all port B bits output
        bcf  status,5       ;select bank 0
;
;The "main" program starts here
        movlw 00            ;clear all bits in port A
        movwf porta
        movwf pointer       ;also clear pointer
loop    movf  pointer,0     ;move pointer to W register
        call  table
        movwf portb         ;move W register, updated from table SR, to port B
        call  delay
        incf  pointer,1
        btfsc pointer,3     ;test if pointer has incremented to 8
        clrf  pointer       ;if it has, clear pointer to start over
        goto  loop
;
;****************************************************************
;Subroutines
;****************************************************************
;Introduces delay of 500ms approx, for 800kHz clock
delay movlw D'100'
      movwf delcntr2
```

Program Example 5.4 Using a look-up table

```
outer movlw D'200'
      movwf delcntr1
inner nop
      nop
      decfsz delcntr1,1
      goto   inner
      decfsz delcntr2,1
      goto   outer
      return

;Holds Lookup Table
table addwf pcl
      retlw 23
      retlw 3f
      retlw 47
      retlw 7f
      retlw 0a2
      retlw 1f
      retlw 03
      retlw 67
;
      end
```

Program Example 5.4 Continued

Programming Exercise 5.2

People often find the concept of the PIC look-up table quite difficult to understand. It is therefore a particularly good idea to simulate an example. Create a project called Flashing LEDs, copy the source code of Program Example 5.4 from the book CD and include it in the project. Then simulate. Open a Watch window with **PCL**, **PORTB**, **WREG** and **pointer** as observed variables. Step through the program and see carefully how the value of the W register changes as the subroutine is entered and left. Use Step Over to avoid getting stuck in the delay subroutine. Ensure that you understand all stages of the program and the new instructions that have been used.

5.5 Introducing logical instructions

So far we have seen a good selection of the 16 Series instructions, but have yet to see any logical ones. These instructions, like **andwf**, **andlw**, **iorwf** or **xorwf**, perform logical operations between the contents of the W register and either a literal value or a value held in a memory location. They do it on a *bitwise* basis. For example, if the **andlw k** instruction was applied, then bit 0 of the literal value is ANDed with bit 0 of the W register, bit 1 is ANDed with bit 1 and so on. These instructions are useful for actual logical operations. Commonly, **and** instructions are used for suppressing unwanted bits in a word and **or** instructions are used for setting individual bits in a word.

As an example, let's look at an alternative way of resetting the pointer in Program Example 5.4. Remember, every time the pointer increments to value 8 (0000 1000), it needs to be reset to 0. Instead of doing this, why not just suppress the higher 5 bits of the word, which are of no use in this application?

The alternative, using logical instruction **andlw**, is shown in Program Example 5.5. Instead of testing the value of **pointer** every time it is incremented, it is now 'ANDed' with the number 07, or 0000 0111$_B$. Now when it increments to 0000 1000$_B$, bit 3 (which has been set to 1) is ANDed to 0 and the value of **pointer** returns to zero.

```
loop   movf   pointer,0      ;move pointer to W register
call table
movwf portb                 ;move W register, updated from table SR, to port B
call delay
incf pointer,0              ;increment pointer, place result in W reg
andlw 07
movwf pointer
goto loop
```

Program Example 5.5 Introducing a logical instruction

5.6   Introducing arithmetic instructions and the Carry flag

The 16 Series instruction set has six arithmetic instructions: **addwf**, **addlw**, **subwf**, **sublw**, **incf** and **decf**. Their use is central to any arithmetic processing that the microcontroller may have to do, as we will see shortly in the next program example.

5.6.1   Using add instructions

The use of the two add instructions is straightforward. Using **addwf** the contents of the W register are added to the contents of the memory location specified; using **addlw** a literal value is added to the contents of the W register. The situation becomes slightly more complex when one realises that two 8-bit results added together can lead to a 9-bit result. This ninth bit is the Carry flag in the Status register.

5.6.2   Using subtract instructions

The subtract instructions follow a similar pattern to the add. The Carry bit now acts as a Borrow, except the polarity is reversed (see the Status register, Figure 2.3). Therefore, if a subtract occurs and the result is positive, then the Carry bit is *set*. If the result is negative, then the Carry bit is *clear*.

5.6.3   An arithmetic program example

Program Example 5.6 demonstrates the use of some simple arithmetic techniques. It generates a Fibonacci series by an adding process, as described in the header. A counter indicates how many numbers in the series have been calculated. When it exceeds the 8-bit range, it reverses the series by subtracting. It detects the range overflow by checking the Carry bit after each addition.

The program starts by preloading the three first numbers in the series into the memory store. It then starts moving up the series, from the label **forward**. The two most recent numbers are added and the Carry bit then checked. If it is set, the 8-bit range has been exceeded and the program will need to reverse. Assuming Carry was not set, the program then increments the counter and shuffles the numbers in the memory store, discarding the oldest. The program then loops up to **forward**. If, however, the Carry had been set earlier, the program branches to **reverse**. Now it works down the series, by subtraction. It tests the counter number to determine when it should return to **forward**.

```
;********************************************************************
;In a Fibonacci series each number is the sum of the two previous
;ones, e.g. 0,1,1,2,3,5,8,13,21.....
;This program calculates Fibonacci numbers within an 8-bit range,
;first going up and then down.
;Program intended for simulation only, hence no input/output.
;The program demonstrates addition, subtraction, compare.
;TJW 17.3.05.                        Tested by simulation 18.3.05
;********************************************************************

    list p=16F84A
;no i/o ports used
status equ 03
c       equ  0
z       equ  2
;these memory locations hold the three highest values of the Fibonacci series
fib0    equ 10        ;lowest number (oldest when going up,
                           ;newest when reversing down)
fib1    equ 11        ;middle number
fib2    equ 12        ;highest number
fibtemp equ 13        ;temporary location for newest number
counter equ 14        ;indicates value reached, opening value is 3

    org 00
;preload initial values
        movlw 0
        movwf fib0
        movlw 1

        movwf fib1
        movwf fib2
        movlw 3
        movwf counter ;we have preloaded the first three numbers,
                        ;so start count at 3
;
forward movf  fib1,0
        addwf fib2,0
        btfsc status,c       ;test if we have overflowed 8-bit range
        goto  reverse        ;here if we have overflowed, hence reverse down
        movwf fibtemp        ;latest number now placed in fibtemp
        incf  counter,1
```

Program Example 5.6 Generating a Fibonacci series

```
;now shuffle numbers held, discarding the oldest
      movf  fib1,0        ;first move middle number, to overwrite oldest
      movwf fib0
      movf  fib2,0
      movwf fib1
      movf  fibtemp,0
      movwf fib2
      goto  forward
;when reversing down, subtract fib0 from fib1 to form new fib0
reverse movf fib0,0
      subwf fib1,0
      movwf fibtemp        ;latest number now placed in fibtemp
      decf  counter,1
;now shuffle numbers held, discarding the oldest
      movf  fib1,0        ;first move middle number, to overwrite oldest
      movwf fib2
      movf  fib0,0
      movwf fib1
      movf  fibtemp,0
      movwf fib0
;test if counter has reached 3, in which case return to forward
      movf  counter,0
      sublw 3
      btfsc status,z
      goto  forward
      goto  reverse
;
      end
```

Program Example 5.6 Continued

Programming Exercise 5.3

Create a project in MPLAB® called Fibonacci. Copy from the book CD the source file of Program Example 5.6 into it and simulate. In the Watch window display **counter, fib0, fib1, fib2, fibtemp, WREG** and **STATUS**. Single-step initially and watch the Fibonacci series develop, in **fib0, fib1** and **fib2**. How many numbers in the series fit into the 8-bit range? Watch the Carry bit being set as the range is exceeded and see the program reverse down the series. Notice now that the Carry bit (now acting as Borrow) is *set* after each subtraction. Try halting the program at **reverse**, and forcing two values for **fib0** and **fib1** that will give a negative result. Single-step through the subtraction, check the result in the W register and notice that the Carry bit is clear. See how the comparison of **counter** with the literal number 3 is achieved, and see the program return to **forward**.

5.6.4   *Using indirect addressing to save the Fibonacci series*

Program Example 5.7 extends the Fibonacci series program to incorporate storage of the series, using indirect addressing, with the main section of the program shown. The extra lines of code are highlighted in bold. The data block is stored in data memory, starting at location 20_H. The first three numbers in the series are entered using conventional addressing (indirect addressing could already be used here, but this would require slightly more lines of code). The FSR is then loaded with 23_H, the address of the next location in the data block to be loaded. Entering the main loop, it can be seen that every time a new number is generated, it is stored in the data block, using the **movwf indf** instruction. The value of the FSR is then incremented until it reaches a predetermined maximum value.

```
;...
(opening lines of program omitted)
;...
;these memory locations hold the most recent numbers in the Fibonacci series
fib0     equ 10   ;lowest number (oldest when going up; newest when reversing down)
fib1     equ 11   ;middle number
fib2     equ 12   ;highest number
fibtemp  equ 13   ;temporary location for newest number
counter  equ 14   ;indicates which value we have reached, opening value is 2

        org 00
;preload initial values
        movlw 0
        movwf fib0
        movwf 20       ;save at start of list
        movlw 1

        movwf fib1
        movwf 21       ;save in list
        movwf fib2
        movwf 22       ;save in list
        movlw 3
        movwf counter ;have preloaded the first three numbers, so start at 3
        movlw 23       ;Initialise File Select Register, with next location for
        movwf fsr      ;list of numbers to be saved
;
forward movf fib1,0
        addwf fib2,0
        btfsc status,c ;test if we have overflowed 8-bit range
        goto  reverse ;here if we have overflowed, hence reverse down the series
        movwf fibtemp ;latest number now placed in fibtemp
        movwf indf    ;save by indirect addressing
        movf  fsr,0   ;test to see if FSR is at top of range
        sublw 30
        btfss status,z
        incf  fsr,1   ;increment FSR, if available range not full
        incf  counter,1
;now shuffle numbers held, discarding the oldest
        movf  fib1,0 ;first move middle number, to overwrite oldest
        movwf fib0
        movf  fib2,0
        movwf fib1
        movf  fibtemp,0
        movwf fib2
        goto  forward
```

Program Example 5.7 Storing the Fibonacci series with indirect addressing

---

Programming Exercise 5.4

Create a project in MPLAB called Fibo+storage, or use a name of your choice. Copy from the book CD the source file of Program Example 5.7 into it and simulate. Display the File Registers window using View > File Registers. Using 'Step Into', single-step through the program and watch the Fibonacci series being built up in the data memory block starting at location 20_H. On completion you should have a window similar to that in Figure 5.6. Notice in this also the two banks of SFRs, and how **PCL, STATUS** and **FSR** are mirrored across both banks. What does the program do when **FSR** has reached its maximum value? Try changing the definition of the top of the block by varying the literal value in the instruction **sublw 30**. Note the change in program action as you do this.

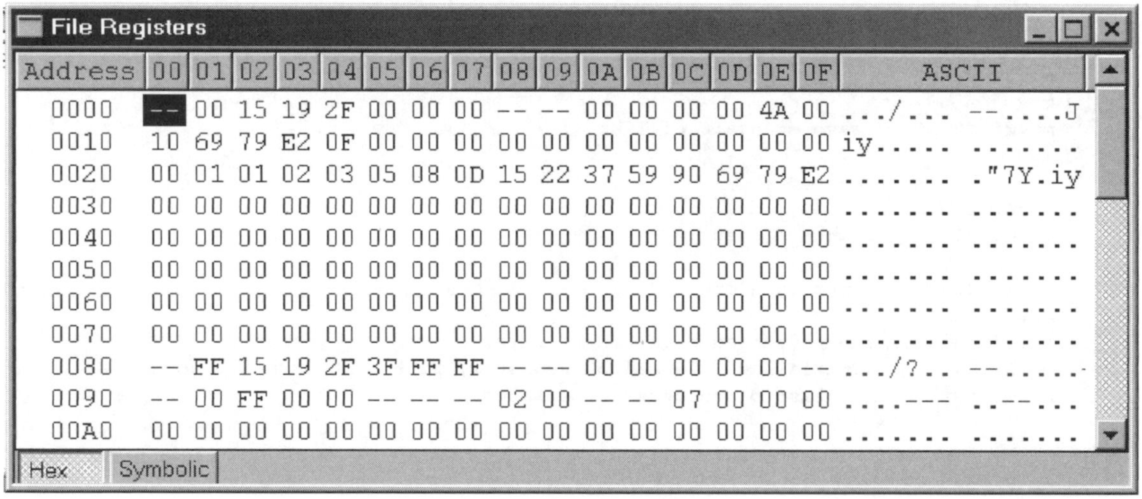

```
■ File Registers                                          _ □ ×

Address │00│01│02│03│04│05│06│07│08│09│0A│0B│0C│0D│0E│0F│      ASCII

  0000   -- 00 15 19 2F 00 00 00 -- -- 00 00 00 00 4A 00  .../... --....J
  0010   10 69 79 E2 0F 00 00 00 00 00 00 00 00 00 00 00  iy..... ........
  0020   00 01 01 02 03 05 08 0D 15 22 37 59 90 69 79 E2  ......."7Y.iy
  0030   00 00 00 00 00 00 00 00 00 00 00 00 00 00 00 00  ....... ........
  0040   00 00 00 00 00 00 00 00 00 00 00 00 00 00 00 00  ....... ........
  0050   00 00 00 00 00 00 00 00 00 00 00 00 00 00 00 00  ....... ........
  0060   00 00 00 00 00 00 00 00 00 00 00 00 00 00 00 00  ....... ........
  0070   00 00 00 00 00 00 00 00 00 00 00 00 00 00 00 00  ....... ........
  0080   -- FF 15 19 2F 3F FF FF -- -- 00 00 00 00 00 --  .../?.. --.....
  0090   -- 00 FF 00 00 -- -- -- 02 00 -- -- 07 00 00 00  ...--- ..--...
  00A0   00 00 00 00 00 00 00 00 00 00 00 00 00 00 00 00  ....... ........

 Hex   Symbolic
```

Figure 5.6 The Fibonacci series stored in a data block starting at address 20_H

5.7 Taming Assembler complexity

You are now beginning to see how complex even a simple Assembler program can become. We need every means possible of keeping the program as short and understandable as possible. A few options are now described.

5.7.1 Include Files

The Assembler directive **#include** allows any file to be embedded within a program, thereby saving the trouble and space of pasting in large program sections which already exist elsewhere. A file so included is called an *Include File*. Initially, the most useful way to use this is to replace all the microcontroller-specific memory definitions that must occur at the start of a program. Like other Assemblers, MPLAB contains Include Files for each microcontroller, containing **equ** statements for all SFRs and their bits. These can be found in one of the MPLAB folders.

Use of an Include File is useful for a small microcontroller like the 16F84A, where the file is several pages long. It becomes almost essential for larger processors, which have a huge array of SFRs and hence very long Include Files. Once an Include File is used, it is of course essential to ensure that the microcontroller-specific labels that are referenced in the program are identical to the ones in the Include File. The advantage of an Include File, even for a very small application, is illustrated in Program Example 5.8.

```
;specify SFRs
timer    equ    01
status   equ    03                      #include p16f84A.inc
porta    equ    05                      OR
trisa    equ    05                      include p16f84A.inc
portb    equ    06
trisb    equ    06
intcon   equ    0B
```

        the above, *and much more besides, can be replaced by:* **the above**

Program Example 5.8 Using Include Files

5.7.2 Macros

We are finding in every program we see that program development for a RISC processor is laborious, due to the limited function of each individual instruction. A CISC instruction set, with its somewhat more powerful instructions, offers some modest advantage, as we saw in Section 4.9, but not much. Is there a way we can get around the minimalist nature of the instruction set while remaining in the Assembler environment?

One answer to this problem is by the use of *macros*. A macro is a grouping of instructions, defined by the programmer and given a name. Once defined, the macro can be used in the program at any time. In some ways a macro offers the convenience of a subroutine, but it is used differently. When the source code is assembled, the macro is expanded out into the original instructions that made it up. Therefore, using macros is a form of shorthand in programming rather than a way of structuring the program.

Program Example 5.9 shows three macros inserted at the start of the ping-pong program (Appendix 2). The macro itself is contained within the directives **macro** and **endm**. *Arguments* are defined for the macro, which are data values that the macro can apply. The macro **movlf** moves a data constant into a memory location. It applies two arguments, **const** and **address**. The macros **bfbset** (branch if file bit set) and **bfbclr** (branch if file bit clear) are similarly defined. All three macros are then applied within the first few lines of program, each time saving one line of code. Thus, the eight original lines of code in loop **wait** are reduced to four.

```
;now ready for action
;macro to move a literal value to a file
movlf          macro const,address
               movlw const
               movwf address
               endm
;macro to branch if a specified bit is set
bfbset macro file,bit,target
               btfsc file,bit
               goto  target
               endm
;macro to branch if a specified bit is clear
bfbclr macro file,bit,target
               btfss file,bit
               goto  target
               endm
wait           movlf  04,porta      ;at rest, "out of play"
               movlf  00,portb      ;all play leds off
;both paddles must initially be clear before play allowed to commence
               bfbclr porta,4,wait   ;go to wait if right paddle pressed
               bfbset porta,3,wait   ;go to wait if left paddle pressed
;
```

Program Example 5.9 Applying macros to the ping-pong program

Programming Exercise 5.5

Enter the code of Program Example 5.8 into a copy of the ping-pong program, removing the lines of code it replaces. Assemble the code and open the list file. Notice how the original macro definition occupies no memory space and observe how the macro is expanded out into its original form whenever it is invoked, thus replicating the original ping-pong program. Continue through the program, applying these two macros wherever you can. How many times can you do this and how many lines of code do you save? Are there other macros that could usefully be defined?

5.7.3 MPLAB special instructions

Microchip further eases the problem of the restrictive RISC instruction set by defining a set of 'special instructions'. These are recognised by the Assembler and expanded out to the equivalent instructions shown. Examples are given in Table 5.1, while a full listing appears in Appendix B.11 of Ref. 4.1. Most are operations using or manipulating the **Z** or **C** bits in the Status register. Some, like **bc** or **bnc**, offer no saving in lines of code, but improve the clarity of programming. Others, like **addcf**, create new and useful functions not originally available in the instruction set, which are very similar to CISC instructions.

Table 5.1 Example MPASM^{TM} 'special' instructions

Mnemonic	Description	Equivalent	Status flags affected
addcf f,d	Add Digit Carry to file	BTFSC 3,1 INCF f,d	Z
bc k	Branch on Carry	BTFSC 3,0 GOTO k	–
bnc k	Branch on No Carry	BTFSS 3,0 GOTO k	–
clrc	Clear Carry	BCF 3,0	–
movfw f	Move file to W	MOVF f,0	Z
subcf f,d	Subtract Carry from file	BTFSC 3,0 DECF f,d	Z
tstf f	Test file	MOVF f,1	Z

5.8 More use of the MPLAB simulator

We have already seen the enormous value of the software simulator as a means of running through a program and observing outputs. We did this just using the simple controls, of single-step, animate or run. As programs grow, however, we need greater sophistication in the way we can run them and how we observe their behaviour.

5.8.1 Breakpoints

Once programs become long, it becomes increasingly tedious to step through them when simulating. We need a means of getting them to run through the code that we may not be interested in, but stopping where we need to take a closer look at what is happening. Breakpoints allow this functionality. In their simplest form, breakpoints allow you to run a program up to a specified instruction. Program execution then stops, and memory and register values can be inspected. In MPSIM^{TM} you can set a breakpoint simply by double-clicking on an instruction in the program window, and remove it in the same way. The number of breakpoints is unlimited, so they can be used freely.

Programming Exercise 5.6

The Fibonacci program is perhaps the longest example we have looked at so far, and it is annoying to have to step through it if we wish to see something happen deep inside the program. Open the Fibonacci project you created earlier (or create it for the first time) and select the MPLAB simulator using Debugger > Select Tool > MPLAB SIM. Scroll through the **.asm** source file and double-click on the line labelled **reverse**. A breakpoint symbol should appear, as seen in Figure 5.7. Check that you can remove this by double-clicking again. Reset the simulator and run. See how the program stops at the breakpoint. You can inspect all windows at this point and then proceed any way you wish, for example by single-stepping. Try setting another breakpoint at the second line shown in Figure 5.7 and running to here.

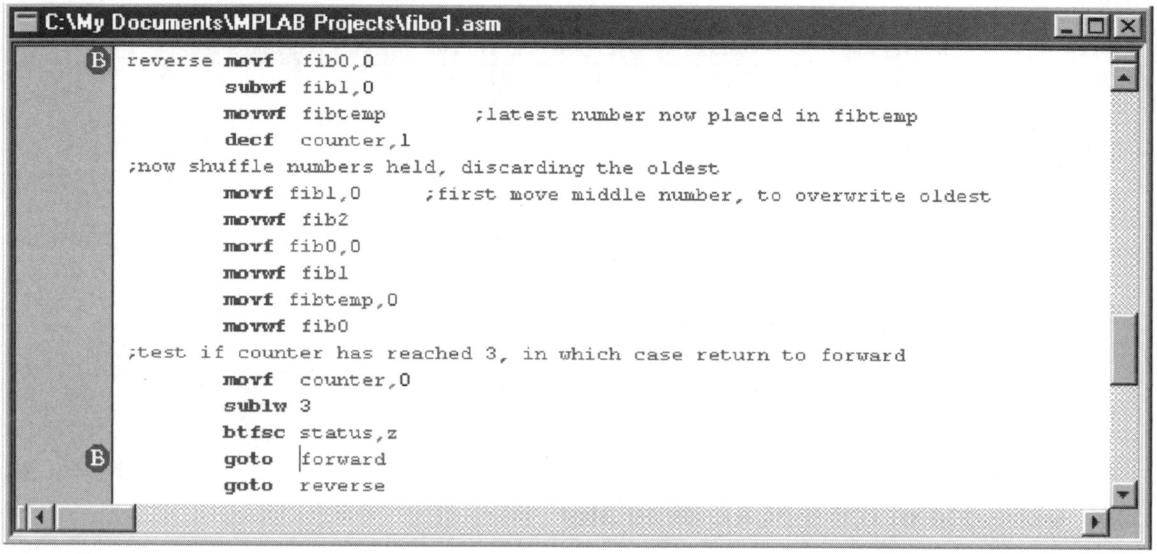

```
C:\My Documents\MPLAB Projects\fibo1.asm                          _ □ ✕
  B  reverse movf   fib0,0
             subwf  fib1,0
             movwf  fibtemp        ;latest number now placed in fibtemp
             decf   counter,1
        ;now shuffle numbers held, discarding the oldest
             movf   fib1,0     ;first move middle number, to overwrite oldest
             movwf  fib2
             movf   fib0,0
             movwf  fib1
             movf   fibtemp,0
             movwf  fib0
        ;test if counter has reached 3, in which case return to forward
             movf   counter,0
             sublw  3
             btfsc  status,z
  B          goto   forward
             goto   reverse
```

Figure 5.7 Breakpoints inserted in Fibonacci program

5.8.2 Stopwatch

A weakness of the software simulator is that it does not run in real time, yet in embedded systems we have a strong desire to understand the timing behaviour of our programs. The Stopwatch facility of the simulator allows accurate time measurements to be simulated. It simply requires that the simulator 'knows' what the oscillator frequency is. As it can record the number of instruction cycles executed, it can then calculate time taken.

In MPSIM the oscillator frequency is set through the 'Simulator Settings' window, found using Debugger > Settings > Osc/Trace (Figure 5.8(a)). The Stopwatch (Figure 5.8(b)) is displayed using Debugger > Stopwatch.

Programming Exercise 5.7

Still with the Fibonacci project open, set the processor frequency to 4 MHz. This usefully gives an instruction cycle time of 1 μs. Leave the breakpoints as set in Programming Exercise 5.6. Press Debugger > Stopwatch, and Zero the Stopwatch. Reset the simulator and run the program to the breakpoint. The Stopwatch should show 166 μs. Can you account for this value?

5.8.3 Trace

The various windows available in MPSIM give a good picture of the state of the processor status and memory locations at any time, but they do not tell us the history of program execution. Even if program

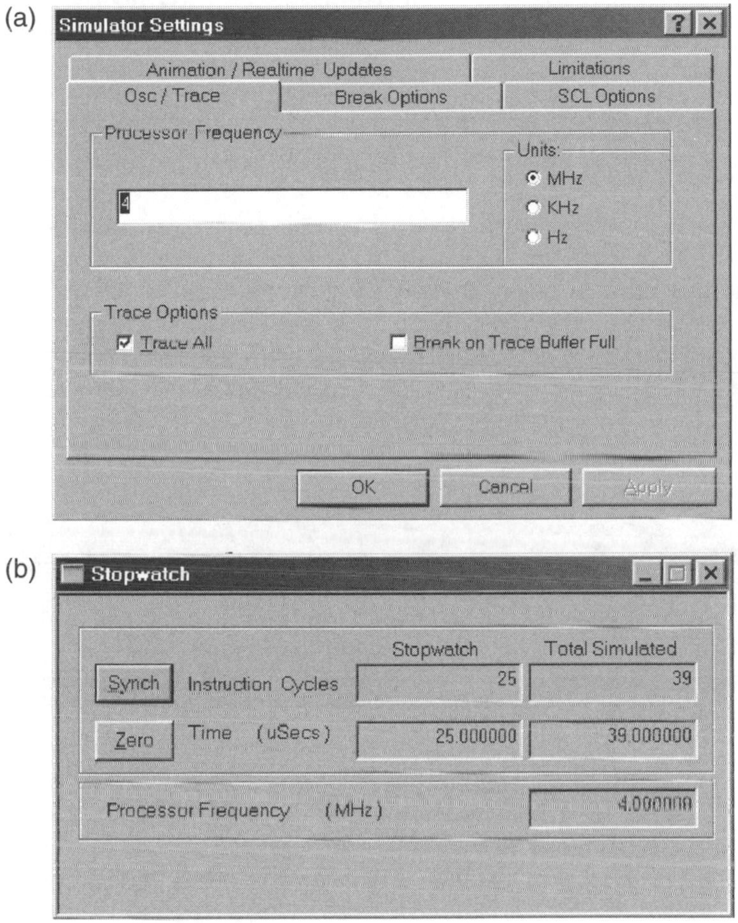

(a)

(b)

Figure 5.8 Using the Stopwatch. (a) Simulator Settings window. (b) Stopwatch window

execution has halted at a breakpoint, there may have been a number of program paths for it to go down to reach that point. The *Trace* function is there to give a record of the recent past of the program execution. In Trace memory the simulator keeps a continuous record of all instructions that have been executed. This can be inspected when program execution stops.

MPLAB has a Trace function, with memory size of 32 767 lines. The Trace function is enabled in the Simulator Settings window (Figure 5.8(a)), found by following Debugger > Settings > Osc/Trace. Note that it slows down simulator speed somewhat if it is enabled. The Trace window is viewed using View > Simulator Trace. Here the columns are all self-explanatory, except for:

SA = Source address − address or symbol of the source data
SD = Source data − value of the source data

DA = Destination address − address or symbol of the destination data
DD = Destination data − value of the destination data.

Programming Exercise 5.8

Return again to the settings of Programming Exercise 5.6. Ensure the Trace is enabled as described above, reset the simulator and run to the breakpoint at label **reverse**. Now open the Trace window, which should appear as in Figure 5.9. See that it is a list of all the instructions recently executed, finishing (in line 142) with the **goto** instruction which takes execution to the breakpoint line. Looking back over the SD and DD columns, we see the Fibonacci series being formed. The value of the Status register in line 141 is 19_H, or 0001 1001_B. This shows that bit 0, the Carry flag, has been set and the **goto** instruction is accordingly invoked.

Line	Addr	Op	Label	Instruction	SA	SD	DA	DD	Cycles
134	000F	0812		MOVF 0x12, W	0012	90	0012	90	0000009B
135	0010	0091		MOVWF 0x11	----	--	0011	90	0000009C
136	0011	0813		MOVF 0x13, W	0013	E9	0013	E9	0000009D
137	0012	0092		MOVWF 0x12	----	--	0012	E9	0000009E
138	0013	2807		GOTO 0x7	----	--	----	--	0000009F
139	0007	0811	forward	MOVF 0x11, W	0011	90	0011	90	000000A1
140	0008	0712		ADDWF 0x12, W	0012	E9	0012	79	000000A2
141	0009	1803		BTFSC 0x3, 0	0003	19	----	--	000000A3
142	000A	2814		GOTO 0x14	----	--	----	--	000000A4

Figure 5.9 Trace window for section of Fibonacci program

5.9   The ping-pong program

It is useful now to look at the full ping-pong program, as seen in Appendix 2. It is never simple looking at assembler code written by someone else (in fact, it's often difficult looking at your own assembler code!), so you should not feel worried if initially it appears difficult.

5.9.1   A structure for the ping-pong program

Let us first of all try to get a feel for the overall structure. For this program a state diagram gives a clear overall representation, which would be difficult to achieve with a flow diagram. This is seen in Figure 5.10.

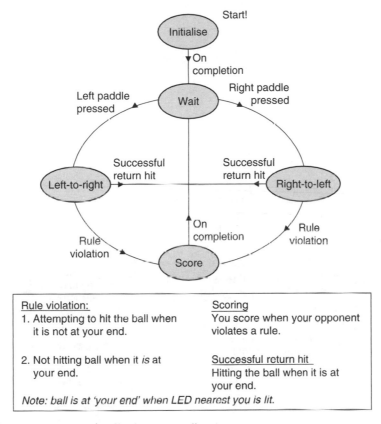

Figure 5.10 The ping-pong program visualised as a state diagram

The program starts in the *Initialise* state. When this has completed it immediately enters a *Wait* state, where it stays until play commences. If the left player presses a paddle then a *Left-to-right* state is entered. In this the 'ball' starts at the left-most position and starts moving towards the right. Exit from this state occurs either if there is a rule violation (for which definitions are given) or if there is a successful return hit, when the ball has reached the right-most position. With no rule violation play continues, with the state alternating between *Left-to-right* and *Right-to-left*. When either player makes a mistake this is classified as a rule violation and the game enters a *Score* state. It leaves this when scoring is complete and returns to the *Wait* state.

Having grasped the ping-pong state diagram, try to find each state in the Assembler listing. Three of the five states, Initialise, Wait and Score, should be easy to follow. Indeed, we have met the first in Chapter 4. The two states where play is actually in progress, Left-to-right and Right-to-left, are a little more difficult to grasp. Each is a mirror image of the other, so when one is understood, the other immediately follows.

While the program overview is best represented as a state diagram, the actual Right-to-left/Left-to-right states are essentially looping structures, and are most easily represented as a flow diagram (Figure 5.11).

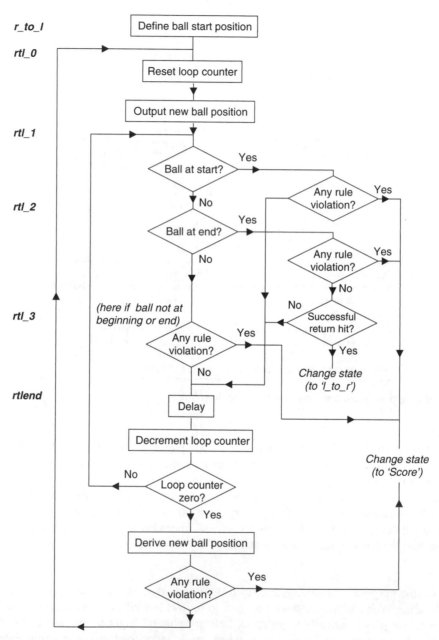

Figure 5.11 Flow diagram of Right-to-left/Left-to-right states

Here we are confronted, perhaps for the first time, with the detailed complexity that such a program requires, even in a product that appears so simple. There are certainly a number of requirements to be met within the state. The 'ball' is to 'move' by lighting a series of LEDs, each to be illuminated for a set period of time. The state of the paddles is to be continuously checked; at certain times a paddle press is a legal action, at others it represents a rule violation. If timing were to be achieved simply by entering a timing loop, the function of input checking could not be carried out. Hence each LED time duration is made up of a certain number of loop iterations – within each the inputs are tested, followed by a short delay.

5.9.2   Exploring the ping-pong program code

As an aid to further understanding, certain sections of the ping-pong code are now described.

Opening section and memory allocation

In the opening comments the program gives details on hardware allocation. This is followed by a section on memory allocation. Here names are given to memory locations in the general-purpose RAM area. The names used, **delcntr1**, etc., are chosen by the programmer and are placed as labels, i.e. starting fully left on the program line. The **equ** directive is used and the memory location is chosen from the memory map of Figure 2.5, which shows that available memory locations range from address $0C_H$ to address $4F_H$.

The Wait state

Let's explore this by looking at the opening part of the actual ping-pong program, which follows the initialisation section.

In the first four lines the program switches on the 'out-of-play' LED and switches off all others. It then tests the state of both paddles. Remember that, when pressed, the switch input bit goes to Logic 0. If neither is pressed then program execution skips forward to **wait1**. However, if one or both are pressed, then program execution just returns to **wait** and loops until the button is released. This is to stop a 'false start' to play, which would otherwise occur if a player switched on the game while a paddle was pressed. As this is the point where play restarts after a score, it also ensures that the previous round of play is completed before starting again. Note that the loop execution includes the setting of the LEDs. This is not strictly necessary but does no harm and minimises the number of labels used.

Program execution then enters another loop, **wait1**. Both buttons have been cleared, so the game can now start. Again both paddles are tested. This time, however, if a button is pressed, instead of looping back play goes forward, to either **l_to_r** or **r_to_l**.

The main play states

Let us start by looking at the **r_to_l** section. This opens with the 'out-of-play' LED being switched off and the opening LED position being defined. The larger loop then starts at the line labelled **rtl_0**. Here the loop counter **loop_cntr** is loaded with the number **led_durn**. This number was defined in the opening section of the program and represents the number of times the inner loop is to be iterated. This inner loop starts at line **rtl_1**. Much of it is concerned with checking for rule violation, the interpretation of which depends on the position of the ball. The general structure is shown in the flow diagram, while the actual rule interpretation can be determined from looking at the source code. Scoring occurs when any rule violation is detected. At the end of the loop the 5 ms delay subroutine is called. The loop counter is decremented. If zero, then a new ball position is set up, by rotating the **led_posn** memory location. A score occurs if this causes the ball to go off the end of the 8-bit number; this happens if there has not been a successful return hit while the ball has been at the end position.

The Score state is divided into two parts and is simple. It lights the appropriate Score LED, calls a half-second delay and switches off the LED. The state is then left and execution returns to the Wait state.

5.10   Simulating the ping-pong program – tutorial

The ping-pong program is not exactly complex, but it is full of loops and delays, and therefore illustrates the problems of thoroughly testing a program with a simulator. The art lies in using each feature where needed. Generally, for a program segment that is to be explored in detail, you will single-step or animate. For the sections of code you want to get through quickly, you will simply run through, heading for a breakpoint you have already inserted. The following is a tutorial that guides you through the simulation of this program.

Ensure that you have created and built a project, which contains a copy of the ping-pong program.

5.10.1   *Setting up input stimulus*

The ping-pong program has two digital inputs, the two player paddles, which need to be simulated. Go to Debugger > Stimulus Controller. This will give you the Stimulus Controller dialogue box. Under Pin select RA3, and under Action select Set High. Repeat this for RA4. Then create two more lines, for RA3 and RA4 again, this time with Pulse Low as the action for each. Set the duration to 50 ms, which is representative of a fast switch push. Can you work out from the program what is the maximum theoretical duration a player can press a button?

5.10.2   *Setting up the Watch window*

Click View > Watch to set up a Watch window. A useful selection for the ping-pong is **PCL** (to track where you are in the program), **PORTA** and **PORTB** using the Add SFR button, and **duration, led_posn** and **delcntr1** using the Add Symbol button. By right-clicking on the title bar near the top of the Watch window, you will see that you can display further columns, for example the data shown in binary. The Watch window you set up will be saved for you at the end of your session.

5.10.3   *Single stepping*

Press F6 to reset the Program Counter and then try single stepping the ping-pong program. You will be able to see registers changing under program instruction, with the changes being highlighted in red.

If one or both of the user paddles are set low (i.e. 'pressed'), then the simulation will get stuck in the first wait loop. (You can see their logic state by inspecting the Port A display in the Watch window.) Set these lines high by pressing the Fire buttons on the Stimulus Controller box. You should see the change reflected in the Watch window.

Now you should be able to single step on to the **wait1** loop. Having looped round here once or twice, Fire the RA3 pulse. You should now exit the loop and move on to **l_to_r**. See the Port B value change to 01 as the ball position is set up. You can continue stepping from here, and either step over, or enter, the **delay5** subroutine. Once in, you can Step Out of it at any time. Clearly it would be tedious to single step all the way through this subroutine. Even if we step over it, the loop repetitions become endless and the limitations of single stepping are revealed.

5.10.4 Animate

Press F6 again and try running the program in Animate mode. Adjust the speed to one you are comfortable with by adjusting the setting in Debugger > Settings > Debugger Animation. Now you can't use the Step Over function and you will find yourself stuck again in the delay subroutine. Here you can watch the **delcntr** value being decremented in the File Registers window.

5.10.5 Run

If you select Run there is not much to watch, as the memory windows are no longer updated as the program runs. Stimulus inputs are, however, still accepted. Hence it is helpful to start using breakpoints.

5.10.6 Breakpoints

Set a breakpoint initially at the **l_to_r** label. Now reset the simulated CPU, set RA3 and RA4 high in the Stimulus Controller box, and enter Run. Fire the pulse on RA3. The program should then halt at your breakpoint. Without resetting, try setting another breakpoint at the **ltr_1** label and press Run again. You have an unlimited number of breakpoints in MPLAB simulator, so use them freely.

5.10.7 Stopwatch

Using Debugger > Settings > Osc/Trace, set the processor frequency to 800 kHz, which is the nominal frequency for the ping-pong. Set a breakpoint at the first line of the **delay5** subroutine and run to there. Now press Debugger > Stopwatch. Zero the Stopwatch and insert a breakpoint at the Return instruction of that subroutine. Run the program to there. Does the Stopwatch value agree with the calculated value of the delay routine? This is a very useful facility for measuring durations of program execution.

5.10.8 Trace

Try enabling the Trace, running the program to a breakpoint and then inspecting Trace memory in View > Simulator Trace.

Table 5.2 Suggested breakpoint locations for ping-pong debug

Breakpoints	Action when breakpoint reached
wait	Check value of Port A – are bits 3 and 4 high? Press Run to loop here and see 'out-of-play' LED being set high by the program. Set bits 3 and 4 high with the Stimulus generator. Press Run.
wait1	Port A should now be 0001 1100. Pulse RA4 low with Stimulus generator (i.e. a player initiates the game); this will stay low for 50 ms of run time. Press Run.
r_to_l l_to_r	Here if RA4/RA3 has been pressed, note that this bit is now low in Port A. Press Run.
rtl_1 ltr_1	Single step a few lines here and see **led_posn** set to new value. This represents the ball position. Press Run. Watch **loop_cntr** step down on each loop iteration. Then watch **led_posn** change.
rtlend ltrend	Have completed one internal cycle, will now call a delay. Press Run. Remove breakpoint for faster running.
score_left	Here if an 'illegal' paddle press forces a Score. Press Run.
score_right	Here if an 'illegal' paddle press forces a Score. Press Run.
delay5	Single step a few times through this loop. See value of **delcntr1** set and then decremented. Press Run. Remove breakpoint for faster running.
delay500	Here if a Score LED is lit. It will be illuminated for 0.5 s. Press Run.

5.10.9 Debugging the full program

Try inserting breakpoints at the positions shown in Table 5.2. Note that all breakpoints must be entered on lines containing instructions. Reset the Program Counter with F6 and run the program. Step from breakpoint to breakpoint, and observe the Watch window values shown in the table.

Having looped around a few times and understood the program, try in the simulator:

- An 'illegal' paddle press to force a score
- Looping until the ball reaches the far end and returning it with a 'legal' paddle press.

5.11   What others do – graphical simulators

These past two chapters have aimed to give a good introduction to MPLAB and its simulator, MPSIM. While powerful in their own way, it is worth reminding oneself that they are free. What if we are willing to spend some money on a simulator?

There are a number of simulators available which go well beyond the simple text-based interface of MPSIM. An example is the simulator found in Ref. 5.2, shown in Figure 5.12. Here a 16F84 microcontroller is being simulated. The W register, pipelined instruction, current instruction, Stack, Status register, ports and program listing are all clearly displayed. The program can be run or be single stepped, with the internal status being clearly updated and displayed.

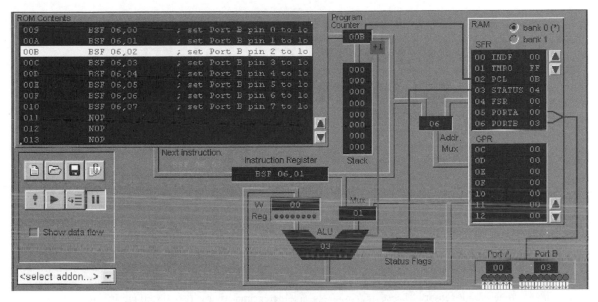

Figure 5.12 The 'Virtual PICmicro' screen, Matrix Multimedia

Summary

- It is important to devise good structures for programs as they are developed. Flow and state diagrams can help with this.
- A number of techniques assist in producing clear and well-structured programs. These include subroutines, look-up tables, macros and the use of Include Files
- The full 16F84A instruction set can be applied to build big and sophisticated programs.
- More complex programs require greater expertise in the use of simulation techniques.

References

5.1. *Implementing a Table Read*. Microchip, Application Note AN556; www.microchip.com.
5.2. Assembly for PICmicro[TM] Microcontrollers, V3.0. John Becker, Matrix Multimedia Ltd; http://www.matrixmultimedia.co.uk or http://www.labvolt.com/

6
Working with time: interrupts, counters and timers

Apart from the lucky few, our daily lives are ruled by time. We have alarm clocks to wake us in the morning, stopwatches to measure time duration, timers to start off single events (like a VCR recording) and timers to maintain periodic events (like a house heating system coming on at the same time every day). For the young and those who teach, the working day is ruled by a school timetable – a complex series of timed events.

For embedded systems, time is similarly of the essence. At a simple level the system needs to respond in a timely manner to external events. It may also need to measure time between external events and generate time-based activity. These requirements are met primarily by two differing, but related, features of a microcontroller: the interrupt and the counter/timer. While each is a stand-alone element in its own right, both are so useful that they have become ubiquitous, finding their way into many other microcontroller features. Interrupts are to be found generated in almost every microcontroller peripheral. Counter/timers provide the timing for a range of activity, from motor control with pulse width modulation to baud rate generation in serial communications. The principles of each are introduced in this chapter and need to be learned in detail. Because they are used both at a simple level and in advanced and sophisticated ways, we return to them on a number of occasions throughout the book. Ultimately, we find that interrupts and counter/timers both form part of the exciting techniques that underpin real-time programming.

In this chapter you will learn about:

- Why we need interrupts and counter/timers
- The underlying interrupt hardware structure
- The 16F84A interrupt structure
- How to program with interrupts
- The underlying microcontroller counter/timer hardware structure
- The 16F84A Timer 0 structure
- Simple applications of the counter/timer
- The Sleep mode.

If you wish you will also:

- Learn about alternative approaches to interrupt strategies, using examples from other microcontroller families
- Get deeper into 16F84A interrupt issues, in particular its interrupt latency.

6.1 The main idea – interrupts

As we know, a computer CPU is a deeply orderly entity, following the instructions of the program one by one, doing what it is told in a precise and predictable fashion. An interrupt disturbs this order. Coming maybe when least expected, its function is to alert the CPU in no uncertain terms that some significant external event has happened, to stop it from what it is doing and force it (at greatest speed possible) to respond to what has happened. Originally this was applied to allow emergency external events to get the attention of the CPU, emergencies like power failure, the system overheating or major failure of a subsystem. But the concept of interrupts was recognised as very powerful, and as time went on, more and more subsystems gained the power to generate interrupts. This forced increasing complexity in interrupt structures and a need to recognise that not all interrupts were equal.

To work successfully with interrupts, we need to understand both the hardware interrupt structure and the programming techniques needed to program successfully with them. An introduction to these now follows.

6.1.1 Interrupt structures

Different microcontrollers have rather different interrupt structures. Inevitably they have more than one interrupt source, usually with some internally generated and others external. A generic structure, which illustrates the main hardware principles, is shown in Figure 6.1. At the left we see one of several sources, 'Interrupt X'. If an interrupt occurs, it sets an S-R bistable. The occurrence of the interrupt, even if it is only momentary, is thus recorded. The output of the bistable, the latched version of the interrupt, is called the *interrupt flag*. This is then gated with an enable signal, *Interrupt X Enable*. If this is high, then the interrupt signal progresses to an OR gate. If it is low, the interrupt signal gets no further. If enabled, it is ORed with other enabled interrupt inputs of the microcontroller. The OR gate output will go high if *any* interrupt input is high. There is then a further gating of the OR gate output, this time with a *Global Interrupt Enable*. Only if that value is high can any interrupt signal reach the CPU. When the CPU has responded to an interrupt,

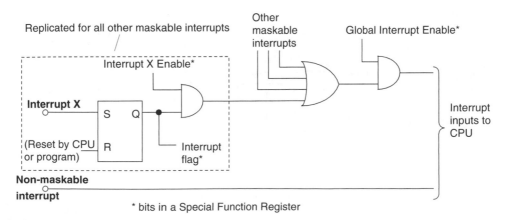

Figure 6.1 A simple generic interrupt structure

it is necessary to clear the interrupt flag. In some processors this is done automatically by the CPU, in others it must be done within the program.

The action of disabling an interrupt is sometimes called *masking*. It seems strange, however, to be able to switch off a capability which is so important and which is meant to be there to report emergencies. Therefore, some microcontrollers have interrupts that cannot be masked. These are always external (i.e. not from an internal peripheral) and are used to connect to external interrupt signals of the greatest importance. A non-maskable interrupt is shown in Figure 6.1. As the CPU always responds if it occurs, there is less point in storing it as a flag, and this is sometimes therefore not done.

6.1.2   The 16F84A interrupt structure

The 16F84A has four interrupt sources, all of which can be individually enabled or disabled:

- *External interrupt*. This is the only external interrupt input. It shares a pin with Port B, bit 0 (Figure 2.1). It is edge triggered.
- *Timer overflow*. This is an interrupt caused by the Timer 0 module, which is the subject of the second half of this chapter. It occurs when the timer's 8-bit counter overflows.
- *Port B interrupt on change*. This interrupts when a change is detected on any of the higher 4 bits of Port B. The mechanism was described in Section 3.4.1.
- *EEPROM write complete*. This interrupts when a write instruction to the EEPROM memory is completed.

The interrupt structure is shown in Figure 6.2 and the SFR that controls it, **INTCON**, in Figure 6.3. It is useful to study the two diagrams in parallel, as every bit in the **INTCON** register appears in the structure logic diagram. The four sources appear labelled on the left of Figure 6.2. When comparing this diagram with Figure 6.1, it is interesting to note the absence of the interrupt flag flip-flop. These exist, but are not shown in the Microchip diagram. Each source has an enable line (labelled ...**E**) and a flag line (labelled ...**F**).

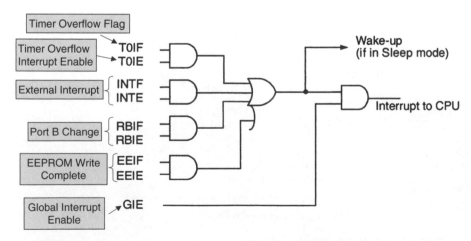

Figure 6.2 The 16F84A interrupt structure (supplementary labels in shaded boxes added by the author)

R/W-0	R/W-0	R/W-0	R/W-0	R/W-0	R/W-0	R/W-0	R/W-x
GIE	EEIE	T0IE	INTE	RBIE	T0IF	INTF	RBIF

bit 7   bit 0

bit 7   **GIE:** Global Interrupt Enable bit
          1 = Enables all unmasked interrupts
          0 = Disables all interrupts

bit 6   **EEIE:** EE Write Complete Interrupt Enable bit
          1 = Enables the EE Write Complete interrupts
          0 = Disables the EE Write Complete interrupt

bit 5   **T0IE:** TMR0 Overflow Interrupt Enable bit
          1 = Enables the TMR0 interrupt
          0 = Disables the TMR0 interrupt

bit 4   **INTE:** RD0/INT External Interrupt Enable bit
          1 = Enables the RB0/INT external interrupt
          0 = Disables the RB0/INT external interrupt

bit 3   **RBIE:** RB Port Change Interrupt Enable bit
          1 = Enables the RB port change interrupt
          0 = Disables the RB port change interrupt

bit 2   **T0IF:** TMR0 Overflow Interrupt Flag bit
          1 = TMR0 register has overflowed (must be cleared in software)
          0 = TMR0 register did not overflow

bit 1   **INTF:** RB0/INT External Interrupt Flag bit
          1 = The RB0/INT external interrupt occurred (must be cleared in software)
          0 = The RB0/INT external interrupt did not occur

bit 0   **RBIF:** RB Port Change Interrupt Flag bit
          1 = At least one of the RB7:RB4 pins changed state (must be cleared in software)
          0 = None of the RB7:RB4 pins have changed state

Figure 6.3 The 16F84A **INTCON** register

Thus, the lines **TOIF, INTF** and so on are actually the interrupt flags, rather than the interrupt inputs themselves. All can be seen as bits in the **INTCON** register, with the exception of the EEPROM write complete flag and enable. Note that the external interrupt is edge triggered. The edge it responds to is controlled by the setting of the **INTEDG** bit of the **OPTION** register (shown later, as it mainly relates to Timer 0, in Figure 6.9).

As in Figure 6.1, each flag is ANDed with a corresponding Enable input (**TOIE, INTE, RBIE** and **EEIE**). The enable bits are located in the **INTCON** register and can be set by the programmer. The outputs of the four AND gates are then ORed together, before passing on to the Global Enable gate. Interrupt flags must be cleared by manipulating their **INTCON** bits in the program. The 16F84A has no non-maskable interrupt input.

6.1.3   The CPU response to an interrupt

Let us assume that an interrupt has occurred, and both its local enable and the global enable are set. The interrupt is therefore detected by the CPU and it executes a special section of program called the *Interrupt Service Routine* (ISR). It is important to understand the underlying detail of what goes on, which is illustrated in the flow diagram of Figure 6.4. The CPU completes the instruction it is currently executing and saves the value of the Program Counter on the top of the Stack. Thus, it will 'know' where to come back to when the ISR is complete. To avoid other interrupts possibly interrupting this interrupt, it also clears the Global Interrupt Enable.

In the PIC® 16 Series, the ISR *must* start at the interrupt vector, program memory location 0004 (Figure 2.4). Therefore, when an interrupt occurs, this value is loaded into the Program Counter and program execution then continues from the reset vector. In any processor, the ISR *must* end with a special 'return from interrupt' instruction. In the 16 Series this is the **retfie** instruction. When this is detected, the CPU sets the **GIE** to 1, loads the Program Counter from the top of the Stack and then resumes program execution. Thus, it returns to the instruction which follows the instruction during which the interrupt was detected.

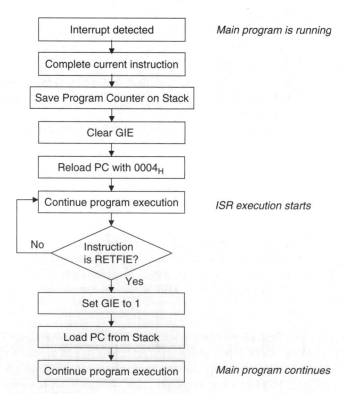

Figure 6.4 The 16F84A interrupt response sequence of events

6.2   Working with interrupts

6.2.1   *Programming with a single interrupt*

It is comparatively easy to write simple programs with just one interrupt. For success, the essential points to watch are:

- Start the ISR at the interrupt vector, location 0004
- Enable the interrupt that is to be used, by setting the enable bit in the **INTCON** register
- Set the Global Enable bit, **GIE**
- Clear the interrupt flag within the ISR
- End the ISR with a **retfie** instruction
- Ensure that the interrupt source, for example Port B or Timer 0, is actually set up to generate interrupts!

Program Example 6.1 gives a very simple interrupt example, intended for simulation. The interrupt guidelines in the list above are applied. The program starts as usual at the reset vector 0000. However, the interrupt vector is now also in use. The first action of the program is to branch over the reset vector to location **start**, where initialisation takes place. Within this we see the **GIE** and **INTE** bits being set. These bit labels can be used because they appear in the 16F84A include file. The main program simply outputs the bit patterns $0A_H$ and 15_H to Port A in turn.

When an interrupt occurs the interrupt vector address is loaded into the Program Counter, from where program execution continues. The first action of the ISR is to jump to location **Int_Routine**. This is placed at program memory location 0080_H to give clarity to the simulation. The ISR simply clears Port A before clearing its interrupt flag and returning to the main program.

```
;******************************************************************
;Interrupt Demonstration 1
;This program demonstrates simple use of single interrupt.
;Intended for simulation.
;Int_Demo1                                    14.4.05
;******************************************************************
;
      #include p16f84A.inc
;Port A all output. Port B: Bit 0 = Interrupt input
;
      org 00
      goto start
       org 04 ;Interrupt Service Routine starts here
       goto Int_Routine
      org 0010
;Initialise
start bsf    status,rp0  ;select bank 1
      movlw 01
      movwf trisb              ;portb bits 1-7 output
                               ;bit 0 input
      movlw 00
      movwf trisa              ;porta bits all output
```

Program Example 6.1 Simple interrupt application

```
        bcf    status,rp0   ;select bank 0
          bsf    intcon,inte  ;enable external interrupt
          bsf    intcon,gie   ;enable global int
;Remove semi-colon from following instruction to change
                                    ;interrupt edge
;       bsf    option_reg,intedg
wait    movlw 0a     ;set up initial port output values
        movwf porta
        movlw 15
        movwf porta
        goto  wait
;
        org 0080
Int_Routine          ;Interrupt Service Routine continues here
        movlw 00
        movwf porta
        bcf    intcon,intf
        retfie
        end
```

Program Example 6.1 Continued

Simulation Exercise 6.1

Copy the program Int_Demo1 from the book CD into MPLAB® and create a project around it. Build the project and enable the simulator. Open a Watch window, displaying **PORTA, PORTB, INTCON** and **PCL**. Open the Hardware Stack window (under **View**) to observe the contents of the Stack. Open the Stimulus Controller and set Pin RB0 to Toggle. Single step through the program, and observe and understand the change to each observed variable, on every instruction. Now 'fire' the RB0 pin, setting RB0 high, and continue stepping. This will cause no change to program execution, as the interrupt edge response will be negative-edge triggered (the **INTEDG** bit has been left at a Reset value of 0). Fire the RB0 pin again and an interrupt sequence should be instigated as you single step further. See the Hardware Stack change, program execution transfer to the ISR and on ISR completion the program resuming after the instruction after it was interrupted. Try changing the **INTEDG** bit, as shown in the program, to change the edge to which the interrupt responds.

When you have implemented the above program successfully, try inserting the errors below into the program. They are commonly made by novices. Observe the effect:

- Fail to clear the interrupt flag by removing the instruction **bcf intcon,intf**
- Terminate the ISR incorrectly by replacing **retfie** with **return**.

6.2.2   Moving to multiple interrupts – identifying the source

Using one interrupt in a program is generally quite easy, but be warned – once we start using more than one, interrupts can interact with each other in ways which are far from simple. Complexity then seems to rise approximately in proportion to the *square* of the number of interrupts used!

As we have seen, the 16F84A has four interrupt sources, but only one interrupt vector. Therefore, if more than one interrupt is enabled, it is not obvious at the beginning of an ISR which interrupt has occurred. In this case the programmer must write the ISR so that at its beginning it tests the flags of all possible interrupts and determines from this which one has been called. An example piece of code that does this, assuming all four interrupt sources are enabled, is shown in Program Example 6.2.

```
interrupt btfsc intcon,0      ;test RBIF
        goto portb_int
        btfsc intcon,1        ;test external interrupt flag
        goto ext_int
        btfsc intcon,2        ;test timer overflow flag
        goto timer_int
        btfsc eecon1,4        ;test EEPROM write complete flag
        goto eeprom_int

portb_int
...
place portb change ISR here
...
        bcf     intcon,0  ;and clear the interrupt flag
        retfie

ext_int
...
place external interrupt ISR here
...
        bcf     intcon,1  ;and clear the interrupt flag
        retfie

timer_int
...
place timer overflow ISR goes here
...
        bcf     intcon,2  ;and clear the interrupt flag
        retfie
eeprom_int
...
place EEPROM write complete ISR here
...
        bcf     eecon1,4  ;and clear the interrupt flag
        retfie
```

Program Example 6.2 Interrupt source identification

6.2.3 Stopping interrupts from wrecking your program 1 – context saving

Because an interrupt can occur at any time, it has the power to be extremely destructive. Program Example 6.3 is written to illustrate this. It applies a 16-bit addition subroutine. For the purposes of the example, the 16-bit number 9999_H is added to itself, with an expected 17-bit result 13332_H. In the subroutine, the lower 2 bytes, **qlo** and **plo**, are added. Any Carry generated is then added into one of the higher bytes, and the higher two bytes are added. An ISR is written which affects both the Carry flag and the W register, as most ISRs would.

Suppose the interrupt occurs immediately after the first subroutine **movf** instruction, where the W register is holding the value of **plo**. The ISR changes the W register, so when program execution returns to the subroutine, it will be with the incorrect W register value. Suppose the interrupt occurs immediately after the first **addwf** instruction. The value of the Carry bit is essential to the success of the addition, but again is lost in the ISR.

```
;****************************************************************
;Int_context
;This program demonstrates the need for context saving.
;Intended for simulation.

;TJW 15.4.05                                Tested 17.4.05
;****************************************************************
;Port A not used. Port B bit 0 used for ext.interrupt ip.
        #include p16f84A.inc
;
rhi             equ    10
rlo             equ    11
phi             equ    12
plo             equ    13
qhi             equ    14
qlo             equ    15


                org 00
                goto start
                org 04 ;here if interrupt occurs
                goto Int_Routine
;
start           org 0010
                bsf    intcon,inte ;enable external interrupt
                bsf    intcon,gie  ;enable global int
loop            movlw 99
                movwf phi    ;preload numbers to be added
                movwf plo
                movwf qhi
                movwf qlo
                call Double_add
                movlw 00     ;clear result
                movwf rhi
                movwf rlo
                goto loop
;This subroutine adds two 16-bit numbers, stored in phi-plo, and qhi-qlo,
;and stores result in rhi-rlo. 16-bit overflow in Carry flag at end.
Double_add
                movf  plo,0       ;move plo to the W reg
                addwf qlo,0       ;add lower bytes
                movwf rlo
                btfsc status,0
```

Program Example 6.3 Impact of interrupts

```
                incf    phi,1           ;add in Carry
                movf    phi,0
                addwf   qhi,0           ;add upper bytes
                movwf   rhi
                return
Int_Routine
                bcf     status,0        ;clear the Carry flag
                movlw   0ff             ;change W reg value
                bcf     intcon,intf
                retfie
                end
```

Program Example 6.3 Continued

Simulation Exercise 6.2

Copy the program Int_Context from the book CD into MPLAB and create a project around it. Build the project and enable the simulator. Open a Watch window, displaying **qhi, qlo, phi, plo, rhi, rlo, STATUS** and **WREG**. Open the Stimulus Controller and set Pin RB0 to Toggle. Single step through the program, and check that the addition works correctly and the expected result is achieved. Now try inserting interrupts at different points in the program. Note how at many points the occurrence of the ISR destroys the validity of the addition result.

The temporary data being used in a particular activity in the CPU is called its *context*. In the PIC 16 Series this includes at least the W register value and the Status register. It is clearly important to save the context when an interrupt occurs. Some microcontrollers do this automatically, but PIC 16 Series microcontrollers do not. Therefore, it is up to the programmer to ensure that whatever context saving that is needed is done in the program.

Program Example 6.4 shows the recommended Microchip method for saving the W register into a pre-designated memory location **W_TEMP** and Status register into a location called **STATUS_TEMP**. The **swapf** and **movwf** instructions are used because they do not affect any Status register bits.

```
PUSH    movwf w_temp          ;Copy W to W_TEMP register,
        swapf status,0        ;Swap status to be saved into W
        movwf status_temp     ;Save status to STATUS_TEMP register
ISR                           ;Interrupt Service Routine
...
        actual ISR goes here
...
POP     swapf status_temp,0   ;Swap nibbles in STATUS_TEMP register
                                            ;and place result into W
        movwf status          ;Move W into STATUS register ;sets bank to original
                                            ;state
```

Program Example 6.4 Context saving

```
        swapf w_temp,1        ;Swap nibbles in W_TEMP and keep result in W_TEMP
        swapf w_temp,0        ;Swap nibbles in W_TEMP and place result into W
...
        clear interrupt flag(s) here
...
        retfie
```

Program Example 6.4 Continued

Simulation Exercise 6.3

Adapt the context saving shown in Program Example 6.4 and insert it into the Int_Context program (Program Example 6.3). You will need to define memory locations for **w_temp** and **status_temp**. Check that the program now operates correctly wherever you force an interrupt to occur.

6.2.4   *Stopping interrupts from wrecking your program 2 – critical regions and masking*

We can resolve *some* of the problems of an interrupt occurring in a program section like the subroutine discussed above by appropriate context saving. Unfortunately, we can't resolve them all, at least not just with context saving.

What if an interrupt occurred in a software delay routine, for example that of Program Example 5.2? The delay length would be increased by the duration of the ISR, which could be disastrous, and no amount of context saving would improve the situation.

Consider a more subtle problem. The ISR shown in Program Example 6.5 takes the word held in **rhi-rlo**, calculated in the subroutine of Program Example 6.3, and outputs it to a 12-bit digital-to-analog converter (DAC) connected to Ports A and B. We assume that the overall program constrains the word in **rhi-rlo** to 12 bits. Suppose the ISR shown in Program Example 6.5 occurs during the subroutine of Example 6.3. Context saving is implemented, so should there be a problem?

Unfortunately, there is a problem. The ISR is making use of a result that is being calculated in a program section that it is interrupting. Suppose **rlo** has just been updated and not **rhi** when the interrupt occurs. The ISR outputs the new value of **rlo** and the old one of **rhi**. Together, they might make a number that has no sense, with potentially disastrous consequences.

```
Int_Routine
        movwf W_temp          ;Copy W to TEMP register,
        swapf status,0        ;Swap status to be saved into W
        movwf status_temp     ;Save status to STATUS_TEMP register
        bcf   status,5        ;ensure we are in Bank 0
```

Program Example 6.5 An interrupt using data calculated in the program

```
movf  rhi,0              ;output higher 4 bits to DAC
movwf porta
movf  rlo,0              ;output lower 4 bits to DAC
movwf portb
swapf status_temp,0      ;Swap nibbles in STATUS_TEMP register
                                   ;and place result into W
movwf status             ;Move W into STATUS register ;set bank to original
                                   ;state
swapf W_temp,1           ;Swap nibbles in W_TEMP and place result in W_TEMP
swapf W_temp,0           ;Swap nibbles in W_TEMP and place result into W
bcf   intcon,intf
retfie
```

Program Example 6.5 Continued

Therefore, we must accept the fact that in certain program areas we will not want to accept the intrusion of an interrupt under any circumstances, with or without context saving. We call these *critical regions*. We can disable, or *mask*, the interrupts for their duration, by manipulating the enable bits in the **INTCON** register. Critical regions generally include all time-sensitive activity and any calculation where the ISR makes use of the result. Time-sensitive activity itself includes timing loops and multi-instruction setting of outputs.

By applying properly the techniques of context saving and critical regions, we can make good use of interrupts, without them displaying the more destructive side of their nature.

6.3   The main idea – counters and timers

6.3.1   *The digital counter reviewed*

It is very easy to make a digital counter using flip-flops. Counters can be made which count up, count down, which can be cleared back to zero, pre-loaded to a certain value, and which by the provision of an overflow output can be cascaded with other counters. A simple example is shown in Figure 6.5. Eight negative edge-triggered J–K bistables are interconnected, so that the Q-output of one drives the clock input of the next. With J and K both tied to Logic 1, the flip-flop toggles on every input negative edge. The counter holds an 8-bit binary number, made up of the eight Q-outputs of the bistables, where Q_7 is the most significant and Q_0 the least significant. It counts up by one on the negative edge of every incoming clock cycle.

The output timing diagram is shown in the lower part of the figure. It can be seen that after one input cycle Q_0 has gone to Logic 1. After 16 input cycles have been completed (i.e. during cycle 17) the 8-bit word forms 00010000_B, i.e. 16_D, and after 31 cycles it forms 00011111_B, i.e. 31_D. When 255 input cycles have been completed the counter holds the word 11111111_B, or FF_H. If another input cycle comes along, then all flip-flops ripple through to 0 and the output returns to 00000000_B. The negative-going edge of Q_7 can be used to indicate that the counter has overflowed.

The counter of Figure 6.5 can be reset to zero if the clear line is activated. With a little more complexity it is possible to add the facility to pre-load the counter with any number desired. By so doing we gain a

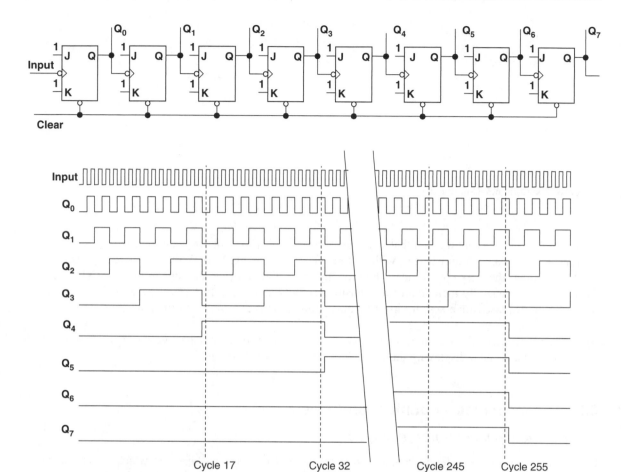

Figure 6.5 A digital counter made of eight flip-flops

versatile digital subsystem which becomes the basis for a microcontroller counter. This can be represented as in Figure 6.6. The only interconnections of significance are the Clock input, the overflow output and the 8-bit Read or Load capability, which can be gated to share a single bi-directional data path.

6.3.2   The counter as timer

It is extremely useful for a microcontroller to be able to count – widgets passing on a conveyor belt for example, or coins in a slot machine, or people going through a door. It is, however, even more useful if it can measure time, and the counter allows us to do this.

Suppose the Input signal of Figure 6.5 was a stable 1 ms clock frequency. Then the counter would increment exactly every 1 ms. After 16 clock cycles, exactly 16 ms would have elapsed, after 31 cycles 31 ms

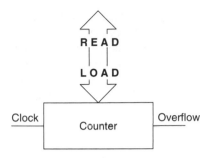

Figure 6.6 The digital counter in block diagram form

and so on. By starting the clock input at a moment of choice, it is therefore possible to measure elapsed time. The resolution of the measurement is determined by the period of the clock. In this example the resolution is 1 ms and we can't measure anything less than that, or a fraction of it! Again, for the 1 ms input period, the 8-bit counter can measure up to 255 ms before overflowing. The use of counters as timers is so important that the counter is often called a counter/timer (C/T), or simply a timer, to reflect this importance.

An obvious application of the counter/timer is to measure the time between two 'events'. These events may both be externally generated. Alternatively, the first is generated by the microcontroller and the second happens some time later, as a response. It may also be necessary to measure the time between two pulses or the duration of a single pulse. The general requirement is illustrated in Figure 6.7. The actual measurement

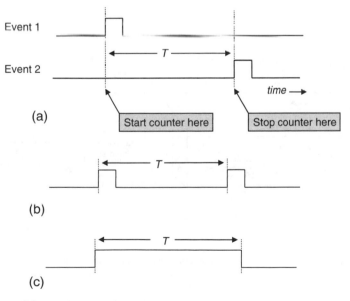

Figure 6.7 The challenge of time measurement

seems easy – start the counter/timer running when the first event occurs and stop it on the second. In practice, this poses a number of challenges. For an accurate measurement, the start and stop of the counter/timer must be perfectly synchronised with the events. The best way of doing this is by using an interrupt. If we don't have an interrupt, then we will have to scan continuously the input to detect when the event occurs – in which case it's hardly worth using the counter/timer, as we might as well do the timing in software. If there are two external events on two different lines then we still have a problem, as with the PIC 16 Series we only have one external interrupt.

We will see a good example of this sort of time measurement in Chapter 10. We will also see enhancements to the counter/timer that get over the problem of accurately synchronising the start and stop of the counter/timer with the events it is measuring.

6.3.3   The 16F84A Timer 0 module

The 16F84A Timer 0 is typical of many simple counter/timers in smaller-scale microcontrollers. It takes an 8-bit counter like the one in Figure 6.5, connects it as an SFR in the memory map and packages it with some useful extra features. Its block diagram representation is shown in Figure 6.8, with the actual 8-bit counter labelled **TMR0**. Looking back to Figure 2.5, we can see that this appears as register **TMR0** at memory location 01 in Bank 0. Like all good microcontroller peripherals, Timer 0 is configurable, controlled by a number of bits that appear in the **OPTION** register, shown in Figure 6.9.

Looking to the left of Figure 6.8, we can see that there are two possible sources of the clock input to the **TMR0** counter. One is the RA4 pin (i.e. pin 3 of the 16F84A – see Figure 2.1). The other is the internal instruction cycle frequency, labelled Fosc/4. The selection of input source is made by the multiplexer

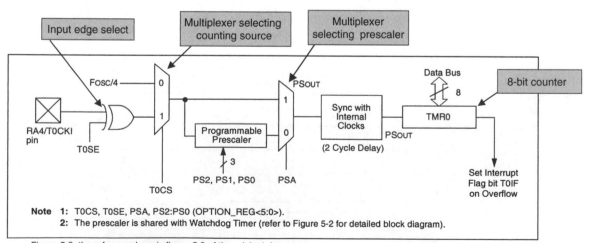

Figure 5.2: the reference here is figure 5.2 of the original document.

Figure 6.8 The 16F84A Timer 0 module (supplementary labels in shaded boxes added by the author)

R/W-1	R/W-1	R/W-1	R/W-1	R/W-1	R/W-1	R/W-1	R/W-1
RBPU	INTEDG	T0CS	T0SE	PSA	PS2	PS1	PS0

bit 7   bit 0

bit 7   **RBPU:** PORTB Pull-up Enable bit

1 = PORTB pull-ups are disabled
0 = PORTB pull-ups are enabled by individual port latch values

bit 6   **INTEDG:** Interrupt Edge Select bit

1 = Interrupt on rising edge of RB0/INT pin
0 = Interrupt on falling edge of RB0/INT pin

bit 5   **T0CS:** TMR0 Clock Source Select bit

1 = Transition on RA4/T0CKI pin
0 = Internal instruction cycle clock (CLKOUT)

bit 4   **T0SE:** TMR0 Source Edge Select bit

1 = Increment on high-to-low transition on RA4/T0CKI pin
0 = Increment on low-to-high transition on RA4/T0CKI pin

bit 3   **PSA:** Prescaler Assignment bit

1 = Prescaler is assigned to the WDT
0 = Prescaler is assigned to the Timer 0 module

bit 2-0   **PS2:PS0:** Prescaler Rate Select bits

Bit Value	TMR0 Rate	WDT Rate
000	1 : 2	1 : 1
001	1 : 4	1 : 2
010	1 : 8	1 : 4
011	1 : 16	1 : 8
100	1 : 32	1 : 16
101	1 : 64	1 : 32
110	1 : 128	1 : 64
111	1 : 256	1 : 128

Figure 6.9 The 16F84A **OPTION** register

controlled by bit **T0CS**, which appears in the Option register. The external input path includes the option of inverting the signal with the Exclusive OR gate, the inversion being controlled by bit **T0SE**. The output of the first multiplexer branches before reaching a second multiplexer. This selects either a direct path or the path taken through a programmable prescaler. The choice is controlled by bit **PSA** of the Option register. A complication here is that the prescaler is actually shared with the Watchdog Timer (WDT), which we meet a little later in this chapter. For now we just need to recognise that if **PSA** is set to 1, then the prescaler is assigned to the WDT and the multiplexer is selecting the input path which avoids the prescaler. The prescaler itself is controlled by bits **PS2**, **PS1** and **PS0** of the Option register. Inspection of these bits in Figure 6.9 shows that they allow a choice of frequency division of the incoming clock signal. The output of the second multiplexer is synchronised with the internal clock, before becoming the input to the actual counter. When the counter overflows, it sets the timer overflow flag, one of the PIC microcontroller's four interrupt sources, which we met in Figure 6.2.

6.4   Applying the 16F84A Timer 0, with examples using the electronic ping-pong

A simple counter/timer like the Timer 0 can be used for many applications. We will look at two examples. Both are based on the electronic ping-pong program and can be readily simulated.

6.4.1   *Object or event counting*

The simplest application of Timer 0 is to use it as a counter, counting pulses entering the microcontroller through the external input. Looking at the electronic ping-pong circuit (Appendix 2, Figure A2.1), we see that the right paddle is connected to pin 3 of the 16F84A. The program of Program Example 6.6 is a very simple counting example. It enables the counter appropriately and uses the right paddle as the counter input, continuously displaying the current value on the LEDs connected to Port B.

To configure Timer 0, we'll need to select its external input, i.e. **T0CS** = 1. The input edge that we trigger from is not too important. As there is a risk of switch bounce, however, we will choose the edge associated with switch release, i.e. the rising edge, as there is less likelihood of bounce. Therefore, **T0SE** = 0. We will not want the prescaler, as we wish to count the exact number of switch presses; therefore, **PSA** = 1. Hence the values of **PS2**, **PS1** and **PS0** do not matter (as this application does not make use of the WDT). All Option register bits that have not been mentioned in this paragraph are not of importance to the ping-pong, so will be arbitrarily set to 0. A final value for the Option register setting is thus 00101000$_B$.

```
;******************************************************************
;cntr_demo                                    Counter Demonstration
;This program demos Timer 0 as counter, using ping-pong hardware
;TJW 15.4.05                                        Tested 15.4.05
;******************************************************************
;Clock freq 800kHz approx (RC osc.)
;Port A 4    right paddle (ip) Counter input.
;       2    "out of play" led (op)
;Port B 7-0  "play" leds (all op)
;Interrupts not used
;Config Word: RC oscillator, WDT off, PU timer on, code protect off
;
      list p=16F84A
      #include p16f84A.inc
;
            org    00
; Initialise
            bsf    status,rp0    ;select memory bank 1
            movlw B'00011000'
            movwf trisa          ;port A according to above pattern
            movlw 00
            movwf trisb          ;all port B bits outout
            movlw B'00101000'    ;set up TMR0 for external input, +ve edge,
                                 ;no prescale
```

Program Example 6.6 Initialising Timer 0 for measuring ping-pong rally length

```
              movwf TMR0         ;as we are in Bank 1, this addresses OPTION
              bcf   status,rp0   ;select bank 0
;
              movlw 04     ;switch on "out of play" led to show power is on
              movwf porta
loop          movf  TMR0,0 ;Continuously display Timer 0 on Port B
              movwf portb
              goto  loop
              end
```

Program Example 6.6 Continued

This program can be run on the ping-pong hardware, in which case every press of the right paddle causes a binary display on the play LEDs to increment by one.

Simulation Exercise 6.4

Copy the program Cntr_Demo from the book CD into MPLAB and create a project around it. Build the project and enable the simulator. Open a Watch window, displaying **PORTB** and **TMR0**. Open the Stimulus Controller and set Pin RA4 to Pulse high, with a pulse width of one cycle. Animate the program, 'fire' the input pulse and see how the Timer 0 and Port B SFRs count up.

6.4.2 Hardware-generated delays

In the original ping-pong program software-generated delays are used to time how long the LEDs are to be illuminated. This is only acceptable in simple programs, as in software-generated delays the CPU is doing nothing useful during the whole of the delay. Now that we have a counter/timer at our disposal, we can use that to generate the delay and if necessary the CPU can busy itself with other things. This seems quite simple, but a small problem presents itself: how do we know when the delay period is up? If we have to keep checking the timer value, then we will have made little progress. This is where the *interrupt on overflow* comes into its own. If things are set up so that an interrupt is generated as the delay ends, then we have a powerful means of creating efficient delays.

As a first step, let's replace the 5 ms software delay subroutine in ping-pong with a delay controlled by Timer 0. The internal clock is approximately 800 kHz and the instruction cycle rate (Fosc/4) is therefore 200 kHz, or a period of 5 µs. Now with this clock frequency, Timer 0 would count up to its maximum value (255) in 255 × 5 µs, or 1275 µs, and would overflow on the next cycle, i.e. after 1280 µs. We can, however, make use of the prescaler here. If the incoming signal is divided by 4 (i.e. **PS2**, **PS1**, **PS0** set to 001), then Timer 0 will overflow after 256 × 4 × 5 µs, or 5.120 ms. This is very close to the 5 ms we're looking for, but it's not quite exact.

Although the ping-pong does not need accurate timing, suppose we genuinely needed a delay very close to 5 ms? Let us divide the incoming clock by 8 instead of 4, which gives a divided frequency of 25 kHz, or period of 40 µs. Now 125 Timer 0 input cycles will cause a delay of 40 × 125 µs, or 5.00 ms, which is exactly our target. If we arrange for this prescaling, and at the start of each delay pre-load Timer 0 with $256 - 125$, i.e. 131_D, then an exact delay, terminated by the interrupt on overflow, is possible.

```
...
;Initialise
      org   0010
start bsf   status,5     ;select memory bank 1
      movlw B'00011000'
      movwf trisa        ;port A according to above pattern
      movlw 00
      movwf trisb        ;all port B bits op
      movlw B'00000010'  ;set up TMR0 for internal input, prescale by 8
      movwf TMR0         ;as we are in Bank 1, this addresses OPTION
      bcf   status,5     ;select bank 0
...
...
;introduces delay of 5ms approx
      delay5 movlw D'131'       ;preload counter, so that 125 cycles, each
                                ;of 40us, occur before timer overflow
      movwf TMR0
del1  btfss intcon,2            ;test for Timer Overflow flag
      goto del1                 ;loop if not set
      bcf intcon,2              ;clear Timer Overflow flag
      return
```

Program Example 6.7 Using Timer 0 in the **delay5** subroutine ·

An implementation of this approach is shown in the program sections in Program Example 6.7. This includes both the initialisation section and the revised delay subroutine. Interrupts are *not* enabled and the subroutine determines when the delay is complete by testing the overflow interrupt flag. The advantage to the programmer is that timing is now achieved by manipulating the Timer 0 settings, rather than by adjusting the software routine. The 'interrupt on overflow' has not been enabled, as it would in this instance offer little advantage. In a more demanding program, however, the interrupt could be enabled and the time spent in the delay used to undertake other CPU activities.

Simulation Exercise 6.5

Modify the ping-pong program to include the changes given in Program Example 6.7. Using **Debugger > Settings** ensure that the clock frequency is set to 800 kHz. Use the Stopwatch facility to check the time duration of the new delay subroutine. How much do the **call**, **return** and Timer loading instructions add to the delay? Can you fine-tune it to improve its accuracy?

6.5 The Watchdog Timer

There is another timer in the 16F84A that we need to take note of, even though it is not normally used in simple applications. This is the Watchdog Timer (WDT). A big danger with any computer-based system is that the software fails in some way and that the system locks up or becomes unresponsive. In a desktop computer such lock-up can be annoying and one would normally have to reboot. In an embedded system it can be disastrous, as there may be no user to notice that there is something wrong and maybe no user interface anyway. The WDT offers a fairly brutal 'solution' to this problem. It is a counter, internal to the microcontroller, which is continually counting up. If it ever overflows, it forces the microcontroller into reset (Figure 2.11). It is up to the programmer to ensure that within the program the WDT is repeatedly cleared. This is done with the instruction **clrwdt**. It is only when the program ceases to run correctly that these instructions are no longer executed and the overflow occurs.

A WDT Reset is generally not good news for an embedded system, as all current settings are of course destroyed, and the program starts over. It is, however, better than a program which is not running at all. Note that the WDT leaves one clue of its action behind, and that is through the $\overline{\text{TO}}$ bit in the Status register (Figure 2.3). It is possible to test this bit towards the beginning of a program and hence distinguish between a Power on Reset and a WDT Reset.

The 16F84A WDT is enabled by one of the configuration bits, seen in Figure 2.6. Thus, it either runs or it doesn't for the duration of the time the microcontroller is switched on. It is driven by an internal RC oscillator, which gives a nominal time-out period of 18 ms. This, however, is to some extent dependent on temperature, supply voltage and variation from device to device. It can be extended by applying the Timer 0 prescaler to it, in which case the time-out period can be stretched up to 128×18 ms, or around 2.3 seconds.

6.6 Sleep mode

Although we are considering timing in this chapter, it is an appropriate moment to consider one aspect of microcontroller operation when time is almost suspended – the Sleep mode. This represents an important way of saving power. The microcontroller can be put into this mode by executing the instruction **SLEEP**, seen in Appendix 1. Once in Sleep mode, the microcontroller goes into almost suspended animation. The clock oscillator is switched off, the WDT is cleared, program execution is suspended, all ports retain their current settings, and the $\overline{\text{PD}}$ and $\overline{\text{TO}}$ bits in the Status register (Figure 2.3) are cleared and set respectively. If enabled, the WDT continues running. Under these conditions, power consumption falls to a negligible amount – Ref. 2.1 quotes a typical value of 1 µA, under specific ideal operating conditions.

Once asleep, the microcontroller requires an explicit event to wake it again. The 16F84A will awake from Sleep in the following situations:

• *External reset through MCLR pin*. While this causes a wake-up, it also resets the microcontroller; therefore, its use seems limited to complete program restarts. It *is* possible, however, to detect that the microcontroller has just been in Sleep mode, due to the state of the $\overline{\text{PD}}$ pin in the Status register.
• *WDT wake-up*. The function of the WDT is a little different in Sleep. Looking at Figure 2.10, it can be seen that the WDT is blocked from causing a reset when in Sleep. Instead, on overflow it just causes a wake-up from Sleep, and the microcontroller continues program execution from the instruction following the Sleep mode.

- *Occurrence of interrupt.* As Figure 6.2 indicates, any individually enabled interrupts cause wake-up from Sleep, regardless of the state of the Global Interrupt Enable. Timer 0 cannot, however, generate an interrupt, as the internal clock is disabled.

On wake-up, the oscillator circuit is restarted. For any crystal oscillator mode this means that the T_{OST} timer, seen in Figure 2.11, is also activated. It must complete its count before program execution can resume. Therefore, like a human being, the 16F84A takes a finite time to wake up and be ready for action.

The Sleep mode is extremely powerful for products that must be designed in a power-conscious way. Many devices are not continuously active when powered. If put into Sleep when not in use, their power consumption can be dramatically reduced.

6.7 What others do

Interrupts are an essential feature of almost every microcontroller, but architectures vary considerably. One difference is the way interrupt vectors are applied.

The Atmel 8051-based microcontrollers have six interrupts, as shown in Table 6.1. Instead of sharing a single interrupt vector, like the PIC 16 Series, each has its own vector. Therefore, for example, the ISR for the external interrupt 0 starts at the address 0003_H, while that for the Timer 0 overflow starts at $000B_H$. Where each interrupt has its own vector, there is no need to poll interrupt flags at the beginning of an ISR to detect the source, as was done in Program Example 6.2. However, once there are multiple interrupt vectors, it is essential to have a means of choosing between interrupts, if two occur at the same time. Therefore, the Atmel device *prioritises* its interrupts, as shown in the table. If two occur simultaneously, the one of higher priority is serviced first. Like the 16F84A, when responding to an interrupt the Atmel device only saves the Program Counter on the Stack.

The Freescale 68HC08, and indeed all previous Motorola products, have adopted a very different approach to interrupt vectors. Instead of the interrupt vector being the start address of the ISR, it *holds* the start address. Freescale places these vectors at the very top of memory space, as Table 6.2 indicates. Clearly, non-volatile memory must be placed here. Each vector is 16-bit, so it occupies two 8-bit memory locations.

Table 6.1 Atmel interrupt sources and vector addresses

Source	Symbol	Priority	Vector address
External interrupt 0	IE0	1 (highest)	0003_H
Timer 0 interrupt request	TF0	2	$000B_H$
External interrupt 1	IE1	3	0013_H
Timer 1 interrupt request	TF1	4	$001B_H$
Serial receive and transmit	RI + TI	5	0023_H
Timer 2 overflow, Timer 2 external	TF2 + EXF2	6 (lowest)	$002B_H$

Table 6.2 Freescale 68HC08 interrupt sources and vector addresses (incomplete)

Source	Symbol	Priority	Vector address
Reset vector (low)	–	–	$FFFF_H$
Reset vector (high)	–	–	$FFFE_H$
Software interrupt (low)	SWI	1	$FFFD_H$
Software interrupt (high)	SWI		$FFFC_H$
Interrupt request (low)	IRQ	2	$FFFB_H$
Interrupt request (high)	IRQ		$FFFA_H$

The reset vector is always placed at the very top of the memory map. At this address is stored the actual start address of the program. Below this the interrupt vectors are placed, in descending order of priority. The table shows only the first few. The advantage to users is that they can place the ISRs anywhere in the memory map they wish, as long as they correctly place their address in the vectors specified. The 68HC08 also differs from both the PIC 16 Series and Atmel, in that when interrupted it stacks *all* CPU registers, that is the Accumulator (equivalent to the PIC W register), the lower byte of the Index register, the Program Counter (2 bytes) and the Condition Code register.

6.8 Taking things further – interrupt latency

The purpose of the interrupt is to attract the attention of the CPU quickly, but actually how quickly does this happen? The time between the interrupt occurring and the CPU responding to it is called the *latency*. The latency is dependent on certain aspects of hardware and ultimately can also depend on the characteristics of the program running. The timing diagram of Figure 6.10 shows how the mid-range PIC family responds to an enabled external interrupt. The interrupt itself can be seen as a positive-going pulse on the **INT** pin line. This causes the interrupt flag **INTF** to be set. This flag is sampled on the Q1 cycle of the internal oscillator clock. Once this is done, the CPU has detected the interrupt and the sequence then follows that of Figure 6.4. Two dummy cycles are needed to save the Program Counter to the Stack, reload it with 0004_H and fetch the instruction at that address.

Simulation Exercise 6.6

Working with the int_demo1 program again, set the clock frequency to 4 MHz using Debugger > Settings. Enable the Stopwatch (under Debugger) and single step through the program. See how the Stopwatch updates elapsed time in a predictable way. Now instigate the interrupt. Notice how the Stopwatch records a latency of two instruction cycles. At the end of the ISR see that the **retfie** instruction also takes two cycles.

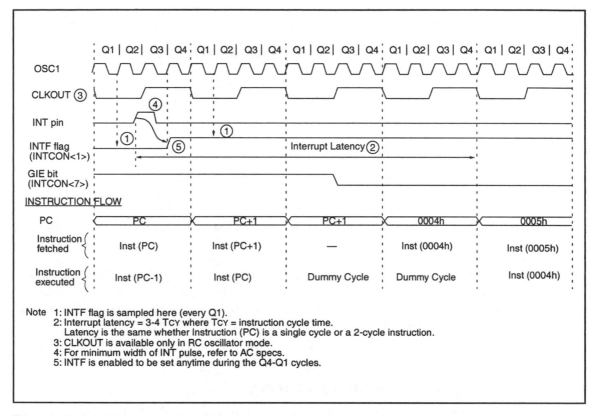

Figure 6.10 16F84A external interrupt latency

Summary

- Interrupts and counter/timers are important hardware features of almost all microcontrollers.
- They both carry a number of important hardware and software concepts, which must be understood.
- The basic techniques of using interrupts and counter/timers have been introduced in this chapter. There is considerably increased sophistication in their use in more advanced applications.

Section 3
Larger Systems and the PIC® 16F873A

This section, also of five chapters, focuses on developing a good understanding of microcontroller peripherals and their underlying principles. It makes use of the same microcontroller core as the earlier chapters, but applies a 'large' PIC 16 Series microcontroller. All peripherals introduced are also directly applicable to the 18 Series. Emphasis is placed on understanding the peripherals and using them in increasingly sophisticated applications. Program examples are in Assembler.

7
Larger systems and the PIC® 16F873A

Over the previous five chapters the PIC® 16F84A has been used as the example microcontroller. It, and other microcontrollers like it, are fine devices for the smaller product. However, there are many things they cannot do, and for more demanding applications we need to look to a more powerful microcontroller. But what does 'more powerful' actually mean? Remember what was said in Chapter 1, that a microcontroller is essentially made up of:

> Microprocessor core *plus* memory *plus* peripherals

More 'power' can be added to a microcontroller by enhancing any of these areas. The core, containing the CPU, can be made more powerful by making it faster, or enhancing the internal architecture or instruction set. The memory can be made 'more powerful' by updating its technology, increasing its capacity and speed. Alternatively, or in addition, more peripherals can be added or the current peripherals enhanced.

In many cases in embedded systems, the most dramatic advance is not made by enhancing the core. Instead, it is the addition of new peripherals, perhaps accompanied by memory upgrades, which give the main sense of progress. With the addition of appropriate peripherals, suddenly analog-to-digital conversion, or serial communication, or complex timing functions, become readily available.

In this section of the book, Chapters 7–11, we stay with the PIC 16 Series. We move, however, from a small member of the family to a large one, the PIC 16F873A. As 16F84A and 16F873A are members of the same family, the core and instruction set remain constant, but the ability to engage in embedded control is dramatically enhanced, by the addition of an excellent range of 'new' peripherals.

In these chapters the material is illustrated by application to the Derbot Autonomous Guided Vehicle (AGV). If you are not building a Derbot, don't worry; it will still act as a perfect case study for introducing the many new concepts that we will meet. We remain with Assembler programming, but will begin to experience the limitations both of using Assembler and of some of the hardware features of the 16 Series. This will lead in a logical way to the final section of the book, where we take on the challenge of an advanced microcontroller, programming in C, and the development of complex, multi-tasking programs.

As we move to larger systems, it is important to develop greater ability to get them working and to get them working reliably. Therefore, this chapter introduces new and important diagnostic tools and techniques, which are applied in the chapters that follow.

By the end of the chapter, you should have a good grasp of:

- The architecture of the 1687XA family, of which the 16F873A is a member
- The 1687XA memory map and interrupt structure

- Some of the more advanced tools used for test and commission of an embedded system
- The use of the Microchip in-circuit debugger.

If you are building the Derbot AGV, this chapter will also give you guidance on the first step of a staged construction, which will continue over several chapters.

7.1   The main idea – the PIC 16F87XA

The 16F873A, our example microcontroller, is part of a family group that was briefly introduced at the beginning of Chapter 2. The group is made up of the 16F873A, the 16F874A, the 16F876A and the 16F877A. Generically, we can refer to them as 16F87XA. Each microcontroller in the group also has an LF version, for example 16LF873A, which can run at a lower power supply voltage than the standard device.

The features of this group are summarised in Table 2.1. Looking back at that table, it is easy to see that the four group members are distinguished simply by different memory sizes and different package sizes, where the larger package allows more parallel input/output ports to be used. This is illustrated in the pin connection diagrams of Figure 7.1. The 'extra' pins on the larger devices are enclosed in a dotted line. It can be seen that a primary difference is the Port D and Port E that the '874A/877A have.

The descriptions that follow in this chapter, and in all chapters up to 11, tend to use the 16F873A as the example device, as this is used in the Derbot AGV project that we will study. The other members of the group are, however, mentioned at times, particularly if they have a feature not found in the '873A.

7.2   The 16F873A block diagram and CPU

The block diagrams of both the 16F873A and the '876A are shown in Figure 7.2. The difference between the two microcontrollers lies in memory size, as detailed in the table at the bottom of the diagram. It is worth studying this diagram carefully and comparing it in detail with the 16F84A block diagram of Figure 2.2. This will show that there are some areas that are identical, others where there has been slight incremental change, others where the diagram is effectively the same but has been drawn in a different way, and a large number of additions.

7.2.1   Overview of CPU and core

Let us note first that the CPU structure, made up essentially of ALU, Working register and Status register, remains as expected like the 'F84A. The addition in this diagram of three lines from ALU to Status register is simply acknowledgement of the three status bits in the Status register, **Z**, **DC** and **C**, which are controlled from the ALU. The lines could be included in Figure 2.2.

One difference that does occur in the CPU is in the Status register. With the 16F84A Status register (Figure 2.3), the upper two bits are not used, while bit 5 is used to select between the two banks of data memory. Figure 7.3 shows the Status register for the 16F87XA. With its much bigger data memory, all the

(a)

(b)

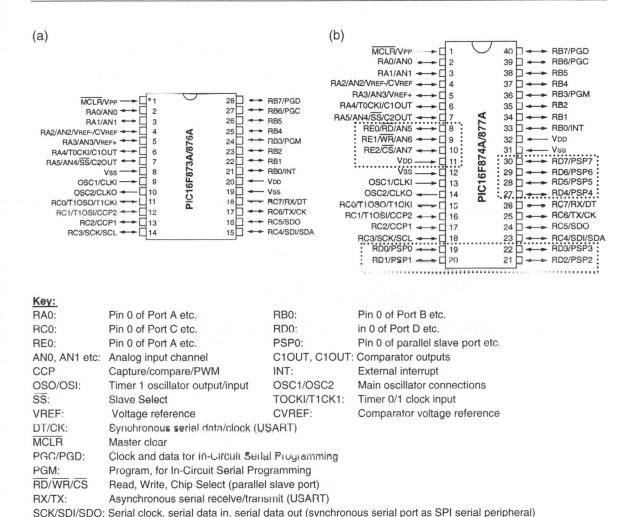

Key:

RA0:	Pin 0 of Port A etc.	RB0:	Pin 0 of Port B etc.
RC0:	Pin 0 of Port C etc.	RD0:	in 0 of Port D etc.
RE0:	Pin 0 of Port A etc.	PSP0:	Pin 0 of parallel slave port etc.
AN0, AN1 etc:	Analog input channel	C1OUT, C1OUT:	Comparator outputs
CCP	Capture/compare/PWM	INT:	External interrupt
OSO/OSI:	Timer 1 oscillator output/input	OSC1/OSC2	Main oscillator connections
$\overline{SS}$:	Slave Select	TOCKI/T1CK1:	Timer 0/1 clock input
VREF:	Voltage reference	CVREF:	Comparator voltage reference
DT/CK:	Synchronous serial data/clock (USART)		
$\overline{MCLR}$	Master clear		
PGC/PGD:	Clock and data for In-Circuit Serial Programming		
PGM:	Program, for In-Circuit Serial Programming		
$\overline{RD}/\overline{WR}/\overline{CS}$	Read, Write, Chip Select (parallel slave port)		
RX/TX:	Asynchronous serial receive/transmit (USART)		
SCK/SDI/SDO:	Serial clock, serial data in, serial data out (synchronous serial port as SPI serial peripheral)		
SCL/SDA:	Serial clock, serial data (synchronous serial port as I²C serial peripheral)		

Figure 7.1 The PIC 16F87XA pin connection diagrams. (a) 16F873A/876A. (b) 16F874A/877A ('extra' pins enclosed in dotted boxes)

higher three bits of the Status register are now used for memory bank selection. Apart from this, the Status register remains unchanged.

7.2.2   Overview of memory

Further to the CPU being similar to the 16F84A, it is easy to see that the whole memory structure is also the same, with some slight adjustments. The 13 program address lines from the Program Counter, which can

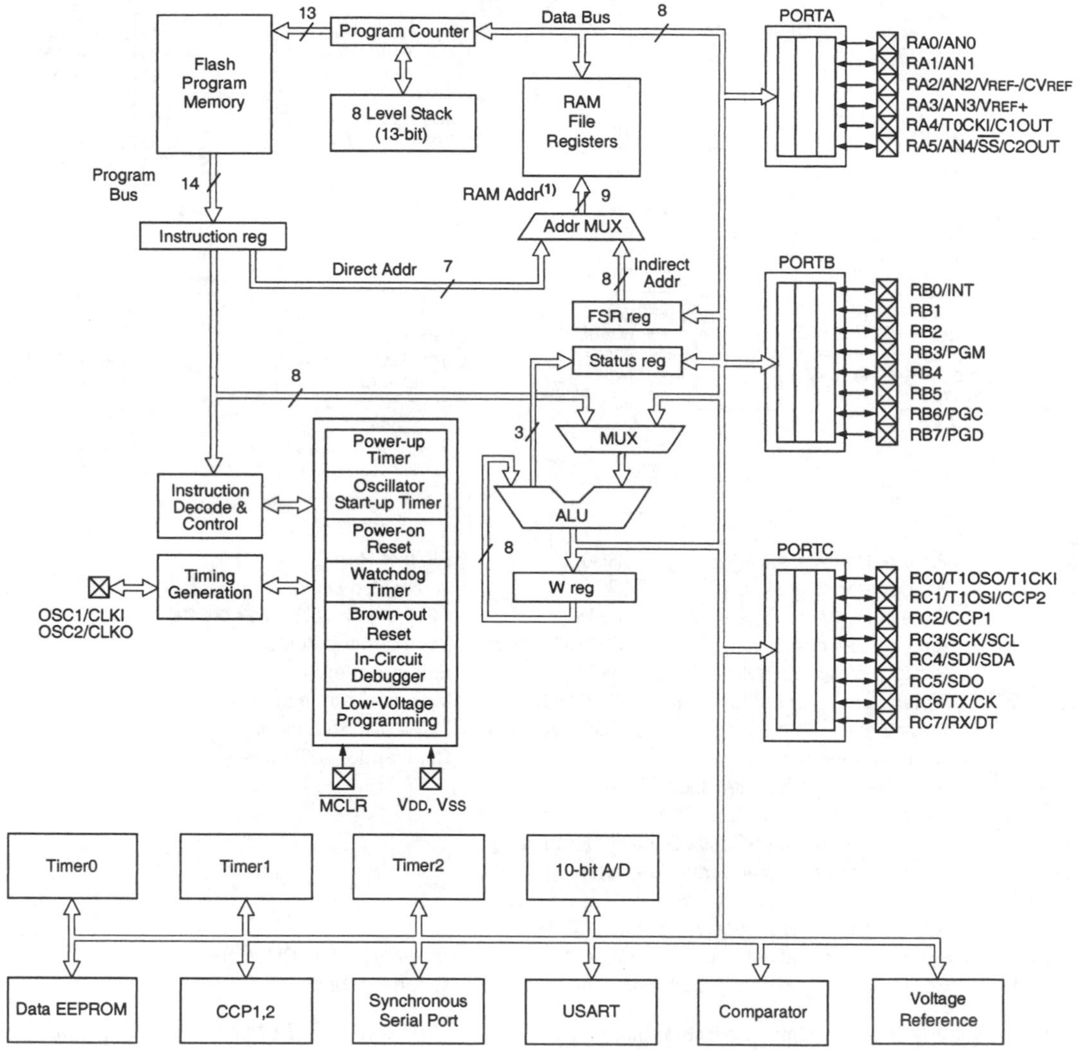

Device	Program Flash	Data Memory	Data EEPROM
PIC16F873A	4K words	192 Bytes	128 Bytes
PIC16F876A	8K words	368 Bytes	256 Bytes

Note 1: Higher order bits are from the Status register.

Key: Peripherals

A/D: Analog-to-digital converter CCP: Capture/compare/PWM module
USART: Universal Synchronous/Asynchronous Receiver/Transmitter

Figure 7.2 The 16F873A/876A block diagram

R/W-0	R/W-0	R/W-0	R-1	R-1	R/W-x	R/W-x	R/W-x
IRP	RP1	RP0	$\overline{\text{TO}}$	$\overline{\text{PD}}$	Z	DC	C

bit 7   bit 0

bit 7   **IRP:** Register Bank Select bit (used for indirect addressing)

           1 = Bank 2, 3 (100h-1FFh)

           0 = Bank 0, 1 (00h-FFh)

bit 6-5   **RP1:RP0:** Register Bank Select bits (used for direct addressing)

           11 = Bank 3 (180h-1FFh)

           10 = Bank 2 (100h-17Fh)

           01 = Bank 1 (80h-FFh)

           00 = Bank 0 (00h-7Fh)

           Each bank is 128 bytes.

bit 4   **$\overline{\text{TO}}$:** Time-out bit

           1 = After power-up, CLRWDT instruction or SLEEP instruction

           0 = A WDT time-out occurred

bit 3   **$\overline{\text{PD}}$:** Power-down bit

           1 = After power-up or by the CLRWDT instruction

           0 = By execution of the SLEEP instruction

bit 2   **Z:** Zero bit

           1 = The result of an arithmetic or logic operation is zero

           0 = The result of an arithmetic or logic operation is not zero

bit 1   **DC:** Digit carry/borrow bit (ADDWF, ADDLW, SUBLW, SUBWF instructions)

           (for borrow, the polarity is reversed)

           1 = A carry-out from the 4th low order bit of the result occurred

           0 = No carry-out from the 4th low order bit of the result

bit 0   **C:** Carry/borrow bit (ADDWF, ADDLW, SUBLW, SUBWF instructions)

           1 = A carry-out from the Most Significant bit of the result occurred

           0 = No carry-out from the Most Significant bit of the result occurred

        Note: For borrow, the polarity is reversed. A subtraction is executed by adding the two's complement of the second operand. For rotate (RRF, RLF) instructions, this bit is loaded with either the high, or low order bit of the source register.

Figure 7.3 The 16F873A Status register

address 2^{13} (i.e. 8192) memory locations, are now fully exploited in the 16F876A with its 8K of program memory, and half used in the '873A.

One bus size which *has* changed is the 'Direct Addr', shown here as 7 bits. This is to accommodate the much larger RAM size, which we see in the next section. The apparent change is not, however, a 'stretching' of the 16 Series structure. Those 7 bits are extracted from the instruction word (Figure 4.13). We can see from this diagram that 7 bits are reserved for the 'file register address', so the larger microcontroller is simply exploiting what is available to it.

7.2.3 Overview of peripherals

The 16F84A and 16F873A share three peripherals that are the same (or almost the same). These are the Timer 0, Port A and Port B. The big difference between the two microcontrollers is of course the extra peripherals, as listed in Table 2.1 and seen in Figure 7.2. With the exception of the parallel ports, which are described in this chapter, we meet all of these peripherals in the subsequent chapters.

The increased number of peripherals brings with it two significant challenges – how do they interface with the CPU and how do they interface with the outside world? We have seen a preliminary answer to the second of these questions in the greater pin count of the 16F873A and the increased sharing of functions on many pins. To provide interfacing with the CPU, we will expect to see a greatly increased number of Special Function Registers (SFRs) and interrupt sources.

Three important extra features found in the 16F873A and seen in Figure 7.2 are brown-out detect, the in-circuit debugger and low-voltage programming. These will be considered later in this chapter.

7.3 16F873A memory and memory maps

The 16F873A has a memory structure very similar to the 16F84A. It is, however, greatly increased in capacity, as has already been noted, *and* there are some important technical developments as well, notably in terms of in-circuit programming.

7.3.1 The 16F873A program memory

The map of the program memory is seen in Figure 7.4. Comparing this to Figure 2.4, we see that it differs from the 16F84A only in size. As it has already been deduced that the 13-bit Program Counter word is adequate to address the whole memory space, it is surprising to see in the figure that there are two 'pages' of program memory. The situation is worse for the 16F876A/877A controllers, each of which have four pages of program memory.

The fact that program memory has to be paged in this way is due to the way the program address, held in the Program Counter, is generated in different situations. The Program Counter is 13 bits long. Its lower 8 bits form the **PCL** register, one of the Special Function Registers (SFRs) that can be accessed and manipulated just like any data memory location. The upper 5 bits of the Program Counter are not readable, but can be written to via the lower 5 bits of the **PCLATH** register, another SFR. The contents of **PCLATH** are transferred to the upper bits of the Program Counter whenever **PCL** is written to.

In normal program execution, the Program Counter is incremented after every instruction. However, there are three other ways by which the program can change the Program Counter value. Each is described below, and also illustrated in Figure 7.5.

By Stack transfers

The Stack is a full 13 bits wide. Therefore any instruction, like **return**, which uses the Stack causes a full 13-bit value to be transferred between Stack and Program Counter – there is no need to worry about looking after any missing bits.

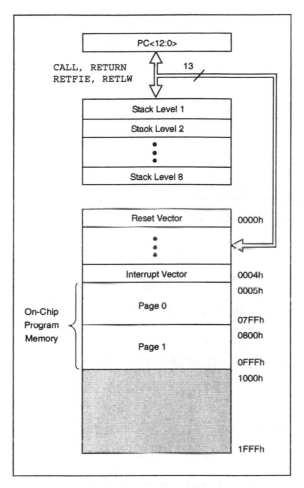

Figure 7.4 PIC 16F873A/874A program memory map and Stack

By **call** and **goto** instructions

As the instruction word format in Figure 4.13 shows, these two instructions are formatted to provide only 11 bits of addressing. These 11 bits can address a range of 2^{11}, or 2K words, and thus represent the pages in Figure 7.4. When these instructions are used, the two 'missing' address bits are taken from bits 4 and 3 of the **PCLATH** register. It is up to the programmer to recognise if a branch over a page is to occur, and to set the bits appropriately. The data sheet [Ref. 7.1] contains an example of how to do this.

By writing to the **PCL** register

The **PCL** is written to directly in situations like the 'computed go to', as described in Section 5.4. Now only the lower 8 bits of the Program Counter, in **PCL**, are adjusted directly and it is as if the programmer is working with 256-word pages. If an address boundary is to be crossed in a computed go to, then **PCLATH**

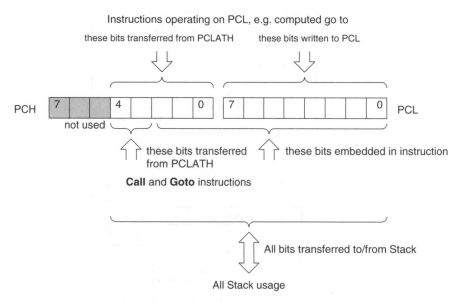

Figure 7.5 Three ways of manipulating the Program Counter

must be set to the right value. For larger look-up tables this can be quite challenging; Ref. 5.1 gives examples of how to do it.

7.3.2 The 16F873A data memory and Special Function Registers

The 16F873A/874A data memory map, including all the SFRs, is shown in Figure 7.6. The sight of all those SFRs is initially unnerving. How is it possible to make sense of them all? Even more unnerving is the thought that most contain 8 active bits, each of which has a function that probably needs to be understood. As always, we will find that a carefully staged approach will allow a good understanding to be developed. Most of these SFRs will be introduced and used, in the right context, over the next few chapters.

Structurally the memory is divided into four banks, which are selected by the values set in bits 6 and 5 of the Status register (Figure 7.3). The 7-bit 'Direct Addr' of Figure 7.2 is drawn from either of the first two instruction patterns of Figure 4.13. It gives the potential to address the 2^7, i.e. 128, memory locations which make up each bank. It is this 7-bit address, concatenated with the two bank select bits in the Status register, which form the 9-bit 'RAM Addr' seen in Figure 7.2, equivalent to the 'File Address' of Figure 7.6.

The first two data memory banks use this memory range to the full, the first one, for example, having 32 SFRs and 96 general-purpose memory locations. The higher two banks are of limited use in the 16F873A/874A. There is no general-purpose memory and most of the SFRs are just mirrored over from the other banks. Check carefully which SFRs are unique to Banks 3 and 4. What are they?

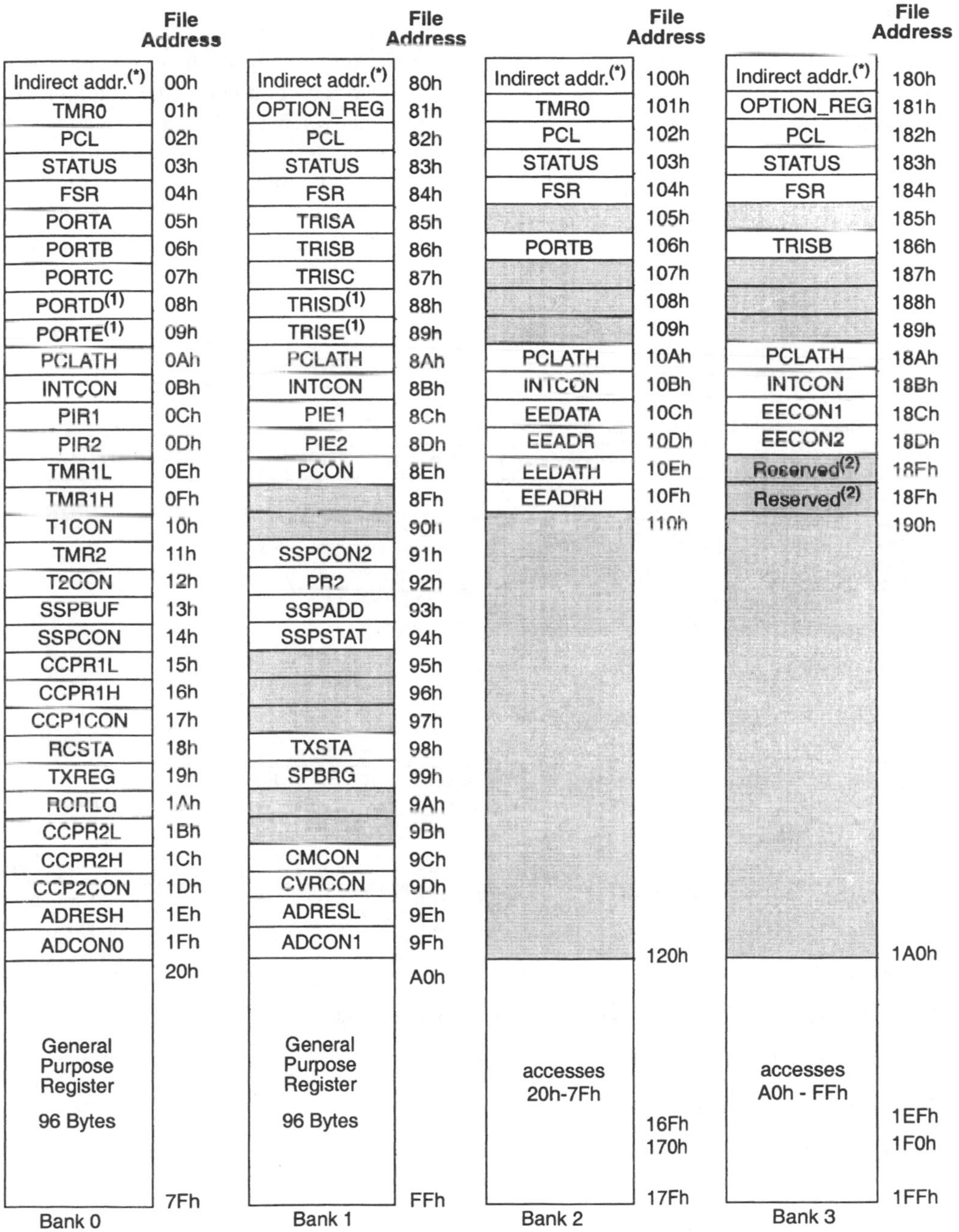

File Address		File Address		File Address		File Address	
Indirect addr.(*)	00h	Indirect addr.(*)	80h	Indirect addr.(*)	100h	Indirect addr.(*)	180h
TMR0	01h	OPTION_REG	81h	TMR0	101h	OPTION_REG	181h
PCL	02h	PCL	82h	PCL	102h	PCL	182h
STATUS	03h	STATUS	83h	STATUS	103h	STATUS	183h
FSR	04h	FSR	84h	FSR	104h	FSR	184h
PORTA	05h	TRISA	85h		105h		185h
PORTB	06h	TRISB	86h	PORTB	106h	TRISB	186h
PORTC	07h	TRISC	87h		107h		187h
PORTD(1)	08h	TRISD(1)	88h		108h		188h
PORTE(1)	09h	TRISE(1)	89h		109h		189h
PCLATH	0Ah	PCLATH	8Ah	PCLATH	10Ah	PCLATH	18Ah
INTCON	0Bh	INTCON	8Bh	INTCON	10Bh	INTCON	18Bh
PIR1	0Ch	PIE1	8Ch	EEDATA	10Ch	EECON1	18Ch
PIR2	0Dh	PIE2	8Dh	EEADR	10Dh	EECON2	18Dh
TMR1L	0Eh	PCON	8Eh	EEDATH	10Eh	Reserved(2)	18Eh
TMR1H	0Fh		8Fh	EEADRH	10Fh	Reserved(2)	18Fh
T1CON	10h		90h		110h		190h
TMR2	11h	SSPCON2	91h				
T2CON	12h	PR2	92h				
SSPBUF	13h	SSPADD	93h				
SSPCON	14h	SSPSTAT	94h				
CCPR1L	15h		95h				
CCPR1H	16h		96h				
CCP1CON	17h		97h				
RCSTA	18h	TXSTA	98h				
TXREG	19h	SPBRG	99h				
RCREG	1Ah		9Ah				
CCPR2L	1Bh		9Bh				
CCPR2H	1Ch	CMCON	9Ch				
CCP2CON	1Dh	CVRCON	9Dh				
ADRESH	1Eh	ADRESL	9Eh				
ADCON0	1Fh	ADCON1	9Fh				
	20h		A0h		120h		1A0h
General Purpose Register 96 Bytes		General Purpose Register 96 Bytes		accesses 20h-7Fh		accesses A0h - FFh	
					16Fh		1EFh
					170h		1F0h
	7Fh		FFh		17Fh		1FFh
Bank 0		Bank 1		Bank 2		Bank 3	

▓ Unimplemented data memory locations, read as '0'.

* Not a physical register.

Note 1: These registers are not implemented on the PIC16F873A.

2: These registers are reserved; maintain these registers clear.

Figure 7.6 PIC 16F873A/874A register file map

7.3.3   The Configuration Word

The Configuration Word of a 16 Series PIC microcontroller determines some of the programmable features of the microcontroller, which can be changed only when the device is programmed. The Configuration Word of the 16F87XA is shown in summary form in Figure 7.7. This reveals some of the underlying features of the microcontroller. The lower 4 bits, and the highest, are already familiar, from the Configuration Word of the 16F84A (Figure 2.6). The two 'new' operating modes, of in-circuit programming and in-circuit debugging, are enabled through the Configuration Word, as are new and flexible code protect features, and the detection of partial loss of power – a 'brown-out' – enabled by bit **BOREN**. These are described in the sections which follow.

R/P-1	U-0	R/P-1	R/P-1	R/P-1	R/P-1	R/P-1	R/P-1	U-0	U-0	R/P-1	R/P-1	R/P-1	R/P-1
CP	—	DEBUG	WRT1	WRT0	CPD	LVP	BOREN	—	—	PWRTEN	WDTEN	Fosc1	Fosc0

bit 13   bit0

Bit 13. **CP:** Flash program memory Code Protection bit

Bit 12. **Unimplemented:** Read as '1'

Bit 11. **DEBUG:** In-circuit debugger mode bit

Bit 10-9. **WRT1, WRT0:** Flash program memory Write Enable bits

These bits determine which sections of program memory can be written to during program execution:

WRT1:WRT0	PIC16F876A/877A		PIC16F873A/874A	
	This area write-protected	This area writeable	This area write-protected	This area writeable
11	none	all	none	all
10	0000h to 00FFh	0100h to 1FFFh	0000h to 00FFh	0100h to 0FFFh
01	0000h to 07FFh	0800h to 1FFFh	0000h to 03FFh	0400h to 0FFFh
00	0000h to 0FFFh	1000h to 1FFFh	0000h to 07FFh	0800h to 0FFFh

.Bit 8. **CPD:** Data EEPROM memory Code Protection bit

Bit 7. **LVP:** Low-Voltage (single-supply) In-Circuit Serial Programming Enable bit

Bit 6. **BOREN:** Brown-out Reset Enable bit

Bits 5-4. **Unimplemented:** Read as '1'

Bit 3. **PWRTEN:** Power-up Timer Enable bit

Bit 2. **WDTEN:** Watchdog Timer Enable bit

Bit 1-0. **FOSC1, FOSC0:** Oscillator Selection bits. 11 = RC, 10 = HS, 01 = XT, 00 = LP.

Note: The erased state of all bits is Logic 1.

Figure 7.7 PIC 16F87XA Configuration Word

7.4   'Special' memory operations

A conventional microcontroller reads its instructions from non-volatile program memory and uses volatile RAM for storing temporary data. With the introduction of Flash memory, these distinctions between traditional memory usages are diminishing. This is because Flash memory technology is non-volatile and hence used for program memory. It can, however, also be very easily written to, so there is a temptation to use it for other forms of storage.

Because of its Flash memory technology, the 16F87XA allows – under certain restrictions – program memory to be written to while the program is running. It also allows the program memory to be programmed serially, if wanted, while the IC is in the target location. It also of course allows its EEPROM data memory to be accessed, through its special control registers. It is these memory functions that are covered in this section.

7.4.1   Accessing EEPROM and program memory

The ability to write to program memory, as just mentioned, is controlled by the settings of the **WRT1** and **WRT0** bits in the Configuration Word. It is worth looking back at these in Figure 7.7. It can be seen that they permit writing access to different blocks of program memory.

Interaction with program memory is through the same data and address registers that are used for EEPROM (**EEDATA** and **EEADR**), which were described in Section 2.4. These are only 8-bit registers, however, and program memory holds 14-bit words. Furthermore, it requires a 13-bit address to access the full 8K words (of the 16F876A and '877A), or 12 bits to access 4K (of the 16F873A and '874A). Therefore, a 'high-byte' register is added to each of **EEDATA** and **EEADR**. These extra registers are called **EEDATH** and **EEADRH** respectively, and are used for accessing program memory only.

The arrangement is depicted in Figure 7.8, with program memory at the top of the diagram and EEPROM at the bottom. The register pair formed by **EEADRH** and **EEADR** addresses program memory, while only **EEADR** addresses EEPROM. The situation is similar for data transfer, with **EEDATH** and **EEDATA** carrying data to and from program memory, and **EEDATA** alone being used for EEPROM data memory.

If a section of program memory is enabled, then blocks of four words must be written at the same time. This process is described in full in Ref. 7.1. Single word reads are, however, possible.

All data transfers are controlled by the **EECON1** register, shown in Figure 7.9. It can be seen first of all that the MSB of this controls whether EEPROM or program memory is to be accessed. The other bits are similar to Figure 2.7 and therefore familiar. A major difference, however, is that the interrupt flag bit, **EEIF**, has been relocated, now being found in the **PIR2** register (Figure 7.13).

Accessing EEPROM and program memory is not entirely simple, and requires care in sequencing the actions correctly. Writing to each requires the codes 55_H followed by AA_H to be sent to the virtual **EECON2** register. This adds further security to the process, aiming to ensure that accidental writes are not made. Code examples are given in Ref. 7.1. These can be adapted as appropriate.

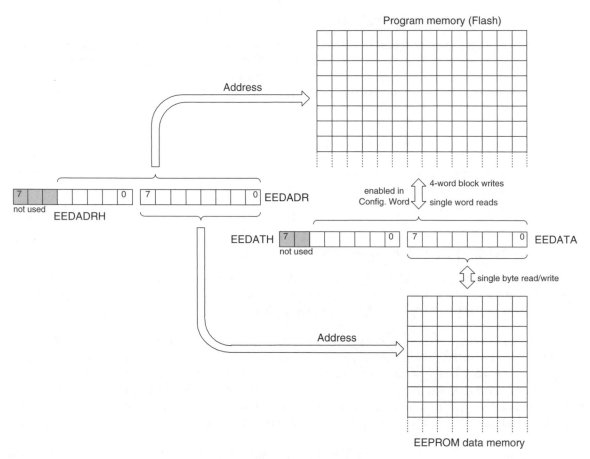

Figure 7.8 Writing to program memory and EEPROM

If the **CP** bit of the Configuration Word is cleared, then it is impossible to read the program memory externally, for example using a programmer like the PICSTART® Plus. This protects the intellectual property of the designer or maker of a product. This bit does not, however, affect internal access to program memory. Similar comments apply to the EEPROM memory, which is protected from external access by the **CPD** bit of the Configuration Word.

7.4.2   In-Circuit Serial Programming (ICSP™)

All the 16F87XA microcontrollers can be programmed while in the target circuit. This is of great importance. In a development environment, it means that test programs can be downloaded straight to the microcontroller in the target system, without the need to remove it and put it in a programmer. In a production environment, it allows the microcontroller to be programmed in place, just before it leaves the factory. The very latest

R/W-x	U-0	U-0	U-0	R/W-x	R/W-0	R/S-0	R/S-0
EEPGD	—	—	—	WRERR	WREN	WR	RD
bit 7							bit 0

bit 7   **EEPGD**: Program/Data EEPROM Select bit

1 = Accesses program memory
0 = Accesses data memory
Reads '0' after a POR; this bit cannot be changed while a write operation is in progress.

bit 6-4   **Unimplemented:** Read as '0'

bit 3   **WRERR**: EEPROM Error Flag bit

1 = A write operation is prematurely terminated (any $\overline{\text{MCLR}}$ or any WDT Reset during normal operation)
0 = The write operation completed

bit 2   **WREN**: EEPROM Write Enable bit

1 = Allows write cycles
0 = Inhibits write to the EEPROM

bit 1   **WR**: Write Control bit

1 = Initiates a write cycle. The bit is cleared by hardware once write is complete. The WR bit can only be set (not cleared) in software.
0 = Write cycle to the EEPROM is complete

bit 0   **RD**: Read Control bit

1 = Initiates an EEPROM read; RD is cleared in hardware. The RD bit can only be set (not cleared) in software.
0 = Does not initiate an EEPROM read

Figure 7.9 The EECON1 register

software can therefore be installed. Once the product is in use, it allows program updates to be made, possibly by the Internet and without the owner even knowing about it!

The disadvantage of in-circuit programming is that some pins have to be committed to the function, or at least must be carefully designed to be dual purpose, in such a way that the normal function of the pin does not impede the programming function. ICSP uses the pins shown in Table 7.1. The Derbot AGV uses ICSP – we will see it designed into the circuit.

Under normal programming mode a high voltage, of around 13 V, is applied to the $\overline{\text{MCLR}}$ pin. A special case of ICSP dispenses with this high voltage, however. It is called the Low-Voltage Programming mode (LVP). This is enabled by the **LVP** bit in the Configuration Word, and allows the memory to be programmed using just the normal power supply V_{DD}. Its disadvantage is that an extra pin, RB3, must be used, as Table 7.1 indicates. The pin is used just to control entry to and exit from this programming mode.

Aside from the hardware interconnection, ICSP requires very specific software routines to transfer the data serially into the microcontroller. These are not described further here, but full information can be found in Ref. 7.2.

Table 7.1 Pins used in In-Circuit Serial Programming

Pin	Description
RB3	Control input in Low-Voltage Programming mode (**LVP** configuration bit is 1)
RB6	Clock input
RB7	Data input/output
$\overline{\text{MCLR}}$	Program mode select, programming voltage connection
V_{DD}	Power supply
V_{SS}	Ground

7.5 The 16F873A interrupts

7.5.1 The interrupt structure

As a member of the 16 Series family of microcontrollers, the 16F873A is expected to fit into the structure of that family. This design strategy works well in places, but elsewhere tests the 16 Series structure to its limits. An example of this is the 16F873A interrupt structure, shown in Figure 7.10. Here we see the minimalist structure of the 16F84A (Figure 6.2) replicated to the right of the diagram. The EEPROM write complete

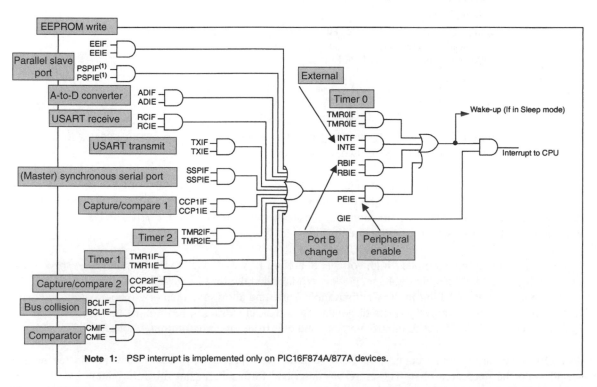

Figure 7.10 PIC 16F87XA interrupt structure (supplementary labels in shaded boxes added by the author)

interrupt is, however, replaced by a line linking across to the huge requirements of the larger microcontroller, with no less than 11 more interrupt sources connected. All of these are routed ultimately through to the single interrupt vector that we are familiar with, seen in the program memory map of Figure 7.4.

7.5.2 The interrupt registers

With 15 interrupt sources, the single **INTCON** register introduced in Figure 6.3 for the smaller 16 Series microcontroller can now only hold a small proportion of the overall number of flags and enable bits. The 16F87XA version is shown in Figure 7.11. Essentially it remains unchanged, *except* that the **EEIE** (EEPROM write complete enable) bit is replaced by the bit **PEIE**, seen in Figure 7.10. **PEIE** is the *Peripheral Interrupt*

R/W-0	R/W-0	R/W-0	R/W-0	R/W-0	R/W-0	R/W-0	R/W-x
GIE	PEIE	TMR0IE	INTE	RBIE	TMR0IF	INTF	RBIF

bit 7   bit 0

bit 7   **GIE:** Global Interrupt Enable bit
        1 = Enables all unmasked interrupts
        0 = Disables all interrupts

bit 6   **PEIE:** Peripheral Interrupt Enable bit
        1 = Enables all unmasked peripheral interrupts
        0 = Disables all peripheral interrupts

bit 5   **TMR0IE:** TMR0 Overflow Interrupt Enable bit
        1 = Enables the TMR0 interrupt
        0 = Disables the TMR0 interrupt

bit 4   **INTE:** RB0/INT External Interrupt Enable bit
        1 = Enables the RB0/INT external interrupt
        0 = Disables the RB0/INT external interrupt

bit 3   **RBIE:** RB Port Change Interrupt Enable bit
        1 = Enables the RB port change interrupt
        0 = Disables the RB port change interrupt

bit 2   **TMR0IF:** TMR0 Overflow Interrupt Flag bit
        1 = TMR0 register has overflowed (must be cleared in software)
        0 = TMR0 register did not overflow

bit 1   **INTF:** RB0/INT External Interrupt Flag bit
        1 = The RB0/INT external interrupt occurred (must be cleared in software)
        0 = The RB0/INT external interrupt did not occur

bit 0   **RBIF:** RB Port Change Interrupt Flag bit
        1 = At least one of the RB7:RB4 pins changed state; a mismatch condition will continue to set
            the bit. Reading PORTB will end the mismatch condition and allow the bit to be cleared
            (must be cleared in software).
        0 = None of the RB7:RB4 pins have changed state

Figure 7.11 PIC 16F87XA INTCON register

Enable bit, acting as a subsidiary Global Enable, to all those interrupt sources upstream of the AND gate it drives. It must be set to 1 if *any* of the 'upstream' interrupts are to be used.

To augment the **INTCON** register, *four* new SFRs are added – **PIE1** and **PIE2**, which hold the enable bits and are located in memory bank 1, and **PIR1** and **PIR2**, which hold the flag bits and are in memory bank 0. These are shown, in summary form, in Figures 7.12 and 7.13. It can be seen that **PIE1** and **PIR1** each follow the same pattern, as do **PIE2** and **PIR2**. As would be expected, all active bits are reset to 0 on any form of power-up.

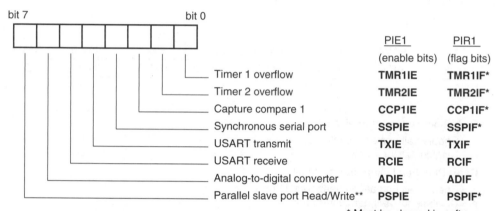

Figure 7.12 16F87XA **PIE1/PIR1** (Peripheral Interrupt Enable/Peripheral Interrupt Request) registers

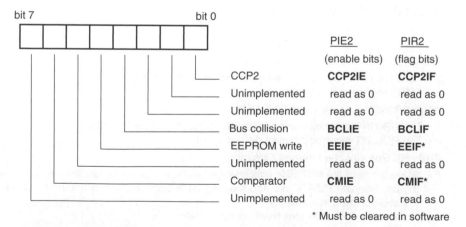

Figure 7.13 16F87XA **PIE2/PIR2** registers

7.5.3   *Interrupt identification and context saving*

If more than one interrupt source is enabled, it will be necessary for the ISR to contain a program section at its beginning to identify which is the calling interrupt. This is exactly similar to the case of the 16F84A, and the principle of Program Example 6.2 can be applied to meet this need.

When it comes to context saving, the 16F87XA again acts like the 16F84A, with only the return address being saved on the stack when an ISR is called. It is up to the programmer to save any other registers that are needed, usually the W and Status registers, at the start of the ISR, and retrieve them at the end. Program Example 6.4 can be adapted for this.

7.6   The 16F873A oscillator, reset and power supply

7.6.1   *The clock oscillator*

The 1687XA has exactly the same oscillator structure and range of options as the 16F84A, described in Section 3.5. The oscillator type is again selected by settings in the Configuration Word (Figure 7.7). It may be of interest at this point to look forward to Figure 7.19 and its accompanying description, to see actual oscilloscope traces of oscillator waveforms.

7.6.2   *Reset and power supply*

The reset structure is extremely similar to that of the 16F84A (Figure 2.10), with the exception that a new source has been added: the Brown-out Reset. A brown-out is a dip in the supply voltage. This form of power loss can be particularly dangerous, because it can pass unnoticed. Part of a circuit or system may keep going, whereas another part may temporarily fail or lose data. This reset is intended to ensure that the device is fully reset if a brown-out occurs. It is enabled by the **BOREN** bit of the Configuration Word. If **BOREN** is set high and a brown-out occurs, then the microcontroller is forced into reset. The device data [Ref. 7.1] quotes 4 V as being the typical voltage that triggers this form of reset. Apart from this the 16F87XA has almost identical power supply requirements to the 16F84A, as seen in Figure 3.16. A difference is that the minimum supply voltage is now clearly defined by the Brown-out Reset, if this is enabled.

With multiple sources of reset now available, it can be useful when programming to know which form of reset has most recently occurred. A Watchdog Timer reset is indicated by the $\overline{\text{TO}}$ bit of the Status register (Figure 7.3). Further information is provided by the $\overline{\textbf{POR}}$ and $\overline{\textbf{BOR}}$ bits of the **PCON** register (which has no other active bits). The first of these is set to 0 if a Power-on Reset occurs, the second if a brown-out occurs.

7.7   The 16F873A parallel ports

It can be seen from Figure 7.2 that the PIC 16F873A has three ports, A, B, and C. Ports A and B are similar to the ports of the 16F84A, except that more alternate functions are crowded onto the pins, and Port A is now 6-bit, a 1-bit advance on the 5 bits of the 16F84A Port A. At reset all port bits are set to input. The port pin output characteristics, for a supply of 5 V, are shown in Figure 7.14. It can be deduced, following the

(a)

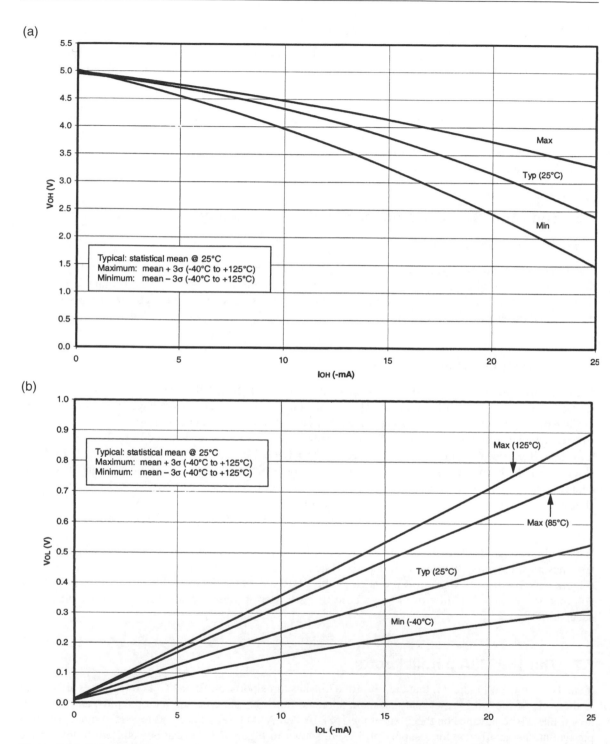

Figure 7.14 PIC 16F87XA port output characteristics. (a) Typical V_{OH} vs. I_{OH} ($V_{DD} = 5$ V, -40 to $125°$C). (b) Typical V_{OL} vs. I_{OL} ($V_{DD} = 5$ V, -40 to $125°$C)

same procedure as applied in Section 3.4.3, that the 'typical' output resistance at Logic 1 is approximately 70 Ω and around 22 Ω at Logic 0.

The port characteristics are now surveyed in turn.

7.7.1   The 16F873A Port A

The 6 bits of Port A appear on pins 2–7 of the 16F873A, as can be seen in Figure 7.1. It is worth checking the accompanying key to work out the connections made to the pins. The port can be used for general-purpose bi-directional digital data. It is also shared with the Analog functions, notably the analog-to-digital converter (ADC) module and the comparators. While we meet both of these in Chapter 11, it is very important to note that on power-up the port bits are set as analog inputs. To use the port for digital purposes the **ADCON0** register, described in Chapter 11, must be set appropriately. As with the 16F84A, the important Timer 0 input is on bit 4 of Port A.

Two of the Port A pin driver circuits are shown in Figure 7.15. It is useful to compare these with the equivalent 16F84A circuits, shown in Figure 3.11. With the added functionality of most pins, it is interesting

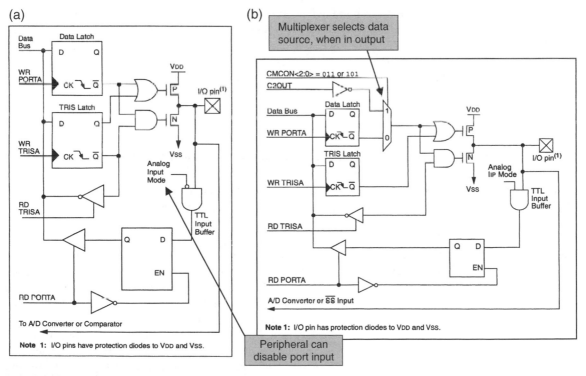

Figure 7.15 PIC 16F87XA Port A output characteristics. (a) Pins RA0, 1, 2 and 3. (b) Pin RA5 (supplementary labels in shaded boxes added by the author)

to see how the peripherals that share the pins begin to 'invade' the pin driver circuit. A simple example is in Figure 7.15(a). This is almost the same as Figure 3.11(a), except the ADC peripheral, via the 'Analog Input Mode' line, can disable the digital input path of the port. A more extreme case is seen in Figure 7.15(b). Here a multiplexer is introduced in the digital output path. Now the peripheral, in this case the comparator circuit, can disable the normal port output and claim the output for its own purposes. In this case the microcontroller pin can be configured as output of comparator 2 (**C2OUT**), rather than as the port bit output.

The pin driver circuit for pin 4 is not shown. It is based on the circuit of Figure 3.11(b), but has added to it the multiplexer arrangement in the output path, as just described, for comparator 1. It keeps the Open Drain output of Figure 3.11(b).

An important point is emerging in this exploration of the pin driver circuits. The port pins are no longer simply controlled by the port 'TRIS' register. Other SFRs can have a major impact, disabling or reallocating a resource. This can lead to considerable programming frustration. A port appears not to work because a peripheral SFR setting has reallocated its function. Let the programmer beware!

7.7.2   The 16F873A Port B

The interconnection of Port B can be seen in Figure 7.1 on pins 21–28 of the 16F873A. It remains a simple port, almost identical to that of the 16F84A. The pin driver circuits are effectively the same as those shown for the 16F84A in Figure 3.10. As seen in Figure 7.1 and Table 7.1, bits 3, 6 and 7 are used for In-Circuit Serial Programming (ICSP). If ICSP is to be used, the designer must either not use these pins for any other purpose or use them in such a way that they are still available for the ICSP function.

7.7.3   The 16F873A Port C

The Port C interconnection can be seen on pins 11–18 of the 16F873A (Figure 7.1). It is the most complex of the 16F873A ports. Its pins can simply be used as general-purpose digital input/output (I/O). Interestingly, *all* inputs now have Schmitt trigger characteristics. Aside from general-purpose digital I/O, Port C pins are shared with some of the more complex microcontroller peripherals, including those dealing with serial communication.

Diagrams of the Port C pin driver circuits are shown in Figure 7.16. At the heart of both of them it should still be possible to find the standard digital I/O port capability, such as we saw in the simple 16F84A pin drivers. To this is added the multiplexer on the output path, as seen just now for Port A. A further feature is that the peripheral can now take over the TRIS function, through the 'Peripheral OE' line, and the OR gate that it drives, as labelled in the diagram. A final modification of the standard port pin circuit is seen in Figure 7.16(b). Because the SMBus (system management bus) has defined input characteristics, there are now alternative input paths, one with standard Schmitt trigger inputs and the other with SMBus characteristic. This is not the end of the added complexity that serial I/O brings. Not appearing in Figure 7.15(b), but available on these pins if needed, is the full I/O interfacing capability of the inter-integrated circuit (I^2C) serial standard. This can be glimpsed by looking forward to Figure 10.19.

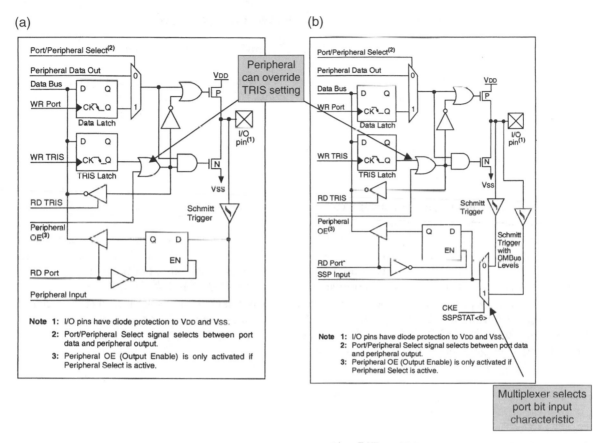

(a)

(b)

Peripheral can override TRIS setting

Multiplexer selects port bit input characteristic

Figure 7.16 Block diagram of Port C pin driver circuits. (a) Pins RC7 to RC5, RC2 to RC0. (b) Pins RC4, 3 (supplementary labels in shaded boxes added by the author)

7.8 Test, commission and diagnostic tools

As we are moving to a distinctly more complicated family of microcontrollers, it follows that we will be developing distinctly more complicated systems. It is therefore necessary to take time to consider here how we will test and commission such systems. The following sections explore equipment and techniques that can be used for such activity, and which will be applied as we develop the Derbot AGV design.

7.8.1 The challenge of testing an embedded system

Section 4.1.3 outlined the program development process, which included the use of a software simulator. When the program is downloaded to the hardware, however, our problems can increase. This is particularly the case if both hardware and program are untested. Then the developer can be left in the unhappy position experienced by all embedded systems developers at some time – of being confronted with a totally inanimate system, not having any idea of where the problem lies.

In the test procedure, the developer is confronted with broadly two types of problem:

- *Design problems.* These are problems in the design of hardware or software, leading to partial or total system non-function.
- *Implementation problems.* These are due to the way the design, which may be perfectly good, has been implemented. They include things like broken PCB tracks, failed or corrupted program download, or configuration bits not set in the download process. These problems can similarly lead to partial or total system non-function.

Each problem type can have the same sort of symptoms, yet it is important to be able to distinguish between the two, as the solution to each can be quite different.

Let us therefore try to establish a commissioning hierarchy, such that a systematic approach to test and commission can be made. The layers of dependency that exist between different elements in an embedded system are shown in an informal way in Figure 7.17. Generally, in a test procedure one moves from the bottom to the top of this diagram. If a car does not have petrol it does not run. Similarly, if a microcontroller does not have the correct power supply, it does not run, however good the circuit or program may be. Nor will it run in any way if its oscillator is not running, or if its Reset pin ($\overline{\text{MCLR}}$ for a PIC) is active, or if the program has not downloaded properly. Once these four essential conditions are met, there is a chance of the program *starting* to run. *If a system is entirely inactive, these must be the ones that are tested first.* Faults in this area are most likely to be in implementation, for example the crystal is not properly soldered to the board or program download has failed for some reason.

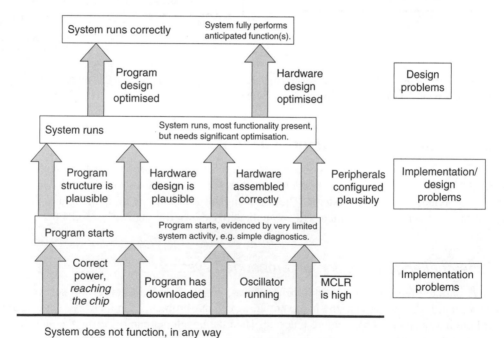

Figure 7.17 Layers of dependence in an embedded system

Once these fundamental conditions have been satisfied, a further set applies if the system is to run continuously and achieve a moderate level of functionality. As indicated, these include plausible circuit and program designs, correct hardware assembly and all the peripherals being configured appropriate to the situation. Here the word plausible is used to indicate that the design does not have fatal flaws, although it might have some features that need improvement. As the conditions indicated are met, the system should progress to a stage of optimisation. Now it shows a good level of functionality, although still imperfect in some areas. From here it is likely that further tests must be accompanied by ongoing incremental design development, which may lie in either hardware or the program. Finally, one expects to see a system functioning to the full anticipated level of performance.

Given an understanding of the layers of dependence in an embedded system, how should the test procedure be implemented, particularly once the basic tests indicated by the bottom layer of Figure 7.17 have been completed? In Ref. 1.1, this author identified three 'golden rules' for test. These were:

- *Divide and rule* Divide the system, both hardware and program, up into modules or subsystems which can be isolated as much as possible from each other, and test these separately. The system can then be built back together from 'known good' subsystems.
- *Guilty until proven innocent.* Make no assumptions that any part of the system is working correctly until you have demonstrated that it is. In other words, assume it doesn't work correctly until you have demonstrated that it does. This is the opposite of the court of law, where the accused is innocent until proven guilty.
- *Work systematically, document ruthlessly.* This can be expanded out to:
  - Plan your test procedure ensuring that things are tested in a sensible and logical order. A starting point for the sequence is to some extent indicated in Figure 7.17. Start at the bottom of the diagram and work upwards.
  - Work from good documents, updating them as necessary. As you work on your embedded system, have the up-to-date circuit diagram and program listing beside you. *Always.* Test by reference to these. Have appropriate data sheets (especially for the microcontroller in use) within easy reach. Update diagrams as needed, ensuring each new version has the revision date on it.
  - Record your test results. This becomes increasingly important as the system becomes more complex. It is too easy, in a series of tests, to forget what the outcome of a single test was. However good your memory, in a week or two you will have forgotten anyway. With the result, record the circuit and program version in use.

To be successful in the test process, we need instruments that allow us to control and see what is going on. A number of these are now surveyed.

7.8.2   Oscilloscopes and logic analysers

An oscilloscope is the most powerful general-purpose instrument available in the electronics world. It allows simple and reasonably accurate measurements of voltage and time to be made, especially if they are DC or periodic. A standard oscilloscope is greatly enhanced by the addition of a storage facility. In this case it is possible to examine aperiodic signals, or single events, with reasonable ease. An oscilloscope is essential in the embedded world for examining the basic conditions of power supply, oscillator and simple port activity. With expertise, and a good 'scope, it can be used for looking at more complex signals.

A logic analyser has some of the characteristics of an oscilloscope, in that it can also display the value of an input signal against time. The input signal, however, is assumed to be digital, so the display can only take one of two values, Logic 0 or Logic 1. The threshold applied is determined by the logic analyser, and is for many instruments adjustable. A usual characteristic of the logic analyser is that it has many inputs, for example 16, 32 or 48. Thus, it can be used to look at activity in data and/or address buses. Logic analysers usually carry a range of extra features, including memories that can store a history of the input values, and complex triggering capability, which allows defined combinations of inputs to be identified and to cause a trigger. Logic analysers were most prominent in the days when systems were built up of multiple ICs and there was a need to study bus activity. Now in many systems the ICs are very complex and the buses no longer accessible. They can still be very useful, however, in looking at complex digital activity, for example a parallel port, or serial data flow.

This book makes use of a wonderful instrument which combines features of both oscilloscope and logic analyser. This is the Agilent Mixed Signal Oscilloscope, with the 54622D model used here and pictured in Figure 7.18. It has two conventional oscilloscope inputs and 16 logic analyser inputs. Any combination of input can be displayed simultaneously. The 'oscilloscope' has fantastic triggering, signal analysis and storage facilities. Its combination of analog and digital capability reflects the mixture of analog and digital in most embedded systems. It thus forms a natural tool to undertake a comprehensive range of tests in the embedded environment.

Two screen images of oscillator signals, using the analog inputs of the Agilent 'scope, are shown in Figure 7.19. Note from information on the screen border that the vertical scale of both channels 1 and 2 are 2 V/cm and the horizontal scale is 1 μs/cm. The two small arrowheads appearing on the left of the screen indicate the 0 V reference position for each trace.

The first screen is of an RC oscillator, applying the circuit of Figure 3.14(b). The upper trace shows the OSC1 microcontroller terminal, i.e. the junction of R_{EXT} and C_{EXT}. The characteristic rising voltage of a capacitor charging through a resistor is seen, followed by its rapid discharge as the oscillator Schmitt trigger output switches to logic high. As Figure 3.14(b) indicates, the signal at the OSC2 pin is Fosc/4. This can be seen in the lower trace. Component values of $R_{EXT} = 8.2\,k$ and $C_{EXT} = 100\,pF$ are chosen to give a nominal oscillator frequency of 700 kHz (i.e. a period of 1.43 μs). The actual signal can be seen to have a period of close to 1.4 μs (just over 700 kHz), with the Fosc/4 having a period of 5.6 μs. This is in reasonably good agreement with what is expected. It should be noted, however, that when the 'scope probe was removed from the OSC1 pin, the Fosc/4 period fell to 5.0 μs. Herein lies a pitfall of oscilloscope use – the probe plus oscilloscope input adds a loading capacitance (and resistance) to the circuit under test. Whether it has any impact of significance depends on the test circuit itself. In this case the loading is 15 pF approximately, which is placed directly in parallel with R_{EXT}, increasing its value by over 10 per cent! The measurement itself has introduced an error.

The second trace is of the microcontroller in HS crystal mode, and applies the circuit of Figure 3.14(a). Both microcontroller pins have similar, approximately sinusoidal, signals of the same frequency. One is the inverse of the other. It can be seen that the signal clearly has a period of 125 ns, confirming a frequency of 4 MHz.

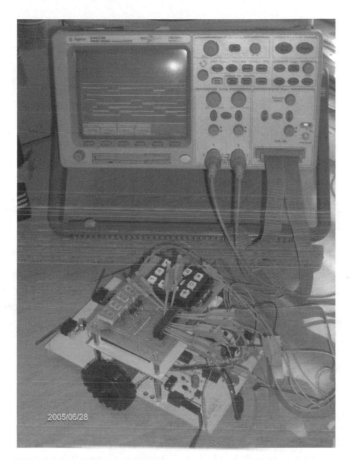

Figure 7.18 An Agilent 54622D Mixed Signal Oscilloscope connected to the Derbot

(a)   (b)

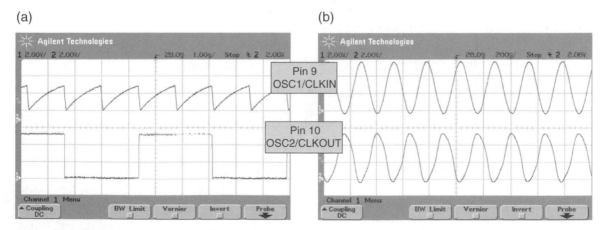

Figure 7.19 Oscillator traces from 16F873A devices. (a) RC oscillator, 1 MHz nominal. (b) Crystal oscillator, 4 MHz

7.8.3   In-circuit emulators

Oscilloscopes and logic analysers are very good for testing signals that are accessible, when the system is running. There are many situations, however, when we need to be able to undertake the same sort of tests we were doing with the software simulator in Chapter 5, for example testing specific sections of code or single stepping, but now with the code running in the target hardware.

The solution to this need has been the *in-circuit emulator* (ICE). This is a device which *replaces* the microcontroller in the circuit, replicates its action as closely as possible, but which remains linked back to a host computer. The host computer has the power to control program execution, in much the same way as the software simulator does.

In-circuit emulators represent one of the most powerful ways of testing and commissioning an embedded system. They come with just a few disadvantages:

- They are very sophisticated pieces of equipment and hence expensive
- They replicate one microcontroller or processor only, and hence a different one is needed for every different microcontroller used
- There are areas where their action is imperfect, for example they are not usually good at replicating the action of the microcontroller in terms of the clock oscillator, and they may have power supply requirements which are less flexible than the microcontroller itself
- They do not allow the genuine final operating condition of the system to be fully replicated – after all, the microcontroller itself has been removed.

7.8.4   On-chip debuggers

Given the benefits of the ICE, but also its disadvantages, it was natural for IC designers to ask themselves if they could actually design features of the ICE into the microcontroller itself. A good part of the ICE, after all, is a replica circuit of the microcontroller itself. Thus, a variety of on-chip test facilities came into being. Motorola (now Freescale) used the terminology *background debug mode* (BDM), while Microchip uses the terminology *in-circuit debugger* (ICD). We will explore this facility in detail, as it is now available on many PIC microcontrollers, including the 16F873A.

The *advantages* of the ICD approach can be summarised as:

- Testing is done with the target system substantially undisturbed.
- The ICD can also act as a programmer, downloading program code to the target microcontroller.
- The ICD approach is more flexible – connection, test and further download can be done 'in the field' as well as in the development lab.

The *disadvantages* of the ICD approach can be summarised as:

- Some microcontroller resources are taken by the ICD function; this includes a few I/O pins, some program memory and other internal resources.
- The target microcontroller must be functioning, with its clock running.

- It is generally less powerful than a fully fledged ICE system.
- Testing of certain aspects may still be imperfect. For example, when microcontroller operation has been halted by the ICD, peripherals may continue running, leading to erroneous results.

7.9 The Microchip in-circuit debugger (ICD 2)

The features of the Microchip ICD 2 system [Ref. 7.3] are illustrated in diagrammatic form in Figure 7.20; the host computer is linked to the target system via a special adaptor unit or pod. This is seen in a real system in Figure 7.21, where an ICD 2 pod is shown linked to a Derbot AGV. The ICD is controlled from the host computer, which must be running MPLAB®. The ICD 2 pod links to the host computer via a USB cable. An RS232 link may also be used. It then connects to the target system via another cable, having five interconnections. These are shown in the diagram and are the same as those for ICSP, except that the low-voltage programming pin, bit 3 of Port B, is not used. The internal microcontroller resources needed by the ICD are also shown in the diagram, and include elements of program memory, data memory and the Stack.

The ICD 2 can be used as a debugger, in which case it can program the microcontroller, at the same time downloading its own 'debug executive', which is loaded into the high end of program memory. Running from MPLAB it can then execute all the functions of the MPLAB simulator, as introduced in Chapters 4 and 5, except that now the program is actually running in the hardware. The ICD 2 can also be used simply as a programmer, in which case it replicates the action of programmers like the PICSTART Plus, described in Chapter 4. The operating mode is selected from within MPLAB.

The normal use of the ICD 2 in the development cycle is to download the program under test to the microcontroller with the ICD in debug mode. The program is then debugged using all the facilities of

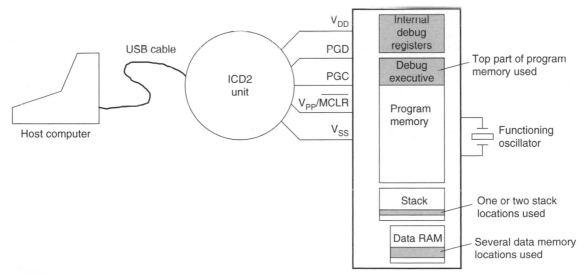

Figure 7.20 The ICD 2 system and internal resource demand

Figure 7.21 ICD 2 connected to a Derbot AGV

the ICD. When a fully functioning version has been developed, the ICD is switched to debugger mode. The program is then downloaded again, this time without the special features of the debugger, and the microcontroller can run in its normal operating mode.

Use of the ICD 2 will be described in a simple Derbot-based tutorial following description of the first Derbot build.

7.10   Applying the 16F873A: the Derbot AGV

Following the descriptions earlier in this chapter, we turn now to see the 16F873A applied in a practical situation, the Derbot AGV. The Derbot block diagram is shown in Figure 1.5 and its circuit diagram in Figure A3.1. Some of the underlying considerations in the design and application of the Derbot are described in Appendix 4.

7.10.1   Power supply, oscillator and reset

The power needs of the Derbot are twofold, to power the motors and to power the microcontroller circuit and all associated sensors and displays. While low-voltage motors, running from 2 or 3 V, are available, it is less easy to arrange the necessary switching circuits to run at this voltage. Experimentation with several motors, however, showed that adequate torque and speed could be obtained from a motor running from around 9 V, with a tolerable supply current. This voltage fitted well with currently available motor interface ICs, like the L293D, which is used. The second power supply need is for the microcontroller circuit. While this could have been supplied unregulated, as in the ping-pong project, there are a number of sensors, particularly the light-dependent resistors and reflective opto-sensors, which require stable operating conditions for correct results. Therefore, a regulated supply was chosen.

The Derbot is therefore normally supplied from six 'AA' Alkaline cells, giving an overall nominal supply of 9 V. This voltage is used 'raw' for the motor supply. For supply to the microcontroller and associated circuit it is regulated down to 5 V, using a National Semiconductors LP2950 low-power voltage regulator. A number of the system elements, for example the opto-sensors, rely on this regulated voltage to operate correctly.

As there are precise timing requirements in the application, a crystal oscillator was selected; 4 MHz was chosen as the oscillator frequency, to allow simple timing functions to be derived from the resulting 1 μs instruction cycle time.

7.10.2 Use of the parallel ports

As can be seen, Derbot has extensive I/O requirements, although not all of these may be implemented in any one version of it. These are listed in Table 7.2. Some simply require general-purpose parallel I/O. Others,

Table 7.2 Derbot I/O needs

Function	Data direction	Microcontroller peripheral used	Port and pin allocation
In-circuit debug	Bi-directional	Parallel port	Port B, 6 and 7
'Bump' microswitches	Input	Parallel port	Port B, 4 and 5
Servo drive	Output	Parallel port	Port B, 3
Opto-sensor enable	Output	Parallel port	Port B, 2
Piezo sounder	Output	Parallel port	Port B, 1
Interrupt	Input	Parallel port/interrupt	Port B, 0
Mode switch, also USART RX	Input	Parallel port	Port C, 7
Diagnostic led, also ultrasound echo, also USART TX	Output Input Output	Parallel port	Port C, 6
Diagnostic led, also ultrasound pulse	Output	Parallel port	Port C, 5
I²C, serial data	Bi-directional	Synchronous serial port	Port C, 4
I²C, serial clock	Bi-directional	Synchronous serial port	Port C, 3
Right motor PWM drive, also PWM demo, TPs 1 and 2	Output	PWM	Port C, 2
Left motor PWM drive	Output	PWM	Port C, 1
Reflective opto-sensor	Input	Timer 1	Port C, 0
Left motor enable	Output	Parallel port	Port A, 5
Reflective opto-sensor	Input	Timer 0	Port A, 4
Light sensor, rear	Input	Analog-to-digital converter (ADC)	Port A, 3
Right motor enable	Output	Parallel port	Port A, 2
Light sensor, left	Input	ADC	Port A, 1
Light sensor, right	Input	ADC	Port A, 0

as indicated, are special purpose functions that link to specific peripherals or I/O, through uniquely identified pins. These functions were therefore allocated immediately to their special-purpose pins. Like the port pins themselves, there are some shared functions apparent in the table. These are mainly used in the short term for demonstration purposes.

The Derbot general-purpose I/O can be allocated with some degree of flexibility. The choice is only influenced by the opportunity of Schmitt trigger input, internal pull-up resistor (Port B) and of course the physical position on the microcontroller IC. The implementation of Derbot switch and LED interfacing is considered here; other interfacing is considered in later chapters.

Switch interfacing

The Derbot AGV uses the mode switch, a user control which can cause the robot to switch between two modes, and a couple of microswitches, which are used for bump sensing. Electrically these switches are the same, being SPST, and so can be connected using the diagram of Figure 3.7(b). A regular pull-up resistor can be used or the Port B pull-up. In the latter case, it's worth noting pull-ups will be switched on for all Port B bits used as inputs, regardless of whether they're wanted or not. As only three pull-ups are required, it was decided to use external components and not activate those of Port B.

LED driving

Two general-purpose diagnostic LEDs are included in the circuit. These can be used for any purpose that the programmer wishes. The high-efficiency red type chosen gives excellent output at very low currents. Tests showed that adequate visibility could be obtained with a current in the region of 3.5 mA. Using the characteristics of Figure 3.8(a), the forward voltages across the LED for this current will be around 1.88 V, and from Figure 7.14(a) the port bit output voltage will be around 4.7 V. Hence, applying equation (3.1):

$$R = \frac{4.7 - 1.88}{0.0035} = 806\,\Omega$$

820 Ω is a close 'preferred value' and was found to provide good visibility.

7.10.3   Assembling the hardware

If you are building up a Derbot AGV, then use the component layout diagram in the book CD and assemble the following:

- All components in the power supply path, including decoupling capacitors
- Reset switch and pull-up resistor
- Crystal and load capacitors

- Diagnostic LEDs with current-limiting resistors
- Two front microswitches with pull up resistors
- Mode switch and pull-up resistor
- (Optionally) the piezo sounder and drive transistor
- The ICD connector, if you have an ICD 2 unit.

Having done this, you will have a PCB assembled with the circuit shown in Figure 7.22.

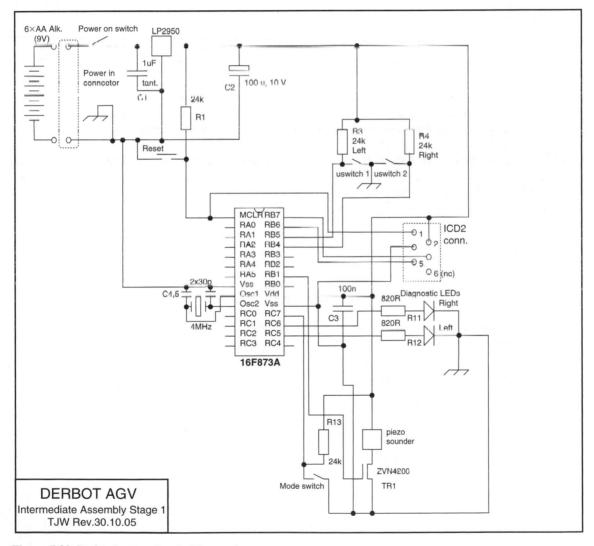

Figure 7.22 Derbot intermediate build stage 1

7.11 Downloading, testing and running a simple program with ICD 2

Let us now attempt running a program for the new hardware design and, if you have access to one, the ICD 2.

7.11.1 A first Derbot program

Program Example 7.1 is a very simple test program for the embryonic Derbot AGV, as just constructed. It simply tests the states of the front microswitches and sets or clears the diagnostic LEDs accordingly. If both switches are pressed, the sounder comes on. Notice that the program uses the 16F873A Include File, in the line:

```
include p16f873a.inc
```

With this larger microcontroller it is no longer practical to consider entering the addresses of each register, as we did, for example, at the head of Program Example 4.2. Errors introduced at this stage can also be very difficult to detect. Using the Include File should become your regular practice, if this is not already the case.

In MPLAB create a project **Dbt_sw2led** (or name of your choice) and copy the program **Dbt_sw2led.asm** from the book CD, or enter it by hand; it is shown in its entirety in the printed example. Build the project in the usual way. If you do not have an ICD 2, then download the program to the PIC, using your usual programmer, for example the PICSTART Plus, as described in Chapter 4.

```
;***********************************************************
;Dbt_sw2led
;Moves state of front microswitches to diagnostic leds.
;If both switches on then buzzer goes on.
;TJW 24.3.05                   Tested & working 20.8.05
;***********************************************************
;
      list p=16f873a
      include p16f873a.inc

;Specify RAM labels
delcntr1 equ 20      ;used in delay SR
delcntr2 equ 21
;
;Set Configuration Word: crystal oscillator HS, WDT off,
;     power-up timer on, code protect off, LV Program off.

      __CONFIG _HS_OSC & _WDT_OFF & _PWRTE_ON & _LVP_OFF
      org 00
;Initialise
start  bsf    status,5     ;select memory bank 1
      movlw B'00000011'  ;all port A bits op
      movwf trisa
      movlw B'11111000'
```

Program Example 7.1 Simple bit moving with the Derbot

```
        movwf trisb          ;port B bits
        movlw B'10000000'
        movwf trisc          ;port C bits
        movlw B'00000110'
        movwf adcon1         ;set port A for digital function
        bcf   status,5       ;select bank 0
;
;The "main" program starts here
;Switch all outputs off
        clrf  porta
        clrf  portb
        clrf  portc
;diagnostic, switch leds on for half a second
        bsf   portc,6
        bsf   portc,5
        call  delay500
        bcf   portc,6
        bcf   portc,5
        call  delay500
;move microswitch states to diag leds
loop bcf     portc,6         ;preclear port C, bit 6 (led off)
        btfsc portb,4        ;jump if switch pressed
        bsf   portc,6        ;led on if switch clear
;
        bcf   portc,5        ;preclear port C, bit 5 (led off)
        btfsc portb,5        ;jump if switch pressed
        bsf   portc,5        ;led on if switch clear
;
        btfsc portb,4
        goto  loop1
        btfsc portb,5
        goto  loop1
        bsf   portb,1        ;switch on sounder if both pressed
        goto  loop
loop1 bcf    portb,1        ;switch off sounder
        goto  loop
;************************************************
;SUBROUTINES
;************************************************
;introduces delay of 1ms approx
delay1  movlw D'250'        ;250 cycles called,
                             ;each taking 4us
        movwf delcntr1
del1    nop         ;4 inst cycles in this loop, ie 4us
        decfsz delcntr1,1
        goto del1
        return
;
;500ms delay (approx)       ;500 calls to delay1
delay500  movlw D'250'
        movwf  delcntr2
```

Program Example 7.1 Continued

```
del5    call   delay1
        call   delay1
        decfsz delcntr2,1
        goto   del5
        return
        end
```

Program Example 7.1 Continued

7.11.2 Applying the ICD 2

The following assumes you have an ICD 2, or later version of it. It is a very brief introduction, written in the expectation that you already have knowledge of the MPLAB simulator, whose features are used by the ICD 2. Further expertise can then be gained by experimentation and reading the full manual [Ref. 7.3].

Ensure, by reading the 'Getting Started' section of the manual, that the ICD 2 is configured correctly for your computer and operating system. Connect it between computer and AGV, as illustrated in Figure 7.20. Open MPLAB and open the project that you have created, containing Program Example 7.1. Set the ICD to operate initially in debug mode, by clicking **Debugger > Select Tool > MPLAB ICD 2**. Then complete the connection to the AGV by pressing **Debugger > Connect**. The ICD runs a self-test, which checks for the presence of power supply, as well as its ability to apply the correct voltages to the lines it controls. If the target system is powered and working, the ICD 2 should pass the self-test and will be able to identify the microcontroller. It displays a message indicating whether it passed the test and issues an **MPLAB ICD 2 Ready** message. The drop-down menu under **Debugger** then becomes active, as does the toolbar, as seen in Figure 4.9 and repeated in Figure 7.23. An extra toolbar, special for the ICD 2, also appears. This appears to the right of the figure, and allows the user to read program memory, download to it and reset the ICD. Notice, however, that if on the main toolbar **Programmer > Select Programmer** is pressed, the menu indicates that no programmer is selected. This is because in this mode the ability to program is only available within the ICD debugger facility.

Ensuring that the Derbot is switched on, download the program memory, using either the toolbar button of Figure 7.23 or **Debugger > Program**. Open a Watch window, as described in Chapter 4, and display registers **PCL**, **PORTB** and **PORTC**. Then start single stepping through the program, using the Step Over function to skip the delay routines. Once in the main loop, try pressing the microswitches in turn, and see the response of the Port values in the Watch window as you step. Unlike when using the simulator, it is actual values from the hardware that are being transmitted back to the display you are seeing. Try actually setting or clearing port bits from your computer, and see how you can actually switch the Derbot leds on

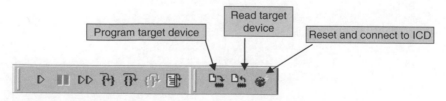

Figure 7.23 Debug toolbars for ICD 2 – Standard debug, with extra ICD program features

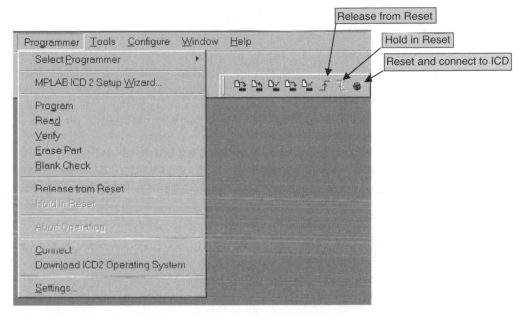

Figure 7.24 Programmer features for ICD 2 – pull-down menu and toolbar

and off in this way. As with the simulator, breakpoints can be set in the program by double-clicking on any line.

To finally program a product, the ICD 2 should be switched to programmer mode. On the top toolbar click **Programmer > Select Programmer > MPLAB ICD 2** and then **Programmer > Connect** (an automatic connection mode is also possible). A check back at the Debugger menu will show that there is now no debugger selected. It is now possible to program and read memory. The pull-down menu and programmer toolbar are available, as shown in Figure 7.24. Notice that Reset can be controlled from the screen. When programming is complete, Reset is left active. It can be released using the toolbar button shown, and the program then runs immediately.

7.11.3 Setting the configuration bits within the program

The 16F873A has more configuration bits than the 16F84A and it is easy to make an error when setting them in the MPLAB window. Incorrect settings can lead to very odd behaviour in the target embedded system or to no behaviour at all! It is, however, possible to set the configuration bits in the program, by using the **_CONFIG** assembler directive. This directive uses symbols that are defined in the microcontroller Include File. The relevant section for the 16F873A is shown as Program Example 7.2. It can be seen that each hexadecimal value that a symbol equates to is derived from the Configuration Word setting shown in summary in Figure 7.7. As combinations of these are ANDed together, a final setting for the Configuration Word is formed. The setting used, seen in the program example shown below, can be applied to all Derbot programs based on the 16F873A.

```
;Set Configuration Word: crystal oscillator HS, WDT off,
;     power-up timer on, code protect off, LV Program off.
        __CONFIG _HS_OSC & _WDT_OFF & _PWRTE_ON & _LVP_OFF
```

Even with this setting in the program, there is a danger that a change is made inadvertently in the MPLAB Configuration Bits window. You can demonstrate this by building Program Example 7.1. Look at the configuration bit settings in the window – they are correct. However, it is possible at this stage to introduce an error by making a change in the window. Whatever is now shown in the window will be downloaded to program memory on the next program download. This danger can be removed by clicking Configure > Settings > Program Loading and checking the 'Clear configuration bits upon loading the program' box. Now, on program download, the window setting is cleared and the configuration bits defined in the program are downloaded.

```
;==========================================================================
;
;          Configuration Bits
;
;==========================================================================
_CP_ALL                   EQU      H'1FFF'
_CP_OFF                   EQU      H'3FFF'
_DEBUG_OFF                EQU      H'3FFF'
_DEBUG_ON                 EQU      H'37FF'
_WRT_OFF                  EQU      H'3FFF'   ;No prog memory write protection
_WRT_256                  EQU      H'3DFF'   ;First 256 prog memory write protected
_WRT_1FOURTH              EQU      H'3BFF'   ;First 1/4 prog memory write protected
_WRT_HALF                 EQU      H'39FF'   ;First half memory write protected
_CPD_OFF                  EQU      H'3FFF'
_CPD_ON                   EQU      H'3EFF'
_LVP_ON                   EQU      H'3FFF'
_LVP_OFF                  EQU      H'3F7F'
_BODEN_ON                 EQU      H'3FFF'
_BODEN_OFF                EQU      H'3FBF'
_PWRTE_OFF                EQU      H'3FFF'
_PWRTE_ON                 EQU      H'3FF7'
_WDT_ON                   EQU      H'3FFF'
_WDT_OFF                  EQU      H'3FFB'
_RC_OSC                   EQU      H'3FFF'
_HS_OSC                   EQU      H'3FFE'
_XT_OSC                   EQU      H'3FFD'
_LP_OSC                   EQU      H'3FFC'
```

Program Example 7.2 Defining configuration bits in 16F873A Include File

7.12   Taking things further – the 16F874A/16F877A Ports D and E

Looking at Figure 7.1(b), one can see that the 16F874A and 16F877A have two extra ports, the 8-bit Port D and the 3-bit Port E. Either port can be used for general-purpose I/O, like any of the other ports. The block diagram of Port D, when configured for normal digital I/O, is shown in Figure 7.25(a). An alternative function for Port E is to provide three further analog inputs. Because of this, Port E is also under the control

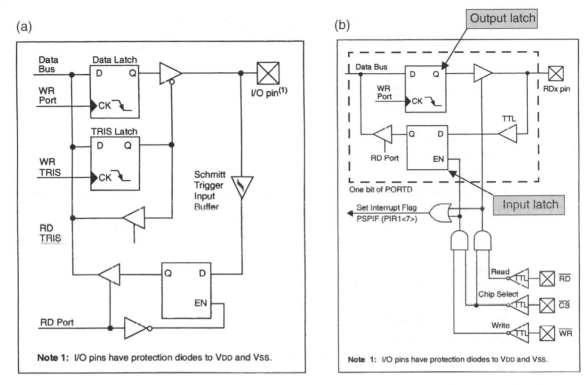

Figure 7.25 Block diagrams of Port D pin driver circuits. (a) In I/O mode. (b) As parallel slave port (supplementary labels in shaded boxes added by the author)

of one of the registers that control the ADC, **ADCON1**. The setting of this determines whether the port is used for digital or analog signals.

Together, Ports D and E can also form the parallel slave port. The ports are put into this mode by setting the **PSPMODE** bit in the **TRISE** register (Figure 7.26). The purpose of this mode is to allow the microcontroller to interface as a slave to a data bus, controlled perhaps by a microprocessor. The Port E bits must be set as inputs (with digital mode selected in **ADCON1**); the state of **TRISD** is, however, immaterial. Ports D and E are then configured as in Figure 7.25(b). The diagram shows 1 bit of Port D, together with the three control lines, $\overline{\text{CS}}$, $\overline{\text{WR}}$ and $\overline{\text{RD}}$. These are the 3 bits of Port E, now configured for this purpose. Notice that there is an output latch and an input latch for each Port D bit.

An example application for the parallel slave port appears in Figure 7.27. Here the port is connected to a data bus and control lines that form part of a larger system, controlled by a microprocessor. The 16F874A program can write data to the port, in which case bit **OBF** of **TRISE** is set. If $\overline{\text{CS}}$ and $\overline{\text{RD}}$ are taken low by the external circuit, then the port outputs the data held on its output latches onto the external bus. This action clears **OBF**. If $\overline{\text{CS}}$ and $\overline{\text{WR}}$ are taken low, the port latches data from the bus into its input latches, and bit **IBF** of **TRISE** is set. **IBF** is cleared when the port is read by the microcontroller program. If, however,

R-0	R-0	R/W-0	R/W-0	U-0	R/W-1	R/W-1	R/W-1
IBF	OBF	IBOV	PSPMODE	—	Bit 2	Bit 1	Bit 0

bit 7   bit 0

Parallel Slave Port Status/Control Bits:

bit 7   **IBF:** Input Buffer Full Status bit
1 = A word has been received and is waiting to be read by the CPU
0 = No word has been received

bit 6   **OBF:** Output Buffer Full Status bit
1 = The output buffer still holds a previously written word
0 = The output buffer has been read

bit 5   **IBOV:** Input Buffer Overflow Detect bit (in Microprocessor mode)
1 = A write occurred when a previously input word has not been read (must be cleared in software)
0 = No overflow occurred

bit 4   **PSPMODE:** Parallel Slave Port Mode Select bit
1 = PORTD functions in Parallel Slave Port mode
0 = PORTD functions in general purpose I/O mode

bit 3   **Unimplemented:** Read as '0'

PORTE Data Direction Bits:

bit 2   **Bit 2:** Direction Control bit for pin RE2/$\overline{CS}$/AN7
1 = Input
0 = Output

bit 1   **Bit 1:** Direction Control bit for pin RE1/$\overline{WR}$/AN6
1 = Input
0 = Output

bit 0   **Bit 0:** Direction Control bit for pin RE0/$\overline{RD}$/AN5
1 = Input
0 = Output

Figure 7.26 The **TRISE** register

the external circuit writes to the port again, before the previous word has been read, then the **IBOV** bit of **TRISE** is set. The interrupt flag **PSPIF** (Figure 7.10) is set when either a slave write or read is completed by the external circuit. This flag must be cleared in software.

Summary

- The 16F87XA group of microcontrollers is an important subset of the 16 Series family. Its added capability is achieved by the inclusion of a powerful set of peripherals, along with memory upgrades.
- The central architecture of the 16F87XA group is the same as other microcontrollers in the PIC 16 Series of microcontrollers, with the same instruction set.

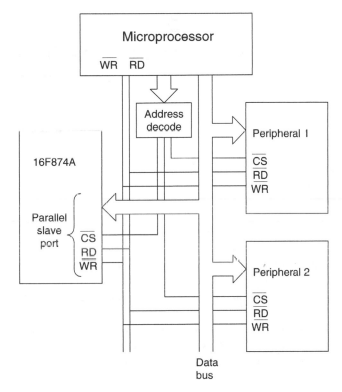

Figure 7.27 Parallel slave port connected to system bus

- Testing of an embedded system must be undertaken systematically, applying the best tools that are available to do the job.
- In-circuit debugging is a powerful technique for testing and commissioning both program and hardware, allowing minimum invasiveness.

References

7.1. PIC 16F87XA Data Sheet (2003). Microchip Technology Inc., Reference no. DS39582B; www.microchip.com

7.2. PIC 16F87XA Flash Memory Programming Specification (2002). DS39589. www.microchip.com

7.3. MPLAB ICD 2 In-Circuit Debugger User's Guide (2005). Microchip Technology, DS51331B.

8
The human and physical interfaces

Figure 1.1 illustrates a typical embedded system. It shows an embedded computer reading in signals, outputting control signals, interacting with a human user and possibly interacting with an external system via a network. All of these activities fall under the broad heading of interfacing, which forms a very large part of the hardware design of embedded systems. In this chapter we look at some aspects of interfacing, including both the human and the physical interfaces.

In order to design the interface we need to know something about the sensors and actuators that can be used. These lie between the wider system and the electronic domain of the microcontroller circuit. There are input devices, sensors for measurement or data entry devices for human interaction. There are also output devices, displays or alarms for the human being, and motors or other actuators for the physical system. There are thousands of these devices to choose from; we take just a few as examples.

Further to this knowledge of the sensors and actuators themselves, we will need to know how they can be connected to the microcontroller. Sensor signals may need amplification to connect to the microcontroller; motor control output signals may need to drive powerful switching circuits to get the motor turning, at the right speed and the right time.

Finally, and crucially, there is a need to understand how programs can be written to effect this interfacing.

In this chapter you will therefore learn about:

- Human interfacing needs and some simple means of meeting these
- Some simple example sensors
- Some ways of interfacing between sensor signals and the microcontroller
- Some simple example actuators
- Some ways of interfacing between the microcontroller and the actuator
- The Derbot application of some of its sensors and actuators.

Many of these topics will be illustrated by examples from the Derbot AGV.

It is important to note that important elements of interfacing do not appear in this chapter. The whole process of the input of analog signals gets its own chapter (Chapter 11). Networking concepts, through an exploration of serial communication, are introduced in Chapter 10.

8.1 The main idea – the human interface

The human has to interface with any machine that he/she works with. This is almost inevitably in some form of closed loop interaction. The user perceives what the current status of the machine and perhaps the wider environment is. In so doing, he/she receives information from the system. Then, based on what he/she wants to happen, the user interacts with the machine to cause some change in action. This interaction may be purely in the form of information exchange. For example, the user of the fridge may read the current temperature on the display, decide he wants it colder and thus enter a new demand temperature on the keypad. Alternatively, the user may have to input some physical actuation of his own. This is very much what happens in driving a car, where the driver receives information from the dashboard but still turns this into physical actuation, in the form of movements of the steering wheel, gear stick or accelerator.

Figure 8.1 shows examples of human interfaces found in everyday products, which are also embedded systems of one form or other. The fridge can operate in one of two modes, 'super' or 'eco'. It displays the

(a)

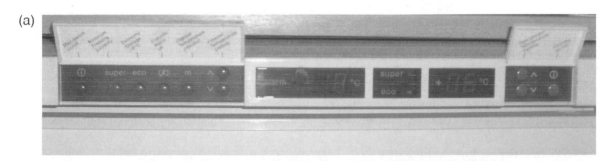

(b)

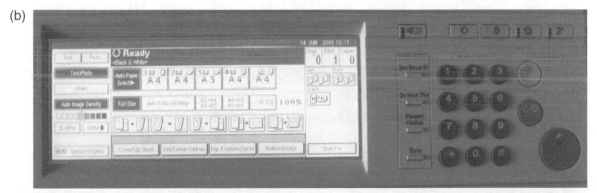

(c)

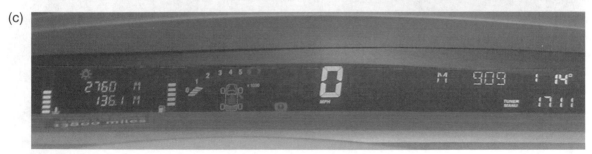

Figure 8.1 The human interface in some familiar products. (a) Domestic fridge. (b) Photocopier. (c) Car dashboard (detail)

actual temperature of its freezer and main compartments. Demand temperatures for each of these can be set. An acoustic alarm sounds if the freezer temperature rises above a certain value. This alarm can, however, be disabled. All the control input is by simple push-buttons, while the display information is by two two-digit numerical displays with polarity indicators.

The photocopier control is somewhat more complex. There is a wide range of operating modes available, with different paper sources and image adjustment capabilities available. Information is conveyed by a customised display. As this is touch sensitive, it also acts as an input of control information. A more conventional numerical keypad allows input of purely numerical data, like codes and the number of copies required.

The detail from the car dashboard represents a further increase in complexity. Its central feature is car speed – as the car in this instance is stationary this shows a big zero. Around this, one can see a range of status information, including things as diverse as engine r.p.m., radio station, temperature, door status and fuel. Despite the complexity and diversity of information, it is all conveyed in simple forms. Again, there is considerable use of numerical display. To this is added bar graphs, simple illuminated symbols and simple diagrammatic information. No user input is seen in this picture.

What is striking in each of these three very diverse interfaces is the strands which unite them. Each has significant need to convey numerical information and status information; means are found to achieve both of these in simple ways. The main sense invoked is sight. Sound is also used, especially in alarm situations.

We will be exploring means of information exchange between the embedded system and user in the section that follows. This will be done through a study of the keypad, seven-segment LED (light-emitting diode) display and the liquid crystal display (LCD), as these are more or less ubiquitous in simple embedded systems. These will be illustrated with the Derbot 'hand controller' module. This is an optional unit that can be hand-held or fitted above the battery pack, as seen in Figure A3.3. The hand controller is available in two versions, one for LED and one for LCD. These are shown in Figure 8.2. The hand controller is

(a)   (b)

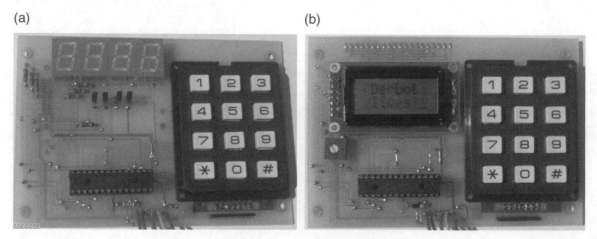

Figure 8.2 The Derbot hand controller. (a) LED version. (b) LCD version

designed as a 'dumb terminal' which interfaces directly with the main Derbot AGV. It can, however, be used in stand-alone form, which is the case in this chapter. In Chapter 10 it is linked to the AGV and used to illustrate serial communications. Its circuit diagram is given in Figure A3.2.

8.2 From switches to keypads

The humble switch, mainstay of so many human interfaces, was introduced in Chapter 3. Switches are good for conveying information of a digital nature – they are two-state, but can be used in multiples, each one taking one port bit. In more complex situations it becomes inappropriate to keep adding switches. For one thing, the demand on port bits becomes excessive, and their relentlessly two-state nature often does not meet the need.

8.2.1 The keypad

A useful step forward from the simple switch is given by the keypad, as seen in the photocopier interface and the Derbot hand controllers in Figure 8.2. The keypad allows numeric or alphanumeric information to be entered. It is widely used in photocopiers, burglar alarms, central heating controllers and so on.

A keypad is based on switches, yet it would be extremely resource-intensive if each of these switches were allocated to a port bit. Instead, to make good use of resource, each switch is connected in a matrix. Figure 8.3 shows the electrical connections for the keypad of Figure 8.2, which has 12 keys. It can be seen that these are arranged in a 4 × 3 matrix, with four rows and three columns. Now only seven interconnections are needed, rather than 12. Whenever a key is pressed, it connects its row with its column.

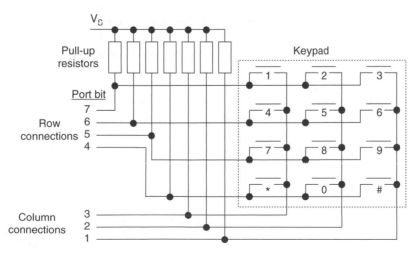

Figure 8.3 Keypad circuit diagram, with pull up resistors

(a)

(b)

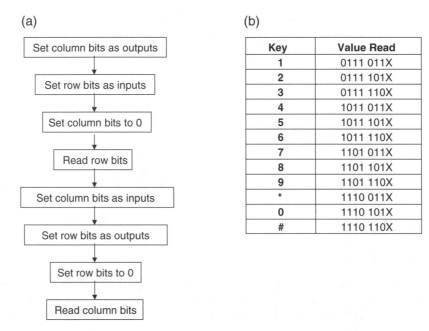

Key	Value Read
1	0111 011X
2	0111 101X
3	0111 110X
4	1011 011X
5	1011 101X
6	1011 110X
7	1101 011X
8	1101 101X
9	1101 110X
*	1110 011X
0	1110 101X
#	1110 110X

Figure 8.4 Reading a keypad with a microcontroller port. (a) Flow diagram. (b) Outputs for keypad of Figure 8.3

In an embedded environment the keypad is usually connected to the bits of a microcontroller input/output (I/O) port. Example port bit connections are shown in the figure, as well as the necessary pull-up resistors. The challenge is how to detect efficiently which key has been pressed.

The technique usually used to read a keypad follows the flow diagram of Figure 8.4(a). First the column bits are set to output, with the row bits as input. The output column bits are set to 0. If no button is pressed all row line inputs will read 1, due to the action of the pull-up resistors. If, however, a button is pressed then its corresponding switch will connect column and row lines, and the corresponding row line will be pulled low. If the same process is repeated instantaneously, but with outputs exchanged for inputs, then the column line for the key pressed can be found and the key fully identified.

Output values for the keypad connection shown in Figure 8.3 are shown in Figure 8.4(b). For example, suppose key 7 was pressed. In the first phase of the flow diagram, with row connections as inputs, the third row line (port bit 5) would read low. In the second phase, with column bits as inputs, the first column (port bit 3) would read low. The final pattern is as shown, with bit 0 showing a 'don't care' condition, as it is unused.

8.2.2   *Design example: use of keypad in Derbot hand controller*

The flow diagram just described is only the starting point for working with keypads. How, for example, do we detect when the keypad is actually pressed and what do we do with the code appearing in the table?

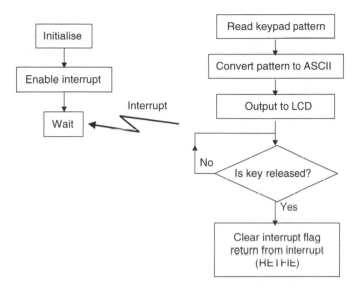

Figure 8.5 Flow diagram of Program Example 8.1

Program Example 8.1 illustrates a practical application of keypad reading, in a program called *keypad_test*. The program is used to illustrate use of both keypad and LCD. It is too long to print in its entirety, so keypad-related sections only are reproduced here. The book CD contains the full program.

The program flow diagram is shown in Figure 8.5. To detect keypad action, it uses the 'interrupt on change' facility, available on the higher 4 bits of Port B. Following initialisation, all program activity is contained in the Interrupt Service Routine, called **kpad to lcd**. This does essentially four things. It reads a pattern from the keypad, converts this to ASCII code, sends the code to the LCD and waits for the key to be released, before leaving the ISR.

Try following the program through, matching it with the flow diagrams shown. In the opening initialisation Port B is set up so that row bits are input and column bits are output. We could start the other way round, but to make use of the 'interrupt on change', the higher 4 bits must be set as input. All output bits are subsequently set to zero and the 'interrupt on change' interrupt is enabled. We have at this stage already entered the flow diagram of Figure 8.4(a).

An interrupt occurs when a key is pressed and the Interrupt Service Routine, labelled **kpad_to_lcd**, is invoked. This immediately calls the subroutine **kpad_rd,** which continues the flow diagram of Figure 8.4(a), started in the initialisation section. The value of Port B is read and stored in memory location **kpad_pat**. Column and row roles are then reversed, and the row bits are then read. These are ORed into **kpad_pat**, ensuring that any unwanted bits are removed by ANDing with 0. The routine then resets Port B to its initial value, ready for the next keypad read.

The value held in **kpad_pat**, on completing the **kpad_rd** subroutine, is not in a very useful format. It is a 7-bit number, with possible values shown in Figure 8.4(b). The program therefore calls subroutine **kp_code_conv**, which converts this number into the keypad code that caused it. The way this is done is described in the next paragraph. This value is then output to the LCD, using subroutine **lcd_write**. This is not shown in this example, but is described later in the chapter. The program then sits in a loop, waiting for release of the key, invoking the **kpad_rd** subroutine again. This is because the user *letting go* the keypad will also cause an 'interrupt on change', which would lead to a second, unwanted keypad read.

```
;***************************************************************
;keypad_test
;Tests keypad, writing key pressed to lcd display on
;Derbot Hand Controller.
;TJW 23.6.05                                    Tested 24.6.05
;***************************************************************
...
(opening program sections omitted)
...
;Initialise
       bsf    status,rp0    ;select memory bank 1
...
       movlw B'11110000'    ;Port B initially Row bits ip, column op
       movwf trisb          ;(port B not used)
       bcf    status,rp0    ;select bank 0
...
(lcd initialisation omitted)
...
       clrf  portb ;initialise keypad value
;enable interrupt
       bcf    intcon,rbif
       bsf    intcon,rbie
       bsf    intcon,gie
loop   goto  loop           ;await keypad entries


;***************************************************************
;Interrupt Service Routine.
;***************************************************************
;Keypad press has been detected through Port B Interrupt on Change.Gets
;keypad pattern, converts to character, stores in kpad_char, sends to lcd,
;and awaits key release,
kpad_to_lcd call    kpad_rd
;now convert code to character, forming address used in lookup table
       call  kp_code_conv
;now send to lcd
       bsf    portc,lcd_RS ;set for character op
       movwf lcd_op
       call  lcd_write
;test now for keypad release
rel_test call kpad_rd
       movf  kpad_pat,0
       andlw 0fe            ;suppress lsb, which is not used
       sublw 0fe            ;test if inactive
       btfss status,z
       goto  rel_test
```

Program Example 8.1 Keypad reading on the Derbot hand controller

```
        bcf    intcon,rbif   ;clear interrupt flag
        retfie
;
;*****************************************************************
;SUBROUTINES
;*****************************************************************
;Reads keypad, places pattern into kpad_pat, and resets keypad interface
kpad_rd movf portb,w         ;read portb value, this will be row pattern
        andlw B'11110000'    ;ensure unwanted bits are suppressed
        movwf kpad_pat
        bsf    status,rp0    ;set row to op, column to ip
        movlw B'00001110'
        movwf trisb
        bcf    status,rp0
        movlw 00
        movwf portb          ;ensure output values still zero
        movf  portb,w        ;read portb value, this will be column pattern
        andlw B'00001110'    ;ensure unwanted bits are suppressed
        iorwf kpad_pat,1     ;OR those results into the pattern
;reset keypad interface
        bsf    status,rp0 ;set row to ip, column to op
        movlw B'11110000'
        movwf trisb
        bcf    status,rp0
        clrf   portb  ;ensure output values still zero
        return

;Converts keypad pattern held in kpad_pat to ASCII character, first forming
;address (in kpad_add) that is used in lu table. Returns with character held
;in kpad_char
kp_code_conv bcf status,c
        rrf    kpad_pat,1    ;discard bit 0 which is not used
        clrf   kpad_add
;deduce row
        btfsc kpad_pat,6
        goto  kp1
        goto  col_find       ;here if row 1, kpad_add stays as is
kp1     btfsc kpad_pat,5
        goto  kp2
        movlw B'00000100'    ;here if row 2
        iorwf kpad_add,1     ;form table address
        goto  col_find
kp2     btfsc kpad_pat,4
        goto  kp3
        movlw B'00001000'    ;here if row 3
        iorwf kpad_add,1     ;form table address
        goto  col_find
kp3     btfsc kpad_pat,3
        goto  kp4
        movlw B'00001100'    ;here if row 3
        iorwf kpad_add,1     ;form table address
        goto  col_find
```

Program Example 8.1 Continued

```
kp4     movlw D'16'          ;no row detected, return "E" via Table
        goto keypad_op
;now deduce column
col_find btfsc     kpad_pat,2
        goto  cf1
        goto   keypad_op    ;here if column 1, kpad_add stays as is
cf1     btfsc kpad_pat,1
        goto  cf2
        movlw B'00000001'    ;here if column 2
        iorwf kpad_add,1     ;form table address
        goto  keypad_op
;assume now column 3
cf2     movlw B'00000010'
        iorwf kpad_add,1     ;form table address
keypad_op movf kpad_add,0
        call  kp_table
        movwf kpad_char      ;save the character
        return
;
;Table called to convert pattern recd from keypad to actual character. Note that
;ASCII codes will be returned, as each digit is in format 'D'.
kp_table addwf pcl,1
        retlw '1'            ;row 1
        retlw '2'
        retlw '3'
        retlw 'A'            ;Error code
        retlw '4'            ;row 2
        retlw '5'
        retlw '6'
        retlw 'B'            ;Error code
        retlw '7'            ;row 3
        retlw '8'
        retlw '9'
        retlw 'C'            ;Error code
        retlw '*'            ;row 4
        retlw '0'
        retlw '#'
        retlw 'D'            ;Error code
        retlw 'E'            ;Error code
...
```

Program Example 8.1 Continued

The **kp_code_conv** subroutine converts the patterns derived from reading the keypad, as seen in Figure 8.4(b), into the ASCII code of the key pressed. It does this by deriving an address, held in memory location **kpad_add**, which is used to access the look-up table **kp_table**. The address follows the format shown in Figure 8.6. The subroutine tests the bits of the pattern in turn, finding out which row and column has been active. It sets up the address bits according to the outcome of its tests. The look-up table is then called, as described in Chapter 5.

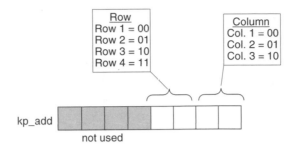

Figure 8.6 Format of look-up table address

8.3   LED displays

The first part of this chapter showed how very important displays are, in almost any system which has a human interface. We look at two types of display here, the seven-segment LED and the liquid crystal.

8.3.1   LED arrays: seven-segment displays

As with the switch, the simple LED was introduced in Chapter 3. We saw that LEDs are visually attractive, are an efficient source of light, can be driven from a logic gate or port bit output and are thus exceptionally useful for conveying simple information. Yet the single LED, or even groups of LEDs, are restricted in the information they can give, and as their number increases they become increasingly complex to drive. There are therefore a number of standard configurations in which LEDs are packaged, including bargraph, seven-segment display, dot matrix and 'star-burst'.

The seven-segment display is a particularly versatile configuration, seen already in Figures 8.1(a, c) and 8.2(a). We explore the LED implementation of it in some more detail here. A single digit, made by Kingbright [Ref. 8.1], is shown in Figure 8.7. This is the display used in the Derbot hand controller (Figure 8.2(a)). By lighting different combinations of the seven segments, all numerical digits can be displayed, as well as a surprising number of alphabetic characters. A decimal point is usually included, as shown. The problem arises that if each segment is illuminated by an LED, then 14 connections are required, and that is just for one digit. If multiple digits are required, then the number of connections soars. Therefore, two clever and simple techniques are used to tame the number of connections used.

The common anode/common cathode connection

When using multiple LEDs in simple configurations, it is almost certain that one 'side' of all the LED terminals will be connected to all others of the same type. In other words, all LED anodes are likely to be connected together, or all LED cathodes. This is what is done in the seven-segment digit, as seen in Figure 8.7(b). The digit is available either in common cathode form or in common anode. There are eight LEDs in the digit (including the decimal point), but instead of 16 connections being needed, there are now only nine, one for each segment and one for the common connection. The actual pin connections in the example shown lie in two rows, at the top and bottom of the digit. There are 10 pins in all, with the common anode or cathode taking two pins.

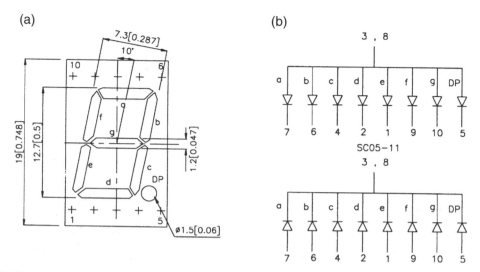

Figure 8.7 The seven-segment display. (a) A seven-segment digit (Kingbright, 12.7 mm). (b) Electrical connection (upper: common anode; lower: common cathode). Reproduced with permission of Kingbright Elec. Co. Ltd

Multiplexing of digits

By making a common cathode or common anode connection, the number of connections to a single digit is reduced. But a digit on its own is rarely used and multiple digits will still require many connections. A four-digit display, for example, with decimal points on each digit, would require 36 connections. Furthermore, a multi-digit display would have power supply requirements that are quite excessive. With all segments on, and each taking 5 mA, it would draw 160 mA. Therefore, a second, important technique is introduced, that of digit *multiplexing*. While this enables us to develop a practical display, it also poses an interesting early challenge in handling some real-time issues. The technique is illustrated in the design example that follows.

8.3.2   Design example: the Derbot hand controller seven-segment display

The circuit diagram of the Derbot hand controller in Figure A.3.2 is comparatively complex, as the controller can be used in LED version or LCD version (but not both). The seven-segment LED display section alone is shown in Figure 8.8(a), and from this an understanding of digit multiplexing can be developed. It can be seen that all equivalent segment lines are connected together, so that, for example, segment 'a' of Digit 1 is connected to segment 'a' of Digit 2 and so on. Each of these segment lines is connected via a resistor (1.2 kΩ; not shown) to a port bit. Kingbright 'high-efficiency red' common cathode displays are used [Ref. 8.1].

Each of the digit common cathode lines is connected to a 'logic-compatible' MOSFET transistor. When the gate voltage of the transistor goes high, the transistor conducts and the common cathode terminal goes low. (Further details on transistor switching is given later in the chapter.) Any segment whose anode is at logic high then illuminates. The value of the series resistors was selected experimentally, to minimise the current,

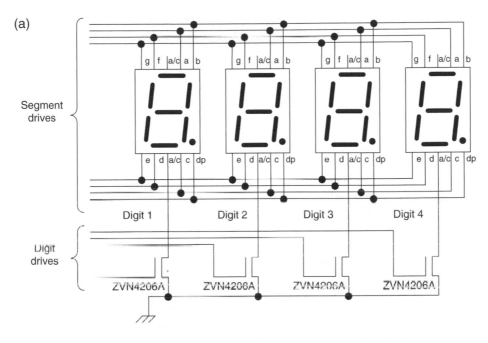

Segment	Port bit	Segment	Port bit	Digit drive	Port bit
a	C, 6	e	A,2	1	C,0
b	C, 7	f	A,3	2	C,1
c	A, 0	g	A,4	3	C,5
d	A, 1	d.p.	A,5	4	B,0

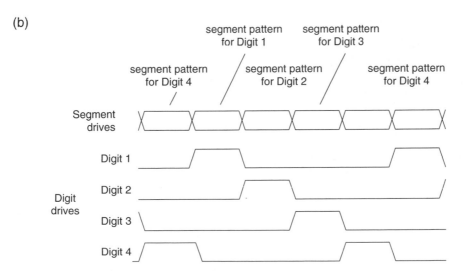

Figure 8.8 Seven-segment display part of the Derbot 'hand controller'. (a) Detail of display circuit diagram. (b) Timing diagram for driving the digits

while maintaining reasonable visibility. Noting from the data that each LED has a forward voltage when in conduction of around 1.9 V, the current I flowing per segment can be calculated as:

$$I = (5.0 - 1.9)/(1200)$$

$$I = 1.3 \, \text{mA}$$

This calculation neglects the 'on' resistance of the switching transistor and the output resistance of the port. Both of these are small compared with the $1.2 \, k\Omega$ value of the series resistance.

The way this display is driven is shown in the timing diagram (Figure 8.8(b)). The digits are activated continuously in turn. If this is done at the right speed, the eye is tricked into thinking that all digits are being continuously lit. The timing diagram shows the segments for Digit 4 being set and the Digit 4 common cathode being set to 1. Digit 4 is therefore switched on, while all other digits are off. This is held for a period of time (around 5–20 ms is generally appropriate) and then Digit 2 is illuminated in a similar way. Each digit is lit in turn and the cycle recommences.

If driven from a microcontroller, a program must be written to recreate the timing diagram. Let us see how this is done in practice, in Program Example 8.2. This is a very simple program, which runs on the LED version of the hand controller, and does nothing but display the word HELP on the digits. Figure 8.7(a) shows that, for the letter H, segments b, c, e, f and g should be illuminated. After a simple initialisation, the bit pattern for the letter H is set up on Ports A and C, by setting high each of the bits connected to the 'on' segments. The common cathode of Digit 1 is also set high and a 5 ms delay called. The H segments, and the Digit 1 common cathode drive, are then replaced by the segments for the letter E and the common cathode drive for Digit 2. The program continues through all four letters, before looping back to start again.

```
;****************************************************************
;led_disp_tst
;Tests led display on Derbot Hand Controller by writing
;word "HELP" to 4-digit display.
;TJW 17.6.05
;****************************************************************
;Clock is 1MHz approx
;Configuration Word: WDT off, power-up timer on,
;                    code protect off, RC oscillator
;
;   Port A                Port B                Port C
;   ------                ------                ------
;0  led seg c             led cc digit 4        led cc digit 2
;1  led seg d             keypad col 3          led cc digit 1
;2  led seg e             keypad col 2          Interrupt op
;3  led seg f             keypad col 1          SCL
;4  led seg g             keypad row 4          SDA
;5  led seg dp            keypad row 3          led cc digit 3
;6       -                keypad row 2          led seg a
;7       -                keypad row 1          led seg b
```

Program Example 8.2 Driving the seven-segment display on the Derbot 'hand controller'

```
        list    p=16F873A
        #include p16f873A.inc

;Specify RAM
delcntr1 equ 20        ;used in delay5
;Specify some port bits
;Port C

dig1_cc equ 1          ;digit common cathode drives
dig2_cc equ 0
dig3_cc equ 5
;Port B
dig4_cc equ 0
;
        org 00
;Initialise
        bcf     status,rp1
        bsf     status,rp0     ;select memory bank 1
        movlw   B'00000000'    ;setall port bits op
        movwf   trisa
        movwf   trisb
        movwf   trisc
        bcf     status,rp0     ;select bank 0
;
;set digit1
loop bcf portb,dig4_cc
        movlw   B'00011101'    ;turn on segments for H
        movwf   porta
        bcf     portc,6
        bsf     portc,7
        bsf     portc,dig1_cc  ;enable digit once segments set
        call    delay5
;digit2
        bcf     portc,dig1_cc
        movlw   B'00011110'    ;turn on segments for E
        movwf   porta
        bsf     portc,6
        bcf     portc,7
        bsf     portc,dig2_cc ;enable digit
        call    delay5
;digit3
        bcf     portc,dig2_cc
        movlw   B'00001110'    ;turn on segments for L
        movwf   porta
        bcf     portc,6
        bcf     portc,7
        bsf     portc,dig3_cc  ;enable digit
        call    delay5
;digit4
        bcf portc,dig3_cc
        movlw   B'00011100'    ;turn on segments for P
```

Program Example 8.2 Continued

```
        movwf porta
        bsf    portc,6
        bsf    portc,7
        bsf    portb,dig4_cc ;enable digit
        call delay5
        goto loop
;
;SUBROUTINE: Introduces delay of 5ms approx
delay5 movlw D'250' ;250 cycles called
        movwf  delcntr1
del1   nop                    ;5 inst cycles in this loop, ie 20us
         nop
        decfsz delcntr1,1
        goto   del1
        return
         end
```

Program Example 8.2 Continued

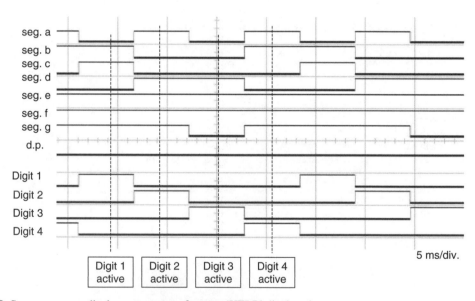

Figure 8.9 Seven-segment display output waveforms – 'HELP' displayed

Figure 8.9 shows a screen print from the logic analyser section of the Agilent Mixed Signal Oscilloscope, with the program running. All segment lines are shown, together with the four digit drives. When any digit common cathode drive is high, the associated common cathode connection itself is taken low and the digit is activated. For Digit 1, it is easy to see the segments b, c, e, f and g are on, forming the letter 'H'. Similarly, for Digit 2, the segments a, d, e, f and g form the letter 'E', and so on. The process repeats continuously and the word 'HELP' is spelt out. It can be seen that segments e and f are continuously on, while the decimal point (d.p.) is always off.

Light-emitting diodes are enormously useful, but they suffer from several drawbacks: they are power hungry, at least for battery-powered designs, and it is difficult, if not impossible, to form them into complex multi-digit or graphical displays. Therefore, it is essential to look elsewhere for sophisticated displays, with the solution being readily found in liquid crystal technology.

8.4 Liquid crystal displays

The liquid crystal display (LCD) has been one of the enabling technologies of the current electronic revolution. It is an essential part of every mobile phone, every laptop and every personal organiser.

Liquid crystal is an organic compound that polarises any light that passes through it. Liquid crystal also responds to an applied electric field by changing the alignment of its molecules, and in so doing changing the direction of the light polarisation that it introduces. Liquid crystal can be trapped between two parallel sheets of glass, with a matching pattern of transparent electrode on each sheet. When a voltage is applied to the electrodes, the optical character of the crystal changes and the electrode pattern appears in the crystal.

A huge range of LCDs has been developed, including those based on seven-segment digits or dot matrix formats, as well as a variety of graphical forms. Many general-purpose displays are available commercially, while customised displays are made for large-volume products. The Derbot hand controller uses an example of a very popular and useful general-purpose format, as seen in Figure 8.2(b). The display shown has two lines, of eight digits each, where each digit is a liquid crystal dot matrix. Larger displays in this format are common, with more digits and more lines.

Driving LCDs directly is not entirely simple. That need not concern us too much here, however, as most displays, like the one shown in Figure 8.2(b), contain their own drive electronics, designed to be interfaced to a microcontroller. The need is then to understand how to interface to the drive electronics.

8.4.1 The HD44780 LCD driver and its derivatives

Some few years back the electronics giant Hitachi developed a microcontroller specially designed to drive LCD alphanumeric modules such as the one shown in Figure 8.2(b). In turn, it had a simple interface that can be connected to general-purpose microprocessors and microcontrollers. This microcontroller, the HD44780 [Ref. 8.2], defined an interface that has become something of an informal standard for this type of display. Many manufacturers of displays integrated it into their products. A generation of derivatives now exists, which have replaced the original Hitachi device but retain most of its features. These include the S6A009 and KS0066U devices made by Samsung. As these derivatives tend to be so similar to the original Hitachi device, the HD44780 features will be described. When designing with a display it is important, however, to ensure that you are working with the correct data for the device.

The HD44780 interface has some peculiarities, partly due to the time-scale laid down by the display itself. It has the following highlights:

- Data is transferred on a 4- or 8-bit data bus, determined by the user. Data may be instruction or character information. Using the 4-bit mode allows the whole interface to be contained within just 7 bits, but the data transfer process is a little slower. Bit 7 of the data bus doubles as a 'busy flag', indicating whether the device is ready to accept new data. This is very important, as many operations of the HD44780 take finite time and must be completed before another instruction can be accepted.

- Control is exercised by three control lines:
    - Register Select (RS), which determines whether an instruction or character data is being transferred
    - Read/Write (R/$\overline{\text{W}}$), which determines data direction
    - Enable (E), which provides a clock function to synchronise data transfer.
- There is a simple instruction set, which allows control of operating characteristics – this includes initialising and clearing the display, and controlling the position and characteristics of the cursor.
- The user can access two registers, depending on the state of the RS line:
    - An instruction register, used to transfer instructions (RS = 0)
    - A data register, used to transfer display data, for example character codes (RS = 1).
- Internal resources include 80 bytes of display RAM and a character generator ROM.

On power-up, the HD44780 must undergo a very specific initialisation process. It is important to get this absolutely right, or any HD44780-driven display may just sit inactive. An instruction sequence is given in the data accompanying any commercially available display based on this interface. An example is reproduced in Ref. 8.3. The early initialisation instructions are independent of whether the display is in 4- or 8-bit mode and the busy flag is not initially available. Therefore, these initialisation instructions are usually separated by delay routines of appropriate duration, to ensure that one action is completed before the next is started.

The HD44780, constrained as it is by the timing requirements of the LCD itself, tends to operate at a slower speed than most microcontrollers. Interfacing with it therefore carries some unique timing problems. An example timing diagram, for an 8-bit interface, is shown in Figure 8.10. Every data transfer to the LCD controller is made by a pulse on the E line. With RS initially set low, the data placed by the microcontroller on the data bus is interpreted as an instruction, which the HD44780 receives and starts to execute. The microcontroller needs to know when that instruction has been completed. R/$\overline{\text{W}}$ is therefore taken high, so on the next cycle of E the LCD controller outputs to the data bus a word made of the busy flag as MSB, with lower bits made up of the internal RAM address counter. No further data can be sent to the LCD controller until the busy flag is cleared, so it is checked repeatedly. When it goes low, RS in this example is taken high, R/$\overline{\text{W}}$ is taken low and the next data transfer is therefore a character code.

8.4.2　Design example: use of LCD display in Derbot hand controller

The Derbot hand controller, in its LCD version, uses a Powertip PC0802-A display [Ref. 8.4]. This display has two lines, each of eight characters. It is controlled by the S6A0069 LCD driver microcontroller, which has the features of the HD44780 just described. It is used in 8-bit interface mode. The full circuit diagram is given in Figure A3.2, with the LCD-only detail given in Figure 8.11.

The program used to demonstrate the LCD display is **keypad_test**, which has already been quoted in Program Example 8.1. When a keypad key is pressed, this program identifies it and transfers the key

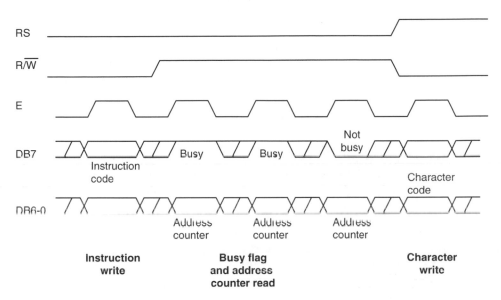

Figure 8.10 HD44780 timing diagram, 8-bit interface

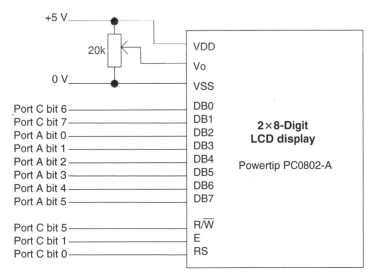

Figure 8.11 Connections to the Powertip PC0802 LCD display, showing connections for the Derbot hand controller

number, in ASCII, to the LCD. The subroutines to write an instruction or character code to the display controller, **lcd_write**, and to check the busy flag, **busy_check**, are shown in Program Example 8.3. The initialisation process can be seen by checking the full program listing on the book CD.

The **lcd_write** subroutine starts with a call to the **busy_check** subroutine, so that no attempt is made to write to the display controller unless it is ready to receive. The subroutine then sets the R/$\overline{\text{W}}$ line low, to indicate a write process is about to occur. The subroutine is rendered a little more complex as the hardware design splits the 8-bit bus to the LCD controller across two ports (Figure 8.11), so two shifts right must be undertaken. Once the correct data value is set up on the port, the enable line, labelled **lcd_E**, is pulsed high to complete the transfer.

The **busy_check** routine first sets Port A to input. The value of RS, which can be in either logic state, is saved. This line is then set low, and R/$\overline{\text{W}}$ set high, which sets the condition for a read of the busy flag. The E line is then strobed high and a test of the busy flag occurs. The subroutine loops until the busy flag is cleared.

```
;Waits until busy clear, and writes word held in lcd_op to display.
;RS must be preset to required value, this status is preserved.
lcd_write call busy_check
        bcf portc,lcd_rw
        bcf status,c
        rrf lcd_op,1     ;form output bits, op word sits
;across ports a & c
        bcf portc,6 ;set value of bit 0 of bus
        btfsc    status,c
        bsf portc,6
        bcf status,c
        rrf lcd_op,1
        bcf portc,7 ;set value of bit 1 of bus
        btfsc    status,c
        bsf portc,7
        movf     lcd_op,0
        movwf    porta
        bsf portc,lcd_E
        bcf portc,lcd_E
        return
;
;Test Busy Flag, and wait till cleared
busy_check bsf status,rp0     ;select memory bank 1
        movlw B'00111111'     ;set port A all ip
        movwf    trisa
        bcf status,rp0
        bcf flags,0
        btfsc    portc,lcd_RS ;save RS bit in flags, 0
        bsf flags,0
        bcf portc,lcd_RS     ;access instruction register
        bsf portc,lcd_RW     ;set to read
```

Program Example 8.3 Keypad Test Program: LCD Drive Subroutines

```
busy_loop    bcf portc,lcd_E
       bsf portc,lcd_E
       btfsc    porta,lcd_busy ;test the busy flag, loop if still busy
       goto    busy_loop
       bcf portc,lcd_E
       bsf status,rp0    ;select memory bank 1
       movlw    B'00000000';set port A all op,
       movwf    trisa
       bcf status,rp0
       bcf portc,lcd_RS
       btfsc    flags,0 ;reinstate RS bit
       bsf portc,lcd_RS
       return
```

Program Example 8.3 Continued

Liquid crystal displays such as the one just described are enormously useful in the world of small to medium embedded systems. They are low power and comparatively flexible in their use. On the downside, interfacing to them can be tiresome, with not insignificant blocks of code required just to transfer simple messages. Because their interface is slow, they can become a limiting factor in high-speed systems.

8.5   The main idea – interfacing to the physical world

Whether or not the embedded microcontroller has a human interface, it will certainly interface with the physical world. To do this it must be able to detect the state of physical variables and it must be able to control those variables. This interaction with the physical world is done by means of transducers, a major field of study in themselves. *Input transducers*, also called sensors, detect and convert physical variables into electrical variables. Examples include light or temperature sensors, or sensors which detect physical position, including measurement of distance or rotary displacement. Output transducers convert electrical variables to physical. Ones which cause physical movement, our main interest here, are also called actuators. Examples include solenoids and motors.

In our study of embedded systems, we need knowledge of what transducers are available, what they can do and how we can interface to them. In the remaining part of this chapter, therefore, some interfacing techniques essential to embedded systems are introduced. The transducers used in the Derbot AGV are also introduced. While this might seem an arbitrary choice, they are as good a selection as any – it would be impossible in this book to attempt to undertake a comprehensive survey of all transducers.

8.6   Some simple sensors

There is an enormous range of sensors available today, some with a long history and others based on very recent technology. These include 'smart' or 'intelligent' sensors, which are integrated onto an IC and have on-chip signal processing. All are based on one or other physical phenomenon that leads to conversion of physical variable to electrical, sometimes via an intermediate variable. The sensors introduced here include electromechanical, optical and ultrasonic. Some are shown in Figure 8.12.

Figure 8.12 Some of Derbot's sensors and actuators – an ultrasonic distance sensor mounted on a Futaba servo, behind a light-dependent resistor and microswitch

8.6.1   The microswitch

The microswitch has been the mainstay of mechanical position sensing over many years, and is likely to retain that position in years to come. Usually, it is in the form of a single-pole, double-throw switch, with a lever or roller for actuation. Microswitches are available from the sub-miniature to the large and rugged, for heavy-duty industrial applications. Electrically they can be interfaced just like a normal toggle switch, using one of the circuits shown in Figure 3.7. In industrial applications further precautions may be taken to minimise electrical interference. On the Derbot two microswitches are used as 'bump' sensors, one of which is seen in Figure 8.12.

8.6.2   Light-dependent resistors

A light-dependent resistor (LDR) is made from a piece of exposed semiconductor material. When light falls on it, it creates hole–electron pairs in the material, which improve the conductivity. When light is removed, the hole–electron pairs recombine and conductivity falls. The overall effect is that as illumination increases, the LDR resistance falls.

The LDR used in the Derbot is the NORP12, made by Silonex [Ref. 8.5], with a resistance when completely dark of around $20\,\text{M}\Omega$, falling to a few hundred ohms when very brightly illuminated. It can be connected

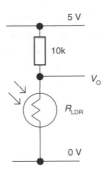

Illumination (lux)	R_{LDR} (Ω)	V_o
Dark	2M	5.00
10	9k	2.36
1000	400	0.19

Figure 8.13 The NORP12 LDR connected in a potential divider, with indicative output values

in a simple potential divider to give a voltage output, as shown in Figure 8.13. The Derbot has three of these sensors, which it uses when in light-seeking mode. These can be seen in Figure A3.1. They are connected to the 16F873A analog-to-digital converter inputs described in Chapter 11.

8.6.3   Optical object sensing

Optical methods are very useful in sensing objects and surfaces. In one configuration the presence of an object can be sensed if it breaks a light beam, in another if it reflects the beam. Many sensors are available with both light source and sensor integrated into the same package. The Derbot AGV uses reflective opto-sensors made by Optek, type OPB608A [Ref. 8.6].

The principle of this sensor is illustrated in Figure 8.14(a). The sensor consists of an infrared LED and phototransistor mounted side by side in the same plastic package. The package material allows infrared light to pass, but filters ambient visible light. When a reflective surface is placed at a suitable distance in front of the sensor, some of the emitted light is reflected back to the phototransistor, which then conducts. If the sensor is connected in the circuit of Figure 8.14(b), then the circuit output V_O is at Logic 1 when no reflection occurs and goes to Logic 0 if a reflective surface is present. Resistor values are dependent on sensor characteristics, the ones given here being indicative. The distance from sensor to reflective surface is critical in many such sensors, with preferred distances around 3 mm being common.

8.6.4   The opto-sensor applied as a shaft encoder

In the Derbot, the reflective opto-sensors just described are used to create very simple shaft encoders, as can be seen in Figure 8.15. A card with a simple black/white pattern (Figure A3.4) is fixed to the wheel, with the sensor face positioned around 3 mm away from it. As the wheel rotates, the sensor produces an approximate square wave, with logic high every time a black section of the pattern goes by. (An actual waveform is shown in Figure 8.20, where the conditioning requirements of the signal are investigated.) If these pulses are counted, they can be used as the basis for distance measurement, or odometry, in the AGV. This is developed further in later chapters. It should be mentioned that the handmade shaft encoder developed here is very crude compared to commercially available units. Whereas the Derbot shaft encoder generates 16 pulses per revolution, a commercial unit can generate hundreds of cycles, giving a far improved resolution.

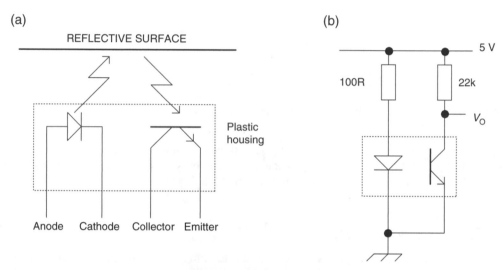

Figure 8.14 The reflective optical sensor. (a) Principle of operation. (b) Electrical connection

Figure 8.15 A reflective opto-sensor used as a shaft encoder on the Derbot

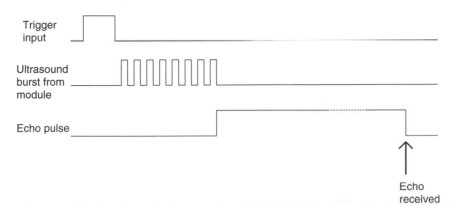

Figure 8.16 Simplified timing diagram for SRF04 ultrasonic ranger

8.6.5   Ultrasonic object sensor

Ultrasound is widely used for sensing and measurement, from simple distance measurement to complex medical imaging. The Derbot AGV uses an ultrasonic reflective sensor to detect obstacles in its path or to allow it to run parallel to a wall. The sensor, a Devantech SRF04 [Ref. 8.7], is seen in Figure 8.12.

The sensor consists of a transmitter and receiver and, to the extent that it is based on a reflective principle, is initially similar to the reflective opto-sensor. The big difference lies in the fact that the ultrasound source is pulsed and the time taken for the echo to return is measured; from this a distance can be calculated. The timing diagram of the sensor is shown in Figure 8.16. A logic pulse is input to the module trigger input. This causes an eight-cycle ultrasonic burst to be generated. The echo output of the module then goes high and remains high until an echo is detected, at which point it goes low. If the duration of the pulse is measured, then the distance away of the object that caused the reflection can be calculated.

8.7   More on digital input

If a microcontroller is to receive logic signals, then it is essential that those signals are at voltage levels which are recognised by it as being either Logic 0 or Logic 1. These voltage levels are usually defined by logic family, for example TTL (Transistor Transistor Logic) or CMOS (Complementary Metal Oxide Semiconductor). When one device is connected to another, and each is supplied by the same voltage and is of the same logic family, then it is usually safe to assume that logic levels will be safely and reliably transferred. However, if signals are generated from a non-logic source, e.g. a sensor, or if they have been received over a long communication link, or have been subject to interference, then it may be that they are not correctly interpreted by the receiver.

8.7.1   16F873A input characteristics

To determine whether a signal will be properly received by a logic device, it is first necessary to understand its input characteristics. The characteristics for a 16F873A port bit, taken from Ref. 7.1, are shown

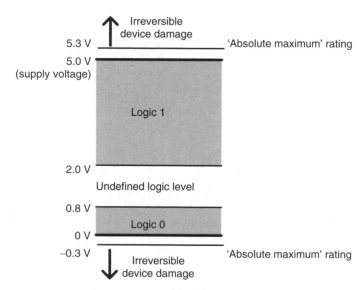

Figure 8.17 Port bit input voltage levels, 5 V supply

diagrammatically in Figure 8.17. From this it can be seen that any input voltage lying between 0 and 0.8 V is interpreted as a Logic 0, and any lying between 2 and 5 V is interpreted as a Logic 1. An input voltage lying between these two regions is not defined.

If the input voltage exceeds 5 V, then there is a danger of damage to the device. Logic inputs, however, almost invariably have internal protection diodes, one connected from input to ground, the other from input to supply rail. This arrangement can be seen by looking forward to Figure 8.19(a). Both diodes are connected so that in normal operation they are reverse-biased. However, if the input voltage exceeds the supply rail by a voltage adequate to cause diode conduction, then the diode connected between them will start to conduct and the input voltage will be clamped at that voltage. Similarly, if the input falls sufficiently below 0 V, then the other diode will start to conduct. This mechanism offers some protection to the input circuit.

For the 16F873A, input protection diodes come into action at +5.3 or −0.3 V. These do not, however, have limitless capability; the maximum input clamp current is specified as ±20 mA. If the absolute maximum voltages are significantly exceeded, then damage will occur, probably starting with the destruction of a protection diode.

8.7.2 Ensuring legal logic levels, and input protection

It is up to the designer to ensure that the input voltage is only ever steady state in one of the recognised logic levels, i.e. one of the shaded zones of Figure 8.17. It can pass quickly through the intermediate undefined zone, but must not linger there. It must never exceed the maximum ratings. It will meet these conditions if signals are generated locally, by another logic device of the same family. If, however, the signal has been

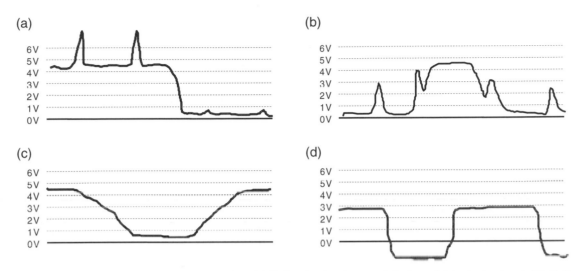

Figure 8.18 Different forms of signal corruption. (a) Spikes in signal, potentially harmful to device input. (b) Spikes in signal. (c) Excessively slow edges. (d) DC offset in signal

transmitted over a long distance, maybe from a remote sensor, or there is interference, or if it was never a proper logic signal in the first place, then problems may arise.

Figure 8.18 shows sketches of a number of forms that a corrupted signal can take. Signal (a) has acquired positive-going spikes, which are potentially damaging to the input circuit. Signal (b) has similarly acquired voltage spikes. While these are not at a level that will damage the input circuit, they may well lead to misleading results, particularly if the signal is to be input to a counter, or is a serial clock or data signal. Signal (c) has very slow edges, perhaps due to the filtering effect of a long cable or because the source is from a sensor. Finally, signal (d) looks like a reasonably healthy logic signal, but it has picked up a voltage offset. This could be due to long-distance transmission, with a voltage differential between the earth references at transmitter and receiver.

A variety of techniques are available to try to correct problems like these, and to ensure legal logic levels. These are found in any good electronics textbook. Figure 8.19 illustrates three of them.

Clamping voltage spikes – with current-limiting resistor

It has been mentioned that if too much current flows in the protection diodes, they will themselves burn out. Therefore, if there is a chance of the protection diodes being invoked, then it is worth including a series resistor to limit the current that may flow in them. This is shown in Figure 8.19(a). Such a method would correct the corrupted signal of Figure 8.18(a), if the magnitude of the spikes was not too great.

Suppose in Figure 8.18(a) that the maximum permissible diode current was 20 mA, R_{prot} was 1 kΩ and the upper protection diode started to conduct when the input voltage was 5.3 V. Then the maximum permissible voltage spike would have a peak value of [(20 mA × 1 kΩ) + 5.3], or around 25 V.

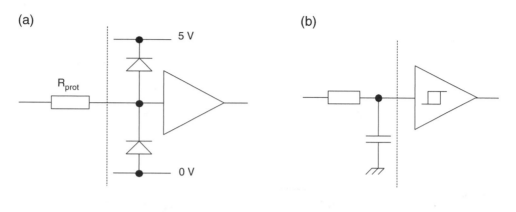

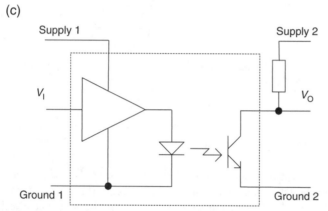

Figure 8.19 Some simple methods to condition a digital input. (a) Invoking protection diodes. (b) Filtering spikes with Schmitt trigger. (c) Isolation or level shifting with the opto-isolator

Schmitt trigger

The Schmitt trigger was described in Chapter 3. It provides an easy means of speeding up slow logic edges, such as are seen in Figure 8.18(c).

Analog input filtering

Sometimes a logic signal picks up interference that is not potentially damaging to the microcontroller, but can cause problems in the system operation. For example, if the signal is the input to a counter, then spurious counts will occur if there are voltage spikes in the signal. A simple RC filter, as seen in Figure 8.19(b), is sometimes enough to remove low-level interference. The edges of the signal, which will have been slowed by the filter, can be recovered by means of the Schmitt trigger. This approach could solve the problem seen in the signal of Figure 8.18(b).

Figure 8.20 shows the oscilloscope trace at the output of one of the reflective opto-sensors on the Derbot AGV. The connections to this run close to the motors, which being brushed DC motors are a rich source of

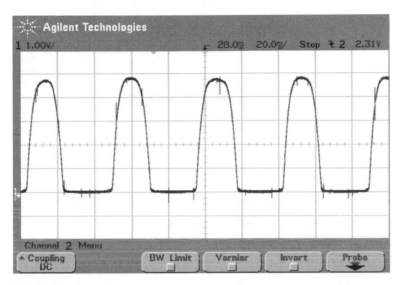

Figure 8.20 Signal from the Derbot reflective opto-sensor

interference. The signals are connected to the inputs of Timer 0 and Timer 1. While the signal does not look severely corrupted, it was found that the little voltage spikes that are present are quite enough to introduce spurious counts.

As the circuit diagram of Figure A3.1 shows, an RC filter having component values $R = 11K$, $C = 10n$ (i.e. cut-off frequency close to 1.4 kHz) is inserted in the signal path. This is well above the maximum frequency of 40 Hz expected from the shaft encoder, but proved adequate to remove the effects of the voltage spikes. The Schmitt trigger inputs of both timers then serve to correct the slow rate of change of the incoming signal.

Opto-isolation

The opto-isolator, as seen in Figure 8.19(c), is a very useful means of protecting a logic input, especially when a signal has been received over a distance. The incoming signal drives an LED, whose light output activates an opto-transistor. There is no electrical connection between input and output, so problems such as the earth differential, seen in Figure 8.18(d), can be resolved.

Digital input filtering

Many ICs designed to accept signals that may carry interference have some form of digital input filtering designed into their input circuit. This is the case with the 16F873A, as we shall see in Chapter 10 with the asynchronous serial input. A simple digital filtering strategy is to sample the input three times in succession and then use a 'majority vote' circuit to determine the logic value to be accepted. Thus, if two ones and one zero are detected, the Logic 1 will be accepted and the zero – possibly representing a glitch due to interference – discarded. The sampling must of course occur at a faster rate than any intended rate of genuine input change.

8.7.3   Switch debouncing

A particular problem with mechanical switches is the property that the switch contacts literally bounce as they close. This leads to a short period of time, generally less than 10 ms, when the switch state bounces between open and closed. When switching on a conventional electrical load, like a light, this is not a problem and is not even noticed. When connected to a digital circuit, particularly one that has a rapid response or is counting, the effect can be disastrous.

A number of standard ways exist to eliminate the effect of switch bounce. Hardware techniques, generally based on bistables or Schmitt triggers, can be found in any good electronics textbook and in Ref. 1.1. It is interesting to explore software techniques briefly here, as these can be implemented in a system without extra cost. Figure 8.21 shows two possibilities. In (a) and (b), a switched input is being polled (i.e. read periodically), with the polling period being greater than the period of switch bounce. If, as in (a), the switch change and bounce occurs between input polls, then a clean transition is perceived. If the input is polled as the switch bounce occurs, as in (b), then the logic value read is unpredictable. Either the previous logic state is detected and retained for another poll interval (dotted line) or the new logic state is detected (solid line), a value repeated at the next poll. In either case the transition is clean.

Sometimes the programmer does not want to poll an input, but may instead use it to force an interrupt. If the interrupt is caused by a single switch, changing to a known state (for example, always switching from high to low), then switch bounce is unlikely to cause a problem. The interrupt could, however, be caused by one of several switches, for example the keypad read described earlier in this chapter. If the switch states were read immediately the interrupt occurred, then the read would probably occur during the period of switch bounce. In this case a short programmed delay, following the first detection of change, can be introduced. Switch states can then be read at the end of the delay, when the bounce has completed.

8.8   Actuators: motors and servos

A common requirement in an embedded system is to cause physical movement. This is usually either linear, i.e. movement in a straight line, or rotary. Many of the actuators used to create these movements are electrical. Solenoids can be used for linear movement, 'servos' for angular only, and DC or stepper motors for angular or rotary. Other actuation methods, particularly for high forces, include pneumatic and hydraulic.

8.8.1   DC and stepper motors

DC and stepper motors are very widely used in embedded systems. While based on very different operating principles, they tend to compete for similar applications. Given the right operating environment, both can be used for continuous rotary motion or for precise angular displacement.

DC motors range from the extremely powerful to the very small. DC motors drive huge electric trains, but also drive tiny mechatronic systems. Their popularity is due to their wide, useful speed range, the ability

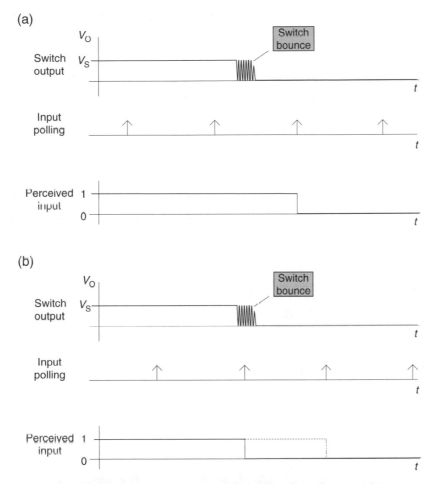

Figure 8.21 Eliminating switch bounce by polling. (a) Input poll misses bounce. (b) Input poll hits bounce

to control this speed and their potentially good efficiency. When used with a feedback potentiometer or shaft encoder, they can be used to provide accurate angular positioning. Small DC motors, such as are used in embedded systems, are usually of the permanent magnet type, i.e. the magnetic field within which the armature rotates is provided by one or more permanent magnets. This leads to one of the attractive features of the DC motor – its operating simplicity. Only the armature winding needs to be driven.

The big attraction of stepper motors is their ability to interface very directly with a digital system. Each digital pulse sent to a stepper controller can be used to advance the motor shaft position by a known angle. Therefore, in theory, a microprocessor or microcontrollers can control speed and angular position of the motor shaft to a high level of accuracy, without feedback. In practice this happy position is not completely achieved. Stepper motors have awkward start-up characteristics, show mechanical resonance in a particular speed range and lose torque at high speed, with a limited top speed. Any of these can cause the motor to lose synchronisation with its digital drive. On top of this, stepper motors tend to be less efficient and more complex to drive than the DC motor.

The choice between stepper and DC motor is not always an obvious one. In general, if precise and limited rotary motion is required, and power consumption is not of primary importance, then stepper motors usually win. If less precision is needed, and perhaps higher speed and/or higher efficiency, then a DC motor may be the choice.

The Derbot AGV needs controllable rotary actuation to drive its wheels. It steers by these, and needs to exercise speed and distance control. All these functions could be achieved by stepper motors. However, the simplicity of driving a DC motor, coupled with their good efficiency – essential for the Derbot battery operation – led to the choice of this type of motor. To effect control some form of feedback was necessary. For this the simple shaft encoder already described was implemented. The version of the Derbot described in this book uses a geared motor made by MFA/Como Drills [Ref. 8.8]. The motor is pictured in Figure 8.22, with basic data provided in Tables A3.1 and A3.2.

8.8.2   Angular positioning: the 'servo'

The 'servo' is a device that has claimed the generic title of servomechanism for its own! Widely used in radio-controlled modelling and robotics, it allows precise angular positioning. The servo output is a shaft that can take an angular position over a range of around 180°. The input to the servo is a pulse stream, generally of repetition rate 50 Hz (i.e. period of 20 ms). The *width* of the input pulse determines the angular position of the output shaft. In the example of Figure 8.23, a pulse width of 1.25 ms leads to an output shaft position of 0°, 1.5 ms to an output shaft position of 90° and 1.75 ms to an output shaft position of 180°. This is an example of pulse width modulation, which we shall meet in later chapters.

Figure 8.22 A Derbot geared motor MFA/Como RE280/1

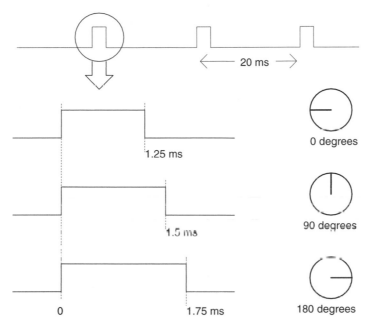

20 ms

0 degrees

1.25 ms

1.5 ms

90 degrees

0   1.75 ms   180 degrees

Figure 8.23 Servo input and output characteristics

In some of its configurations the Derbot uses the Futaba S3003 servo [Ref. 8.9] to rotate the ultrasonic sensor, thus allowing a single sensor to make measurements in different directions.

8.9 Interfacing to actuators

8.9.1 Simple DC switching

Only very small electrical loads, like LEDs, can be driven directly by a microcontroller port bit. Larger loads, drawing beyond 10 or 20 mA, or powered from a voltage higher than the logic supply voltage, need to be interfaced via power switching devices.

Transistor switches provide an easy way of switching DC loads. Figure 8.24 shows two types applied, MOSFET and bipolar, interfacing a controller port bit or logic gate output to a resistive load R_L. In both circuits a logic high from the microcontroller causes current to flow in the load. As with the Open Drain output shown in Figure 3.6, the load supply voltage V_S does not have to be the same as the microcontroller supply. In many cases it is greater, for example a microcontroller powered from 5 V can drive a load powered from 12 or 24 V.

A bipolar transistor (Figure 8.24(a)) requires a small base *current* to switch a much larger collector current. MOSFETs (Figure 8.24(b)) require a modest gate *voltage*, with negligible current, to switch a large drain current. As the MOS device is purely voltage controlled, its gate can be connected directly to the port bit output. This output must then just comfortably exceed the gate-to-source threshold voltage necessary for

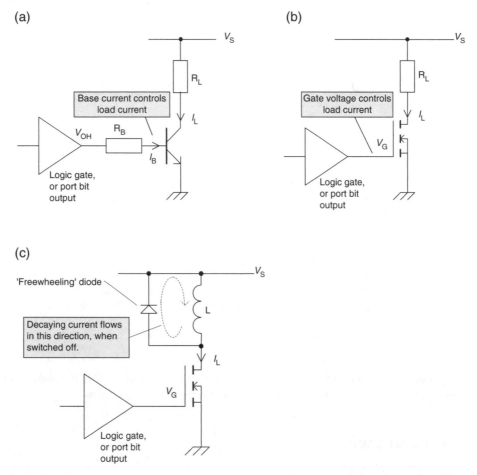

Figure 8.24 Transistor switching of DC loads. (a) Resistive, bipolar transistor. (b) Resistive, MOSFET. (c) Inductive, MOSFET

the MOSFET to switch on. This simplicity of connection makes the MOSFET a very attractive option for load switching in the microcontroller environment, as long as the threshold voltage just mentioned can be exceeded.

Special families of MOSFETs, designed for direct interface to logic levels, are available. Two examples, made by the company Zetex [Ref. 8.10], are shown in Table 8.1. For either transistor, it can be seen that if their gate-to-source voltage V_{GS} exceeds 3 V, then their drain-to-source resistance falls from near infinity to a low value. How low it goes depends on the type and internal construction. The ZVN4306A has the lower maximum 'on' resistance of 0.33 Ω. Its input capacitance is, however, 350 pF. The slightly lower-priced ZVN4206A has an 'on' resistance of 1.5 Ω, but a lesser input capacitance, of 100 pF. This input capacitance has to be charged by the drive circuit as it switches from Logic 0 to 1. For high current-capacity MOSFETS it becomes a significant factor, in many cases needing its own drive circuit.

Table 8.1 Characteristics of two popular logic-compatible MOSFETs

Characteristic	ZVN4206A	ZVN4306A
Maximum drain-to-source voltage, V_{DS} (V)	60	60
Maximum gate-to-source threshold, $V_{GS(th)}$ (V)	3	3
Maximum drain-to-source resistance when 'on', $R_{DS(on)}$ (Ω)	1.5	0.33
Maximum continuous drain current, I_D	600 mA	1.1 A
Maximum power dissipation (W)	0.7	1.1
Input capacitance (pF)	100	350

If the load is inductive, like a DC motor, the winding on a stepper motor, a solenoid or electromechanical relay, then special precautions must be taken. The inductance stores energy when current is flowing in its magnetic field. If the applied voltage is switched off, it is essential for a path to be provided for the current to decay to zero, by which process the inductance returns the stored energy to the circuit. This is normally done by the inclusion of a 'freewheeling' diode, as seen in Figure 8.24(c).

8.9.2 Simple switching on the Derbot

The Derbot AGV uses transistor switching in a number of places, both on the main AGV as well as on the hand controller, to switch loads on and off. Two of these are shown in Figure 8.25, taken from the main Derbot circuit diagram of Figure A3.1.

The piezo sounder is rated at 9 mA, 3–20 V. As such, it can be driven directly from the 5 V supply, so does not demand its own drive transistor. Nevertheless, to minimise loading on the microcontroller, it was

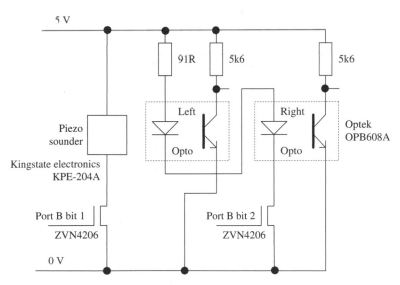

Figure 8.25 Section of Derbot circuit diagram – opto-sensors and piezo sounder

decided to include one. The opto-sensors tend to be a little more power hungry. To minimise overall current consumption, the two sensor LEDs are connected in series, as seen in the figure. Experimentally, the sensors were found to operate well if the resistor in series with the LEDs was 91 Ω. With (from device data) a forward voltage across each diode of around 1.7 V, this indicates a current of:

$$I = (5 - 3.4)/91$$

$$I = 17.6 \, \text{mA}$$

As with the piezo sounder, a PIC® microcontroller port output would be able to switch this, although it is closer to the 25 mA 'absolute maximum' limit. Again, however, it was decided to use transistor switching to minimise microcontroller loading.

8.9.3   Reversible switching: the H-bridge

As we have seen, it is easy to switch loads on and off, with the current always going in the same direction. Some loads, however, for example DC or stepper motors, need to have a reversible voltage applied, even if only a unipolar supply voltage is available. The way this is usually achieved is by a simple yet ingenious circuit connection called the H-bridge, shown in Figure 8.26.

In the H-bridge two pairs of switching devices, usually transistors, are connected between supply rail and 0 V. For simplicity the switching devices are shown in the diagram as switches, with a control input assumed to close the switch if the control is at Logic 1. The switch pairs are labelled A and B. Each pair has a 'high-side' and a 'low-side' switch. The load is connected between the two pairs to form an overall H configuration. Clearly, the switches in a pair must never be on at the same time or the supply will be shorted to ground. Therefore, it is common to drive them through a logic inverter, as shown, to ensure that only one can be on at any time.

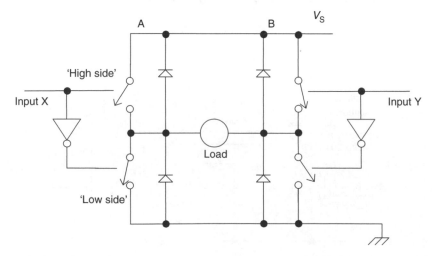

Figure 8.26 The principle of the H-bridge

If input X is switched low and input Y high, then the switch positions will be as shown in the diagram, with upper right and lower left closed, and the other two open. Current will then have a path through the high side of Pair B, through the load and through the low side of pair A. If the two inputs are reversed, then all switches will change state and current flows in the opposite direction. Reversible current drive has been achieved. If the load is inductive, freewheeling diodes must be connected as shown. This circuit, and derivations of it, is applied in many applications – from low power to very high power indeed.

A low-power, practical realisation of the H-bridge is available in the L293D IC, made by ST Microelectronics [Ref. 8.11]. This contains four half-bridges, so that two full H-bridges can be configured. A simplified diagram is shown in Figure 8.27. This is drawn so that each half-bridge is depicted as a logic buffer. This may not seem immediately obvious, but consider in the previous diagram that when input X is high, then the voltage output from transistor pair A is high, or low when input X is low.

Buffers 1 and 2 can form one complete H-bridge, as can buffers 3 and 4. Two power supplies, V_{LS} and V_{OS}, are used. The former supplies the input logic, while V_{OS} supplies the bridge itself. These two supplies

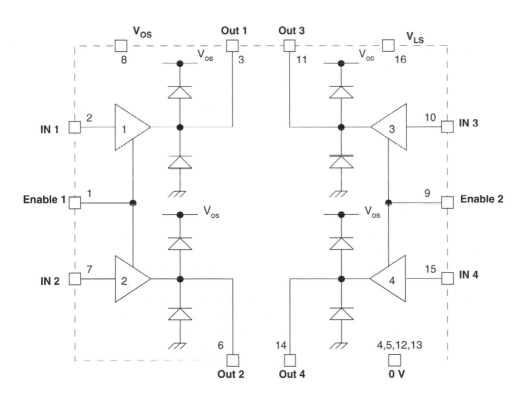

Operating Conditions – Highlights
600 mA output current per channel
Over temperature protection

1.2 A peak output current (non-repetitive) per channel
High noise immunity (Logic 0 input voltage to 1.5 V)

Figure 8.27 The L293D dual H-bridge

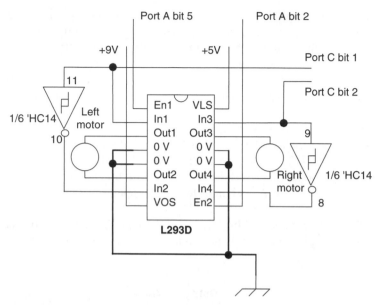

Figure 8.28 The L293D applied in the Derbot motor drive circuit

can be of different values, although V_{OS} must not be less than V_{LS}. The logic supply could, for example, be at 5 V, with the load supply at 12 V. An enable is provided to pairs of buffers, so that all switches in the bridge can be easily switched off. Internal freewheeling diodes are included. The L293D has four ground pins, which can be soldered to a copper plane on the PCB to provide a limited amount of heat sinking.

8.9.4   Motor switching on the Derbot

The Derbot AGV uses the L293D to drive its two motors, with the circuit detail shown in Figure 8.28. Instead of having separate drives to each logic input, it uses an inverter between each. Thus, the motor is always driven in one direction or the other. However, the enable lines are each driven from a port bit, so that each motor can be disabled. The logic supply is connected to the main 5 V supply, while the output supply is taken directly from the incoming battery voltage, meaning that 9 V is available for driving the motors. Internal voltage drops in the IC, however, reduce this to closer to 7 V. The enable lines each have a pull-down resistor of value 10 kΩ, connected between them and ground, not shown in this diagram (but seen in Appendix 3). These have the important function of disabling the motors when the microcontroller is not driving these lines, for example during initialisation.

8.10   Building up the Derbot

If you are building up the Derbot it is suggested that you now add the motors and their drive, based around the L293D, leading to the circuit shown in Figure 8.29. If you want a running AGV, then you will also need

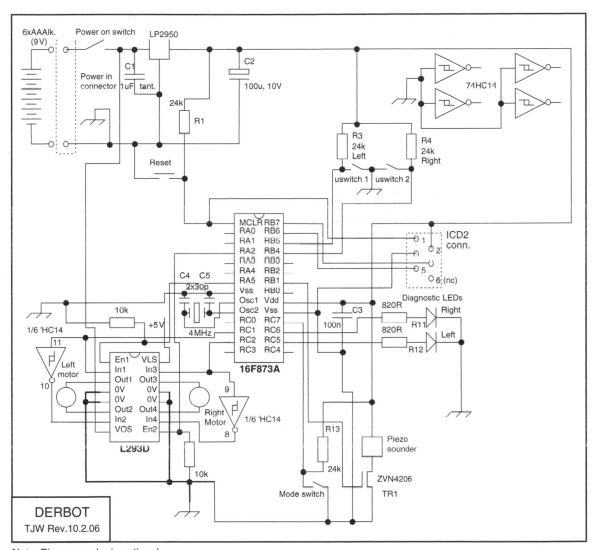

Note: Piezo sounder is optional.

Figure 8.29 Derbot intermediate build stage 2

to build the battery pack using the separate PCB and mount the spacing pillars. Further build guidance is given on the book CD.

You can also at this time build up a hand controller card, either in its LED or its LCD version. The programs used in the first sections of this chapter can then be downloaded and tested. These programs do not yet cause the hand controller to interact with the AGV; this is coming in later chapters.

8.11   Applying sensors and actuators – a 'blind' navigation Derbot program

With the Derbot build as just described, it is possible to run the program **Dbt_blind_Nav**, as found on the book CD, with the main features appearing in Program Example 8.4. This program sends the AGV running forward until it hits an obstacle, detected by its front microswitches. At this point it reverses, turns and runs forward again in a new direction.

The main program runs from the label **start,** by setting the motors running forward. It should be easy to do this simply by switching the motors on. However, they are chosen to give a good top speed, which is excessive for this simple application. Therefore, pulse width modulation is applied to set a slower speed. The detail of this is hidden in the subroutines **leftmot_fwd** and **rtmot_fwd**, which are called in turn. These are not included in the program example below, but can be found in the full book listing and will be fully explored in a later chapter. The program then enters a loop, at label **loop**. Here it repeatedly tests its microswitches and continues running until one or other is detected as being activated. When a microswitch is hit, the AGV stops, sounds the piezo sounder and reverses for 1.5 s approximately, again calling pulse width modulation subroutines to do this. It then turns, driving one motor forward while the other continues to reverse, before returning to running forward in the main program loop.

```
;****************************************************************
;Dbt_blind_Nav
;Derbot moves by "blind" navigation.
;Moves forward, and reverses and turns on bump.
;Fixed rate PWM applied to set reasonable speeds.
;
;TJW 5.5.05                              Tested 9.5.05
;****************************************************************
...
(Memory Allocation and Initialisation omitted)
...
;start motors
start call leftmot_fwd       ;sets left motor running forward
      call rtmot_fwd         ;sets right motor running forward
;test for bumps - reverse and turn if either microswitch closes
loop btfss  portb,us_rt      ;test right microswitch
     goto   rev_rt
     btfss  portb,us_left    ;test left microswitch
     goto   rev_left
     call   delay100
     goto   loop
;
rev_rt bsf  portc,led_rt
     bcf    porta,mot_en_left ;stop motors
     bcf    porta,mot_en_rt
     bsf portb,sounder       ;small bleep from sounder
     call delay200
     bcf portb,sounder
```

Program Example 8.4 Derbot blind navigation (section)

```
;reverse both motors
      call leftmot_rev
      call rtmot_rev
      call delay500
      call delay500
      call delay500
      call leftmot_fwd      ;left motor forward to turn
      call delay500
      call delay500
      bcf   portc,led_rt
      goto start
;
rev_left bsf portc,led_left
      bcf   porta,mot_en_rt ;stop motors
      bcf   porta,mot_en_left
      bsf portb,sounder   ;small bleep from sounder
      call delay200
      bcf portb,sounder
;reverse both motors
      call leftmot_rev
      call rtmot_rev
      call delay500
      call delay500
      call delay500
      call rtmot_fwd        ;right motor forward to turn
      call delay500
      call delay500
      bcf   portc,led_left
      goto start
...
(subroutines omitted)
...
```

Program Example 8.4 Continued

Summary

- An embedded microcontroller must be able to interface with the physical world and possibly the human world as well.
- Much human interfacing can be done with switches, keypads and displays.
- To interface with the physical world, the microcontroller must be able to interface with a range of transducers. The designer needs an understanding of the main sensors and actuators available, and must be ready to keep abreast of current technology in the field.
- Interfacing with sensors requires a reasonable knowledge of signal conditioning techniques.
- Interfacing with actuators requires a reasonable knowledge of power switching techniques.

References

8.1. 12.7 mm (0.5 inch) Single Digit Numeric Displays (2003). Kingbright, Spec. no. DSAD0006; http://www.kingbright.com.tw

8.2.   HD44780 Data. Available from a number of websites, including http://www.electronic-engineering.ch/microchip/index.html

8.3.   Interfacing PICmicro MCUs to an LCD Module (1997). Microchip Technology Inc., DS00587B.

8.4.   Powertip; http://www.powertip.com.tw/

8.5.   Silonex; http://www1.silonex.com/

8.6.   Optek; http://www.optekinc.com/

8.7.   Devantech. SRF04 data available on http://www.robot-electronics.co.uk/ and other supplier sites.

8.8.   MFA/Como Drills; http://www.comodrills.com/

8.9.   Futaba; http://www.futaba-rc.com/

8.10. Zetex; http://www.zetex.com/

8.11. STMicroelectronics; http://www.st.com/stonline/

9
Taking timing further

We began to see in Chapter 6 how important counting and timing are in the embedded environment. We also saw how easy it is to count digitally and to convert that counting ability to an ability to measure time. We need now to take this capability much further, especially in the timing arena. Once we have good tools for timing, we can use them to underpin other microcontroller functions, like the generation of serial data or of pulse width modulation. We can also use those timing tools to facilitate complex external activity, generating, for example, the timing signals for an engine management system.

This chapter will explore counting and timing needs in the embedded environment, and develop capabilities to meet those needs, to a comparatively sophisticated level. While counting remains the underlying technique, it is its timing function that emerges as the predominant activity.

A central theme of this chapter will be an exploration of enhanced counter/timer structures. This will lead to the microcontroller being able to hand over much of its time-based activity to this hardware. Program execution can then continue, doing other things, while the hardware looks after timing issues. This activity will include:

- Maintaining continuous counting functions
- Recording ('capturing') in timer hardware the time an event occurs
- Using timer hardware to trigger events at particular times
- Setting up an environment whereby repetitive time-based events can be generated
- Measuring frequency, and hence physical variables which can be expressed as frequencies, for example motor speed.

The structures examined will be those of the 16F873A microcontroller. This will lead both to increased expertise in using this device and its relatives, as well as to an understanding of the underlying concepts, applicable to any microcontroller system. Many aspects will be illustrated with examples from the Derbot AGV. The necessary build for this is outlined in the final section of this chapter.

Note that Microchip tend to use the terminology 'timer' when referring to their counter/timer modules. In this chapter, 'timer' and 'counter/timer' will be used more or less interchangeably.

9.1   The main ideas – taking counting and timing further

Building on the material of Chapter 6, we need now to move on to more advanced applications of counting and timing, as shown in the bullet-point list above. The ability to do these things depends very much on extending the basic counter/timer hardware. Therefore, in this chapter each new counting or timing technique will be introduced alongside a description of the hardware that enables it to happen.

The Timer 0 of the PIC® 16 Series was introduced in Section 6.3 of Chapter 6. Take a moment to look at Figure 6.8 and remind yourself of its principle features, if there is any chance you have forgotten. At the heart is a digital counter, which is memory mapped, and can be written to and read from. The clock input to the counter can be selected from two sources. Either an external input can be chosen, in which case the module is often viewed as being in 'counter' mode. Alternatively, the internal oscillator signal can be connected to the counter, in which case it is viewed as being in 'timer' mode. A prescaler can be applied, which divides down the frequency of the incoming clock signal. When the counter counts up to its maximum value and then overflows to zero, an interrupt can be generated.

The PIC 16F873A has three timers, with some important accessories, which we now explore. The comparatively simple structure just reviewed forms the basis of the more advanced designs that we shall see.

9.2   The 16F87XA Timer 0 and Timer 1

9.2.1   Timer 0

The 16F873A uses the standard mid-range Timer 0 module. This is therefore the same counter/timer design as used by the 16F84A. Hence the description of Chapter 6, Section 6.3.3 fully applies.

9.2.2   Timer 1

The PIC mid-range Timer 0 is limited by being only 8 bits. Timer 1, shown in Figure 9.1, builds directly on the concepts of Timer 0, but has a number of important differences. First of all, it is 16-bit. As the figure shows, it is made up of two 8-bit registers, **TMR1H** and **TMR1L**. These are Special Function Registers (SFRs), which in the usual way are readable and writeable. They can be seen in the register file map

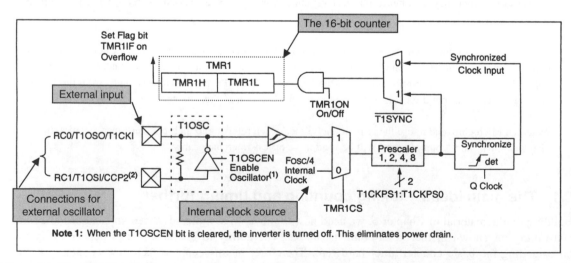

Figure 9.1 The 16F87XA Timer 1 block diagram (supplementary labels in shaded boxes added by the author)

of Figure 7.6. Together, they can count from 0000 to FFFF$_H$, or 65536$_D$. When the count rolls over from FFFF$_H$ back to 0, the interrupt flag **TMR1IF** (seen in the interrupt structure diagram of Figure 7.10) is set. This can be enabled to form an interrupt on overflow.

Timer 1 is controlled by the **T1CON** register, shown in Figure 9.2. The timer is switched on and off with bit **TMR1ON**. The timer has three distinct clock sources, which can be traced on Figure 9.1. For counting functions, the external input **T1CKI**, shared with pin 0 of Port C, must be used. For timing functions, there remains the option of the internal clock oscillator, Fosc/4. Selection between these two is made by the **TMR1CS** bit in the control register. A third possible clock source is made available due to the possibility of setting up a dedicated Timer 1 external oscillator, connected to the two external pins shown. This removes the dependence of working with the main oscillator frequency. The external oscillator can operate at a frequency entirely distinct from the main oscillator, and can also run when the main one is shut down in Sleep mode. The external oscillator is enabled with the **T1OSCEN** bit. The oscillator input for Timer 1 is identical to the main LP oscillator (Section 3.5.3 of Chapter 3). It is intended for lower frequency oscillation, up to around 200 kHz. Typically it is used with a 32.768 kHz crystal, which can be divided down to provide a one-second time base.

U-0	U-0	R/W-0	R/W-0	R/W-0	R/W-0	R/W-0	R/W-0
—	—	T1CKPS1	T1CKPS0	T1OSCEN	T1SYNC	TMR1CS	TMR1ON

bit 7   bit 0

bit 7-6   **Unimplemented:** Read as '0'

bit 5-4   **T1CKPS1:T1CKPS0:** Timer1 Input Clock Prescale Select bits

                  11 = 1:8 prescale value
                  10 = 1:4 prescale value
                  01 = 1:2 prescale value
                  00 = 1:1 prescale value

bit 3   **T1OSCEN:** Timer1 Oscillator Enable Control bit

                  1 = Oscillator is enabled
                  0 = Oscillator is shut-off (the oscillator inverter is turned off to eliminate power drain)

bit 2   **T1SYNC:** Timer1 External Clock Input Synchronization Control bit

                  When TMR1CS = 1:
                  1 = Do not synchronize external clock input
                  0 = Synchronize external clock input

                  When TMR1CS = 0:
                  This bit is ignored. Timer1 uses the internal clock when TMR1CS = 0.

bit 1   **TMR1CS:** Timer1 Clock Source Select bit

                  1 = External clock from pin RC0/T1OSO/T1CKI (on the rising edge)
                  0 = Internal clock (Fosc/4)

bit 0   **TMR1ON:** Timer1 On bit

                  1 = Enables Timer1
                  0 = Stops Timer1

       Note: In the 18 Series register of this name, bit 7 is called **RD16.** If set to 1 it enables the '16-bit Read/Write' mode

Figure 9.2 The Timer 1 control register, **T1CON** (address 10$_H$)

Whichever clock source is chosen, there is the option of applying the prescaler. With only two control bits, **TICKPS1** and **TICKPS0**, this does not have the range of scaling options offered by Timer 0, with only three effective division values, of 2, 4 and 8. Finally, synchronisation of the external input is enabled by bit $\overline{\text{T1SYNC}}$. The timer must run synchronously if either the Capture or Compare modes, described in the coming sections, are to be used. If operating asynchronously, however, it can continue to run when the microcontroller is in Sleep mode.

When the external clock source is chosen, the counter is always incremented on a rising edge (on Timer 0, a rising *or* falling edge can be chosen). However, the counter must first have a falling edge before it starts to count. There is a danger, therefore, that a first pulse might be lost. We will see in the Derbot program that follows how to get over this little problem.

9.2.3   Application of Timer 0 and Timer 1 as counters for Derbot odometry

Having suggested that timing is to be the dominant theme of this chapter, our first example is of counting! A fundamental need of any vehicle is an ability to measure how far it has travelled, a technique known as *odometry*. This is done in the Derbot by using its handmade optical shaft encoders, as described in Section 8.6.4 of Chapter 8. By counting pulses from the shaft encoder, and calculating from the wheel geometry the distance represented by each pulse, a measure of total distance travelled can be obtained.

Program Example 9.1 applies the shaft encoder to implement simple odometry with the Derbot. The outputs of the two optical sensors are connected on the Derbot PIC to the inputs of Timer 0 and Timer 1, as shown in Figure A3.1. Both of these are used in Counting mode. For the wheel diameters used in the prototype, and applying the shaft encoders described in Appendix 3, the program drives the Derbot forward for 1 m. It then completes a 180° turn on the spot and runs forward for 1 m again. The program loops continuously in this manner.

As before, to save space, the full program listing is not reproduced in the book, but can be found in the book CD. It is instructive to look at a number of features of the program. Port bits are set up in the normal way, to reflect Derbot bit usage. **ADCON1** controls whether Port A bits are to be used for analog input or digital input/output. It is initialised here for the final expected analog/digital distribution of Port A bit, even though the analog input is not used in this program.

By checking the Option register diagram in Figure 6.9, the Timer 0 settings described in the comment can be verified. Similarly, a look at Figure 9.2 should verify the setting for Timer 1.

At the program section **opto_move**, both timers are cleared to zero. The input to Timer 1 is, however, tested. If it is zero, then the timer value is incremented to one, as the first rising edge is not detected in the hardware. The two motors are then set running forward, using the same subroutines used in Program Example 8.4. Program execution then moves to the loop **opto_loop**. Here the timer values are continuously tested. A count of 91 represents a distance travelled of 1 m, for an encoder pattern of 16 black/white cycles per wheel revolution, as described in Appendix 3. When a timer value reaches 91, the respective motor enable bit is cleared to zero and the motor stops.

When both motors have stopped, the Derbot executes a turn on the spot, running the motors in opposite directions. The distance that each wheel must cover is calculated by applying the geometry described in Appendix 4, and found to equate to the count of 23 that is found in the program. In program execution, the arc travelled by each wheel in the turn is measured with the shaft encoder, and the motors stop running when the distance is complete. The AGV then runs the same fixed distance back to its starting point and the process continues.

```
;*****************************************************************
;odometry_test
;Runs forward a fixed distance, turns by 180 degrees,
;and returns - looping continuously.
;
;TJW 19.5.05                             Tested 20.5.06
;*****************************************************************
...
(comments, and memory definition omitted)
...
      org 00
;set up SFRs in Bank 1
...
(set up Tris A, B, C)
...
      movlw  B'01000100'
      movwf  adcon1 ;set port A for right analog/digital mix
      movlw  B'11101000' ;set up Timer 0: external input, low to high
                                  ;transition,no prescale
      movwf  option_reg
      movlw  D'250'       ;set PWM prd
      movwf  pr2
;set up SFRs in Bank 0
      bcf    status,rp0   ;select bank 0
      movlw  B'00000011' ;set up Timer 1: no prescale, oscillator
      movwf  t1con                 ;disabled, external sync input
...
further initialisation, and opening section of program
...
;*********************************************
;run forward fixed distance, then turn and return
;*********************************************
opto_move clrf tmr0          ;clear timers
      clrf   tmr1l
      clrf   tmr1h
      clrf   flags
      btfss  portc,0        ;increment T1 if ip is zero, as first rising edge
                                          ;isn't detected
      incf   tmr1l
      call   leftmot_fwd    ;start motors running
      call   rtmot_fwd
opto_loop call opto_to_led ;transfer opto states to diagnostic leds
;move forward set distance (1m)
      movlw  D'91'          ;test if counter has reached this value
```

Program Example 9.1 Application of Odometry with the Derbot

```
        subwf   tmr0,0
        btfsc   status,z
        bcf     porta,mot_en_left  ;disable motor if value reached
        movlw   D'91'
        subwf   tmr11,0
        btfsc   status,z
        bcf     porta,mot_en_rt
;if both motors stopped, proceed to turn, otherwise loop
        btfsc   porta,mot_en_left
        goto    opto_loop
        btfsc   porta,mot_en_rt
        goto    opto_loop
;now turn
        call    delay500       ;ensure AGV is at rest
        movlw   00
        movwf   tmr0           ;clear timers and flags
        movwf   tmr11
        btfss   portc,0        ;increment T1 if it is zero,
        incf    tmr11
        call    leftmot_fwd    ;turn on spot, left motor forward, right back
        call    rtmot_rev
;execute the turn
opt_loop1 call opto_to_led ;transfer opto states to diagnostic leds
;rotate by 180 degrees
        movlw   D'23'          ;test if counter has reached this value
        subwf   tmr0,0
        btfsc   status,z
        bcf     porta,mot_en_left  ;disable motor if value reached
        movlw   D'23'
        subwf   tmr11,0
        btfsc   status,z
        bcf     porta,mot_en_rt
;if both motors stopped, proceed to straight line, otherwise loop
        btfsc   porta,mot_en_left
        goto    opt_loop1
        btfsc   porta,mot_en_rt
        goto    opt_loop1
        call    delay500       ;ensure we're at rest
        goto    opto_move
;************************************************
;SUBROUTINES
;************************************************
...
motor control and delay subroutines
...
;transfers opto sensor state to leds
opto_to_led bcf portc,led_left   ;preclear left led
        btfss   porta,4
        bsf     portc,led_left     ;but set it if opto on
        bcf     portc,led_rt ;preclear right led
        btfss   portc,0
```

Program Example 9.1 Continued

```
        bsf    portc,led_rt   ;but set it if opto on
        return
        end
```

Program Example 9.1 Continued

This program demonstrates the application of counting to odometry, and also its weaknesses. The resolution of the distance measurement using this simple home-made shaft encoder is poor and suitable for demonstration purposes only – interesting though these are. The AGV can execute only approximate 180° turns and an approximate return home. If it is allowed to continue its back and forth trajectory, its return becomes increasingly approximate, as a cumulative error builds up. Moreover, there is no speed control on the motors. Although driven from the same voltage, any two motors are rarely identical and the AGV tends to run in a curve, albeit slight. We will be able to resolve this problem when speed measurement is implemented later in this chapter, and speed control in Chapter 11.

9.2.4 Using Timer 0 and Timer 1 to generate repetitive interrupts

An important use of a counter/timer is often for it to generate a continuously running series of accurately timed interrupts. These can be used for a host of timing applications, including forming the basis of a sophisticated programming framework known as a *Real Time Operating System* – but that's the business of Chapter 18! Sometimes such a series of interrupts is called a *clock tick*. This can be a little misleading, as we already have a clock oscillator and now we are talking about something different. However, such a clock tick can become almost as fundamental as the microcontroller clock oscillator itself, and can be used as the basis for almost as many time-based operations.

Timer 0 and Timer 1 can readily be used to generate such a clock tick, as both can produce an interrupt on overflow. The principle is illustrated in Figure 9.3. The timer is set to run continuously, in Timer mode, with interrupts enabled. Its value is represented by the stepped waveform in the diagram. It repeatedly counts up to its maximum value and overflows back to zero. Every time it does this, an interrupt is generated.

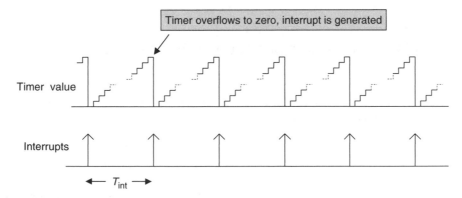

Figure 9.3 Generating a 'clock tick' – a repetitive interrupt stream

In general, if the timer is n-bit, then it will count 2^n cycles, from one zero to the next. The time between the interrupts, T_{int}, is then given by:

$$T_{int} = 2^n \times t_{clk},$$

where t_{clk} is the period of the clock input to the timer.

Design Example 9.1

Assuming an oscillator frequency of 4 MHz, what is the slowest 'clock tick' rate that can be obtained from Timer 0 and Timer 1? What is the next fastest in each case?

For Timer 0, refer to Figures 6.8 and 6.9. To achieve the slowest interrupt rate, the prescaler is set to $\div 256$. The input clock frequency to the timer itself is therefore 1 MHz/256, or 3.906 kHz. The action of the 8-bit timer itself is to divide this frequency by 256 to produce the clock tick frequency, which will be 3.906 kHz/256, or 15.26 Hz. The next frequency up would be if the prescaler were set to $\div 128$. In this case, the interrupt rate would be 30.52 Hz.

For Timer 1, refer to Figures 9.1 and 9.2. The maximum prescale rate is $\div 8$. Using this, the clock input to the timer itself will be 125 kHz. The action of the timer is to divide this by 2^{16}, leading to an interrupt frequency of 1.91 Hz. The next highest frequency would be if the prescaler were set to $\div 4$, leading to an interrupt frequency of 3.81 Hz.

It is interesting to see that there is not a major difference between the slowest rates of each timer. Although Timer 0 is only 8-bit, the fact that it has a very effective prescaler compensates in this sort of application for its smaller size.

An example of interrupt on overflow appears in Program Example 9.3.

9.3 The 16F87XA Timer 2, comparator and PR2 register

9.3.1 Timer 2

The 16F87XA Timer 2 is a simple 8-bit device. It is shown in block diagram form in Figure 9.4. Only two of the elements in the diagram, the **TMR2** register and the prescaler, actually relate to the conventional timer function. Timer 2 is a pure timer, so is driven only from the internal oscillator, shown at the right of the diagram. No external input is possible. The 8-bit Timer register is readable and writeable, and can be seen at memory location 11_H in Figure 7.6. Modest prescaling opportunities are possible.

The control register for Timer 2, **T2CON**, appears in Figure 9.5. The explanations in the figure give a good description of the role of each bit.

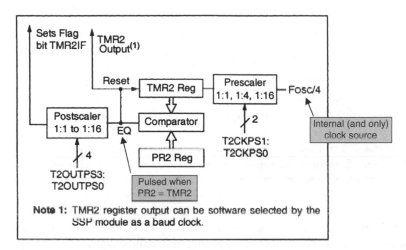

Figure 9.4 The 16F87XA Timer 2 block diagram (supplementary labels in shaded boxes added by the author)

U-0	R/W-0	R/W-0	R/W-0	R/W-0	R/W-0	R/W-0	R/W-0
—	TOUTPS3	TOUTPS2	TOUTPS1	TOUTPS0	TMR2ON	T2CKPS1	T2CKPS0

bit 7   bit 0

bit 7   **Unimplemented:** Read as '0'

bit 0-3   **TOUTPS3:TOUTPS0**. Timer2 Output Postscale Select bits

                        0000 = 1:1 postscale

                        0001 = 1:2 postscale

                        0010 = 1:3 postscale

                        •

                        •

                        •

                        1111 = 1:16 postscale

bit 2   **TMR2ON:** Timer2 On bit

                        1 = Timer2 is on

                        0 = Timer2 is off

bit 1-0   **T2CKPS1:T2CKPS0**: Timer2 Clock Prescale Select bits

                        00 = Prescaler is 1

                        01 = Prescaler is 4

                        1x = Prescaler is 16

Figure 9.5 The Timer 2 control register, **T2CON**

Despite its seeming simplicity, Timer 2 begins to emerge as a powerful module when its add-on elements, described now, are brought into play. Its usefulness is increased further when the Capture and Compare register, with its pulse width modulation (PWM) capability, is introduced.

9.3.2 The PR2 register, comparator and postscaler

Timer 2 has a Period register **PR2**, at memory location 92_H in the memory map (Figure 7.6), which can be preset by the programmer. When the timer is running, its value is continuously compared with the **PR2** register by the comparator. When it reaches the value held in **PR2**, it resets to 0 on the next input cycle.

This process is illustrated in Figure 9.6. The **PR2** value is shown fixed, although it can be changed within the program. The Timer 2 value is now just represented as a sawtooth, although its underlying stepped nature (as shown in Figure 9.3) should still be recognised. It counts up to the **PR2** value. When they are equal, a reset is generated and Timer 2 is cleared to zero. This occurs on the increment cycle *after* the equality has been detected, so there are (**PR2** + 1) cycles between each reset. Figure 9.4 shows that this reset forms an output of the module, labelled 'TMR2 output'. This can be selected as a baud rate generator by the synchronous serial port (SSP), as described in the following chapter.

Design Example 9.2

The Derbot AGV has an oscillator frequency of 4 MHz. What is the slowest Timer 2 interrupt rate that it can generate?

To achieve the slowest rate, **PR2**, and post- and prescalers must be set to maximum. If the prescale is set to ÷16, then the frequency of the clock driving Timer 2 will be 1 MHz ÷ 16, or 62.5 kHz. If **PR2** is set to 255, then Timer 2 will be able to clock up to its maximum value (i.e. 255) before returning to 0. Therefore, the reset frequency will be 62.5/256 kHz, or 244.14 Hz. If this is further postscaled by 16, then the interrupt frequency will be 15.26 Hz. This is the slowest possible interrupt frequency for this source.

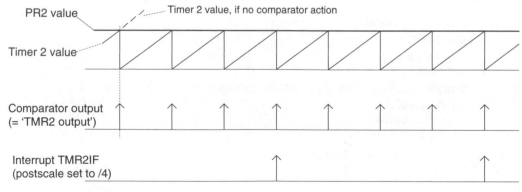

Figure 9.6 PR2 and comparator action

The series of resets just described also forms an input to the postscaler, whose output can be used as an interrupt source. The postscaler is controlled by the **TOUTPSX** bits of the control register. Notice that it can take any division value up to 16 – it is not just limited to binary powers. Figure 9.6 shows the case when the postscaler is dividing by four. The outcome of this is a stream of interrupts, whose frequency can be adjusted across a range of values.

9.4   The capture/compare/PWM (CCP) modules

9.4.1   A capture/compare/PWM overview

At the beginning of Chapter 6 it was suggested that embedded systems have timing needs which are equivalent in some ways to various timing needs in everyday life. The embedded system does need the equivalent of setting an alarm and of recording the time of an event. These and other requirements are neatly resolved by the addition of one or more registers to a basic counter/timer. A register that can record the time of an event is called a 'Capture' register. One that can generate an alarm does this by holding a preset value and comparing it with the value of a running timer (as we have seen already with **PR2**). The alarm occurs when the two are equal. Such registers are called 'Compare' registers. The PIC 16 Series microcontrollers combine these different functions, and more besides, in their capture/compare/PWM (CCP) modules.

The CCP modules are very versatile, and interact with both Timer 1 and Timer 2. The 16F873A has two such modules and they are well worth understanding. Each module has two 8-bit registers, called **CCPRxL** and **CCPRxH**, where x is 1 or 2. Together they form a 16-bit register that can be used for capture, compare or to form the duty cycle of a PWM stream. The CCP modules are controlled by their respective **CCPxCON** registers (Figure 9.7). It is the least significant 4 bits of these that determine the mode of operation of the module. It can be seen that they can be switched off, set for PWM or set for one of four possible configurations in either Capture or Compare mode. We will examine each operating mode in turn.

9.4.2   Capture mode

A Capture register operates something like a stopwatch. When an event occurs, the value indicated by a running clock is recorded. In a stopwatch, the watch then stops running. In a Capture register, the clock goes on running, but its value at the instant when the event occurred is recorded.

The CCP1 in Capture mode is shown in Figure 9.8. CCP2 will act the same as CCP1 in this mode, but will share its input with Port C bit 1. Because the input pin is shared with Port C, it must be set as an input in **TRISC**. An external signal, representing an 'event', is connected to the microcontroller pin RC2/CCP1. The purpose of the Capture mode is to record in **CCPR1L** and **CCPR1H** the value of Timer 1 when the event occurs. Further flexibility can be built in, as the external signal can be prescaled by 4 or 16, and rising or falling edge can be selected. The possible options can be seen in Figure 9.7. As well as causing a capture to occur, the occurrence of the event also causes an interrupt, represented by the **CCP1IF** flag in the interrupt structure diagram of Figure 7.10. Module CCP2 has an equivalent interrupt, represented by **CCP2IF**.

U-0	U-0	R/W-0	R/W-0	R/W-0	R/W-0	R/W-0	R/W-0
—	—	CCPxX	CCPxY	CCPxM3	CCPxM2	CCPxM1	CCPxM0

bit 7   bit 0

bit 7-6   **Unimplemented:** Read as '0'

bit 5-4   **CCPxX:CCPxY**: PWM Least Significant bits

               <u>Capture mode:</u>
               Unused.

               <u>Compare mode:</u>
               Unused.

               <u>PWM mode:</u>
               These bits are the two LSbs of the PWM duty cycle. The eight MSbs are found in CCPRxL.

bit 3-0   **CCPxM3:CCPxM0**: CCPx Mode Select bits

               `0000` = Capture/Compare/PWM disabled (resets CCPx module)
               `0100` = Capture mode, every falling edge
               `0101` = Capture mode, every rising edge
               `0110` = Capture mode, every 4th rising edge
               `0111` = Capture mode, every 16th rising edge
               `1000` = Compare mode, set output on match (CCPxIF bit is set)
               `1001` = Compare mode, clear output on match (CCPxIF bit is set)
               `1010` = Compare mode, generate software interrupt on match (CCPxIF bit is set, CCPx pin is unaffected)
               `1011` = Compare mode, trigger special event (CCPxIF bit is set, CCPx pin is unaffected); CCP1 resets TMR1; CCP2 resets TMR1 and starts an A/D conversion (if A/D module is enabled)
               `11xx` = PWM mode

Figure 9.7 The **CCP1CON/CCP2CON** registers (addresses 17_H and $1D_H$ respectively)

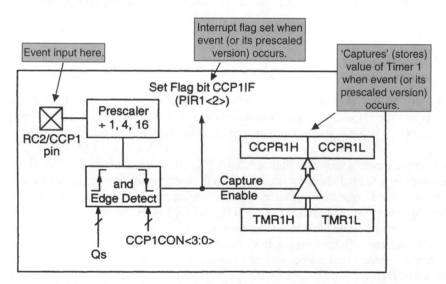

Figure 9.8 Capture mode block diagram (CCP1) (supplementary labels in shaded boxes added by the author)

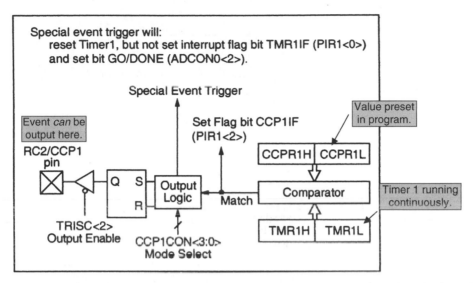

Figure 9.9 Compare mode block diagram (CCP1) (supplementary labels in shaded boxes added by the author)

9.4.3 Compare mode

The CCP1 module configured in Compare mode is shown in Figure 9.9. Here a digital comparator is designed into the hardware, which continuously compares the value of Timer 1 and the 16-bit register made up of **CCPR1H** and **CCPR1L**. Timer 1 must be running in Timer mode or in synchronised Counter mode. The module is connected to an external pin, which must be configured as an output in **TRISC**. When a match occurs, one of several things may happen, depending on the setting of the control register (Figure 9.7). The associated output bit can be set or cleared (with the interrupt flag bit set in both cases). This allows external events to be started or terminated. Alternatively, the interrupt may be set with *no* change to the output. Finally, a 'special event' may be triggered. For both modules this includes clearing Timer 1 and leaving the output pin unchanged. Module CCP2 also initiates an analog-to-digital conversion, if the converter is enabled.

9.5 Pulse width modulation

Pulse width modulation is a very powerful technique, which allows analog variables to be controlled from a purely digital output, and just with a single data connection at that!

9.5.1 The principle of PWM

An understanding of a typical application of PWM requires a little bit of background electronics. This is so important that we will review it briefly, by reference to Figure 9.10.

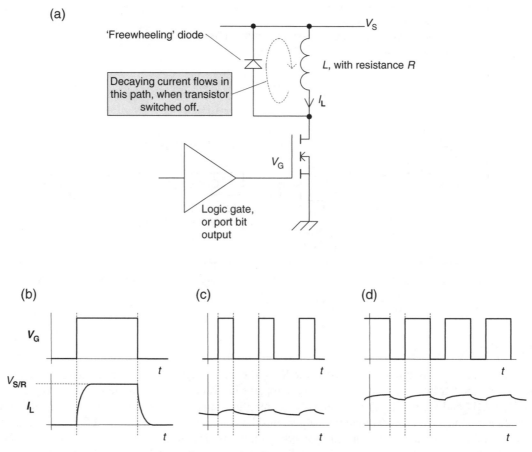

Figure 9.10 The principle of PWM. (a) Example circuit. (b) Time constant small compared to 'on' time. (c) Time constant large compared to 'on' time, narrow pulse. (d) Time constant large compared to 'on' time, wide pulse

The voltage–current relationship in an inductor is given by the equation:

$$V = L\,di/dt,$$

where V is the voltage across the inductor, i the current through it and L its inductance. Derived from this, if a step change in voltage is applied, then the current rises exponentially to a steady state value of V/R, where R is the resistance of the inductor. The rate of rise depends on a circuit quantity known as the *time constant*, which for an inductor is given by L/R. It can be shown that, for a step change in applied voltage, the current rises from 10 to 90 per cent of its final value in time $2.2L/R$. If the voltage is removed, the time taken for it to fall (assuming there is a circuit for it to flow in) is the same.

Let us imagine an inductive load is switched on and off by a transistor switch, as shown in Figure 9.10(a). When the transistor is switched off, the decaying current flows in the freewheeling diode. If the time duration it is on is much greater than the circuit time constant, then the current rise and fall times will just be a small proportion of the overall time, and the current waveform will appear as shown in Figure 9.10(b). If, however, the switching is repetitive, with a period less than the inductor time constant, then the current has no chance to reach steady state. It rises a little when the switch is on and falls a little (flowing through the freewheeling diode) when it is off. If the switching is continuous, the current rises to an average value which is dependent on the Mark : Space ratio of the switching waveform. This is represented in Figure 9.10(c, d).

The great thing about this is that the average current flowing in the inductor can be controlled by the Mark : Space ratio of the switching waveform. PWM is born!

A PWM waveform generally has a fixed period T, with a variable pulse width, t_{on}. It can be shown, for the circuit of Figure 9.10(a), that the average current flowing in the inductor, I_{ave}, is given by:

$$I_{ave} = \frac{t_{on}}{T} \times \frac{V}{R} \tag{9.1}$$

9.5.2 Generating PWM signals in hardware – the 16F87XA PWM

The beauty of PWM lies in its simplicity, and the way in which it acts as a gateway between the digital and analog worlds. It is easy to generate a PWM waveform in digital hardware. We will use the PIC 16 Series as an example to explain this. Although the principle is straightforward, the useful 'extras' introduced by Microchip add to the complexity. A little care is therefore needed in its understanding.

The simplified block diagram of the CCP configured to generate PWM is shown in Figure 9.11, with example waveforms in Figure 9.12. It can be seen that Figure 9.11 contains the central features of Figure 9.4, with Timer 2, comparator and the **PR2** register working together. Although not now shown, Timer 2 is still driven from the on-chip oscillator, through its own prescaler. With Timer 2 free-running, it is reasonable to expect the waveforms of Figure 9.6 to be found in the PWM process. This is indeed the case – the waveforms of Figure 9.6 are extended to become those of Figure 9.12. The comparator output, however, now drives an R–S flip-flop. When the value in **PR2** equals Timer 2, then the comparator clears the timer and *sets* the flip-flop, whose output goes high. This is seen in Figure 9.12. This action sets the PWM period.

Having established the PWM period, let us consider how the pulse width is determined. A second Compare register arrangement is introduced to do this. This is made up of the **CCPR1H** register, plus a second comparator. As the logic of the diagram shows, every time this comparator finds equal input values, it *resets* the output flip-flop, clearing the output to zero. It is this comparator that determines the pulse width. Again, this is shown in Figure 9.12. To change the pulse width, the programmer writes to the **CCPR1L** register, which acts as a buffer. Its value is transferred to **CCPR1H** only when a PWM cycle is complete, to avoid output errors in the process.

The block diagram is made more complex because three of the registers are 'stretched', to make them potentially 10-bit instead of 8-bit. This increases the resolution. **CCPR1L** uses 2 bits of the **CCP1CON** register (as already seen in Figure 9.7). **CCPR1H** is extended with an internal 2-bit latch, while the extension to Timer 2 is as described in Note 1 of Figure 9.11. Because of these two extra bits, in its 10-bit version it is effectively clocked direct from the internal oscillator signal, undivided. If the prescaler is used, then it

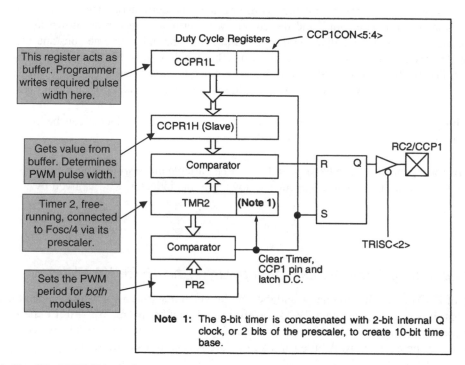

Figure 9.11 Simplified PWM block diagram (for CCP1 – CCP2 is equivalent) (supplementary labels in shaded boxes added by the author)

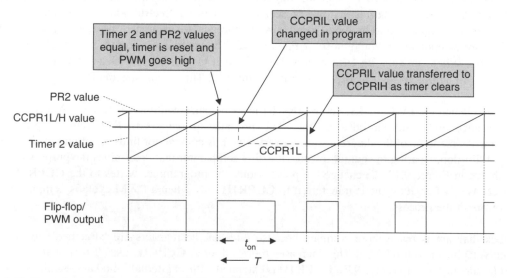

Figure 9.12 Example waveforms for the 16F873A PWM generator

acts on this frequency, not the usual Fosc/4. Notice, however, that the **PR2** register remains at 8 bits. This means that the PWM period has only an 8-bit equivalent resolution.

The PWM period T is determined by the interaction of the **PR2** register and the 8 bits of Timer 2. It may be calculated as follows:

$$T = (\mathbf{PR2} + 1) \times (\text{Timer 2 input clock period})$$

$$= (\mathbf{PR2} + 1) \times \{\text{Tosc} \times 4 \times (\text{Timer 2 prescale value})\} \qquad (9.2)$$

The PWM pulse width t_{on} is determined by the interaction of the extended **CCPR1H** register (all 10 bits of it) and the extended (10-bit) Timer 2. It may be calculated as follows:

$$t_{on} = (\text{pulse width register}) \times (\text{PWM timer input clock period}),$$

where 'PWM timer input clock period' is the period of the clock input to the extended Timer 2 and 'pulse width register' is the value in the extended **CCPR1H** register. Hence,

$$t_{on} = (\text{pulse width register}) \times \{\text{Tosc} \times (\text{Timer 2 prescale value})\} \qquad (9.3)$$

Note that there is *not* here a factor of 4 with the Tosc term, as explained above.

9.5.3   PWM applied in the Derbot for motor control

Let us now explore an application of the 16F873A PWM to the Derbot AGV. We need first of all to understand how PWM is applied in the circuit design. We saw in Chapter 8 how an H-bridge could be used for reversible drive. If PWM, rather than just on/off drive, is applied to an H-bridge, then continuously variable reversible drive can be achieved.

The Derbot circuit diagram for the left motor is shown in Figure 9.13(a). Half of the L913D IC (Figure 8.27) is used to form an H-bridge, with outputs driving the motor. The two inputs of the H-bridge are connected with a logic inverter between them, so that whenever one is at Logic 1, the other is at Logic 0, and vice versa. If the input labelled 'Bridge Drive' is held at Logic 1, the motor will run in one direction, if at Logic 0 in the other.

If now the Bridge Drive is connected to a PWM source, interesting things happen. If it is an exact square wave of sufficiently high frequency, the motor stands still, as the average current is zero. This is illustrated in Figure 9.13(b). If the Bridge Drive changes to a Mark : Space ratio of less than 1 the current takes up a negative average value. With a Mark : Space ratio of greater than 1 it takes up a positive average value. This is also seen in Figure 9.13(b). By varying the PWM pulse width across its full range, from absolute minimum to absolute maximum, the motor speed can be continuously varied, in both directions.

Looking at the full circuit diagram in Figure A3.1, it should be possible to work out that CCP1 drives the right motor, while CCP2 drives the left. Enable signals come from bits 2 and 5 of Port A.

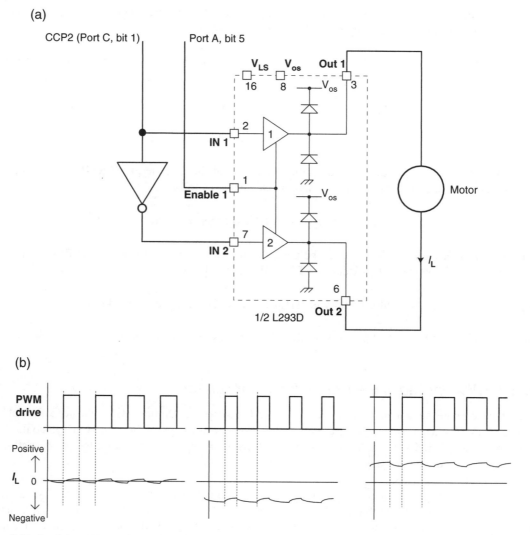

Figure 9.13 Applying PWM with an H-bridge to the Derbot AGV – achieving continuously variable bi-directional control. (a) Circuit diagram (left motor). (b) Example waveforms

The simplest program that we have available to look at PWM use in the Derbot AGV is the 'blind navigation' program of Chapter 8, appearing in part as Program Example 8.4. The parts of the program relating to PWM, omitted in the earlier chapter, now appear as Program Example 9.2.

The example program sections start with initialisation of the PWM. By writing to the **T2CON** register (Figure 9.5), Timer 2 is switched on, with neither pre- nor postscale (see Figure 9.4). Applying the **CCPxCON** registers, the two CCP control registers are set up in PWM mode. Finally, the **PR2** register is loaded with 249$_D$. Noting that the clock oscillator frequency is 4 MHz, we can deduce from equation (9.2) that the PWM period is:

$$T = (\mathbf{PR2} + 1) \times \{\text{Tosc} \times 4 \times (\text{Timer 2 prescale value})\}$$

$$= 250 \times (250\,\text{ns} \times 4 \times 1)$$

$$= 250\,\mu\text{s}$$

i.e. PWM frequency = 4.00 kHz.

This is a convenient frequency value, which we shall use for other things later. It retains almost the full resolution of the PWM system.

Also shown in Program Example 9.2 are the subroutines used to set the motors running forward and back, **leftmot_fwd**, **rtmot_fwd** and so on. Each of these starts by enabling the respective motor drive, by setting the port bit **mot_en_rt** or **mot_en_left** high. A number is then loaded into **CCPR1L** or **CCPR2L**, which determines the PWM pulse width. For simplicity, the two least significant bits of the pulse width setting, held in **CCPxCON**, are preset to zero and not further used in this program. Therefore, the pulse width setting can range from 00 to 250$_D$ loaded into **CCPRxL**, where 00 gives full reverse, 250$_D$ gives full forward and 125$_D$ gives no rotation. The numbers actually loaded in the program, 176$_D$ and 80$_D$, were found experimentally to give gentle forward and reverse speeds. Later in this chapter, and then later in the book, programs are developed to exploit the PWM capability to give variable speed drive.

```
;^^^^^^^^^^^^^^^^^^^^^^^^^^^^^^^^^^^^^^^^^^^^^^^^^^^^^^^^^^^^^^^^^^^
;Dbt_blind_Nav
;Derbot moves by "blind" navigation.
;Moves forward, and reverses and turns on bump.
;Fixed rate PWM applied to set reasonable speeds.
;
;TJW 5.5.05                              Tested 9.5.05
;*****************************************************************
...
(opening program sections omitted)
...
;Specify some port bits
;For Port A
mot_en_rt     equ   2      ;right enable input to L293D ic
mot_en_left   equ   5      ;left enable input to L293D ic
...
      bcf     status,rp0         ;select bank 0
;set up PWM
      movlw   B'00000100'   ;switch on Timer2, no pre or postscale
      movwf   t2con
      movlw   B'00001100'   ;enable PWM
```

Program Example 9.2 PWM – related sections of 'blind navigation' program

```
        movwf   ccp1con
        movwf   ccp2con
        movlw   0f9
        movwf   pr2
...
(main program omitted - appears as Program Example 8.4)
...
;motor drive subroutines
leftmot_fwd                     ;sets left motor running forward
        bsf     porta,mot_en_left
        movlw   D'176'
        movwf   CCPR2L
        return

rtmot_fwd bsf porta,mot_en_rt
        movlw   D'176'
        movwf   CCPR1L
        return

leftmot_rev bsf porta,mot_en_left
        movlw   D'80'
        movwf   CCPR2L
        return

rtmot_rev bsf porta,mot_en_rt
        movlw   D'80'
        movwf   CCPR1L
        return
```

Program Example 9.2 Continued

9.6 Generating PWM in software

While hardware PWM sources are versatile and simple to use, they are not always available. Perhaps a cost-conscious application has chosen a microcontroller without a PWM source. Alternatively, an application may have used up all its PWM sources and still need more. PWM is a classic case of a capability that can be developed *either* in hardware *or* software – if using a software solution, the only hardware resource that is needed is a single port I/O pin, set to output!

PWM outputs can be generated based on software delay loops only, such as those described in Chapter 5. A possible flow diagram appears in Figure 5.4 of Ref. 1.1. This approach, however, ties up the CPU almost completely, so is likely to be of only limited use in practice.

An alternative to using a purely programming approach to generating a PWM stream, still without requiring a PWM hardware module, is to use a timer interrupt to set the period. This can simplify programming complexity, and allows the CPU to be used in part for other activities. All that is needed is to set a timer interrupt on overflow running at an appropriate speed, as described in Section 9.3.3 of this chapter. On every interrupt the PWM output should be set high and a programmed delay initiated. At the end of the delay the PWM output is cleared. This is illustrated in Figure 9.14.

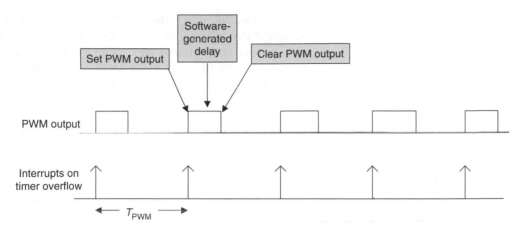

Figure 9.14 Generating PWM with timer interrupt

9.6.1 An example of software-generated PWM

A good example of the technique just described is found in Program Example 9.3. This drives the Derbot servo, which is used to rotate the ultrasound detector. The 16F873A has only two PWM sources and these are committed to the motors. The servo requires the pulse waveform shown in Figure 8.23. This is particularly appropriate for a program-generated technique, as the pulse width is always small compared to the period, so the CPU is committed for only a comparatively short period of time. The program moves the servo shaft position in 45° steps, from a reference 0° to 180°. Because the current consumption is very high if an instantaneous step movement is demanded of the servo, a ramp is generated between each step position.

With Timers 0 and 1 committed to the shaft encoder function, only Timer 2 is possibly available for timer overflow use. However, it appears to be tied up with the PWM function. Here now we see how a timer can be used for two apparently unrelated things. The timer can be left in its PWM role, but its interrupt on overflow, via the postscaler (Figure 9.4), can be evaluated for this new application. The two applications are in some conflict – the servo needs a period of 20 ms, while the PWM frequency needs to remain in the region of several kilohertz. The flexibility of the postscaler is an advantage here. As with the motor PWM setting, it can be seen that if the contents of **PR2** are set to 249$_D$, then the Timer 2 overflow rate is 4 kHz. With the postscaler set to ÷16, the interrupt frequency becomes 250 Hz, or a period of 4 ms. This is not the 20 ms we want, but it is easy in the program to output a pulse every five interrupts.

The settings just described can be seen in the program example. Register **PR2** is loaded with 249$_D$ and the Timer 2 postscaler set to ÷16. Three software counters are then preset – **int_cntr** counts the incoming interrupts, **pulse_cntr** counts the number of pulses sent at each pulse width, while **past_pulse** holds the pulse width of the previous pulse, and is used to determine whether ramping is needed between one pulse width and the next. The Timer 2 interrupt is then enabled and the program moves to sit in a wait loop. All subsequent program action is then in the ISR (Interrupt Service Routine).

Within the ISR, the value of **int_cntr** is first decremented to test whether a pulse is to be output. If it is, the value of **pulse_cntr** is then decremented to test if the pulse width is to be changed. In this case, the value of **servo_dirn** is incremented. This acts as a pointer to the look-up table, from which the new period is taken. At label ISR1, the look-up table is accessed and the 'demand' pulse width is read. This is stored in memory location **pw_cntr**. The value stored determines how many times a 20 μs delay routine is called, setting the pulse width.

The ramp, if needed, is now generated. The value in **pw_cntr** is compared with the previous value, in memory location **past_pulse**. If they are equal, the pulse is immediately output. If not, a test of the Carry flag determines which was the greater. The value of **pw_cntr** is then incremented or decremented accordingly, to set the new pulse width. The pulse is then output using a simple delay loop.

```
;*****************************************************************
;Dbt_servo_tst
;Rotates servo position in 45 degree steps from 0 degs to 180 degs.
;Servo PWM drive period, is generated with interrupt on overflow
;(Timer 2), with pulse width in software.
;Ramp between each step is generated.
;TJW 26.5.05                                         Tested 30.8.05
;*****************************************************************
...
(opening comments omitted)
        cblock 20
delcntr1        ;used in delay5 & delayADC SRs
delcntr2
temp            ;a temp location, to be used only in consecutive instructions
int_cntr        ;counts incoming interrupts
pulse_cntr      ;counts how many pwm pulses sent to servo, before moving to
                    ;next posn.
pw_cntr         ;counter used to set width of pulse to servo
servo_dirn      ;little counter which determines direction servo points, used
                    ;as look-up table pointer
past_pulse      ;holds value of most recent pulse width, used with demand value
                    ;to determine actual
        endc
...
        org  00
        goto  begin
;
        org  04
        goto  ISR
;Initialise
;set up SFRs in Bank 1
begin bcf     status,rp1
        bsf     status,rp0    ;select memory bank 1
...
        movlw  B'11110001'   ;set port B bits, bit 3 is servo PWM op
        movwf  trisb
...
```

Program Example 9.3 PWM generated in software

```
        movlw   D'249'              ;set Timer 2 Overflow period
        movwf   pr2
;set up SFRs in Bank 0
        bcf     status,rp0          ;select bank 0
        movlw   B'01111100'  ;switch on Timer2, no prescale, /16 postscale
                                ;giving 4ms interrupt prd (with pr2=250)
        movwf   t2con
;
;Initialise counter values
        movlw   D'5'   ;preset interrupt counter
        movwf   int_cntr
        movlw   D'200'
        movwf   pulse_cntr   ;preload pulse counter
        movlw   D'60' ;start with intermediate pulse width
        movwf   past_pulse
...
(opening program section omitted)
...
;Enable interrupts
        bcf     pir1,tmr2if     ;clear pending interrupts
        bsf     status,rp0      ;select memory bank 1
        bsf     pie1,tmr2ie     ;enable Timer 2 interrupt
        bcf     status,rp0      ;select memory bank 0
        bsf     intcon,peie     ;enable peripheral interrupts
        bsf     intcon,gie
wait goto wait          ;let ISR do the work
;
;********************************************************************
;ISR is here. Interrupts occur every 4ms. Count 5, and on 5th
;emit pulse to Servo. Pulse length will depend on servo_dirn setting
;********************************************************************
ISR     decfsz int_cntr     ;decrement interrupt counter. Action occurs
                                ;only if it is 0.
        goto    intend
        decfsz pulse_cntr   ;here if a pulse to be output, test if pulse
                                ;width is to change
        goto    ISR1
        incf    servo_dirn,1 ;here if pulse width changing, point to new
                                ;duration
        movlw   D'200'      ;reload pulse counter, 200 pulses will be emitted
        movwf   pulse_cntr      ;at new pulse width
;now determine pulse width, calculating ramp value if needed
ISR1    movf    servo_dirn,0 ;get demand pulse duration from Table
        andlw   07          ;use only 3 lsbs of servo_dirn
        call    int_table
        movwf   pw_cntr     ;this is demand value, may need to ramp towards it
;now determine ramp value, if needed
        subwf   past_pulse,0 ;compare demand value with most recent pulse
        btfsc   status,z
        goto    pulse_op        ;if equal, go direct to pulse
        btfsc   status,c        ;if carry clear, demand>past
        goto    $+3
```

Program Example 9.3 Continued

```
        incf   past_pulse,1 ;here if demand>past, hence ramping up
        goto   $+2
        decf   past_pulse,1 ;here if demand<past, hence ramping down
        movf   past_pulse,0 ;and save new value for next time round
        movwf  pw_cntr
;now send pulse
pulse_op bsf  portb,3       ;set pulse high
pw_loop call  delay20u
        decfsz pw_cntr
        goto   pw_loop
        bcf    portb,3       ;clear pulse
        movlw  D'5'   ;reload interrupt counter
        movwf  int_cntr
intend bcf pir1,tmr2if ;clear interrupt flag
        retfie
;Table for servo positions
int_table addwf pcl
        retlw D'20'           ;400us delay, 0 degrees
        retlw D'40'           ;800us delay, 45 degrees
        retlw D'60'           ;1200us delay, 90 degrees
        retlw D'80'           ;1600us delay, 135 degrees
        retlw D'100'          ;2000us delay, 180 degrees
        retlw D'80'           ;1600us delay, 135 degrees
        retlw D'60'           ;1200us delay, 90 degrees
        retlw D'40'           ;800us delay, 45 degrees
;
;*************************************************************
;SUBROUTINES
;*************************************************************
;introduces delay of 20us
delay20u  movlw 5   ;5 cycles called, each taking 3us,
              ;plus call, return (2 ea), and 2 move insts
;             less one cycle lost when last goto is hopped
        movwf        delcntr1
dela   decfsz        delcntr1,1 ;3 inst cycles in this loop, ie 3us
       goto dela
       return
;
...
(other delay subroutines omitted)
...
        end
```

Program Example 9.3 Continued

This little program was found to work well, with the position change being incremented smoothly.

9.6.2 *Further Assembler directives for memory definition and branching*

Notice the use in the program of two useful additions to our Assembler coding repertoire. Early in the program, the directive pair **cblock** and **endc** are used to define the list of data memory locations. This saves the use of a long list of **equ** statements. In the ISR, branching is achieved with instructions such as the following:

```
btfsc   status,c      ;if carry clear, demand>past
goto    $+3
incf    past_pulse,1   ;here if demand>past, hence ramping up
goto    $+2
```

Here the dollar symbol $ represents the current Program Counter value. Its usage in the line **goto $+3** therefore forces a forward jump of three lines. Negative values can also be used. This technique is invaluable for local branches, of a few lines forward or back, reducing the need for line labels. It is less useful with longer jumps, as a program change may be made some distance from the **goto** instruction, which renders the indicated jump value invalid.

9.7   PWM used for digital-to-analog conversion

While PWM is perhaps primarily used for load control, it allows a simple but very effective digital-to-analog converter (DAC). If a PWM stream of fixed Mark : Space ratio is low-pass filtered, with a suitable cut-off frequency, then the digital stream becomes a DC voltage, with a little residual ripple. If the PWM Mark : Space ratio is modulated, then a varying output voltage can be produced.

This idea is illustrated in simple form in Figure 9.15. The PWM stream is taken through an RC filter, which has the smoothing effect shown. The filter characteristic, in Figure 9.15(b), is chosen so that the PWM frequency is well into the filter stop-band. If the modulation frequency is in the filter pass-band, then the PWM is effectively blocked and the modulating frequency passed. The effect can be quite dramatic.

9.7.1   An example of PWM used for digital-to-analog conversion

The Derbot has two low-pass RC filter sections in its circuit, with outputs at TP1 and TP2 (Figure A5.1). These are simply there to demonstrate analog voltage generation with PWM and are nothing whatever to

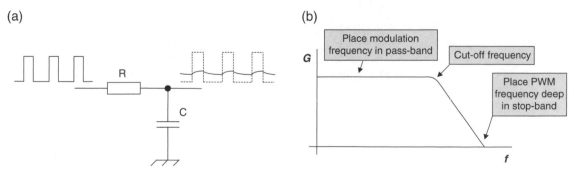

(a)   (b)

Figure 9.15 Filtering a PWM stream to produce an analog voltage. (a) The low-pass filter with input and output signals. (b) Filter characteristics

250 Taking timing further

do with the operation of the AGV. If used experimentally for this purpose, the motor enable bits should be set to 0!

The program **dbt_pwm_qrtr_sinwave** on the book CD is a simple illustration of this technique. It is reproduced in part in Program Example 9.4. The program uses the full 10-bit width available for setting the PWM time period. It generates a quarter sine wave, by taking samples in turn from a look-up table called **Sin_Table**. A memory location called **pointer** indicates where in the look-up table the program currently is. There are 45 samples, held as 16-bit values, in 2 bytes each. These give a binary, scaled sine value for $0°$, $2°$, $4°$ and so on up to $90°$. Only the most significant 2 bits of the lower byte are used, however.

Once initialisation is complete, the program is structured as a simple loop, starting at **sin_loop**. The more significant sample byte is transferred directly to **ccpr1l**. The less significant byte has to be adjusted to fit the two target bits in **ccp1con**. At the same time, the other bits of **ccp1con** must not be affected. It should be possible to follow through in the program listing the way this is done. A 1 ms delay is inserted for each sample, using the **delay1** subroutine. It is this, along with execution time of the remainder of the program, which determines the sine wave frequency.

```
;****************************************************************
;dbt_pwm_qrtr_sinwave
;Demonstrates quarter sin wave output, on CCP1
;Uses full ten bits of PWM period setting
;TJW 9.7.05                                 Tested 11.7.05
;****************************************************************
...
(all opening sections of program omitted)
...
        clrf   pointer
sin_loop movf pointer,0
        call   sin_table      ;get more significant byte of sample
        movwf  ccpr1l         ;move it to the PWM output
        incf   pointer,1      ;increment the pointer
        movf   pointer,0
        call   sin_table      ;get the less significant byte
        andlw  B'11000000'    ;we only use ms 2 bits of this
        movwf  temp
        bcf    status,c       ;adjust for CCP1CON
        rrf    temp,1
        rrf    temp,0
        iorlw  B'00001100'    ;ensure other bits of CCP1COn are set
                                ;correctly, ie PWM mode remains selected
        movwf  ccp1con        ;and output
        incf   pointer,1
        movf   pointer,0
        sublw  D'92'          ;test for end of table
        btfsc  status,z
        clrf   pointer        ;reset pointer
        call   delay1
        goto   sin_loop
;
```

Program Example 9.4 Generating a quarter sine wave with PWM

```
Sin_Table
     addwf pcl,1
     retlw 00       ;0 degrees, higher byte
     retlw 00       ;0 degrees, lower byte
     retlw 03       ;2 degrees, higher byte
     retlw 5A       ;2 degrees, lower byte
     retlw 06       ;4 degrees, higher byte
     retlw 0B2      ;4 degrees, lower byte
     ...
```

Program Example 9.4 Continued

Also on the book CD is a program called **dbt_pwm_full_sinwave**. This uses the same look-up table to generate the full sine wave, but is slightly more complex than the quarter sine wave version. It adds all samples to a midway value of 125_D. This represents the offset zero value of the sine wave. In the first quadrant it steps up through the table, in the second it steps backwards. In the third it steps forward again, but subtracts the samples from 125_D. In the fourth it steps backwards and subtracts. This can be seen in the full program listing.

The output of the **dbt_pwm_full_sinwave** program can be seen in Figure 9.16(a), with the output taken through the 100 nF/20 kΩ filter section (cut-off frequency of 80 Hz approximately). Notice the vertical and horizontal settings in each. Channel 2 displays the sine wave in both cases. Image (a) appears to show a very satisfactory sine wave. Image (b) shows a detail of the very same waveform, along with the PWM stream that creates it. The PWM period can be seen to be exactly 250 µs (i.e. a frequency of 4 kHz) in image (b). The characteristic (but small) sawtooth-like rise and fall of the analog waveform that accompanies it is very evident in the analog trace, with the overall rise in the signal voltage being just evident. With a PWM frequency of 4 kHz and the **delay1** subroutine in use, there will be just four or five PWM cycles per sample. The overall sine wave period can be seen to be approximately 3.8 divisions, at 50 ms each, i.e. 190 ms. This is accounted for by 180 samples making up the sine wave, each with a 1 ms delay, plus further program execution time.

(a)   (b)

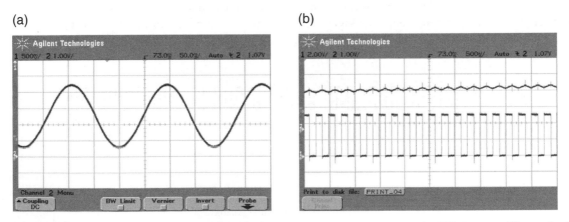

Figure 9.16 A sine wave generated with PWM. (a) The output sine wave. (b) Lower: the PWM stream. Upper: detail of analog output

The use of PWM for digital-to-analog conversion, as just described, can be simple and effective. It does, however, come with some weaknesses, which need to be understood if the technique is to be used. In brief, these are:

- The output analog voltage is directly dependent on the logic levels of the PWM stream. These in turn are dependent on the accuracy of the power supply voltage, cleanness of the ground path and logic technology in use. Overall, accurate D-to-A conversion cannot normally be expected from this technique.
- The impact of the low-pass filter is such that the technique is of little use to generate fast-changing signals. The situation can be improved by use of a higher order filter and/or running the PWM faster. This can be achieved by reducing the value stored in register **PR2**. By increasing the PWM speed in this way, however, it should be noted that the resolution is decreased. In general, we can see that any demand for high PWM frequency conflicts with a demand for high resolution.
- There will always be some residual ripple on the analog output.

9.8   Frequency measurement

9.8.1   The principle of frequency measurement

Frequency measurement is a very important application of both counting and timing. Fundamentally, frequency measurement is a measure of how many times something happens within a certain known period of time, as illustrated in Figure 9.17. The use can be as diverse as how many counts are received per minute in a Geiger counter, how many cycles per second (i.e. Hertz) in an electronic or acoustic measurement, or how many wheel revolutions there are per unit of time in a speed measurement. Both a counter and a timer are needed, the timer to measure the reference period of time and the counter to count the number of events within that time.

9.8.2   Frequency (speed) measurement in the Derbot

The Derbot AGV applies frequency measurement techniques to measure the speed of travel of the machine, later applying this to speed control. As Figure 9.17 indicates, what is needed first is an accurate timebase against which measurement can be made. Although any of the 16F873A timers could be used for this,

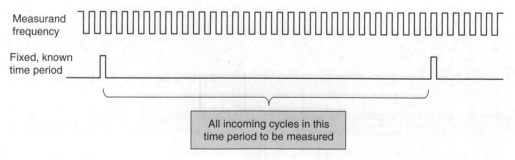

Figure 9.17 The principle of frequency measurement

Timers 0 and 1 are both used in counting mode for the wheel shaft encoders, so are not available. Timer 2 appears to be committed to the PWM function, but just as with the software-generated PWM in Section 9.6, further inspection shows that it can still be used. It was explained in that section how a 4 ms interrupt period could be achieved without impeding the PWM action. This is too low a period for most frequency measurements, but if we build a measurement period based on a certain number of iterations of this clock tick period, then we have the basis of a good time period for frequency measurement.

Program Example 9.5 shows sections of the program **Dbt_speed_meas**, a simple speed measurement program, which can be found in full on the book CD. The program simply switches on the motors and runs them at fixed PWM rate. The program is structured according to the simplified diagram of Figure 9.18. The system is initialised, and motors and Timer 2 interrupt are enabled. The Timer 2 interrupt occurs every 4 ms, and every 250th occurrence of this (i.e. every second), the count accumulated in Timers 0 and 1 is measured.

The timer settings applied can be explored through the program listing. The settings of Timers 0 and 1 are the same as for Program Example 9.1, and the setting of Timer 2 the same as for Program Example 9.3. Following the initialisation, the enabling of interrupts and motors can be seen, and the setting of the interrupt counter **int_cntr**. The motors are set running forward at modest forward speed.

At the beginning of the ISR, the value of **int_cntr** is first of all decremented. If it is non-zero, then the ISR is terminated. However, if the counter has decremented to zero, then one second exactly has elapsed, and the values of Timer 0 and 1 are read and saved. A frequency measurement is thus made. In this simple demonstration program nothing further is done with that measurement. It is used, however, in the next section of this chapter to implement Derbot speed control.

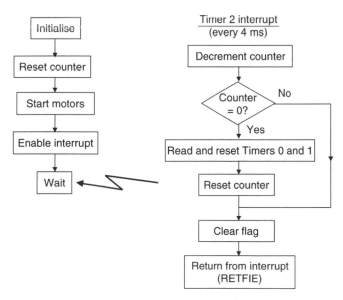

Figure 9.18 Derbot speed measurement program: simplified structure

```
;*********************************************************************
;dbt_speed_meas
;Derbot wheel speed is measured by frequency counting. This is a demo
;program. The frequency measurement concept can be embedded into larger
;programs, eg for speed control.
;TJW 7.7.05                                          Tested 7.7.05
;*********************************************************************
...
        #include p16f873A.inc
...
(opening program sections omitted)
...
;Initialise
;set up SFRs in Bank 1
main bcf status,rp1
...
        bsf     status,rp0     ;select memory bank 1
        movlw   B'11101000'    ;set up Timer 0: external input,
        movwf   option_reg     ;low to high transition, no prescale

        movlw D'249'           ;set PWM prd
        movwf   pr2
        bsf     pie1,tmr2ie    ;enable Timer 2 interrupt
;set up SFRs in Bank 0
        bcf     status,rp0     ;select bank 0
        movlw   B'00000011'    ;set up Timer 1: no prescale, oscillator
        movwf   t1con          ;disabled, external sync input
        movlw   B'01111100'    ;switch on Timer2, no prescale, /16 postscale
        movwf   t2con
        movlw   B'00001100'    ;select PWM
        movwf   ccp1con
        movwf   ccp2con
...
(further initialisation and startup omitted)
...
;clear timers
        clrf    tmr0
        clrf    tmr1l
        clrf    tmr1h
;enable interrupts
        movlw   D'250' ;load interrupt counter
        movwf   int_cntr
        bcf     pir1,tmr2if    ;clear pending interrupt
        bsf     intcon,peie    ;enable peripheral interrupt
        bsf     intcon,gie     ;enable global interrupt
;
;start running, then wait for Timer interrupt
        movlw   D'200' ;set PWM rate for reasonable forward speed
        movwf   ccpr1l
        movwf   ccpr2l
        bsf     portb,2        ;switch on optos
        bsf     porta,mot_en_left ;enable motors
        bsf     porta,mot_en_rt
```

Program Example 9.5 Frequency measurement applied to the Derbot

```
wait goto wait
;
;************************************************************
;ISR is here. Interrupts occur every 4ms. Count 250 (i.e.1.0s), then
;measure pulse count on each wheel.
;************************************************************
Timer2_Int    bsf portc,5           ;diagnostic
        decfsz int_cntr
        goto   int_end
;here if making a measurement
        movf   tmr0,0 ;save counter values
        movwf  tmr0_temp
        movf   tmr11,0
        movwf  tmr1_temp
        clrf   tmr0           ;clear counters
        clrf   tmr11
        btfss  portc,0        ;but increment T1 if it is zero, as first
        incf   tmr11              ;rising edge won't be seen
        movlw  D'250' ;reload interrupt counter
        movwf  int_cntr
int_end bcf pir1,tmr2if
        bcf    portc,5         ;diagnostic
        retfie
...
```

Program Example 9.5 Continued

9.9   Speed control applied to the Derbot

Embedded systems are about control, and now for the first time we will apply some closed loop control!

The **Dbt speed control** program, shown in part in Program Example 9.6, uses the frequency measurement just discussed to implement closed loop speed control on the AGV's motors. It is based on the principles of a simple closed-loop control system [Ref. 9.1], in that an output variable (motor speed, as represented by shaft encoder frequency) is compared with a demand value. The difference between the two is amplified and used to modify the drive to the motors. The control algorithm is identical for each motor and is embedded within the program ISR, following the comment line **adjust left motor speed**.

The program is effectively an extension of Program Example 9.5, except that the contents of **PR2**, governing the maximum PWM period, have been to 255_D. This just makes the data manipulation simpler and the program easier to follow. It does, however, mean that the Timer 2 interrupt period is now 4.096 ms, not 4.000 ms, so the frequency measurements made are not exactly per half second or second. As data is not displayed in human-friendly form, this does not really matter.

To understand settings within this program, an understanding of how the motor speed relates to shaft encoder frequency is needed. A shaft encoder of 16 cycles per wheel revolution was used. With a supply of 9 V applied to the L293 drive IC, the free-running gearbox output speed was measured to be approximately 154 r.p.m. This is in reasonable agreement with Table A3.1, bearing in mind the losses that occur in the drive electronics. This speed converts to a maximum shaft encoder frequency of $(154 \times 16)/60$, or 41 Hz.

To test this program a speed of approximately half this maximum was chosen, implying that over a period of 0.512 s a count of 10 cycles (nearest integer) is expected. This value is embedded within the program.

```
;*****************************************************************
;dbt_speed_control
;Derbot wheel speed is measured by frequency counting. Derbot
;runs forward at fixed speed - demonstrates simple speed control.
;
;TJW 7.7.05                                        Tested, 24.7.05
;*****************************************************************
...
        bsf     status,rp0              ;select bank 0
...
        movlw   B'11101000'   ;set up Timer 0: external input, increment
        movwf   option_reg      ; on low to high transition,no prescale
        movlw   0ff           ;set PWM prd to its maximum.
        movwf   pr2
        bsf     pie1,tmr2ie   ;enable Timer 2 interrupt
;set up SFRs in Bank 0
        bcf     status,rp0    ;select bank 0
        movlw   B'00000011'   ;set up Timer 1: no prescale, oscillator disabled,
        movwf   t1con         ;external input, sync with int clock
        movlw   B'01111100'   ;switch on Timer2, no prescale, /16 postscale.
        movwf   t2con            ;Timer int prd is 4.096ms, freq is 244.14Hz
...
;*****************************************************************
;ISR is here. Interrupts occur every 4.096ms. Count 125 (0.512s), then
;measure pulse count on each wheel, and calculate new PWM setting
;*****************************************************************
Timer2_Int   bsf     portc,5        ;diagnostic
        decfsz int_cntr
        goto    int_end
;here if checking speed
        movf    tmr0,0 ;save counter values
        movwf   tmr0_temp
        movf    tmr11,0
        movwf   tmr1_temp
        clrf    tmr0   ;clear counters
        clrf    tmr11
        btfss   portc,0        ;but increment T1 if it is zero, as first
        incf    tmr11              ;rising edge isn't seen
;Adjust left motor speed. Find "error" frequency by subtraction
        movf    tmr0_temp,0
        sublw   D'10'   ;this is demand speed
        movwf   tmr0_temp
        btfss   status,c ;check for polarity of result, skip if +ve
        goto    left_err_neg
        bcf     status,c ;here if result positive,
;"Amplify" result by shifting left.
;Check for over-range, bit 7 is initially 0 (from numbers used)
        btfss   tmr0_temp,6  ;check for possible over-range,
                             ;don't shift if bit 6 set
```

Program Example 9.6 Derbot speed control program

```
        rlf     tmr0_temp,f
        bcf     status,c
        btfss   tmr0_temp,6   ;shift again
        rlf     tmr0_temp,f
        bcf     status,c
        btfss   tmr0_temp,6   ;shift again
        rlf     tmr0_temp,f
        movf    tmr0_temp,w
        addwf   ccpr2l,0      ;add in current PWM value, result to W
        movwf   tmr0_temp     ;test and correct for over-range
        btfsc   status,c
        goto    set_l_max     ;carry set, so set max op
        movf    tmr0_temp,0   ;output calculated result to PWM
        movwf   ccpr2l
        goto    rt_adj        ;and proceed
set_l_max movlw     D'255'    ;here if output value has saturated,
        movwf   ccpr2l          ;hence output max value

        goto    rt_adj        ;and proceed
left_err_neg comf tmr0_temp,0 ;this will be holding a negative no,
        addlw 1                 ;therefore correct by taking 2's comp.
        movwf   tmr0_temp
        btfss   tmr0_temp,6   ;check for possible over-range,
        rlf     tmr0_temp,f   ;don't shift if bit 6 set
        bcf     status,c
        btfss   tmr0_temp,6   ;shift again
        rlf     tmr0_temp,f
        bcf     status,c
        btfss   tmr0_temp,6   ;shift again
        rlf     tmr0_temp,f
        movf    tmr0_temp,w
        subwf   ccpr2l,1      ;and subtract from current PWM value
                                ;Over range testing not included
;
;now adjust right motor speed
...
```

Program Example 9.6 Continued

The control algorithm itself follows the flow diagram of Figure 9.19. It should be possible to follow the program listing through, by cross-reference to the flow diagram. The motors' PWM rates are initially set to mid-range (i.e. zero speed) at the start of the program and the ISR adjusts them thereafter. The program listing shows the left motor only being controlled in this way. The setting for the right motor can be found in the complete listing on the book CD.

This program was found to run well, with the controlled speed being measured first 'on the bench' by testing the frequency of the output waveform of the reflective opto-sensors. With the Derbot running on the floor, for the first time it was seen to run in a reasonably true straight line, rather than the gradual curve that occurs when the two motors just run freely.

It is worth reminding ourselves that all features of this program, from setting the motor speed to measurement and control of the same, have been achieved using counting and timing techniques.

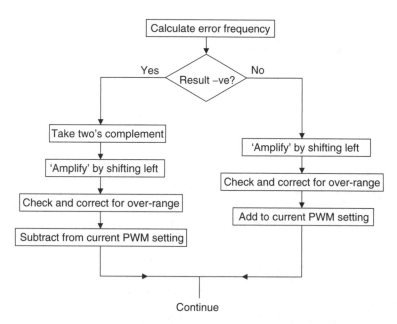

Figure 9.19 Flow diagram of motor control algorithm

9.10 Where there is no timer

While we have seen the power of the hardware timer, situations will continue to arise where they are not available. Maybe there are no timers on the microcontroller in use or maybe they have all been used up. This is the case with the Derbot AGV, where all timers are committed, yet if the ultrasound sensor is applied then a further timer is needed. The solution adopted here is to return to the software timing loop described in Chapter 5. The disadvantage of such a loop is that it wholly commits the CPU. It is justified in this case, as the ultrasound measurement is not made continuously and the measurement time is a comparatively small proportion of time overall.

Program Example 9.7 shows part of a program which measures the distance between the ultrasound sensor described in Section 8.6.5 of Chapter 8 and an object in its range. When we get to Chapter 11, the distance will be displayed in centimetres on the hand controller LCD and the circuit can be adapted as an ultrasonic tape measure. This second section of the program is not shown here, as it depends on data manipulation routines described in that chapter. The whole program is on the book CD. The ultrasound sensor shares pins with the diagnostic LEDs. Therefore, if the sensor is used, the LEDs are no longer available for independent use, although they still light in response to the ultrasound drive pulses.

The main program applies the timing diagram of Figure 8.16 and is structured as a continuous loop, starting at the label **us_loop**. The ultrasound pulse is initiated by the pulse on Port C bit 5. A blanking period follows, to allow the echo pulse to rise. The program then enters a timing loop, within which the counter **echo_time** is incremented on every cycle. The duration of this loop is designed to be exactly 30 μs. With sound travelling at 1 mm/3 μs, this implies that the sound will travel 10 mm for every iteration of the loop.

As the sound must travel to the target and return, this implies in turn a measurement resolution of 5 mm. The state of the echo pulse is tested on every loop iteration, and if and when it is found to fall to zero, the loop is terminated. If the counter overflows, however, then no object has been detected. A short bleep on the AGV sounder is emitted if an object is detected and a long one if none is detected. The full program goes on to convert the reading in **echo_time** to a distance value expressed in centimetres.

```
;********************************************************************
;Dbt_US_tst
;Tests derbot ultrasound sensor.
;Measures distance between sensor and object, converts and sends
;measured distance to lcd display,
;Sound travels 1mm/3us. For range of 1m require timing up to 6ms,
;2m up to 12ms.
;
;TJW 10.9.05                              Tested 12.9.05
;********************************************************************
...
;
;Specify RAM
       cblock 20
...
echo_time      ;counter measuring time for echo to return
...
       endc
;
       org  00
       goto  begin
;  (opening program sections omitted)


;***********************************
;The "main" program loop starts here
;***********************************
us_loop clrf   echo_time     ;clear counter used to measure time
       bsf    portc,5        ;output us pulse. 10us minimum required
       call   delay20u
       bcf    portc,5
       call   delay300u      ;pause for op to be set high; ie blank for 5cm
;this loop takes 30us per cycle, ie 10mm there and back, or 5mm one way;
;hence 8-bit range is 255x5mm = 1275mm. Max duration in loop is 30x255=7.6ms
echo    call   delay24u
       incf   echo_time,1
       btfsc  status,z       ;test for overflow,which indicates no target found
       goto   no_tgt
       btfsc  portc,6        ;test echo pulse, skip if cleared (ie echo recd.)
       goto   echo
;here if target detected
;      bsf    portb,1        ;indicate with short bleep
       call   delay200
       bcf    portb,1
       call   delay200
```

Program Example 9.7 Ultrasound timing in software.

```
...
(data processing, and transfer to lcd inserted here)
...
        goto    us_loop
no_tgt  bsf     portb,1      ;indicate with long bleep
        call    delay500
        call    delay500
        bcf     portb,1
        call    delay200
        goto    us_loop
...
(most subroutines omitted)
...
;This subroutine introduces delay of 24us
delay24u movlw 6    ;6 cycles called, each taking 3us,
        ;plus call (2), & 2 opening insts (2) + 2 at end
        ;less one cycle lost when last goto is hopped
        movwf delcntr1
del21 decfsz delcntr1,1 ;3 inst cycles in this loop, ie 3us
        goto  del21
        nop
        return
```

Program Example 9.7 Continued

It is suggested that this program is just looked at as a theoretical example at this time, as the subroutines necessary to complete it are introduced in Chapters 10 and 11.

9.11   Sleep mode

As with Chapter 6 we embed Sleep mode, when time appears suspended, in a chapter on timing. The Sleep mode of the 16F87XA follows exactly the principles of the Sleep mode of the 16F84A, described in Section 6.6 of Chapter 6. A possible limitation on the 16F84A Sleep mode was the limited range of wake-up opportunities. A significant example of this is the lack of a periodic timed wake-up. (Although the WDT seems to offer this, its frequency of operation is imprecise and the range of overflow periods not particularly wide.)

The 16F87XA in contrast offers a wide range of wake-up opportunities, including a number from peripherals. The interrupt structure diagram of Figure 7.10 suggests that *all* peripherals can cause a wake-up from Sleep. In practice, of course, some of these stop functioning, as they depend on the clock oscillator, which is turned off during Sleep. An interesting case now is Timer 1, which can run with its own oscillator. This can be left running while in Sleep, with a periodic wake-up derived from its overflow interrupt. Peripheral wake-up opportunities are available from the following:

- Timer 1, when operating in asynchronous mode
- The CCP Capture mode interrupt
- The special event trigger, with Timer 1 operating in asynchronous mode with external clock
- The synchronous serial port
- The USART (Addressable Universal Synchronous Asynchronous Receiver Transmitter) port

- The analog-to-digital converter, when running with RC clock oscillator
- EEPROM write complete
- The comparator output change
- The parallel slave port read or write (for 16F874A or '877A).

9.12 Where do we go from here?

We have seen in the last few pages a wealth of tools and techniques that allow time-based activities to be developed. Life has been kept simple, however, as most examples have been introduced as if the activity

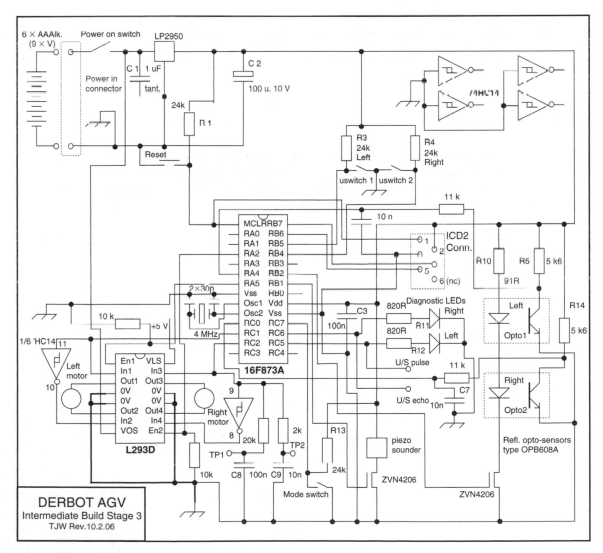

Figure 9.20 Derbot intermediate build stage 3

in question will be the only one it will do. In practice this is not the case, of course – the Derbot AGV will need to run its motors while it measures the motor speed while it controls the servo and so on. The number of time-based tasks will become many, and possibly conflicting. We have already seen, for example, the case of one timer being used for two different and potentially conflicting tasks. This challenge of marshalling time-based activity in an embedded system is common in many systems and needs a systematic approach to resolve it. That systematic approach will be seen in the Real Time Operating System, a subject for Chapter 18.

9.13   Building up the Derbot

To run most of the programs used as examples in this chapter, you will need to add the reflective opto-sensors to the Derbot and the shaft encoder patterns to the wheels. The circuit will then have reached the stage shown in Figure 9.20. To run Program Example 9.3, the servo will need to be added. This is not shown in Figure 9.20, but can be seen in Figure A3.1.

Summary

- Timing is an essential element of embedded system design – both in its own right and to enable other embedded activities, like serial communication and pulse width modulation.
- A range of timers is available, with clever add-on facilities which extend their capability to capture, compare, create repetitive interrupts or generate PWM pulse streams.
- In applications of any complexity, a microcontroller is likely to have several timers running simultaneously, for quite different and possibly conflicting applications. The question remains open at this stage: how can these different time-based activities be marshalled and harmonised?

Reference

9.1.   Bolton, W. (1998). *Control Engineering Pocket Book*. Newnes, Oxford, UK. ISBN 0 75063928 8.

10
Starting with serial

An essential activity within any microprocessor system is the movement of data, between say CPU and memory. Equally important is the transfer of data between subsystems, between say computer and keyboard. Data is sent by ordering bits into words (often bytes, but not always) and then sending those words.

Broadly speaking, there are two ways of transferring the data. In parallel transfer, all the bits of the word are transmitted at the same time, each bit over its own connection. The alternative is to send each bit in turn, over a single connection. This is called serial communication. It is easy to see the first relative advantages. Parallel takes more wires and connections, but is faster. Serial needs fewer wires, is slower, and generally requires more complex hardware to transmit and receive.

Therefore, in the past, over short distances, data transfer has been mainly parallel. Over long distances, where running a large number of wires would be bulky and expensive, it has been serial. Now, however, that distinction has been challenged, and nowhere more so than in the field of embedded systems, where things must be very small. The advantage of serial transmission, with its fewer wire links, has in many situations become overwhelming. Serial communication has therefore become very important and great ingenuity has been applied to overcoming its main apparent disadvantages – its low speed and more complex hardware.

This chapter introduces the main ideas of serial data communication, both in principle and practice. This includes:

- Describing the principles of synchronous serial communication
- Exploring the implementation of synchronous serial communication with the PIC® 16F873A, notably with the SPI (Serial Peripheral Interface) and I²C (Inter-Integrated Circuit) protocols
- Describing the principles of asynchronous serial communication
- Exploring the implementation of asynchronous serial communication with the PIC 16F873A.

As always, a good proportion of the material is illustrated with example code. This is written for the Derbot AGV, and its hand controller board, and will be of interest whether or not one has the hardware to run the programs. A number of oscilloscope traces of serial data are also included. At the time of writing it is effectively not possible to simulate the programs in this chapter with the MPLAB® simulator, as the serial ports are not supported in the simulator.

10.1 The main idea – introducing serial

The introduction above has just identified one of the main advantages of serial communication, the fact that it saves space. Figure 10.1 shows one way that this happens. Both of the memory ICs pictured can store the same amount of data. However, the one with parallel interconnect needs 13 address lines and eight

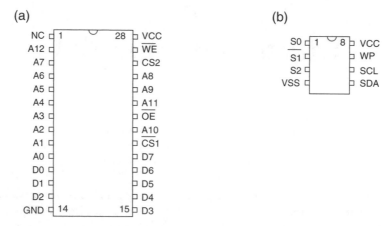

Figure 10.1 Example dual-in line packages required for 8 Kbyte memory, approximately to scale. (a) Parallel address and data buses. (b) Serial address and data

for data, which dominate the chip size. The one using serial interconnect has *two* lines (SCL – serial clock; SDA – serial data) for data address and *three* (S0, $\overline{S1}$, S2) for address. This incidentally is an example of an I^2C chip, of which we shall learn much more soon. The space saving at the IC level is clearly dramatic. To this is added the space saving in PCB tracks, ribbon cable and so on.

While serial communication has a number of clear advantages, it is important to recognize the challenges it also presents. With just a single wire to carry the data how do we know when one bit starts or ends, or where a word starts or ends? These questions are resolved in a number of interesting ways. Two different ways are used to identify the individual bits. The simple way is to send a clock signal to accompany the data; every clock cycle is used to indicate 1 bit of data. This is called *synchronous* serial data communication. Instead of sending a clock signal everywhere, certain timing requirements can be placed on the data itself. In this case, it can be possible to do without the clock and the data bits can still be properly identified. This is called *asynchronous* serial data communication. We look at both of these in some detail during this chapter.

To identify the start and end of a whole word, it is common to package the data in a certain format. Data synchronisation and formatting implies certain sets of rules will be needed, in order to ensure coherent communication. These sets of rules are called *protocols* and are very important in the serial world. Some are comparatively simple, while others are very complicated. We will come across a number in this chapter.

In a serial link, we generally use the idea of a *transmitter*, the device outputting data onto a serial link, and *receiver*, the device receiving the data. In general terms, any device on a serial link is sometimes called a *node*.

To illustrate our study of serial communication, we will be looking at the serial capabilities of the 16F873A. This has two serial ports, as mentioned in Chapter 7. Both are extremely flexible and can be configured in different ways. Therefore, both are quite complex! The Master Synchronous Serial Port (MSSP) is designed for different forms of serial communication, while the Addressable Universal Synchronous Asynchronous Receiver Transmitter (USART) can operate in both synchronous and asynchronous modes.

10.2 Simple serial links – synchronous data communication

10.2.1 Synchronous basics

To understand how data can be sent serially, it is helpful to explore the underlying hardware. Data within a microprocessor or memory is still likely to be used and formatted in parallel. Therefore, a serial transmitter needs to be able to accept data in parallel format, but then transmit it serially; a serial receiver needs to be able to do the reverse. The classic way of doing this is with a shift register.

A simple 8-bit shift register is shown in Figure 10.2. It is made up of eight flip-flops connected together so that the Q-output of one becomes the data input of the next. All are driven from the same clock, and on every clock cycle the data is moved one flip-flop to the right. If a new bit of data appears at the D_{IN} input on every clock cycle, then in eight clock cycles 1 byte will be clocked through such that it can be read as a parallel word from outputs Q_A to Q_H. This simple circuit can act as a *receiver* of serial data.

It is not difficult to enhance the circuit so that parallel data can not only be read from the shift register, it can also be *loaded* with a parallel word. It's not worth drawing the full logic diagram of such a shift register, but instead we could represent it in block diagram form, as in Figure 10.3. With this general-purpose shift register, we can clock serial data into it or out of it, and can read parallel data or load parallel data.

At this point we already have the basis for a serial communication link. If two of the shift registers shown in Figure 10.3 are taken and clocked from the same source, then data from one can be transferred serially to the other using the connection shown in Figure 10.4. Effectively, the two 8-bit shift registers act as one 16-bit shift register, with output connected back to input. After eight clock cycles the data in one shift register is moved to the other. Therefore, *either* can be viewed as the transmitter and either as the receiver. The action of data transfer is actually controlled by the clock. The link is called a *synchronous* link, because the data transfer is synchronised to a common clock signal.

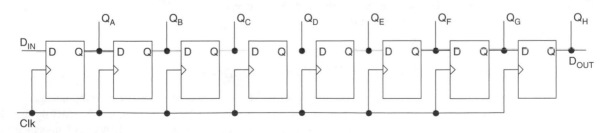

Figure 10.2 An 8-bit shift register – a possible *receiver* of serial data

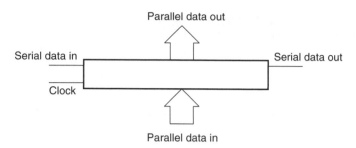

Figure 10.3 Block diagram of a general-purpose shift register

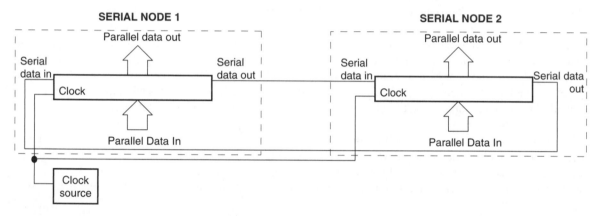

Figure 10.4 A general-purpose serial communication link

10.2.2   Implementing synchronous serial I/O in the microcontroller

The synchronous serial link just described can be readily implemented in a microcontroller, as shown in Figure 10.5. The clock source is placed in the microcontroller, and as the clock controls data flow, this node is called the *master*. The other node is called the *slave*. The slave device could be another microcontroller, a memory device or one of a range of other peripherals.

This sort of connection is the basis of many simple embedded serial links. The shift register is memory mapped, and can be read from and written to.

10.2.3   Microwire and SPI (Serial Peripheral Interface)

In the late 1970s and early 1980s, National Semiconductor, Motorola and others were developing micro-processors and microcontrollers that had built-in synchronous serial capability. They needed to define the associated operating characteristics so that other manufacturers could make devices that could interface reliably with their products. Both National and Motorola were producing serial ports that worked in the

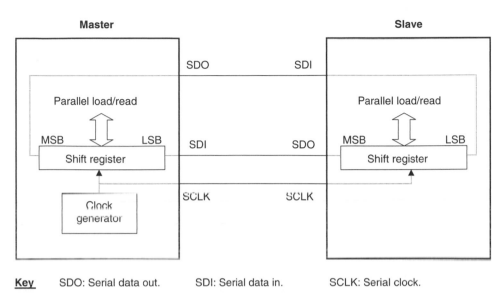

Key   SDO: Serial data out.   SDI: Serial data in.   SCLK: Serial clock.

Figure 10.5 The synchronous serial link implemented in a microcontroller

way described above. From their designs, two standards were produced. National called theirs Microwire and Motorola called theirs SPI. These are similar to each other and can communicate with each other. Each has the flexibility to adjust its characteristics, for example to determine whether data is transferred on rising or falling edge of the clock. Both have now both been going for many years and are well established, with new chips still being produced which interface with them. We will see an implementation of SPI in the following sections.

10.2.4   Introducing multiple nodes

The diagram of Figure 10.5 gives an effective serial link, but it only connects two nodes. How could we extend it? The answer is simple. If a means is introduced of selecting which slave device the master is to communicate with, then more than one slave can be connected to the serial data lines, as shown in Figure 10.6. Each slave input now requires a means of enabling it, sometimes labelled *Slave Select* (SS) or *Chip Select*. The exact function of the SS line varies somewhat from one device to another, but to a greater or lesser extent it causes the slave to disconnect all or part of itself from the serial connection. The master now has to have a dedicated line for each slave with which it communicates; these can be port bit outputs.

10.3   The 16F87XA Master Synchronous Serial Port (MSSP) module in SPI mode

The MSSP module is designed for synchronous communication and can be configured as a simple synchronous port (called SPI mode, but compatible with both SPI and Microwire), or as an I²C port

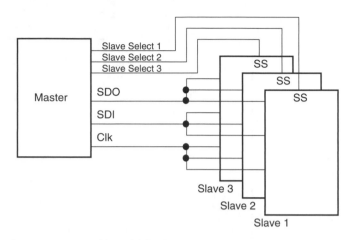

Figure 10.6 Single synchronous master with multiple slaves

(Inter-Integrated Circuit). It has three SFRs dedicated to it, **SSPCON1**, **SSPCON2** and **SSPSTAT**, which can be found in the register file map diagram of Figure 7.6. It also has a register for data transfer, **SSPBUF**, and is the source of an interrupt, as seen in Figure 7.10. In this section we look at the module in SPI mode.

10.3.1   Port overview

Figure 10.7 shows the MSSP configured as an SPI port. It can be configured as master or slave, with a variety of clock speeds if master. Let us see how it builds on the simple serial concepts we have discussed above. At the heart of the serial port is the shift register **SSPSR**. When clocked, it transfers serial data to pin **SDO** (if the output buffer gate is enabled) and transfers serial data in from pin **SDI**. If the port is set up as a slave, it will receive the clock from the system master through pin **SCK**. If the port is set up as master, it will generate the clock, which it now *outputs* through the **SCK** pin. This clock is derived either from the internal clock oscillator, or from Timer 2.

An important enhancement to our earlier simple serial port is the addition of the buffer register **SSPBUF**. This holds a data byte on its way to or from the shift register, and is actually the addressable register that the program writes to or reads from. This makes the serial port much more flexible in use. Data can be moved to or from the buffer while the shift register is in operation. This, for example, allows a received byte to be held temporarily, while the next one is already being clocked in.

10.3.2   Port configuration

The two SFRs that control the action of the port in SPI mode are **SSPCON1** and **SSPSTAT**. Their use differs somewhat between SPI and I^{2}C modes. They are shown, when used in SPI mode, in

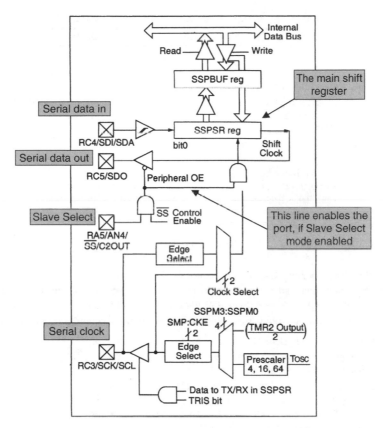

Figure 10.7 The MSSP block diagram, in SPI mode (supplementary labels in shaded boxes added by the author)

Figures 10.8 and 10.9 respectively. In them there are bits to do the following:

- Enabling and configuring the port
- Setting clock rate and clock characteristics
- Managing data transfer and buffering.

The port is enabled with bit 5 (**SSPEN**) of **SSPCON1**, and its operating mode with the lower 4 bits of the same register. It can be seen that these bits determine whether the port is to work as master or slave. If in Master mode, four clock sources are available. As can be seen, these are either the clock oscillator signal divided by 4, 16 or 64, *or* the Timer 2 output, as described in Chapter 9 and seen in Figure 9.4.

If in Slave mode, the Slave Select input pin $\overline{SS}$ can be enabled, through the four lower bits of **SSPCON1**. In this case, an external $\overline{SS}$ signal can control the tristate buffer that drives the **SDO** pin, and the clock to the shift register. The $\overline{SS}$ input then effectively enables the serial port action and the port can be used in a multi-node configuration, such as in Figure 10.6.

R/W-0	R/W-0	R/W-0	R/W-0	R/W-0	R/W-0	R/W-0	R/W-0
WCOL	SSPOV	SSPEN	CKP	SSPM3	SSPM2	SSPM1	SSPM0

bit 7   bit 0

bit 7   **WCOL:** Write Collision Detect bit (Transmit mode only)

1 = The SSPBUF register is written while it is still transmitting the previous word. (Must be cleared in software.)
0 = No collision

bit 6   **SSPOV:** Receive Overflow Indicator bit

<u>SPI Slave mode:</u>

1 = A new byte is received while the SSPBUF register is still holding the previous data. In case of overflow, the data in SSPSR is lost. Overflow can only occur in Slave mode. The user must read the SSPBUF, even if only transmitting data, to avoid setting overflow. (Must be cleared in software.)
0 = No overflow

> **Note:** In Master mode, the overflow bit is not set, since each new reception (and transmission) is initiated by writing to the SSPBUF register.

bit 5   **SSPEN:** Synchronous Serial Port Enable bit

1 = Enables serial port and configures SCK, SDO, SDI, and $\overline{SS}$ as serial port pins
0 = Disables serial port and configures these pins as I/O port pins

> **Note:** When enabled, these pins must be properly configured as input or output.

bit 4   **CKP:** Clock Polarity Select bit

1 = Idle state for clock is a high level
0 = Idle state for clock is a low level

bit 3-0   **SSPM3:SSPM0:** Synchronous Serial Port Mode Select bits

0101 = SPI Slave mode, clock = SCK pin. $\overline{SS}$ pin control disabled. $\overline{SS}$ can be used as I/O pin.
0100 = SPI Slave mode, clock = SCK pin. $\overline{SS}$ pin control enabled.
0011 = SPI Master mode, clock = TMR2 output/2
0010 = SPI Master mode, clock = FOSC/64
0001 = SPI Master mode, clock = FOSC/16
0000 = SPI Master mode, clock = FOSC/4

> **Note:** Bit combinations not specifically listed here are either reserved or implemented in I^{2}C mode only.

Figure 10.8 **SSPCON1** register (address 14_H) in SPI mode

For all pins through which data or clock transfer is to take place, the data direction bits must be set as needed. Therefore, for the **SDO** pin, which is shared with Port C bit 5, bit 5 of **TRISC** must be cleared to make the pin an output. Similarly, for **SCK**, in Master mode bit 3 of **TRISC** must be cleared (to make it an output), while in Slave mode it should be set (to make it an input). The **SDI** pin, however, is under direct control of the MSSP module.

10.3.3 Setting the clock

Figure 10.10 shows the relationship between clock and data waveforms available when the module is in Master mode. If bit **CKP** is set to 1, then the clock idles at Logic 1. The clock edge on which data is

R/W-0	R/W-0	R-0	R-0	R-0	R-0	R-0	R-0
SMP	CKE	D/$\overline{\text{A}}$	P	S	R/$\overline{\text{W}}$	UA	BF

bit 7   bit 0

bit 7   **SMP:** Sample bit

SPI Master mode:
1 = Input data sampled at end of data output time
0 = Input data sampled at middle of data output time

SPI Slave mode:
SMP must be cleared when SPI is used in Slave mode.

bit 6   **CKE:** SPI Clock Select bit

1 = Transmit occurs on transition from active to Idle clock state
0 = Transmit occurs on transition from Idle to active clock state

  Note: Polarity of clock state is set by the CKP bit (SSPCON1<4>).

bits 1 to 5 not used in SPI mode

bit 0   **BF:** Buffer Full Status bit (Receive mode only)

1 = Receive complete, SSPBUF is full
0 = Receive not complete, SSPBUF is empty

Figure 10.9 SSPSTAT register (address 94$_\text{H}$) in SPI mode

transmitted is determined by bit **CKE**. For incoming data, the instant when data is sampled is determined by bit **SMP**. The way these are set may be determined by the particular requirements of a slave device that is being used. It is, of course, essential to maintain consistency throughout a single interconnected system.

10.3.4 *Managing data transfer*

A synchronous serial port such as the one we are looking at can be set as master or slave. Within either of these, the application software can use it as receiver or transmitter, or both. Whatever the application, data is always clocked out from one end of the shift register and in at the other. It is up to the user to determine which data is to be used.

Use of a serial port therefore brings with it some interesting challenges in terms of timing and control. If the port is set as Slave, then an external device causes the transfer, but the slave port must be alerted to this. It must move data into and/or out of the port buffer, according to which direction data is moving. If it is a master, it initiates the transfer, and must also move data to and/or from the buffer. In either case, the serial port hardware can be undertaking a data transfer while the program is doing something completely different. To assist in managing the process, the port has a number of Status bits in its SFRs, as well as the interrupt.

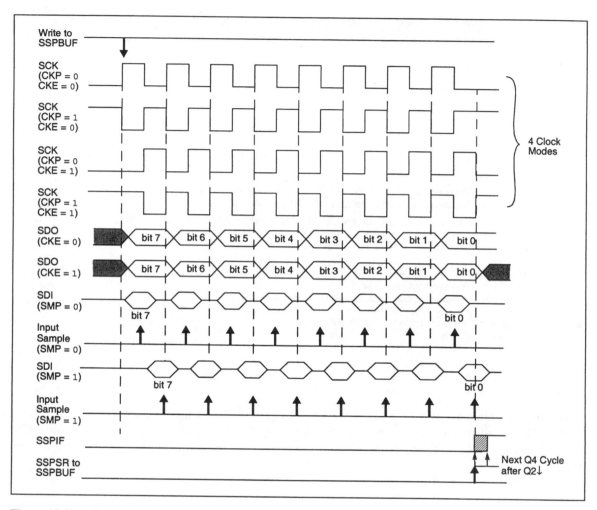

Figure 10.10 SPI timing diagram, Master mode

If the port is set as Master, a write to the buffer register **SSPBUF** automatically starts a transfer, clocking out whatever data has been loaded into **SSPBUF** and clocking in whatever data is present at the **SDI** pin. On completion of eight clock cycles, the interrupt flag **SSPIF** is set and data in the shift register **SSPSR** is automatically transferred to the buffer **SSPBUF**. The **SSPIF** flag can be used as an interrupt to alert the CPU that the transfer is complete. If there is a write to **SSPBUF** before the previous word has been completely sent, then the write collision bit **WCOL** is set.

If set as Slave, then when the **SCK** input starts switching, the port clocks data into the **SSPSR** shift register through the **SDI** pin. At the same time, data is clocked out of it from the other end through the **SDI** pin. It is, of course, up to the system designer to ensure that valid data is ready in the **SSPSR** register and/or is available at the **SDI** input, according to the requirement. When eight cycles are complete, the interrupt flag **SSPIF** flag is set and again data in **SSPSR** is automatically transferred to the buffer **SSPBUF**. If the previous byte has *not* been read from **SSPBUF**, then the **SSPOV** bit is set, indicating a receive overflow.

10.4 A simple SPI example

The Derbot AGV, in the version described in this book, does not use SPI data communication. Nevertheless, Program Example 10.1 provides a simple example program that runs on the Derbot hardware. As always, a full listing appears on the book CD, while only the features of direct interest are shown in the example. In this case, it is such a short program that almost all of it is reproduced. It should be possible to cross-check all initialisation settings with the control registers in Figures 10.8 and 10.9.

The program enables the Derbot microcontroller SPI by writing to **SSPCON1** (for reasons of backward compatibility the Assembler Include File calls this **SSPCON**) and sets it up as a master, with clock frequency F_{osc} /16. Clock control bits **SMP** and **CKE** are both set to zero here. Clock and data output pins are set up as outputs via **TRISC**. The program then transmits 2 bytes in turn repeatedly on the serial link, with a 40 µs delay in between.

```
;*****************************************************************
;sync_ser_demo
;Program to demonstrate MSSP serial output.
;Program sends same two digits repeatedly from serial port, with delay.
;serial data appears on Port C bit 5, serial clock on Port C bit 3.
;2.7.05. TJW                                Tested 2.7.05
;*****************************************************************
...
(early comments and initialisation omitted)
...
;
        bsf     status,rp0    ;select memory bank 1
...
        movlw   B'10000000' ;Set port C bits, SDO and SCK set as op.(SDI line,
        movwf   trisc         ;bit 4, is controlled by SPI module, so leave)
...
        bcf     status,rp0    ;select memory bank 1
        movlw   B'00000000'
        movwf   sspstat       ;SMP=0, CKE=0, other bits don't apply
        movlw   B'00110001'   ;enable serial port, master mode, clock is fosc/16
        movwf   sspcon           ;& idles high.
;Switch all  outputs off
        clrf    porta
        clrf    portb
        clrf    portc
loop    movlw   B'11010101'
        movwf   sspbuf
        call    delay40u
        movlw   B'00101010'
        movwf   sspbuf
        call    delay40u
```

Program Example 10.1 SPI demonstration program

```
        goto    loop
;
;Subroutine: introduces delay of 40us approx
delay40u  movlw     D'10'      ;10 cycles called, each taking 4us
        movwf   delcntr1
del1    nop                     ;4 inst cycles in this loop, ie 4us
        decfsz  delcntr1,1
        goto    del1
        return
        end
```

Program Example 10.1 Continued

Figure 10.11 shows oscilloscope traces of the data and clock lines of Program Example 10.1, for **CKE** = 1 and **CKE** = 0. These are a practical confirmation of some of the waveforms of Figure 10.10. With the horizontal scale set at 10 μs per division, the clock period can be seen to be exactly 4 μs, i.e. a frequency of 250 kHz. This corresponds with the clock setting made in **SSPCON1**, of Fosc (4 MHz) divided by 16. The two data bytes 11010101, followed by 00101010, can clearly be seen, being transmitted MSB first. With bit **CKP** (in **SSPCON1**) in both cases set high, the clock is seen to idle at Logic 1. When bit **CKE** (in **SSPSTAT**) is set high, we see the output data changing on the positive-going edge of the clock, i.e. from its 'active' (Logic 0 in this case) to its idle state. The reverse is true when **CKE** is zero. Notice that the idle state of the data line is not fixed.

The timing between bytes is also of interest. Notice that the program calls a delay of 40 μs between sending bytes, but that the apparent gap is less than 20 μs. This is because the data transmission process is initiated by the program when it writes to **SSPBUF**. The delay subroutine is then called by the program and much of it runs *while* the data transmission is taking place.

(a)   (b)

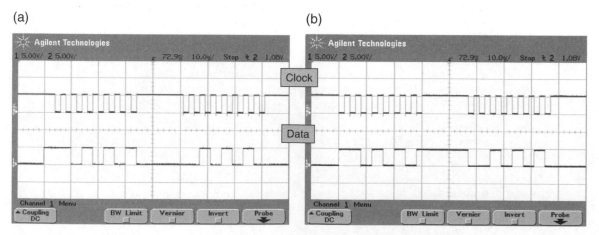

Figure 10.11 Synchronous serial output. (a) **CKP** = 1, **CKE** = 1. (b) **CKP** = 1, **CKE** = 0

10.5 The limitations of Microwire and SPI, and of simple synchronous serial transfer

From what we have seen so far, synchronous links like Microwire and SPI provide a simple and reliable data connection, yet they have limitations. These include:

- They don't cater well for situations where more than one master may be required
- They don't address
- They don't acknowledge – the transmitter simply doesn't know whether the message has been received
- They are not very flexible – it may be not that easy to add another node, even just one more slave, as for each new slave (at least in the configuration of Figure 10.6) an extra Slave Select line is needed.

10.6 Enhancing synchronous serial, and the Inter-Integrated Circuit bus

The Inter-Integrated Circuit (I^2C) protocol was developed by Philips to provide a serial communication standard that overcame some of the shortcomings of Microwire or SPI. As its name suggests, it is meant to provide communication between ICs within a single system. It is intended to be flexible, and tolerant of different technologies and speeds. Like all good standards, it has been exploited well beyond its original intended application. The full I^2C specification can be found in Ref. 10.1, with a useful commentary on it in Ref. 10.2.

10.6.1 Main I^2C features and physical interconnection

Like SPI or Microwire, I^2C is based on a master–slave relationship between nodes. The master controls all bus usage. There is a standard I^2C mode (with clock rate up to 100 kHz), a fast one (with clock rate to 400 kHz) and a high-speed mode (with maximum clock rate of 3.4 Mb/s).

The I^2C bus uses *only* two lines for *all* interconnection, called SDA (serial data) and SCL (serial clock). The output of each node connects to the bus using an Open Drain or Open Collector output, as shown in Figure 10.12, while the node input is through a standard logic buffer. The two interconnect lines each have a pull-up resistor. Both lines are bi-directional, but the SCL clock signal is always generated by the current master.

When no node is accessing the bus the Open Drain outputs are all inactive and the lines idle at Logic 1. The number of nodes connected to the bus is limited just by the number of addresses that are available (see below) and the loading capacitance that each adds. Too much capacitance ultimately will mean that the clock or data rise time will exceed the specified maximum value.

10.6.2 The pull-up resistor

With neither line having an active pull-up, it is the pull-up resistor in conjunction with the line capacitance that determines the rise time. The specification states a maximum line capacitance of 400 pF and a maximum

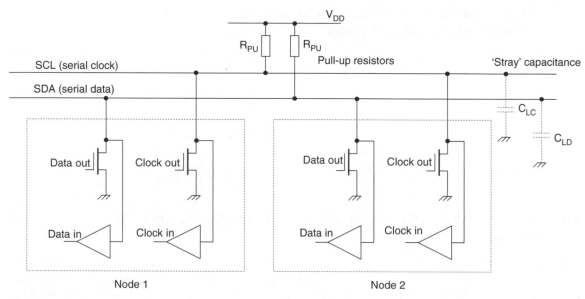

Figure 10.12 The basis of I^{2}C interconnection

rise time (from Logic 0 to Logic 1) in standard mode of 1000 ns. The value of the resistor is chosen to achieve the rise time requirement, dependent on this line capacitance. A low value of resistor reduces the rise time, but increases current consumption. A value of 4.7 kΩ is a widely used; the value must, however, be lowered if the line capacitance is high or can be increased if line capacitance is low. Reference 1.1 gives example calculations for this.

10.6.3 I^{2}C signal characteristics

The I^{2}C protocol follows a very clear format for data transfer, as shown in Figure 10.13. It is initiated by a Start condition, in which the SDA line is taken low, while the SCL line stays high. It is terminated by a

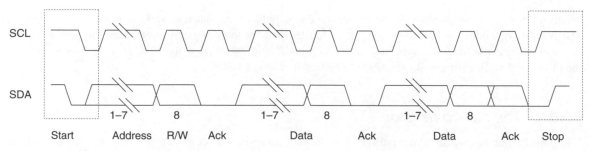

Figure 10.13 I^{2}C signal characteristics

Stop condition, in which the SDA line goes high while the clock is held high. The Start and Stop conditions are asserted by the current master, as is the clock.

Between the Start and the Stop, data is transferred in bytes. During this time, the **SDA** value can only change when **SCL** is low; data must remain stable when the clock line is high. This allows data transfer and Start or Stop to be distinguished. The first byte of any transfer contains address information. The standard allows for either a 7-bit address, within a single byte, or a 10-bit address spread across 2 bytes. The figure shows the 7 bit version. In either mode the eighth bit of the first byte is a Read/Write bit. This determines direction of data flow for the message that follows. At the end of every byte, the transmitter releases the **SDA** line and the receiver must send an acknowledge bit, pulling the **SDA** line low. Any number of bytes can be sent within one message (i.e. between single Start and Stop bits).

There are two slight exceptions to the pattern just described. A 'general call' address is allowed, for which all address bits are zero. This is used to address all nodes on the bus simultaneously. If a master wants to start a new message, when it is still within a message, then a 'Repeated Start' condition is available.

Importantly, the I^2C protocol allows for more than one master, and a node can switch from being a slave to being a master. If the bus is idle, then any of the potential masters can take control of it. If two try to take control of the bus at the same time, an arbitration process is applied, as described in Ref. 10.1.

10.7   The MSSP configured for I^2C

The 16F873A MSSP can be configured for I^2C operation. In this case the **SCL** line is shared with bit 3 of Port C and **SDA** with bit 4 of Port C. In I^2C mode the port is significantly more complex than when in SPI mode, and care needs to be taken to understand it. The first indication of the increased complexity is the whole extra control register, **SSPCON2**, needed for I^2C mode.

We will aim to introduce this complex but interesting serial application incrementally, and illustrate with programs for the Derbot AGV.

10.7.1   The MSSP I^2C registers and their preliminary use

As with the MSSP in SPI mode, the two registers central to the module hardware are the shift register **SSPSR** and the buffer **SSPBUF**. To these are added an address register, **SSPADD**. This is used to hold the slave address when in Slave mode, while in Master mode it forms part of the baud rate generator. Block diagrams of the module hardware, one for each of slave and master, follow shortly.

When in I^2C mode, the MSSP uses the two control registers already introduced, **SSPCON1** and **SSPSTAT**. Most bits in these are, however, used for different functions, so they must effectively be viewed almost as different SFRs, from the point of view of learning about them. They are reproduced in Figures 10.14 and 10.15. To cope with the greater I^2C complexity, there is a further control register, **SSPCON2**, shown in Figure 10.16. There is thus a total of six registers that the programmer uses directly for I^2C operation, in addition to the registers relating to Port C and interrupts.

R/W-0	R/W-0	R/W-0	R/W-0	R/W-0	R/W-0	R/W-0	R/W-0
WCOL	SSPOV	SSPEN	CKP	SSPM3	SSPM2	SSPM1	SSPM0

bit 7   bit 0

bit 7   **WCOL:** Write Collision Detect bit

<u>In Master Transmit mode:</u>

1 = A write to the SSPBUF register was attempted while the I^{2}C conditions were not valid for a transmission to be started. (Must be cleared in software.)

0 = No collision

<u>In Slave Transmit mode:</u>

1 = The SSPBUF register is written while it is still transmitting the previous word. (Must be cleared in software.)

0 = No collision

<u>In Receive mode (Master or Slave modes):</u>

This is a "don't care" bit.

bit 6   **SSPOV:** Receive Overflow Indicator bit

<u>In Receive mode:</u>

1 = A byte is received while the SSPBUF register is still holding the previous byte. (Must be cleared in software.)

0 = No overflow

<u>In Transmit mode:</u>

This is a "don't care" bit in Transmit mode.

bit 5   **SSPEN:** Synchronous Serial Port Enable bit

1 = Enables the serial port and configures the SDA and SCL pins as the serial port pins

0 = Disables the serial port and configures these pins as I/O port pins

    Note: When enabled, the SDA and SCL pins must be properly configured as input or output.

bit 4   **CKP:** SCK Release Control bit

<u>In Slave mode:</u>

1 = Release clock

0 = Holds clock low (clock stretch). (Used to ensure data setup time.)

<u>In Master mode:</u>

Unused in this mode.

bit 3-0   **SSPM3:SSPM0:** Synchronous Serial Port Mode Select bits

1111 = I^{2}C Slave mode, 10-bit address with Start and Stop bit interrupts enabled

1110 = I^{2}C Slave mode, 7-bit address with Start and Stop bit interrupts enabled

1011 = I^{2}C Firmware Controlled Master mode (Slave Idle)

1000 = I^{2}C Master mode, clock = F$_{OSC}$/(4 * (SSPADD + 1))

0111 = I^{2}C Slave mode, 10-bit address

0110 = I^{2}C Slave mode, 7-bit address

    Note: Bit combinations not specifically listed here are either reserved or implemented in SPI mode only.

Figure 10.14 The **SSPCON1** register (address 14$_H$) in I^{2}C mode

R/W-0	R/W-0	R-0	R-0	R-0	R-0	R-0	R-0
SMP	CKE	D/$\overline{\text{A}}$	P	S	R/$\overline{\text{W}}$	UA	BF

bit 7   bit 0

bit 7   **SMP:** Slew Rate Control bit

In Master or Slave mode:

1 = Slew rate control disabled for standard speed mode (100 kHz and 1 MHz)

0 = Slew rate control enabled for high-speed mode (400 kHz)

bit 6   **CKE:** SMBus Select bit

In Master or Slave mode:

1 = Enable SMBus specific inputs

0 = Disable SMBus specific Inputs

bit 5   **D/$\overline{\text{A}}$:** Data/Address bit

In Master mode:

Reserved.

In Slave mode:

1 = Indicates that the last byte received or transmitted was data

0 = Indicates that the last byte received or transmitted was address

bit 4   **P:** Stop bit

1 = Indicates that a Stop bit has been detected last

0 = Stop bit was not detected last

     Note: This bit is cleared on Reset and when SSPEN is cleared.

bit 3   **S:** Start bit

1 = Indicates that a Start bit has been detected last

0 = Start bit was not detected last

     Note: This bit is cleared on Reset and when SSPEN is cleared.

bit 2   **R/$\overline{\text{W}}$:** Read/Write bit information (I²C mode only)

In Slave mode:

1 = Read

0 = Write

     Note: This bit holds the R/$\overline{\text{W}}$ bit information following the last address match. This bit is only valid from the address match to the next Start bit, Stop bit or not $\overline{\text{ACK}}$ bit.

In Master mode:

1 = Transmit is in progress

0 = Transmit is not in progress

     Note: ORing this bit with SEN, RSEN, PEN, RCEN or ACKEN will indicate if the MSSP is in Idle mode.

bit 1   **UA:** Update Address (10-bit Slave mode only)

1 = Indicates that the user needs to update the address in the SSPADD register

0 = Address does not need to be updated

bit 0   **BF:** Buffer Full Status bit

In Transmit mode:

1 = Receive complete, SSPBUF is full

0 = Receive not complete, SSPBUF is empty

In Receive mode:

1 = Data Transmit in progress (does not include the $\overline{\text{ACK}}$ and Stop bits), SSPBUF is full

0 = Data Transmit complete (does not include the $\overline{\text{ACK}}$ and Stop bits), SSPBUF is empty

Figure 10.15 The **SSPSTAT** register (address 94$_{\text{H}}$) in I²C mode

R/W-0	R/W-0	R/W-0	R/W-0	R/W-0	R/W-0	R/W-0	R/W-0
GCEN	ACKSTAT	ACKDT	ACKEN	RCEN	PEN	RSEN	SEN

bit 7   bit 0

bit 7   **GCEN:** General Call Enable bit (Slave mode only)

1 = Enable interrupt when a general call address (0000h) is received in the SSPSR
0 = General call address disabled

bit 6   **ACKSTAT:** Acknowledge Status bit (Master Transmit mode only)

1 = Acknowledge was not received from slave
0 = Acknowledge was received from slave

bit 5   **ACKDT:** Acknowledge Data bit (Master Receive mode only)

1 = Not Acknowledge
0 = Acknowledge

> **Note:**   Value that will be transmitted when the user initiates an Acknowledge sequence at the end of a receive.

bit 4   **ACKEN:** Acknowledge Sequence Enable bit (Master Receive mode only)

1 = Initiate Acknowledge sequence on SDA and SCL pins and transmit ACKDT data bit. Automatically cleared by hardware.
0 = Acknowledge sequence Idle

bit 3   **RCEN:** Receive Enable bit (Master mode only)

1 = Enables Receive mode for I^{2}C
0 = Receive Idle

bit 2   **PEN:** Stop Condition Enable bit (Master mode only)

1 = Initiate Stop condition on SDA and SCL pins. Automatically cleared by hardware.
0 = Stop condition Idle

bit 1   **RSEN:** Repeated Start Condition Enabled bit (Master mode only)

1 = Initiate Repeated Start condition on SDA and SCL pins. Automatically cleared by hardware.
0 = Repeated Start condition Idle

bit 0   **SEN:** Start Condition Enabled/Stretch Enabled bit

In Master mode:
1 = Initiate Start condition on SDA and SCL pins. Automatically cleared by hardware.
0 = Start condition Idle

In Slave mode:
1 = Clock stretching is enabled for both slave transmit and slave receive (stretch enabled)
0 = Clock stretching is enabled for slave transmit only (PIC16F87X compatibility)

Figure 10.16 The **SSPCON2** register (address 91$_H$) in I^{2}C mode

As in SPI mode, the MSSP is enabled for I^{2}C by setting the **SSPEN** bit in the **SSPCON1** register. The mode of operation, notably whether master or slave, and the address length used, is then determined by the setting of the least significant 4 bits of **SSPCON1**. It can be seen from Figure 10.14 that there are six possible I^{2}C modes of operation.

While the bits of the **SSPSTAT** register mostly give information of the current status of the port, the bits in the new **SSPCON2** register (Figure 10.16) initiate one or other of the I^{2}C activities. Setting **SEN**, for example, initiates a Start condition, **PEN** a Stop condition and **RSEN** a Repeated Start. We shall see examples of this soon.

To gain an insight into how these bits are used, and their timing, it is more or less essential to study the timing diagrams that appear in the data sheets. There are many of these, one for each of the possible modes of operation. Two of these are shown a little later in this chapter. The art of developing software to drive the MSSP in I^{2}C mode is very much a case of ensuring that these diagrams are satisfied – completely. That does not mean that every bit displayed in the diagram has to be used, sometimes one does not need to use them all. The flow of events depicted must, however, be followed. The diagrams are not entirely simple and in many cases it is preferable to use or adapt software already written, rather than start from scratch.

10.7.2   The MSSP in I^{2}C Slave mode

When configured in I^{2}C Slave mode, the MSSP operates as shown in Figure 10.17. This diagram is not too complex and indeed has many similarities to the SPI mode diagram of Figure 10.7. The central features of the **SSPSR** shift register and the **SSPBUF** buffer are there. The shift register clock is provided by the external pin **SCL** and data is shifted out or in via the **SDA** pin.

The role of the slave is brutally simple, to wait until it is addressed and then to do what it is told to do. Therefore, a special logic circuit detects when a Start condition occurs, and an address match comparator indicates whether the address that follows matches the internal address of the node, held in the **SSPADD** control register. If there is an address match, then the MSSP interrupt flag **SSPIF**, in register **PIR1**, is set. The programmer uses this to initiate the slave response. This will depend primarily on the value of the **R/$\overline{\text{W}}$** bit in the address word, which is transferred to the **R/$\overline{\text{W}}$** bit in the **SSPSTAT** register, and can be tested in the program.

If a Write is detected, the slave is to act as receiver. The relevant timing diagram, for a 7-bit address, is shown in Figure 10.18(a). The buffer full flag **BF**, set because the address byte has been received, must be cleared by a dummy read of **SSPBUF**. If used, the **SSPIF** flag should be cleared. The slave then clocks in a byte of data, under control of the master clock. If all is operating correctly, an Acknowledge is automatically generated by the slave. Figure 10.18(a) goes on to illustrate the case when the first data byte is not read from **SSPBUF**. The **BF** flag (in **SSPSTAT**) remains high, leading to the overflow flag, **SSPOV**, being set. An Acknowledge is then *not* sent by the slave, and the master terminates the message with a Stop condition.

If a Read is detected in the first byte of the message, the slave must act as a transmitter. The sequence is shown in Figure 10.18(b). When the slave recognises that it has been addressed and a Read demanded, it must write a byte of data into the **SSPBUF** register, which automatically sets the **BF** flag high.

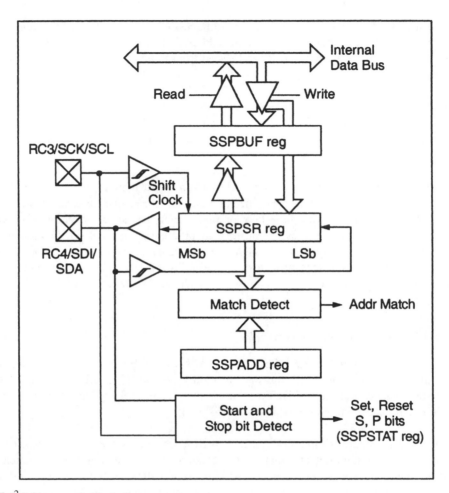

Figure 10.17 I^2C Slave mode block diagram

Of course, it may take a little time to respond to this, so it holds the clock line low until this transfer is complete. This blocks *any* further action on the serial link – the slave has power at last! It releases the serial line, in the software, by setting bit **CKP** high. The hardware automatically returns this low (if **BF** is low) on completion of 9 bits of transmission. This is an example of 'clock stretching' – one of the few ways that a slave can influence the activity of the serial link. It is automatically implemented in Transmit mode, and optionally implemented in Receive mode, by the setting of the **SEN** bit in the **SSPCON2** register.

With the SCK released, the master clocks the data out and should generate an Acknowledge in response. Further status information can be derived from the state of flags in the Status register **SSPSTAT**. The I^2C message is terminated when a stop bit from the master is detected. It should be possible to follow this full sequence in the figure.

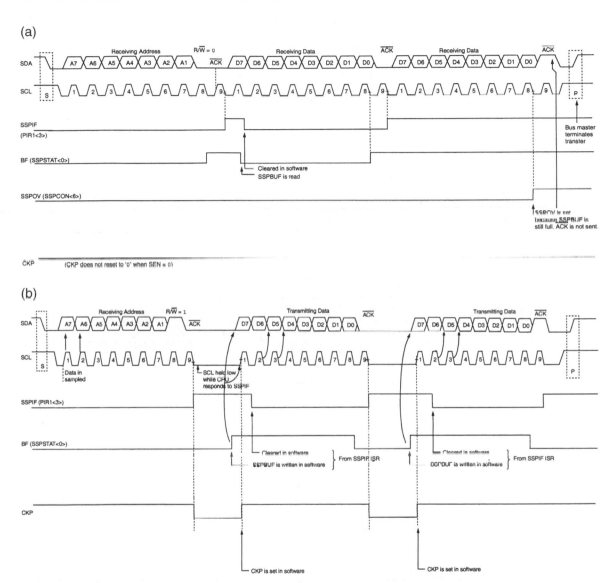

Figure 10.18 I²C Slave mode timing, 7-bit address, **SEN** = 0. (a) Reception. (b) Transmission

10.7.3   The MSSP in I²C Master mode

The MSSP in Master mode is altogether more complex, as can be seen in Figure 10.19. The master must after all control all bus transactions. At the heart of things, however, we still find the **SSPSR** shift register and the **SSPBUF** buffer. Electrical interface with the two serial lines shows the classic I²C connection already seen, with Open Drain for output and Schmitt trigger for input. The clock is now internally generated,

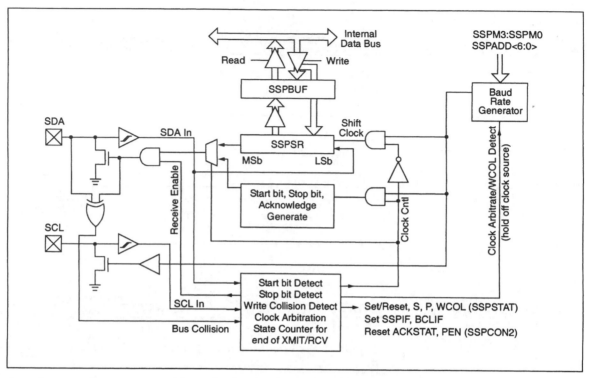

Figure 10.19 The MSSP block diagram, I^2C Master mode

by the baud rate generator. It is routed both to the shift register and to the external bus via the **SCL** pin. The node must be able to generate and detect Start, Stop and Acknowledge conditions. It also, by means of the Exclusive OR gate connected to the **SDA** output, detects a bus collision. This occurs when the master is transmitting, but the logic state on the bus does *not* accord with the intended transmitted value. A collision sets the bus collision interrupt flag, **BCLIF** (Figure 7.10). The master port can also engage in arbitration, if it finds itself in contention with another master.

The baud rate generator, which appears at the top right of Figure 10.19, is shown in further detail in Figure 10.20. It consists of a Down Counter that is reloaded with the value held in the **SSPADD** register when it has counted down to zero. Note clearly that the **SSPADD** register has nothing to do with addresses when the module is in Master mode – masters don't have addresses. The user selects the desired baud rate by the value placed in **SSPADD**. The formula which determines the value, quoted from Ref. 10.3, is:

$$[\textbf{SSPADD}] = \frac{F_{\text{osc}}}{4 \times F_{\text{SCL}}} - 1 \tag{10.1}$$

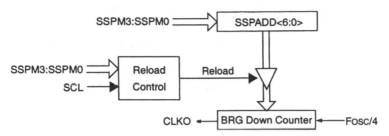

Figure 10.20 The I^{2}C baud rate generator

where [**SSPADD**] is the value loaded into the **SSPADD** register, F_{osc} is the microcontroller clock frequency and F_{SCL} is the desired I^{2}C clock frequency.

Figure 10.21 shows the timing diagram for the port as I^{2}C master, when transmitting a byte of data. The **SEN** bit is seen being used to initiate a Start condition and the **PEN** bit to initiate a Stop. Both are in the **SSPCON2** register. The **R/$\overline{\text{W}}$** bit, in **SSPSTAT**, is used to indicate that transmit is in progress. This sequence of events will be illustrated in the forthcoming Derbot example.

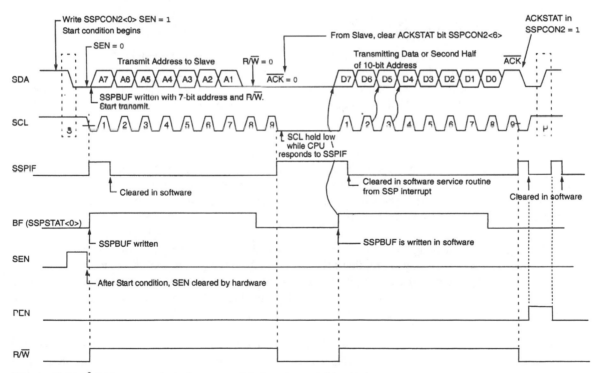

Figure 10.21 I^{2}C Master mode timing – transmission, 7- or 10-bit address

10.8 I²C applied in the Derbot AGV

10.8.1 The Derbot hand controller as a serial node

The Derbot AGV is designed to interface with its hand controller via an I²C link, with the AGV being set as master and the hand controller as slave. Many data transfers are, however, initiated by the hand controller when the user presses a key on the keypad. An I²C slave cannot, however, initiate a transfer! Therefore, an interrupt line is connected from controller to AGV, which forms the external interrupt input of the AGV microcontroller. The full interaction is programmed so that when a keypad press is detected, the controller sends an interrupt to the AGV, which then requests data, via the I²C link.

The two program examples that follow, 10.2 and 10.3, are written respectively for the Derbot AGV acting as master and the hand controller acting as slave. When working together, a keypad press on the hand controller causes an interrupt to be sent to the AGV. This then initiates an I²C message, in which the AGV as master requests a byte of data (the character) from the hand controller. The AGV then echoes the character back to the controller, which sends it to the display. Full program versions can be found on the book CD. The subroutines and ISRs can be used as the basis of any communication between hand controller, AGV and any other I²C peripheral the user may wish to design. It is worth noting, however, that they are comparatively simple routines and do not test or respond to all possible fault conditions.

10.8.2 The AGV as an I²C master

Program Example 10.2 runs on the Derbot AGV and applies the I²C port as a master. The key SFR settings are Port C (where I²C port bits must be set as inputs), **SSPADD** (which determines the clock frequency) and **SSPCON1** (which determines the overall setting). Equation (10.1) is used to determine the value for **SSPADD**. The detail of the **SSPCON1** setting should be explored by comparing it with the register details in Figure 10.14.

All major I²C actions in the program are undertaken with the use of subroutines and it is instructive to look at these. The program starts, from comment **;send opening string**, by transmitting the character message 'Derbot' to the hand controller. This is done in a single multiple-byte message. First the address is sent, using subroutine **I2C_send_add**. The address 52_H is arbitrarily chosen; this must be shifted left by 1 to fit into the address word. With $R/\overline{W}$ set to 0, the transmitted word becomes $A4_H$.

Within the ongoing I²C message, the characters are then read in turn from Table 1 (not shown, but on the book CD) and sent serially, using subroutine **I2C_send_word**. This transfer applies the timing diagram of Figure 10.21. A delay is called between each character, to allow time for the hand controller to write to the LCD display. The end of the character string is marked with the code FF_H. When this is detected, the Stop condition is asserted with subroutine **I2C_send_stop**.

All subsequent program activity is in the ISR, which is initiated by receipt of an external interrupt from the hand controller. The ISR starts an I²C message, addressed to the hand controller, requesting a read. This is done with subroutines **I2C_send_add** and **I2C_rec_word**. With the $R/\overline{W}$ bit now set to 1, the address word is $A5_H$. It should be possible to follow the sequence of the subroutines and their use of the control registers. This message ends with a Stop condition, implemented by subroutine **I2C_send_stop**. The return message, where the received word is echoed back to the hand controller, is done with the subroutine sequence **I2C_send_add**, followed by **I2C_send_word**, followed by **I2C_send_stop**.

```
;*****************************************************************************
;Dbt_kybd_echo_mstr
;This program exercises I2C bus in several ways:
; * sends an opening multi-byte message to the hand controller
; * on interrupt receives a single digit from the Hand Controller,
; * stores it, and echoes it back.
;Routines can be embedded into any program to provide user control of AGV.
;TJW 20.7.05                                     Tested and working 21.7.05
;*****************************************************************************

(opening program sections omitted)
...
;Specify RAM
I2C_RX_word    equ     23      ;holds most recent I2C word recd
I2C_add        equ     24      ;holds address used in I2C message
I2C_TX_word    equ     25      ;holds word to be transmitted on I2C
...
        org  00
        goto start
        org  04
        goto    Interrupt_SR
;Initialise SFRs in Bank 1
start   bcf     status,rp1
        bsf     status,rp0      ;select memory bank 1
...
        movlw   B'10011000'     ;set port C bits, I2C bits are both set as ip
        movwf   trisc
        movlw   07              ;set up 125kHz baud rate
        movwf   sspadd
;Initialise SFRs in Bank 0
        bcf     status,rp0
        movlw   B'00101000'     ;SSPCON1:MSSP on, I2C Master
        movwf   sspcon
...
;Send opening string
        clrf    pointer
        movlw   0a4             ;send slave address, R/W is write
        movwf   I2C_add
        call    I2C_send_add
loop_str1 movf pointer,0
        call    table1
        movwf   I2C_TX_word
        sublw   0ff                     ;test and move on if end marker reached
        btfsc   status,z
        goto    string_end
        call    I2C_send_word
        incf    pointer,1
        call    delay1                  ;give LCD time to write
```

Program Example 10.2 Derbot I^{2}C interchange AGV to hand controller — master (excerpts)

```
        call    delay1
        call    delay1
        goto    loop_str1
string_end    call I2C_send_stop
;Enable interrupts
        bcf     intcon,intf     ;clear pending interrupts
        bsf     intcon,inte     ;enable external interrupt
        bsf     intcon,gie
;Wait for interrupts from Hand Controller
loop    goto    loop
;
;****************************************************************
;ISR. On external interrupt, SSP reads byte from Hand Controller,
;and echoes it back, ie two I2C messages.
;Received Byte stored in I2C_RX_word for further action.
;****************************************************************
Interrupt_SR
        bsf     portc,6         ;diagnostic
;Start new I2C message, requesting word from slave.
        movlw   0a5             ;this is slave address, R/W is read
        movwf   I2C_add
        call    I2C_send_add
;now wait for byte to come in
        call    I2C_rec_word
        call    I2C_send_stop
        bcf     status,rp0
        call    delay20u
;Now echo byte - start new message
        movlw   0a4             ;this is slave address, R/W is write
        movwf   I2C_add
        call    I2C_send_add
;send the echoed character
        movf    I2C_RX_word,0   ;move received word to transmit store
        movwf   I2C_TX_word
        call    I2C_send_word
        call    I2C_send_stop
        bcf     status,rp0
        bcf     portc,6         ;clear diag led
        bcf     intcon,intf
        retfie
;****************************************************************
;SUBROUTINES
;****************************************************************
;initiates I2C message, by sending the word found in I2C_add, which
;must include R/W bit. Waits for all acknowledgement and completion
;states. Leaves RAM in Bank 0.
I2C_send_add
        bsf     status,rp0
        bsf     sspcon2,sen     ;force start bit
        btfsc   sspcon2,sen     ;check for its completion
        goto    $-1
        bcf     status,rp0
        movf    I2C_add,0       ;load address and data dirn bit
```

Program Example 10.2 Continued

```
        movwf   sspbuf          ;and send
        bcf     pir1,sspif      ;will test this soon
        bsf     status,rp0
        btfsc   sspstat,bf      ;test for write complete
        goto    $-1
        btfsc   sspcon2,ackstat ;wait for 0 acknowledge bit
        goto    $-1
        bcf     status,rp0
        btfss   pir1,sspif      ;test for int flag to show completion
        goto    $-1
        bcf     pir1,sspif
        return
;
;Receives (single) word from I2C bus, and stores in I2C_RX_word. Returns Ack of 1,
;signalling this is last byte. Leaves RAM in Bank 0
I2C_rec_word  bsf status,rp0
        bsf     sspcon2,rcen    ;set receive enable bit
        btfss   sspstat,bf      ;wait for buffer full
        goto    $-1
        bcf     status,rp0      ;read the data
        movf    sspbuf,0
        movwf   I2C_RX_word     ;store it for use somewhere
        bcf     pir1,sspif      ;preclear int flag, as we are about to use it
        bsf     status,rp0
        bsf     sspcon2,ackdt   ;set required acknowledge state, 1 as it's
                                ;last byte
        bsf     sspcon2,acken   ;and enable it
        bcf     status,rp0
        btfss   pir1,sspif      ;use interrupt flag to test for end of ack
        goto    $-1
        bcf     status,rp0
        return
;
;Sends word on I2C bus, and awaits acknowledgement. Leaves RAM in Bank 0.
I2C_send_word bcf status,rp0
        movf    I2C_TX_word,0   ;get the word
        movwf   sspbuf          ;this starts the transfer
        bsf     status,rp0
        btfsc   sspstat,r_w     ;test for write complete
        goto    $-1
        btfsc   sspcon2,ackstat ;check for 0 acknowledge bit
        goto    $-1
        bcf     status,rp0
        return
;
;Sends I2C stop bit, and awaits completion. Leaves RAM in Bank 0.
I2C_send_stop bsf status,rp0
        bsf     sspcon2,pen     ;force stop bit
        btfss   sspstat,p       ;test for stop bit completion
        goto    $-1
        bcf     status,rp0
        return
...
```

Program Example 10.2 Continued

10.8.3   The hand controller as an I²C slave

Program Example 10.3 runs on the Derbot hand controller. It is an extension of the program **keypad_test**, which appeared in Program Example 8.1, with the I²C transfer to the AGV now inserted. Notice first the way the control registers are used, compared to the Master mode. **SSPCON1** sets the node as slave and **SSPADD** is now used to hold the slave address. **SSPADD** holds the slave address, 52$_H$, rotated left by 1 bit. The I²C bits of Port C again must be set as input. Two interrupts are enabled – Port B interrupt on change, to detect the keypad being pressed, and the MSSP interrupt, to alert the microcontroller to the arrival of the first byte of an I²C message. Interrupt control bits **GIE**, **PEIE**, **SSPIE** and **RBIE** are accordingly set.

An MSSP interrupt occurs when an address match is detected on an address byte. The interrupt routine first determines the interrupt source. If it is an I²C interrupt, the **D/A̅** bit in **SSPSTAT** is first tested, to determine if the byte just received was address or data. If address, the **SSPBUF** register (which will be holding the address byte) is read – simply to clear the **BF** flag in **SSPSTAT**. A test is then made of the **R/W̅** bit in **SSPSTAT**. If the master is requesting a Write, then the ISR is quit. The program will wait for the next interrupt, which will indicate the expected incoming data byte. If the master is requesting a Read, then program execution stays in the ISR and goes to the label **Send_I2C**. The timing diagram of Figure 10.18(b) is followed. The byte held in **kpad_char** is moved to **SSPBUF**, the **CKP** bit in **SSPCON1** is set high to release the clock, and the interrupt flag is monitored to determine completion.

If in the test of the **D/A̅** bit an incoming data byte had been detected, then program execution moves to the label **ISR1**. The data will by then already be in **SSPBUF**. This is accordingly read and the character is sent to the display, using the **lcd_write** subroutine. A subroutine **dig_pntr_set** is also invoked, seen only in the full listing on the CD. This manages the LCD pointer, ensuring that characters are not sent to positions which do not appear on the size of display used.

```
;******************************************************************
;dbt_kypd_echo_slave                    for Derbot Hand Controller
;Reads keypad value when pressed and sends interrupt to main AGV.
;Transmits on I2C keypad character when asked, and receives echo back.
;Displays anything sent from AGV.
;TJW 13.7.05                                    tested 15.7.05
;******************************************************************
...
(opening program sections omitted)
...
;Initialise SFRs in Bank 1
main    bcf     status,rp1
        bsf     status,rp0      ;select memory bank 1
...
        movlw B'00011000'       ;I2C bits of Port C to ip
        movwf trisc
        movlw B'10100100'
        movwf sspadd            ;our address to be 52H
                                  ;(it's shifted by one in sspadd)
        bsf   pie1,sspie        ;enable I2C interrupt
;
```

Program Example 10.3 Derbot I²C interchange AGV to hand controller – slave (excerpts)

```
;Initialise SFRs in Bank 0
        bcf     status,rp0    ;select bank 0
        movlw   B'00110110'   ;SSPCON1:MSSP on, I2C Slave, 7 bit address,
                              ;interrupts off, no clock stretch on Receive
        movwf   sspcon
...
;enable global interrupts
        clrf    portb   ;initialise keypad value
        bsf     intcon,gie
        bsf     intcon,peie
loop    goto    loop          ;await keypad and I2C interrupts
;
;********************************************************************
;This is ISR, caused by keypad or I2C address match.
;Does not context save, as all action is in ISRs.
;********************************************************************
Interrupt_SR btfsc intcon, rbif ;is it keypad interrupt?
        goto    kpad_ISR
;Here if interrupt is I2C, either address match (Ack sent automatically)
;OR further received byte has been detected.
;check whether this byte was address or data
        bsf     status,rp0
        btfsc   sspstat,d_a
        goto    ISR1            ;go if word was data
        bcf     status,rp0
        movf    sspbuf,0        ;dummy read of the address byte, to clear flag
;check if read, if so load and send byte
        bsf     status,rp0
        btfsc   sspstat,r_w
        goto    Send_I2C
        bcf     status,rp0      ;otherwise exit ISR, to await incoming data byte
        bcf     pir1,sspif      ;clear interrupt bit, and end ISR
        retfie
;Here if data byte has been detected, word is hence already in buffer.
ISR1 call dig_pntr_set          ;sort display pointer
        bcf     status,rp0      ;read word
        movf    sspbuf,0
        movwf   I2C_RX_word     ;save word
        movwf   lcd_op          ;prepare to send word to display
        bsf     portc,lcd_rs
        call    lcd_write
;transfer to lcd is done, end ISR.
        bcf     pir1,sspif      ;clear interrupt bit
        retfie
;here if sending I2C word. Send byte held in kpad_char
Send_I2C bcf    status,rp0
        bcf     pir1,sspif
        movf    kpad_char,0     ;move character to sspbuf
        movwf   sspbuf
        bsf     sspcon,ckp      ;release clock
        btfss   pir1,sspif      ;wait for completion of transfer
        goto    $-1
;transfer is complete, end ISR.
        bcf     pir1,sspif      ;clear interrupt bit
```

Program Example 10.3 Continued

```
retfie
;
;Keypad press has been detected through Port B Interrupt on Change.
;Gets value, converts to character, stores in kpad_char, awaits key release,
;and sends interrupt to AGV
kpad_ISR    call    kpad_rd
...
```

Program Example 10.3 Continued

10.8.4   Evaluation of the Derbot I²C programs

Figure 10.22 shows some of the waveforms of the I²C exhange, when the programs above are running. In (a), we see the characteristic nine clock cycles per word of an I²C data exchange. The clock frequency can be seen to be just a little lower than the expected frequency of 125 kHz. The master starts a message, with the address byte 10100100_B (i.e. $A4_H$) being sent. As we know, the LSB of this is the **R/$\overline{W}$** bit, indicating a Write is requested. The slave then replies with the word 00110111_B (i.e. 37_H, or the ASCII character 7, meaning that the keypad key '7' has just been pressed). The master is seen to Acknowledge and then exert a Stop condition.

The detail of Figure 10.22(b) is characteristic of an Open Drain output driving a logic line. The transition from Logic 1 to 0 is fast, and caused by the transistor output switching on. When it switches off, however, the line only rises to Logic 1 through the action of the pull-up resistor, and the slower, exponential rise of this is clearly seen.

The two programs just discussed are useful, and show the use of I²C in an application that is reasonably constrained. It's worth noting that they don't take account of every state an I²C node can get into, nor do they take corrective action if the serial link is not working properly. In the master program, for example, if no Acknowledge is received, the program simply loops indefinitely.

(a)   (b)

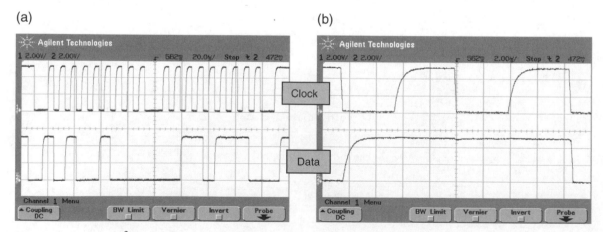

Figure 10.22 Actual I²C waveforms. (a) A complete, single-byte message. (b) Detail, showing signal edges

10.9 Evaluation of synchronous serial data communication and an introduction to asynchronous

Synchronous serial communication as discussed is an incredibly useful way of moving data around, but the question remains: do we really need to send that clock signal wherever the data goes? Although it allows an easy way of synchronising the data, it does have these disadvanges:

- An extra line is needed to go to every data node
- The bandwidth needed for the clock is always twice the bandwidth needed for the data; therefore, it is demands of the clock which limit the overall data rate
- Over long distances, clock and data themselves could lose synchronisation.

10.9.1 Asynchronous principles

For the reasons just listed, a number of serial standards have been developed which do not require a clock signal to be sent with the data. This is generally called *asynchronous* serial communication. It is now up to the receiver to extract all timing information directly from the signal itself. This has the effect of laying new and different demands on the signal, and making transmitter and receiver nodes somewhat more complex than comparable synchronous nodes.

A common approach (but not the only one – the diversity of asynchronous links has been limited only by human ingenuity) to resolving the challenges of asynchronous communication is based on this:

- Data rate is predetermined – both transmitter and receiver are preset to recognise the same data rate. Hence each node needs an accurate and stable clock source, from which the baud rate can be generated. Small variations from the theoretical value can, however, be accommodated.
- Each byte or word is *framed* with a Start and Stop bit. These allow synchronisation to be initiated before the data starts to flow.

An asynchronous data format, of the sort used by such standards as RS-232, is shown in Figure 10.23. The line idles in a predetermined state. The start of a data word is initiated by a *Start* bit, which has polarity opposite to that of the idle state. The leading edge of the Start bit is used for synchronisation. Eight data bits are then clocked in. A ninth bit, for parity checking, is also sometimes used. The line then returns to the idle state, which forms a Stop bit. A new word of data can be sent immediately, following the completion of a single Stop bit, or the line may remain in the idle state until it is needed again.

10.9.2 Synchronising serial data – without an incoming clock

In order to receive correctly an incoming data stream with no accompanying clock, the receiver must be able to detect the start of the byte or word and the moment in each bit when the data is valid. With the data rate predetermined, one might think that this is not a difficult problem, but it is impossible for different microcontrollers to have clock frequencies that are precisely the same.

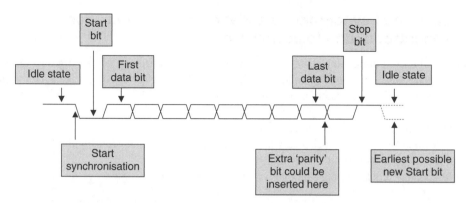

Figure 10.23 A common asynchronous serial data format

The principle of how this timing can be done is shown in Figure 10.24. The receiver runs an internal clock whose frequency is an exact multiple of the anticipated bit rate. Usually, a multiple of 16 is chosen, but this is not essential. The receiver monitors the state of the incoming data on the serial receive line. When a Start bit is detected, a counter begins to count clock cycles until the midpoint of the anticipated Start bit is reached, i.e. eight clock cycles when a ×16 clock is being used. It tests the state of the incoming data line again, to confirm a Start bit is present. If not, the receive is aborted. If the Start bit is confirmed as present, the clock counter counts a further 16 cycles, to the middle of the first data bit. At this point, it clocks that bit into the main receive shift register. In this manner it continues to clock in bits, waiting each time for a bit width before clocking in the next. Depending on the setting of the receiver, it clocks in 8 or 9 bits (or whatever the expected word length is). After a further bit width, it tests for a Stop bit. If this is of the correct value, then a valid reception can be flagged and the received data can be latched into a buffer.

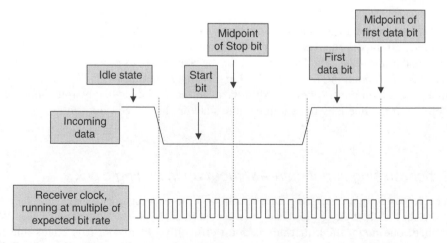

Figure 10.24 Synchronising the asynchronous data signal

If the Stop bit is not present, then a 'framing error' is flagged. The receiver can then be readied to receive the next word.

10.10   The 16F87XA Addressable Universal Synchronous Asynchronous Receiver Transmitter (USART)

10.10.1   Port overview

The second serial port that the 16F87XA family has is an Addressable Universal Synchronous Asynchronous Receiver Transmitter (USART). This rather forbidding title tells us that it can operate in both synchronous and asynchronous modes, and that it can receive and transmit. The inclusion of the word 'Universal' reflects traditional titles and simply implies that it can be configured in all major operating modes needed. The 'Addressable' term indicates a mode of use whereby an incoming byte can be designated and interpreted as an address.

The USART can be configured as synchronous master, synchronous slave or in asynchronous mode. In the latter case it is full duplex – that is, it can transmit and receive both at the same time. Thus, it has both a receive shift register and a transmit shift register, which can operate simultaneously. Both sections share the same baud rate generator and have the same data format. As can be seen from Figure 7.10, it has interrupt sources for both receive and transmit. The USART shares pins with Port C, the Receive line being on bit 7 and the Transmit on bit 6.

Operation of the USART is controlled by two registers, **TXSTA** (Figure 10.25) and **RCSTA** (Figure 10.26). The port is enabled by the **SPEN** bit of **RCSTA**, and selection of synchronous or asynchronous modes is by the **SYNC** bit of the **TXSTA** register.

We will consider each operating mode in turn, with a small example from the Derbot AGV.

10.10.2   The USART asynchronous transmitter

The block diagram of the USART transmitter section is shown in Figure 10.27. It is controlled mainly by the **TXSTA** control register. To start with the familiar, we see in the block diagram that central feature of a serial port – a shift register, 'TSR register'. Notice that data is transmitted LSB first, unlike the MSSP port. The shift register is buffered by the **TXREG** register, an addressable SFR linked to the data bus. It is to this register that the program writes. The shift register is driven by a clock, 'baud rate CLK', which is enabled by the **TXEN** bit. The clock frequency is set by the baud rate generator, depending on the value held in the **SPBRG** register. The output of the shift register is connected to the microcontroller pin via the 'Pin Buffer and Control' circuit. This is enabled by the Serial Port Enable bit, **SPEN**, in the **RCSTA** control register.

Data to be transmitted must be loaded into the **TXREG** buffer by the program. It is transferred immediately to the **TSR** shift register if no transmission is taking place, *or* after the Stop bit if a transmission is already under way. Status information is provided by two bits, the interrupt flag **TXIF** and the **TMRT** bit. The former indicates the status of **TXREG**. When the data transfer between **TXREG** and shift register occurs, then the interrupt flag **TXIF** (in register **PIR1**, Figure 7.12) is set. This cannot be cleared in software and is

R/W-0	R/W-0	R/W-0	R/W-0	U-0	R/W-0	R-1	R/W-0
CSRC	TX9	TXEN	SYNC	—	BRGH	TRMT	TX9D

bit 7   bit 0

bit 7   **CSRC:** Clock Source Select bit
        Asynchronous mode:
        Don't care.
        Synchronous mode:
        1 = Master mode (clock generated internally from BRG)
        0 = Slave mode (clock from external source)

bit 6   **TX9:** 9-bit Transmit Enable bit
        1 = Selects 9-bit transmission
        0 = Selects 8-bit transmission

bit 5   **TXEN:** Transmit Enable bit
        1 = Transmit enabled
        0 = Transmit disabled

        Note: SREN/CREN overrides TXEN in Sync mode.

bit 4   **SYNC:** USART Mode Select bit
        1 = Synchronous mode
        0 = Asynchronous mode

bit 3   **Unimplemented:** Read as '0'

bit 2   **BRGH:** High Baud Rate Select bit
        Asynchronous mode:
        1 = High speed
        0 = Low speed
        Synchronous mode:
        Unused in this mode.

bit 1   **TRMT:** Transmit Shift Register Status bit
        1 = TSR empty
        0 = TSR full

bit 0   **TX9D:** 9th bit of Transmit Data, can be Parity bit

Figure 10.25 The transmit status and control register, **TXSTA** (address 98_H)

only cleared when **TXREG** is reloaded. The bit **TRMT** monitors the state of the shift register and can be polled by the program. It is set when the shift register is empty, i.e. a transmit has been completed.

A ninth data bit, **TX9D**, enabled by bit **TX9**, can be inserted into the transmitted word. Both these bits appear in the **TXSTA** control register. This ninth bit *can* be used as a parity bit. Unlike some serial ports, however, the parity value is not generated automatically in hardware – it is up to the programmer to do this within the program. If the ninth bit is to be used, it should be set up *before* its associated data word is written to **TXREG**. If this is not done, then a transfer to **TXREG** will start a serial data transfer before the

R/W-0	R/W-0	R/W-0	R/W-0	R/W-0	R-0	R-0	R-x
SPEN	RX9	SREN	CREN	ADDEN	FERR	OERR	RX9D

bit 7   bit 0

bit 7   **SPEN:** Serial Port Enable bit

1 = Serial port enabled (configures RC7/RX/DT and RC6/TX/CK pins as serial port pins)
0 = Serial port disabled

bit 6   **RX9:** 9-bit Receive Enable bit

1 = Selects 9-bit reception
0 = Selects 8-bit reception

bit 5   **SREN:** Single Receive Enable bit

<u>Asynchronous mode:</u>
Don't care.

<u>Synchronous mode – Master:</u>
1 = Enables single receive
0 = Disables single receive
This bit is cleared after reception is complete.

<u>Synchronous mode – Slave:</u>
Don't care.

bit 4   **CREN:** Continuous Receive Enable bit

<u>Asynchronous mode:</u>
1 = Enables continuous receive
0 = Disables continuous receive

<u>Synchronous mode:</u>
1 = Enables continuous receive until enable bit CREN is cleared (CREN overrides SREN)
0 = Disables continuous receive

bit 3   **ADDEN:** Address Detect Enable bit

<u>Asynchronous mode 9-bit (RX9 = 1):</u>
1 = Enables address detection, enables interrupt and load of the receive buffer when RSR<8> is set
0 = Disables address detection, all bytes are received and ninth bit can be used as parity bit

bit 2   **FERR:** Framing Error bit

1 = Framing error (can be updated by reading RCREG register and receive next valid byte)
0 = No framing error

bit 1   **OERR:** Overrun Error bit

1 = Overrun error (can be cleared by clearing bit CREN)
0 = No overrun error

bit 0   **RX9D:** 9th bit of Received Data (can be parity bit but must be calculated by user firmware)

Figure 10.26 The **RCSTA** register (address 18_H), receive status and control register

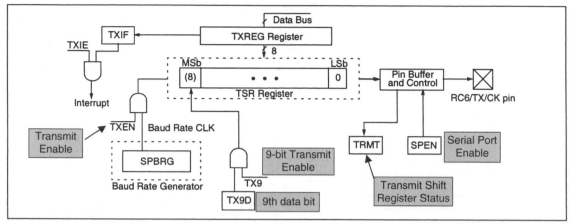

Key: For TXIF, TXIE, see Figure 7.12.
All other control bits in TXSTA (Fig. 10.25) except for SPEN.

Figure 10.27 The USART transmit block diagram (supplementary labels in shaded boxes added by the author)

ninth bit is in place. An alternative use for the ninth bit is to indicate an address in the accompanying byte, as described in Section 10.10.6.

10.10.3   *The USART baud rate generator*

The baud rate generator is used in both synchronous and asynchronous modes of the USART. It is built round a free-running 8-bit counter, controlled by the **SPBRG** register. The counter is clocked by the internal microcontroller oscillator frequency, and the action of the counter is effectively to divide this frequency down by the amount predetermined by the value in **SPBRG**. The division rate is further modified in asynchronous mode by a prescaler bit, **BRGH**, in register **TXSTA**. The resulting baud rate frequencies are as follows, where [**SPBRG**] is the value in the register of the same name:

Asynchronous

$$\text{For } \mathbf{BRGH} = 0 \qquad \text{Baud rate} = \frac{f_{\text{osc}}}{64([\mathbf{SPBRG}] + 1)} \tag{10.2}$$

$$\text{For } \mathbf{BRGH} = 1 \qquad \text{Baud rate} = \frac{f_{\text{osc}}}{16([\mathbf{SPBRG}] + 1)} \tag{10.3}$$

Synchronous

$$\mathbf{BRGH} = \textit{don't care} \qquad \text{Baud rate} = \frac{f_{\text{osc}}}{4([\mathbf{SPBRG}] + 1)} \tag{10.4}$$

10.10.4 The USART asynchronous receiver

The block diagram of the USART receiver section is shown in Figure 10.28. It is controlled mainly by the **RCSTA** register (Figure 10.26). In some ways the diagram is a mirror image of the transmitter diagram of Figure 10.27. Some features are, however, distinctly different due to the differing requirements of the receiver and transmitter. Data enters the Port C bit 7 pin of the microcontroller. The data recovery circuit is included to minimise interference. Every time a data value is to be determined three samples are taken and the majority value transferred. Central to the receiver we see a shift register, 'RSR register'. This is driven by a clock from the baud rate generator. The shift register is *double*-buffered by a FIFO (first in, first out) buffer. The upper byte in this is the **RCREG** register, an addressable SFR. This double-buffer can hold the most recent two data bytes that have been received and thereby allow a third incoming byte to be clocked in to the shift register. The port is enabled overall by the **SPEN** bit. Subservient to this is a Single Receive Enable bit **SREN** (not to be confused with **SPEN**!) or a Continuous Receive Enable bit **CREN**. This allows choice between accepting a single incoming word or a continuous series.

When a complete word has been received correctly, including the Stop bit, the main 8 data bits are transferred to the FIFO, if it is clear to receive, and interrupt flag bit **RCIF** is set. If a ninth bit is used (determined by

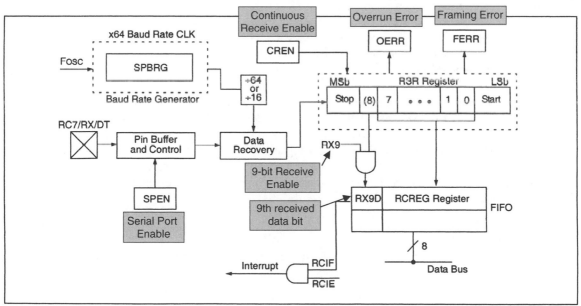

Key: FIFO: First in first out.   For RCIF, RCIE, see Figure 7.6
        For RCIF, RCIE, see Figure 7.12.
        All other control bits in RCSTA (Figure 10.26).

Figure 10.28 The USART receive block diagram (supplementary labels in shaded boxes added by the author)

the state of **RX9**), then it is transferred to bit **RX9D**. This bit has the same level of buffering as the main incoming data bytes.

If an incoming data byte is clocked in, but the Stop bit is detected as Logic 0, then a framing error is indicated by bit **FERR**. As with **RCREG** and **RX9D**, this bit is double-buffered, so the framing status of each buffered bit can be tested. If both buffers of **RCREG** are full and the RSR register has received a further complete byte, then the overrun error bit, **OERR**, is set. The word in the RSR register is lost and further transfers from RSR register to **RCREG** are then blocked. The **OERR** flag must be cleared in software before further reception can take place. When reading a data value, the action of reading **RCREG** causes the **FERR** and **RX9D** bits to be updated with new values. These bits should therefore be read first.

10.10.5 An asynchronous example

In this section we extend Program Example 10.2, described earlier in this chapter, to include an asynchronous communication link. Now when the keypad is pressed on the Derbot hand controller, the keypad character is sent to the AGV, which then sends it out on its asynchronous transmitter. This is looped back to its asynchronous receiver and the word is finally echoed back to the hand controller. While this extra serial link provides no extra functionality whatsoever, it does give a good opportunity to develop a simple test program and look at more serial data! Program Example 10.4 shows the detail relating to the asynchronous communication. The full program is on the book CD.

It should be possible to follow the initial setting of the **TXSTA** and **RCSTA** registers as shown, by reading the program comments and comparing with the control registers in Figures 10.25 and 10.26. The baud rate of 50 kbps was chosen arbitrarily.

As with Program Example 10.2, all data transfer occurs within the main ISR. For this reason the asynchronous interrupts are not used. The ISR starts with a byte being read from the hand controller on the I^2C port and stored in **I2C_RX_word**. The point where data is sent on the asynchronous transmitter is shown. The word just received is transferred to the **TXREG** register. As it transfers out, the asynchronous receiver will simultaneously begin to clock it in. The program simply waits for the receive interrupt flag to be set, indicating that a word has been received and a transfer is complete. The received word is then echoed back to the hand controller, via the I^2C link.

```
;*****************************************************************************
;Dbt_kybd_echo_async
;This program receives a digit from the Hand Controller on the I2C
;bus,stores it, sends it through the asynchronous serial link,
;and echoes it back to the I2C. Each I2C message one byte only.
;Routines can be adapted and embedded into any Derbot program.
;TJW 18.7.05                        Tested and working 19.7.05
;*****************************************************************************
...
(early program sections omitted)
...
```

Program Example 10.4 Asynchronous data transfer on the Derbot

```
        bcf         status,rp0
;Initialise USART in both banks
        movlw  B'10010000'     ;set up async channel: port is on, 8-bit transfer,
        movwf  rcsta           ;continuous receiving, no address detect
        bsf    status,rp0
        movlw  B'00100100'     ;set up async channel:transmit enabled, 8-bit,
        movwf  txsta           ;high speed baud rate
        movlw  04              ;set up baud rate of 50k
        movwf  spbrg
        bcf    status,rp0
...
(program sections omitted)
...
;*********************************************************************
;
;ISR. On external interrupt, SSP reads byte from Hand Controller,
;sends it out on USART, receives it back through USART
;and echoes it back to keypad.
;Received Byte stored in I2C_RX_word for further action.
;*********************************************************************
Interrupt_SR
...
;send out via async comm channel
        bcf    pir1,rcif       ;preclear receive interrupt flag
        movf   I2C_RX_word,0   ;get word, and move to txreg
        movwf  txreg
        btfss  pir1,rcif       ;test for receive interrupt flag,
                               ;indicating receive complete
        goto   $-1
        movf   rcreg,0         ;get and store received word
        movwf  async_RX_word
...
;send the echoed character
        movf   async_RX_word   ;move async received word to transmit store
        movwf  I2C_TX_word
        call   I2C_send_word
...
```

Program Example 10.4 Continued

Waveforms from this program example can be seen in Figure 10.29. Figure (a) shows a byte of asynchronous data, with the display in oscilloscope mode. With the horizontal time base at 50 µs/division, a bit period of exactly 20 µs can be seen, relating exactly to the selected baud rate of 50 kbps. The idle state of Logic 1 and the opening Start bit at Logic 0 can clearly be seen. The data runs 'backwards', LSB first, and can be seen to be 00110101_B, or 35 in hexadecimal. This is ASCII character 5, indicating that this was the keypad button pressed. The Stop bit merges into the next idle period, being simply its first 20 µs, so it cannot be seen explicitly.

Figure 10.29(b) shows the same asynchronous word, but now with part of the preceding and following I^2C messages. The oscilloscope is here in logic analyser mode. The first I^2C message shows just the data byte, again 00110101_B, which is repeated by the asynchronous link, although of course in its own format

(a)   (b)

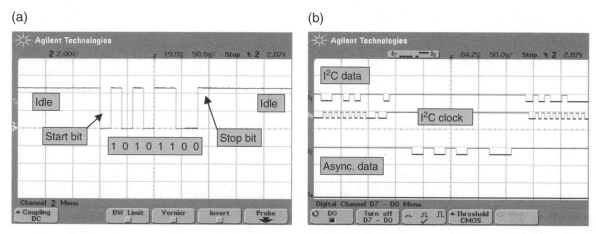

Figure 10.29 Serial waveforms. (a) Single byte, asynchronous. (b) Single byte, passed from I^2C to asynchronous

The following I^2C message is just the address byte of the hand controller, which can be read as 10100100_B. The $R/\overline{W}$ bit is low (i.e. the master will write to the slave) and an Acknowledge can just be seen on the ninth clock cycle.

10.10.6   Using address detection with the USART receive mode

The USART may be used in a way that allows an address to be embedded in the received data. Multiple nodes can then be connected to the serial line and a node can recognise its own address. The USART receiver must be set in 9-bit mode (bit **RX9** = 1) and the address enable bit **ADDEN**, in **RCSTA**, should be set to 1. Once in this mode, a Logic 1 in the ninth bit indicates that the accompanying byte is an address. Initially, all data whose ninth bit is zero is ignored. When a byte is detected whose ninth bit is 1, the program can read the accompanying byte and determine (in software) whether an address match has occurred. If this is the case, the program can revert the port to normal reception by setting **ADDEN** to 0, and further words will be read as data. This continues until a further address word is detected, which may be for another node.

10.10.7   The USART in synchronous mode

Aside from its essential asynchronous capability, the USART can also be used in synchronous mode, which is selected by setting the **SYNC** bit in the **TXSTA** register. The port must be enabled with **RCSTA** bit **SPEN**. Port C bit 7 is then used for serial data and bit 6 for serial clock. Given an understanding of the SPI mode of the MSSP port, as described earlier in this chapter, the underlying concepts of this mode of USART operation will not present any major difficulty. It is therefore left to the reader to read the data on this, if you wish to use it.

10.11 Implementing serial without a serial port – 'bit banging'

The foregoing pages seem to imply that to make use of serial communication, it is absolutely essential to use a microcontroller with one or more serial ports. This is not absolutely true, as it is possible to generate and receive serial data streams in software only, using standard port bits for input/output. This can be a comparatively simple, even attractive (from a cost-saving point of view), option for a simple synchronous link, especially if it is only occasionally used. It remains a possibility for more advanced protocols like I^2C, but becomes increasingly difficult to implement. Chapter 6 of Ref. 1.1 gives further information and an example for the PIC 16F84 microcontroller.

10.12 Building up the Derbot

To run most of the programs used as examples in this chapter, you will need to have a working LCD version of the hand controller. The 'bus' connector on the AGV will have to be in place, as well as the I^2C pull up resistors. The circuit will then be very close to that shown in Figure A3.1, except without the light-dependent resistors and the ultrasound detector.

Summary

- Serial communication is an increasingly important aspect of embedded systems. A good understanding is essential to the aspiring designer.
- There are two broad types of serial communication: synchronous and asynchronous.
- There are a very large number of different standards and protocols for serial communication, ranging from the very simple to the seriously complicated. It is important to match the right protocol with the right application.
- The 16F873A microcontroller has two extremely flexible serial ports. The cost of flexibility is a significant level of complexity in grasping their use. Therefore, it is often worth adapting publicly available routines to use, rather than starting from scratch in writing new code.

References

10.1. The I^2C Bus Specification, Version 2.1 (2000). Philips Semiconductors, Document number 9398 393 40011.

10.2. I^2C Manual (2003). Philips Semiconductors, Application Note AN10216–01.

10.3. Using the PICmicro® MSSP Module for Master I^2C^{TM} Communications (2000). Microchip Technology, Application Note AN735, Ref. no. DS00735A.

10.4. Using the PICmicro® SSP Module for Slave I^2C^{TM} Communications (2000). Microchip Technology, Application Note AN734, Ref. no. DS00734A.

10.5. Asynchronous Communications with the PICmicro® USART (2003). Microchip Technology, Application Note AN774, Ref. no. DS00774A.

11
Data acquisition and manipulation

In the early chapters of this book we limited ourselves to a world that is almost entirely digital. While we want to benefit from the advantages that digital signals can offer us, we need to recognise that most real variables are analog in nature. They are continuously variable and can take an infinite range of different values, whether we are talking about temperature, sound level, frequency or other variables. It is necessary, therefore, for the microcontroller to be able to read values that are analog and if necessary generate output values that are analog, even though internally the microcontroller is relentlessly a digital device. The process of converting an analog signal to digital, along with all the attendant signal manipulation, is usually called 'data acquisition'.

Once data from the outside world has been acquired, it needs to be processed and put to use. It may also need to be averaged, scaled, linearised or stored. Quite possibly it will be used for some form of control purpose and it may need to be displayed, or transmitted to another device.

Data acquisition, and the use of the data acquired, is the business of this chapter. In the chapter you will learn about:

- The main features of a data acquisition system
- The characteristics of an analog-to-digital converter
- The characteristics of the 16F873A analog-to-digital converter
- How the 16F873A analog-to-digital converter can be applied
- Some simple data manipulation techniques
- The use of comparators and the 16F873A comparator capability.

Once we have the ability to acquire data and manipulate it in simple ways, we are in the powerful position of being able to make a variety of measuring devices. The chapter therefore ends with a number of illustrative projects. These use the Derbot either as an AGV or else simply use the core design as the basis for other projects, which have no need for wheels!

11.1 The main idea – analog and digital quantities, their acquisition and use

Most transducers produce output signals that are an *analog* of the quantity they represent. Thus, the voltage output from a temperature sensor represents the temperature as faithfully as it can, increasing or decreasing as the temperature does the same. Similarly, a microphone output signal represents the precise characteristics of the sound wave as best it can, in amplitude, frequency and waveform. Analog signals are fine things, *but* they suffer from a number of big disadvantages, as Table 11.1 shows. *Digital* signals, on the other hand,

Table 11.1 Some properties of analog and digital quantities

Property	Analog	Digital
Means of (electrical) representation	A continuously variable voltage, or current, represents the variable.	Variable is represented by a binary number.
Precision of representation	Can take infinite range of values; absolute precision is theoretically possible, as long as signal is kept completely uncorrupted.	Only a fixed number of digit combinations are available to represent measure; for example, an 8-bit number has only 256 different combinations. 'Continuously variable' quality of analog signal cannot be replicated.
Resistance to signal degradation	Almost inevitably suffers from drift, attenuation, distortion, interference. Cannot completely recover from these.	Digital representation is intrinsically tolerant of most forms of signal degradation. Error checking can also be introduced and with appropriate techniques complete recovery of a corrupted signal can be possible.
Processing	Analog signal processing using op amps and other circuits has reached sophisticated levels, but is ultimately limited in flexibility and always suffers from signal degradation.	Fantastically powerful computer-based techniques available.
Storage	Genuine analog storage for any length of time is almost impossible.	All major semiconductor memory technologies are digital.

as the table indicates, perform better on most counts and with today's technology are easier to work with. In many cases the advantage is dramatic and overwhelming.

It is possible fairly readily to convert a signal from analog to digital form, using an analog-to-digital converter (ADC). The circuits available to do this conversion are comparatively complex. Their design is a mature art form, however, and they are available as ready-to-use integrated circuits or modules within a microcontroller.

As embedded designers, we will need to understand the characteristics of the ADC, so that we can choose the right one and use it effectively.

11.2 The data acquisition system

When converting an analog signal to digital form, it is usually not enough just to find a suitable ADC. Usually, more than one input is required and the signal needs processing before it can be converted. In most cases, therefore, it is necessary to build up a complete *data acquisition system*. The elements of such a system are shown in Figure 11.1. This shows, in block diagram form, a system with multiple inputs, amplification, filtering, source selection, sample and hold, and finally the ADC itself. The different elements are outlined in the sections which follow.

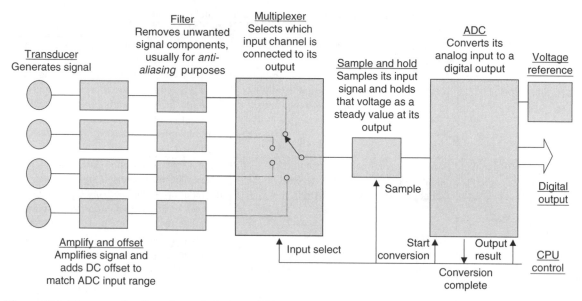

Figure 11.1 Elements of a (four-channel) data acquisition system

11.2.1   *The analog-to-digital converter*

The task of the ADC is to determine a digital output number that is the equivalent of its input voltage. The design of such circuits is a non-trivial task. Many very different ADC circuits have been developed, targeted towards different applications. Some, like the dual ramp ADC, are slow but with very high accuracy, and useful for precision measurements such as digital voltmeters. Others, like the Flash converter (not to be confused with Flash memory technology), are fast but of lesser accuracy, and are used to convert high-speed signals like video or radar. Others, like the successive approximation ADC, are of medium speed and medium accuracy, and useful for general-purpose industrial applications. This is the type most commonly found in embedded systems. Descriptions of how this type of ADC circuit works can be found in most electronics textbooks (see Ref. 1.1).

An ADC is characterised principally by the following features.

Conversion characteristic

The ADC accepts an input voltage that is infinitely variable. It converts this to one of a fixed number of output values. An example ADC conversion characteristic is shown in Figure 11.2, where the input voltage is represented on the horizontal axis and digital output on the vertical. If the ADC is converting continuously and the input voltage is gradually increased from zero, the output is also initially zero. At a certain value of input, the output changes to ...001. It stays at this same value as the input increases further, until at another input value the output switches to ...010. If the input voltage increases continuously, the output at some point reaches its maximum value. The input has then traversed its full *range*. The output will have

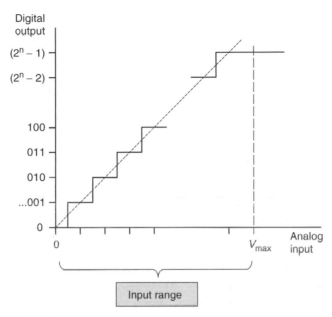

Figure 11.2 The ideal ADC input/output characteristic

moved stepwise up to its maximum value. For an n-bit ADC, the maximum output value will be $(2^n - 1)$. For example, for an 8-bit ADC, the final value will be $(2^8 - 1)$, or 11111111_B, or 255_D.

The input range shown in Figure 11.2 starts from zero and goes up to the value V_{max}. This is placed a little to the right of where one might expect it, at the centre of where a step for 2^n would occur. This positioning allows the horizontal axis to be divided into exactly 2^n equal segments, each centred on an output transition.

Many ADCs have a characteristic like in Figure 11.2, for example with an input range of 0–5 V. Others, however, have a bipolar range, with the input voltage taking both positive and negative values, for example -5 to $+5$ V. In every case the input range V_r is the difference between maximum input voltage and minimum input voltage. The range usually relates in a direct way to the value of the voltage reference, which forms part of the ADC.

It can be seen intuitively from the diagram that the more the number of output bits, the more will be the number of output steps and the finer is the conversion. A measure of the fineness of conversion is called the *resolution*. This is the amount by which the input has to change to go from one output value up to the next. In the diagram, the resolution is the width of one step in the conversion characteristic. An ADC with n output bits can take 2^n possible output values, from 0 up to $2^n - 1$. It therefore has a resolution of $V_r/2^n$, where V_r is the input voltage range. An incoming signal should use as much of the input range as possible, without exceeding it. If it only uses a part of it, then the effective resolution is degraded and the ADC is not being put to best use.

Conversion speed

An ADC takes time to do its work. That time is called the conversion time. A slow ADC, with a high conversion time, will only be able to convert low-frequency signals, as Nyquist's criterion (Section 11.2.2) must always be satisfied. The conversion time of an ADC defines which type of signal it can be used to convert. As suggested earlier, high-accuracy ADCs generally take longer to complete a conversion.

Digital interface

The digital interface is made up of the control signals and the data output. Typical control signals are shown in Figure 11.1. Generally, there is a signal to the ADC that causes a conversion to start. When the conversion is complete, the ADC signals that completion with an output signal. A further signal causes the ADC to output its data. Depending on the type of interface required, the ADC has a parallel or serial data interface.

An ADC always works in conjunction with a *voltage reference*. This is a device or circuit that maintains a very precise and stable voltage, and is based around a zener diode or a band-gap reference. The ADC effectively uses the voltage reference as the ruler, with which it measures the incoming voltage. An ADC is only as good as its voltage reference. For accurate A-to-D conversion, a good ADC must be used with a good reference.

11.2.2   Signal conditioning – amplification and filtering

To make best use of the ADC, the input voltage should traverse as much of its input range as possible, without exceeding it. Yet most signal sources, say a microphone or thermocouple, produce very small voltages. Therefore, in many cases amplification is needed to exploit the range to best effect. Voltage level shifting may also be required, for example if the signal source is bipolar while the ADC input is unipolar (voltage is positive only).

If the signal being converted is periodic, then a fundamental requirement of conversion is that the conversion rate must be at least twice the highest signal frequency. This is known as the Nyquist sampling criterion. If this criterion is not met, then a deeply unpleasant form of signal corruption takes place, known as *aliasing* (see Ref. 1.1 or signal processing text for further details). Anti-aliasing filtering may therefore be required to ensure that the Nyquist criterion is satisfied.

11.2.3   The analog multiplexer

If there are to be multiple inputs, then an analog multiplexer is used. The alternative, of multiple ADCs, is both costly and space consuming. The multiplexer acts as a selector switch, choosing which input out of several is connected to the ADC at any one instant. The multiplexer is built around a set of semiconductor switches. It is important to know that the semiconductor switch is an imperfect device. In particular, when switched 'on', it has internal series resistance, which can range from tens to thousands of ohms. This can impact on the data acquisition process, as we shall see.

11.2.4 Sample and hold, and acquisition time

Because most ADCs are unable to convert accurately a changing voltage, a *sample and hold* (S&H) circuit is often found. This takes a sample of the voltage, like a snapshot, and holds it steady for the duration of the conversion. A circuit of a simple but practical S&H is shown in Figure 11.3. At its heart are just a semiconductor switch and a capacitor. When the switch is closed, the capacitor charges up to the input voltage V_S. At this moment, ideally $V_O = V_C = V_S$, as the buffer amplifier just has unity gain. When the switch opens, the charge is left on the capacitor and V_C (and hence V_O) remains at a fixed value. In practice there is some leakage from the capacitor, so the output voltage drifts. This circuit is sometimes also called *track and hold*, as when the switch is closed the output voltage follows, or tracks, the input.

One problem with this simple circuit is that there is inevitably series resistance in the signal path. This is represented by the resistor in the circuit. When the switch closes, therefore, the capacitor voltage V_C does not take on the signal voltage immediately, but rises towards it exponentially. This is shown in Figure 11.4. The voltage rise is given by:

$$V_C = V_S\{1 - \exp(-t/RC)\}. \tag{11.1}$$

Our interest from a data acquisition point of view is to ensure that the voltage has risen sufficiently close to its final value with the switch closed, before the switch is opened (the signal is then 'held'), and a conversion allowed to start. The time that V_C (and hence V_O) takes to reach a value deemed to be acceptable is called the *acquisition time*.

Let us suppose that V_C must rise to 90 per cent of its final value, V_S. Then, substituting into equation (11.1):

$$0.9V_S = V_S\{1 - \exp(-t/RC)\}$$

$$\exp(-t/RC) = 1 - 0.9$$

$$-t = RC\ln(0.1)$$

$$t = 2.3RC.$$

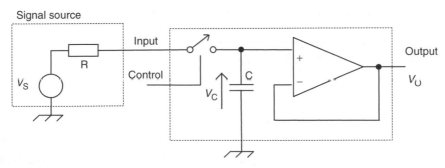

Figure 11.3 A simple form of sample and hold circuit

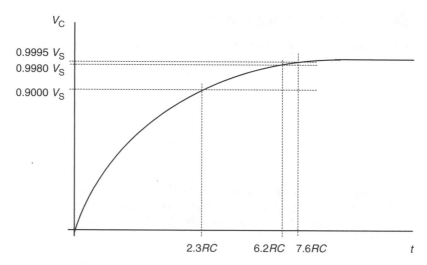

Figure 11.4 Exploring acquisition time (*not to scale*)

This is shown in Figure 11.4. It is, however, an undemanding requirement. To ensure good accuracy in data conversion, the error introduced by this process should be less than the equivalent of half of 1 LSB. Hence, for 8-bit conversion, this implies that the acquired voltage value V_C must reach $\geq (511/512)V_S$, or $0.9980V_S$. For 10-bit conversion it must be $\geq (2047/2048)V_S$, or $0.9995V_S$.

Following the calculation above, but substituting the 10-bit value in, we get:

$$-t = RC\ln(1/2048)$$

$$t = 7.6RC.$$

The resulting acquisition times are shown in Figure 11.4. It is clear that acquisition time increases with increasing resistance, capacitance *and* with accuracy required. We will meet practical application of this calculation later in the chapter.

It is worth noting that the multiplexer circuit and S&H circuit can be merged into one, with the multiplexer switches forming the S&H switch. This is common practice.

11.2.5 Timing and microprocessor control

Usually, a data acquisition system is under the control of a microprocessor or microcontroller. This can control the overall system timing, including which input is being selected, when the selected signal is sampled and when the conversion starts.

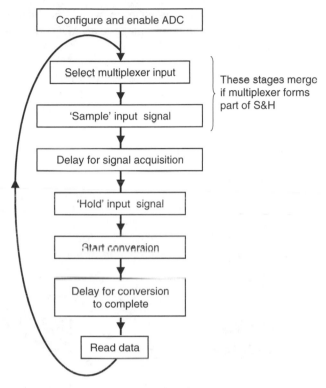

Figure 11.5 Typical timing requirement of one A-to-D conversion

The process of a single conversion can be represented as a flow diagram, as shown in Figure 11.5. Two major time requirements need to be satisfied – the acquisition time (of the S&H) and the conversion time (of the ADC).

Once the system is initialised, the multiplexer switch can be set. The S&H can then start its sample process. A period of time equal to or greater than the required S&H acquisition time must elapse. The ADC can then start its conversion. Again, this takes finite time. The ADC flags when it has completed a conversion and the microprocessor can read the output data.

11.2.6   Data acquisition in the microcontroller environment

Embedded systems need ADCs, so it is natural to expect to find an ADC integrated onto a microcontroller as one of its peripherals. It is important, however, to realise that ADCs and microcontrollers do not make happy bedfellows. To operate to a good level of accuracy, an ADC needs a quiet life (electronically speaking), with excellent and clean power supply and ground, and freedom from electromagnetic interference. A microcontroller, on the other hand, being a digital device, tends to corrupt its power supply and ground with a voltage spike on every switching edge. As a consequence, with all its intensive internal digital activity, it radiates a smog of local interference. Therefore, to integrate an ADC onto a microcontroller is at best a compromise and high accuracy is not usually possible.

Despite this, ADCs are widely available in the microcontroller environment, with many microcontrollers having an on-chip ADC. These are mostly 8- or 10-bit.

11.3 The PIC® 16F87XA ADC module

11.3.1 Overview and block diagram

The 16F87XA has a versatile and powerful 10-bit ADC module, shown in Figure 11.6. This provides a subset of the overall data acquisition system shown in Figure 11.1, having an ADC, a multiplexer and the possibility of using the supply voltage as the voltage reference. The particular ADC design used incorporates in an interesting way the function of sample and hold, discussed further in Section 11.3.3.

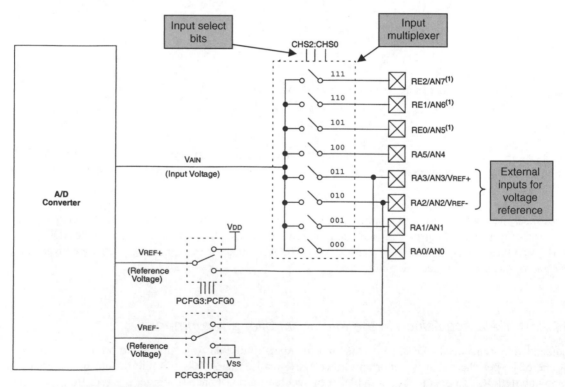

Note 1: Not available on 28-pin devices.

Figure 11.6 The 16F87XA analog-to-digital converter (supplementary labels in shaded boxes added by the author)

The input multiplexer, seen to the right of the diagram, has five channels for the 16F873A and F875A, and eight for the 16F874A and F876A. The inputs are shared with five of the six Port A bits and three of the eight Port E (for 16F874 and F876) bits. Only Port A bit 4 is not used, as it already shares with the important Timer 0 input. Port bits can be allocated in a flexible way to analog or digital input, according to settings in an SFR.

An external voltage reference can be used for applications requiring reasonable accuracy, with terminals for both positive and negative connections provided. Provision of the negative connection means that the reference does not have to be referred to system ground. For lower-cost, lower-accuracy conversions, the power supply voltage can be used as reference. The input range is equal to whatever voltage reference is chosen.

11.3.2 Controlling the ADC

The ADC is controlled by two SFRs, **ADCON0** (Figure 11.7) and **ADCON1** (Figure 11.8). The result of the conversion is placed in two further SFRs, **ADRESH** and **ADRESL**. These four registers can all be seen in Figure 7.6. Other SFRs also have an important impact on the ADC. These include **TRISA** and (for the 40-pin devices) **TRISE**. Any bits used for analog input must be set as inputs in these. Registers **PIR1** and **PIE1**, which contain the ADC interrupt flag and interrupt enable bits respectively, are also used.

The control possibilities are now described, in the approximate sequence they would be used.

Switching on

The ADC is switched on and off by the **ADON** bit of **ADCON0**. Switching it off when not needed offers a slight power-saving advantage.

Setting the conversion speed

Operation of the 16F87XA ADC is governed by the ADC clock, which has a period T_{AD}. A full 10-bit conversion takes around 12 T_{AD} cycles, depending slightly on which clock source is chosen. The user can select the clock frequency from a number of options. Although one generally wants a conversion to take place as quickly as possible, there is an upper limit to the clock frequency. For the 16F87XA the minimum clock period for correct operation is specified as 1.6 μs (from the *Electrical Characteristics* of Ref. 7.1), or a frequency of 625 kHz. This implies a fastest conversion time of 19.2 μs. At the other extreme, if conversion is too slow, charge leaks from the storage capacitance and the conversion becomes inaccurate. Best practice is therefore to set the ADC clock frequency such that it has a period equal to or just less than 1.6 μs.

Selection of the ADC clock source is controlled by bits **ADCS2** in **ADCON1**, and **ADCS1** and **ADCS0** in **ADCON0**, as seen in Figure 11.7. This shows that various divisions of the main clock frequency are possible. There is also a dedicated RC oscillator which can be chosen. This has a typical period of 4 μs, but may range from 2 to 6 μs.

R/W-0	R/W-0	R/W-0	R/W-0	R/W-0	R/W-0	U-0	R/W-0
ADCS1	ADCS0	CHS2	CHS1	CHS0	GO/$\overline{\text{DONE}}$	—	ADON

bit 7   bit 0

bit 7-6   **ADCS1:ADCS0:** A/D Conversion Clock Select bits (ADCON0 bits in **bold**)

ADCON1 <ADCS2>	ADCON0 <ADCS1:ADCS0>	Clock Conversion
0	00	Fosc/2
0	01	Fosc/8
0	10	Fosc/32
0	11	F$_{RC}$ (clock derived from the internal A/D RC oscillator)
1	00	Fosc/4
1	01	Fosc/16
1	10	Fosc/64
1	11	F$_{RC}$ (clock derived from the internal A/D RC oscillator)

bit 5-3   **CHS2:CHS0:** Analog Channel Select bits

000 = Channel 0 (AN0)
001 = Channel 1 (AN1)
010 = Channel 2 (AN2)
011 = Channel 3 (AN3)
100 = Channel 4 (AN4)
101 = Channel 5 (AN5)
110 = Channel 6 (AN6)
111 = Channel 7 (AN7)

> **Note:**   The PIC16F873A/876A devices only implement A/D channels 0 through 4; the unimplemented selections are reserved. Do not select any unimplemented channels with these devices.

bit 2   **GO/$\overline{\text{DONE}}$:** A/D Conversion Status bit

When ADON = 1:
1 = A/D conversion in progress (setting this bit starts the A/D conversion which is automatically cleared by hardware when the A/D conversion is complete)
0 = A/D conversion not in progress

bit 1   **Unimplemented:** Read as '0'

bit 0   **ADON:** A/D On bit

1 = A/D converter module is powered up
0 = A/D converter module is shut-off and consumes no operating current

Figure 11.7 The **ADCON0** register (address 1F$_{\text{H}}$)

If the system clock is fast, it is usually appropriate to use it to derive the clock source. If the system clock is slow, however, it is better to use the RC oscillator. The dividing line between a 'slow' and 'fast' clock oscillator here is around 500 kHz. With an internal oscillator running at this speed, the fastest ADC clock that can be derived from it is 250 kHz. This gives a period of 4 μs, equal to the typical RC oscillator period. If the main oscillator is lower than this frequency, it will then generally be advisable to use the RC oscillator.

R/W-0	R/W-0	U-0	U-0	R/W-0	R/W-0	R/W-0	R/W-0
ADFM	ADCS2	—	—	PCFG3	PCFG2	PCFG1	PCFG0

bit 7   bit 0

bit 7   **ADFM:** A/D Result Format Select bit

1 = Right justified. Six (6) Most Significant bits of ADRESH are read as '0'.
0 = Left justified. Six (6) Least Significant bits of ADRESL are read as '0'.

bit 6   **ADCS2:** A/D Conversion Clock Select bit (ADCON1 bits in shaded area and in **bold**)

ADCON1 <ADCS2>	ADCON0 <ADCS1:ADCS0>	Clock Conversion
0	00	$F_{OSC}/2$
0	01	$F_{OSC}/8$
0	10	$F_{OSC}/32$
0	11	F_{RC} (clock derived from the internal A/D RC oscillator)
1	00	$F_{OSC}/4$
1	01	$F_{OSC}/16$
1	10	$F_{OSC}/64$
1	11	F_{RC} (clock derived from the internal A/D RC oscillator)

bit 5-4   **Unimplemented:** Read as '0'

bit 3-0   **PCFG3:PCFG0:** A/D Port Configuration Control bits

PCFG <3:0>	AN7	AN6	AN5	AN4	AN3	AN2	AN1	AN0	VREF+	VREF-	C/R
0000	A	A	A	A	A	A	A	A	VDD	VSS	8/0
0001	A	A	A	A	VREF+	A	A	A	AN3	VSS	7/1
0010	D	D	D	A	A	A	A	A	VDD	VSS	5/0
0011	D	D	D	A	VREF+	A	A	A	AN3	VSS	4/1
0100	D	D	D	D	A	D	A	A	VDD	VSS	3/0
0101	D	D	D	D	VREF+	D	A	A	AN3	VSS	2/1
011x	D	D	D	D	D	D	D	D	—	—	0/0
1000	A	A	A	A	VREF+	VREF-	A	A	AN3	AN2	6/2
1001	D	D	A	A	A	A	A	A	VDD	VSS	6/0
1010	D	D	A	A	VREF+	A	A	A	AN3	VSS	5/1
1011	D	D	A	A	VREF+	VREF-	A	A	AN3	AN2	4/2
1100	D	D	D	A	VREF+	VREF-	A	A	AN3	AN2	3/2
1101	D	D	D	D	VREF+	VREF-	A	A	AN3	AN2	2/2
1110	D	D	D	D	D	D	D	A	VDD	VSS	1/0
1111	D	D	D	D	VREF+	VREF-	D	A	AN3	AN2	1/2

A = Analog input   D = Digital I/O
C/R = # of analog input channels/# of A/D voltage references

Figure 11.8 The **ADCON1** register (address $9F_{H}$)

Configuring the input channels and selecting the voltage reference

The way the input port bits are used is defined by the setting of bits **PGFC3** to **PGFC0** of **ADCON1**. It is worth looking at this closely in Figure 11.8. The variety of opportunity is impressive, both in terms of input channels and voltage reference. We can see that it ranges from just a single Port A channel used for input (**PGFC3:PGFC0** = 1110) to all eight analog inputs in play (**PGFC3:PGFC0** = 0000). Many combinations which include an external reference are also possible. Note again that any port pin that is to be used as an analog input must be set as an input in its **TRIS** register. Otherwise, the pin will act as an output and the (unintended) digital output value will be converted!

Selecting the input channel

The input channel is selected by the channel select bits **CHS2** to **CHS0** in **ADCON0**. These bits determine which switch in Figure 11.6 is closed. Making this selection is usually the first step in the data acquisition process, as we shall see below.

Starting a conversion and flagging its end

A conversion is initiated by setting bit **GO/$\overline{\text{DONE}}$** in register **ADCON0**. When the conversion is complete the bit is returned to zero by the hardware. Completion of conversion is also signalled by an ADC interrupt flag **ADIF**, as seen in Figure 7.10. Completion of conversion may therefore be detected by testing either of the bits **GO/$\overline{\text{DONE}}$** or **ADIF**, or by enabling the interrupt and responding to it in an ISR.

Formatting the result

The result of the conversion is placed in registers **ADRESH** and **ADRESL**. Two possible result formats are possible, as shown in Figure 11.9. The result can be left justified, in which case the 8 most significant bits

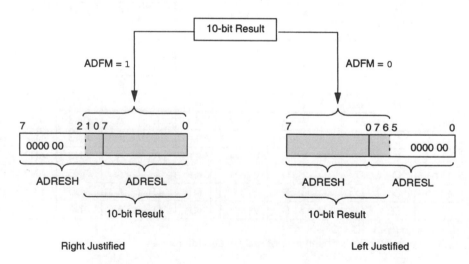

Figure 11.9 Formatting the ADC conversion result

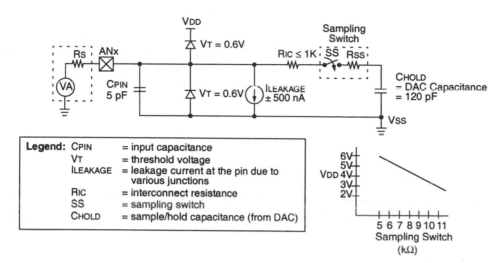

Figure 11.10 The 16F87XA ADC input model

appear in **ADRESH**. This is useful if only an 8-bit result is required, as the contents of **ADRESL** can then be ignored. In most other cases a right-justified result will be the more useful. The formatting is controlled by bit **ADFM** in **ADCON1**.

11.3.3   The analog input model

It was demonstrated earlier in this chapter that an understanding of the actual signal path is necessary in order to understand and predict system characteristics. Figure 11.10 is a diagram of the signal path for this ADC – and what an obstacle course it appears to be! This diagram is effectively a real-life representation of parts of Figure 11.6, shown from the signal's point of view.

The signal source, together with internal resistance, is depicted to the left of the diagram, modelled as voltage source in series with internal resistance. The signal voltage enters the microcontroller through the pin labelled ANx. There is a small input capacitance (5 pF), and the input protection diodes and other input circuitry clearly have the potential to leak current into the signal path. The signal then passes through the interconnect resistance, R_{IC}, before reaching the multiplexer switch. This is one of the switches of the analog input multiplexer in Figure 11.6. The internal resistance of this switch, R_{SS}, is shown. The approximate value of this is dependent on supply voltage and is given by the small graph on the bottom right of Figure 11.10. From this we see that the switch resistance is a sobering 7 kΩ approximately, when the supply voltage is 5 V.

The ADC itself is a so-called switched capacitor type (Ref. 1.1, Chapter 5). First of all, that means that the ADC has internal capacitance, which must be charged up to the input voltage before a conversion can start. Neatly, however, this capacitance takes on the function of the S&H capacitor. On the downside, the capacitance, all 120 pF of it, must be charged up in the first place.

11.3.4 Calculating acquisition time

In Ref. 7.1 Microchip define three sources of time delay in their calculations for acquisition time, t_{ac}, as shown:

$$t_{ac} = \text{Amplifier settling time} + \text{Hold capacitor charging time} + \text{Temperature coefficient.} \qquad (11.2)$$

The reference specifies the amplifier settling time as a fixed 2 μs. The temperature coefficient applies only when temperature is above 25°C and is specified as:

$$\text{Temperature coefficient} = (\text{Temperature} - 25°\text{C})(0.05 \, \mu\text{s/}°\text{C}).$$

It can be seen that this creates a time delay of only 0.5 μs for every 10° above 25°, so its impact in most cases is slight.

It is the capacitor charging time that dominates the acquisition time, which we now explore. The analog input model of Figure 11.10 can be related back to the S&H diagram of Figure 11.3 and equation (11.1). To analyse this, we neglect the effects of the input leakage current and the small input capacitance. Actual values for R and C in Figure 11.3 for the 16F87XA ADC can be extracted from Figure 11.10. R is made up of $(R_{SS} + R_{IC} + R_S)$, or $(1\text{k} + 7\text{k} + R_S)$, for a supply voltage of 5 V. C is the 120 pF shown. Calculations made for Figure 11.4 showed us that, for 10-bit accuracy, an acquisition time of $7.6RC$ was needed. Substituting values in, assuming at first negligible source resistance, gives:

$$t_{ac} = 7.6RC$$

$$= 7.6(1\text{k} + 7\text{k})120 \, \text{pF}$$

$$= 7.3 \, \mu\text{s.}$$

If amplifier settling time is added to this, as it must be, the acquisition time rises to 9.3 μs.

This represents a best possible value. To determine the overall time needed to complete a single conversion, this acquisition time must be added to the conversion time, discussed in Section 3.2. There, a best possible conversion time of 19.2 μs was deduced. Adding this to the best possible acquisition time leads to a total time to complete a conversion of $(2 + 7.3 + 19.2)$ μs, i.e. 28.5 μs.

In many cases the source resistance is not negligible and any external series resistance will degrade the acquisition time calculated above. In Ref. 7.1 Microchip *recommend* a maximum source resistance of 2.5 kΩ. In this case:

$$t_{ac} = 7.6(1\text{k} + 7\text{k} + 2.5\text{k})120 \, \text{pF}$$

$$= 9.6 \, \mu\text{s.}$$

Note that this is not the highest acquisition time that may be encountered, as the maximum *allowed* source resistance is specified in Ref. 7.1 as being 10 kΩ.

11.3.5 Repeated conversions

When a conversion is complete, the converter waits for a period of $2 \times T_{AD}$ before it is available to start a new conversion cycle. Once this time is up, either the same input channel may be converted again, or a new one (which may already have been selected).

A best possible conversion time of 28.5 μs was calculated in Section 11.3.4. If a period of $2 \times T_{AD}$ is added to this, i.e. 3.2 μs for fastest possible, then the complete conversion cycle time becomes 31.7 μs. If successive conversions are intended, this implies a maximum sampling rate of around 30 kHz. Note, however, that this figure takes no account of software overheads, which would tend to slow the conversion rate.

11.3.6 Trading off conversion speed and resolution

The conversion times deduced above are not particularly fast by today's standards and there will be occasions when a faster conversion time is needed. While one option is to use an external ADC, another is to consider whether the full 10-bit resolution is needed. If it is not, then the conversion time can be reduced. One technique, described in Ref. 11.1, is to start a conversion with a valid ADC clock frequency and then to switch it during the conversion to a faster speed, which violates the clock specification. The higher order bits converted before the switch will be valid and can be used. Those converted after will not be valid. A lower resolution conversion, at higher speed, has thus been achieved. It is up to the program, however, to determine when the switch should take place.

An alternative approach is to reduce the acquisition time, as suggested in Figure 11.4, so that, for example, the acquisition is only to 8-bit accuracy. The conversion can then be allowed to run its full course. The switching of the ADC clock frequency, as just described, can still be implemented.

11.4 Applying the ADC in the Derbot light meter program

The Derbot AGV has three light-dependent resistors, used as light sensors, as seen in Figure A3.1. Two are at the front and one at the rear of the vehicle. The Derbot uses the 16F873A ADC to measure light intensity, both for a light meter program and a light-seeking program.

Program Example 11.1 shows sections of the program **Dbt_light_meter**, reproduced in full in the book CD. The program reads the LDR output values and displays these on the LCD of the hand controller unit. There are, however, several data manipulation challenges on the way. The result of the ADC must be scaled to a true voltage reading and then converted to a format suitable for the LCD. Therefore, the reading is converted to Binary Coded Decimal and then to ASCII. The resulting characters are then transferred to the LCD. Let us explore how this is done.

11.4.1 Configuration of the ADC

The initial setting of the ADC control registers, **ADCON0** and **ADCON1**, can be seen in the program example. Figure A3.1 shows that the right LDR is connected to Port A bit 0, the left to Port A bit 1 and the rear to Port A bit 3. No external reference voltage is available, so the power supply is used. These connections lead to the **PCFG** bits in **ADCON1** (Figure 11.8) being set to 0100.

With a main clock period of 250 ns (4 MHz), but a minimum specified T_{AD} time of 1.6 µs, it is necessary to divide the clock frequency by at least 8. This value is chosen, giving a T_{AD} of 2 µs. A right-justified result is selected.

11.4.2 Acquisition time

The operation of the data acquisition can be examined by looking further down the program example.

The acquisition time needs to be calculated for the worst-case condition, which is when ambient light is low and the LDR resistance high. This may be checked in the LDR data [Ref. 8.5]. In the worst case the LDR resistance will be very high and the source resistance will tend towards the value of the 10 kΩ resistor in series with the LDR. Note that this is at the limit of the maximum allowed source resistance. However, under normal lighting conditions the source resistance will be considerably lower. From Figure 11.4, acquisition time is given by:

$$t = 7.6RC$$

$$= 7.6(10k + 1k + 7k)120\,pF$$

$$= 16.4\,\mu s.$$

Adding in the amplifier settling time gives 18.4 µs.

```
;******************************************************************************
;Dbt_light_meter
;Dbt reads LDR values to 10-bit resolution, scales to a voltage reading,
;converts to 4 digits of BCD, and displays on Hand Controller LCD.
;Rear LDR is scaled to be a true millivolt reading, displayed to tens of mV.
;Requires Hand Controller loaded with corresponding program.
;TJW 19.7.05                                            Tested 20.7.05
;******************************************************************************
...
(early program sections omitted)
...
        bsf     status,rp0
        movlw   B'00001011'  ;set port A bits according to their function,
        movwf   trisa          ;ADC channels set as inputs
        movlw   B'10000100'  ;select port A bits 0,1,3 for analog input
        movwf   adcon1         ;right justify result
...
        bcf     status,rp0
        movlw   B'01000001'  ;set up ADC: clock Fosc/8, switch ADC on but not
                                                 ;converting,
```

Program Example 11.1 Applying the ADC in the Derbot light meter program

```
        movwf   adcon0          ;input channel selection currently irrelevant
...
;read and store ldrs
main_loop movlw B'01000001' ;select channel 0 as input (left front ldr)
        movwf   adcon0
        call    delay20u        ;acquisition time
        bsf     adcon0,go       ;start conversion
        btfsc   adcon0,go_done ;wait for conversion to complete
        goto    $-1
        movf    adresh,0        ;read and store ADC output data, high byte
        movwf   ldr_left_hi
        bsf     status,rp0
        movf    adresl,0        ;read and store ADC output data, low byte
        bcf     status,rp0
        movwf   ldr_left_lo
;select channel 1 (right ldr)
        movlw   B'01001001'
        movwf   adcon0          ;select channel 1 as input (right front ldr)
        call    delay20u
...
(samples other two LDRs)
...
(scales, converts to BCD, and outputs values to diplay)
...
        goto    main_loop
...
```

Program Example 11.1 Continued

11.4.3   Data conversion

The actual operation of the ADC can be seen by checking further down the program example, from the label
main_loop. The three LDRs are sampled in turn. For the first (front left), it can be seen that the appropriate
input channel is selected, with the **GO/DONE** bit of **ADCON0** left low. A delay of 20 μs is introduced
for signal acquisition, which just covers the calculated worst-case value of 18.4 μs. The actual conversion
is then initialised by setting the **GO/DONE** bit high. Completion of conversion is awaited by testing the
same bit. The result of the conversion is then transferred from **ADRESH** and **ADRESL** to the two stores in
memory, and the program moves on to sample the next input. The data manipulation that follows is omitted
in this program example, but is described later in the chapter.

11.5   Some simple data manipulation techniques

Almost as soon as data is acquired in a program there follows the need to manipulate it in some way.
This could include simple addition and subtraction, as we have already done, or other arithmetic oper-
ations like multiplication and division. As the data processing demand rises, so most certainly does the
Assembler complexity, creating a strong motivation for a move to a high-level programming language.

Nevertheless, it should not be necessary to write a program or subroutine for any standard mathematical operation in Assembler. Many are already available, for example in Ref. 11.2.

11.5.1   Fixed- and floating-point arithmetic

We have already repeatedly used the fact that an *n*-bit binary number can represent any integer number value from 0 to $2^n - 1$. For example, an 8-bit number can represent from 0 to 255_D, a 12-bit number from 0 to 4095_D, and a 16-bit number from 0 to $65\,535_D$. For larger numbers, more bits just need to be added. We could even represent fractional numbers in this way, by inserting a *binary* point. Then the digits to the right of the binary point represent negative powers of 2. This is seen in the example of Figure 11.11, where the binary number 1101.11 is evaluated to 13.75_D. As long as the binary point remains in the same fixed place in all numbers being used (in fact, we only need to imagine it there), then it is possible to undertake a range of arithmetic operations with a set of numbers.

This type of binary number representation is called *fixed point*. The binary point is not used, or else is assumed to be in a fixed place in the number. Such representation can solve the need to represent integers and non-integer numbers. It does not, however, solve the problem of dealing with number values that range from the very small to the very large. The smallest non-zero number that the 6-bit fixed-point number of Figure 11.11 can represent, for example, is 0.25_D, while the maximum number is 15.75_D. It could neither represent 0.0004_D nor 2.3×10^6.

The solution to this problem lies in *floating-point* representation. This represents the number with a *sign bit*, *mantissa* and *exponent*. A huge range of numbers can be represented, but processing complexity is greater. While floating-point routines are available in Assembler, it is most commonly found in high-level languages, where it is used for all precision calculation. In this chapter we apply only fixed-point arithmetic.

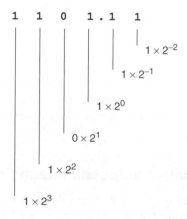

$$1101.11_B = 8+4+0+1+0.5+0.25 = 13.75_D$$

Figure 11.11 A fractional binary number

11.5.2   Binary to Binary Coded Decimal conversion

While all fixed-point arithmetic is done in binary, where there is human interaction, there will be a marked preference for decimal. How do we move data between the binary domain and the decimal?

A simple halfway house between binary and decimal is *Binary Coded Decimal* (BCD). This uses a 4-bit number to represent a decimal digit, as shown in Table 11.2. The only thing that distinguishes BCD from hexademical is that in BCD the binary equivalents of Hex. A_H, B_H, C_H, D_H, E_H or F_H are not allowed. Thus, Table 11.2 shows all legal BCD codes for a single decimal digit. The table also shows the ASCII (*American Standard Code for Information Interchange*) code for each of the numbers. For numeric characters, it can be seen that this is simply a single-byte code, in which the number itself occupies the lower nibble and the number 3 forms the higher.

Given this simple coding, multi-digit decimal numbers can be represented in BCD. In *packed BCD*, which is commonly used, a byte is used to represent two decimal digits. An example is shown in Figure 11.12.

Despite this simple representation, it is not completely straightforward to convert between BCD and binary. Standard algorithms are available, however, such as are used in Ref. 11.3. The Derbot light meter example uses a subroutine taken from this reference that converts a 16-bit number into a five-digit BCD number. This is used to prepare the data for the display. The underlying algorithm can be found in many books on computer arithmetic and in Ref. 1.1.

Table 11.2 Binary, BCD and ASCII

Decimal	Binary (BCD)	ASCII (in hex.)	Decimal	Binary (BCD)	ASCII (in hex.)
0	0000	30	5	0101	35
1	0001	31	6	0110	36
2	0010	32	7	0111	37
3	0011	33	8	1000	38
4	0100	34	9	1001	39

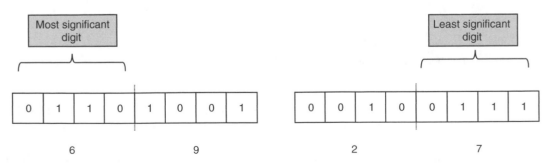

Figure 11.12 Packed BCD representation of the decimal number 6927

11.5.3   Multiplication

After addition and subtraction, multiplication is the next most common arithmetic requirement that is usually needed in computer arithmetic. Some processors have hardware multipliers in them and hence a multiply instruction. For the simpler ones, like the PIC 16 Series, multiplication must be achieved by software routines. The standard algorithm for this is a repeated shift and add process [Ref. 1.1]. A very wide range of standard routines are available, in both fixed point and floating point. The fixed-point ones come in different sizes, depending on the number size to be used. Inevitably, those with longer word length take more time to execute. Reference 11.2 has fixed-point multiply routines for 8-bit × 8-bit (with 16-bit result), 8-bit × 16-bit (with 24-bit result), 16-bit × 16-bit, up to 32-bit × 32-bit (with 64-bit result).

The Derbot light meter example uses a 16-bit × 16-bit multiply subroutine taken from Ref. 11.3, as is now described.

11.5.4   Scaling and the Derbot light meter example

A common application of multiplication is for scaling of input data, perhaps to convert it into a standard unit. The Derbot example is interesting in this respect.

The Derbot uses the power supply voltage of 5 V as its reference. Hence the 10-bit resolution is $(5/1024) = 4.883$ mV. Thus, the ADC least significant bit, when using this voltage reference, is 'worth' 4.883 mV. If the ADC output value is multiplied by 4.883, then the result will give a true millivolt reading. Yet it is awkward to introduce and track a binary point. A possible workaround is to scale the fractional number up by a binary power and then – after the multiplication – divide by the same number. It's easy to do this, as binary division is a straightforward process of shifting a number right, effectively discarding less significant bits.

For example, for the Derbot light meter, we want to do the calculation:

$$\text{ADC output} \times 4.883 = \text{Millivolt reading.}$$

While we can't multiply by 4.883, we note that $4.883 \times 256 = 1250.048$ and we *can* do the following multiplication using integers only:

$$\text{ADC output} \times 1250 = \text{Intermediate result.}$$

With the ADC output being a 10-bit number and 1250_D (i.e. $04E2_H$) an 11-bit number, the result will be at most 21 bits. This defines the type of multiply routine needed.

If the intermediate result is then divided by 256, which can be done simply by discarding its least significant byte, then our scaling process is complete. The outcome is the voltage input to the ADC, given in millivolts.

One is justified in asking how the multiplying factor of 256 was chosen, rather than say 128 or 512. This depends on the rounding error introduced. The rounding error in taking 1250 instead of 1250.048 is very small, well below 0.05 per cent, and so is acceptable when dealing with 10-bit numbers. In fact, it would have been acceptable to use 128 as the multiplying factor, but then the convenience of dropping the least significant byte for the division stage would have been lost.

Remember that all of this is a workaround. We could alternatively convert 4.883 directly to binary, which turns out to be 100.11100010, or $4.E2_H$. With a bit of thought, and comparing this number with the multiplier we found above, it can be seen that the two methods are essentially equivalent.

The process described above is applied in Program Example 11.2, which scales the value of the rear LDR. It uses two subroutines: **mult16x16** to multiply two 16-bit numbers and **Bin2BCD16** to convert a 16-bit number to a five-digit BCD output. The program section starts after the ADC conversions have been made. The stored output of the ADC conversion, held in memory locations **ldr_rear_hi** and **ldr_rear_lo**, is transferred into **aargb0** and **aargb1**, one of the 16-bit inputs to the multiplier subroutine. The other input to the subroutine is the word formed by **bargb0** and **bargb1**. Into this is loaded the word $04E2_H$, the hexadecimal equivalent of 1250_D. The **mult16x16** subroutine is then called, with the result being placed in **aargb0:aargb1:aargb2:aargb3**. As the result can only be 21-bit, the most significant byte **aargb0** will be empty and is ignored. The least significant byte is similarly ignored, as the result needs to be divided by 256. The next byte up, **aargb2**, is the less significant byte of the required result and is therefore transferred into the lower byte (**templ**) of the 16-bit input to the **Bin2BCD16** subroutine. Similarly, **aargb1** is transferred to the upper byte, **temph**. The **Bin2BCD16** subroutine is then called. The output bytes of this are immediately used for transfer to the LCD, by the subroutine **four_dig_disp**. This sends the BCD characters in turn on the I^2C link, for display on the hand controller LCD.

All subroutines appear in the complete program listing in the book CD.

```
(ADC conversions have been undertaken for all three LDRs)
. . .
;scale rear
      movf    ldr_rear_hi,0   ;get higher byte of ADC conversion result
      movwf   aargb0          ;this is multiplier higher byte
      movf    ldr_rear_lo,0
      movwf   aargb1          ;this is multiplier lower byte
      movlw   04              ;higher byte of scaling factor
      movwf   bargb0
      movlw   0e2             ;lower byte of scaling factor
      movwf   bargb1
      call    mult16x16       ;call the multiply subroutine
      movf    aargb1,0        ;discard highest and lowest bytes
      movwf   temph           ;second highest byte for BCD conversion
      movf    aargb2,0        ;third highest byte for BCD conversion
      movwf   templ
      call    Bin2BCD16       ;call Binary to BCD conversion subroutine
      call    four_dig_disp   ;call subroutine which sends BCD bytes to display
. . .
```

Program Example 11.2 Data processing sequence for rear LDR, Derbot light meter program

11.5.5   Using the voltage reference for scaling

The fact that the ADC output in the Derbot example immediately needs to be scaled, in order to provide a true voltage reading, can be annoying, and of course use of the multiply routine itself is time-consuming in program execution.

In some cases a judicious choice of voltage reference value can significantly simplify further calculations that need to be made. For example, if the Derbot ADC was fitted with a reference of 4.096 V, then 1 LSB of the output would represent 4 mV exactly and complex scaling would not be needed. Similarly, if the reference voltage was 1.024 V, then 1 LSB of the output would represent exactly 1 mV. Voltage references with values such as these are readily available for this very purpose.

11.6   The Derbot light-seeking program

This program gives another example of use of the 16F873A ADC. By comparing measurements between the three LDRs, the AGV finds the source of brightest light and moves towards it. It moves at a speed dependent on the light difference between its sensors, so comes to a halt when all three are at similar levels of illumination. As only comparative measurements are being made, the application does not require high accuracy. Therefore, only an 8-bit ADC result is used in the program. It is therefore more convenient to left-format the result (Figure 11.9), so that the result can be taken from just the 8 bits of register **ADRESH**. This can be seen in the changed setting for **ADCON1** in this example.

With only an 8-bit result in use, the opportunity presents itself to shorten the acquisition time, notionally to $6.2RC$ (Figure 11.4). As the program will only operate in reasonable levels of ambient light, the highest LDR resistance is estimated to be $10\,k\Omega$, giving a revised maximum source resistance of $5\,k\Omega$. The ADC input capacitance remains at 120 pF, as do the internal series resistances of the ADC. The acquisition time is therefore $6.2 \times 13k \times 120\,pF$, or $9.7\,\mu s$. An $11\,\mu s$ delay subroutine is used.

```
;*******************************************************************
;Dbt_light_seek
;Derbot seeks light. PWM applied. Speed is dependent on
;light difference (front to back), so Derbot comes to a
;halt when light difference is minimal. Microswitches used
;for bump detection.
;TJW 19.5.05                              Tested 19.5.05
;*******************************************************************
...
(initial program sections omitted)
...
      bsf     status,rp0   ;select memory bank 1
      movlw   B'00001011'  ;set port A bits so that ADC bits are input
      movwf   trisa
      ...
      movlw   B'00000100'  ;Set up ADC, left justified result,
      movwf   adcon1            ;port A bits 0,1,3 for analog input
```

Program Example 11.3 Applying the ADC in Derbot light-seeking program

```
        . . .
        bcf     status,rp0   ;select bank 0
        . . .
        movlw   B'01000001'  ;Set up ADC, clock Fosc/8, ADC on but not running,
        movwf   adcon0           ;input channel selection currently irrelevant.
        . . .
;initiate a conversion
        movlw   B'01001001'  ;select channel 1 (right ldr)
        movwf   adcon0
        call    delay11u
        bsf     adcon0,go    ;start conversion
adc2    btfsc   adcon0,go    ;wait for conversion to complete
        goto    adc2
        movf    adresh,0
        movwf   ldr_rt
        . . .
```

Program Example 11.3 Continued

The actual control algorithm is represented in Figure 11.13 and is repeated approximately every 200 ms. The full listing can be found on the book CD. Having read each LDR value, it makes some preliminary calculations of average and difference values, used later if a forward speed is to be calculated. It then determines the brightest LDR and takes appropriate action. If front left or front right are brightest, it moves forward, turning in the brighter direction, with a speed and turn dependent on the relative light intensities. If the rear is brightest, however, it rotates on the spot for a fixed period, in the direction of the front LDR that is brighter. As it approaches a location where light intensity in all LDRs is similar, the speed slows. The AGV halts if all LDRs are at similar levels of light intensity.

11.7 The comparator module

Having explored the intricacies of the ADC, we end the chapter with one of the simplest interfaces between the analog and digital world – the comparator. This important circuit element acts a little like a 1-bit ADC. It is usually used to compare an input voltage with a reference. If the input is higher than the reference, the comparator output goes high, if it is lower, the output goes low. There are many applications for comparators in embedded systems. These include cleaning up corrupted digital signals (in which case their use is very close to a Schmitt trigger), testing battery voltage, setting an alarm if a temperature or other variable reaches a certain value and so on.

11.7.1 Review of comparator action

A comparator is shown in Figure 11.14(a). The comparator simply compares its two inputs, shown in the diagram as V_+ and V_-. If V_+ is greater than V_-, then the comparator output goes to a positive voltage value, usually the maximum 'saturated' voltage level. If V_+ is less than V_-, then the output goes to a negative or zero voltage value. With suitable design of output circuit, the output voltage levels can easily be made to be recognisable logic levels.

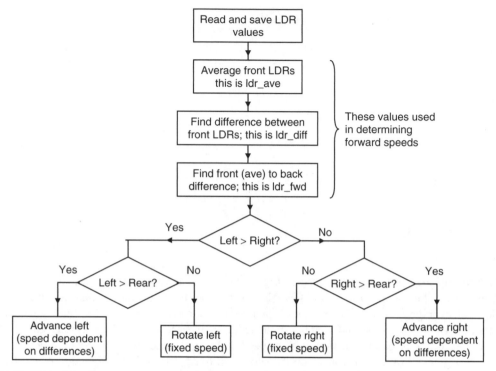

Figure 11.13 Flow diagram – Derbot light-seeking program

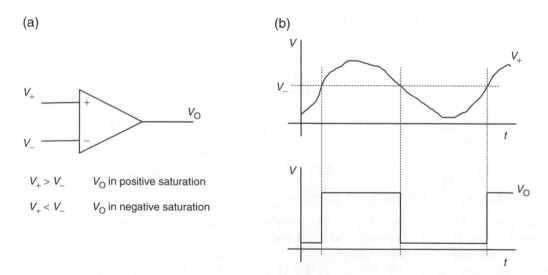

Figure 11.14 Comparator. (a) The comparator symbol. (b) Example signals

An example of input and output voltages is shown in Figure 11.14(b). V_- has been set to a fixed voltage, shown by the dotted line, while V_+ is varying. Whenever V_+ is greater than V_-, the output goes to Logic 1, when it is less, the output goes to Logic 0.

11.7.2   The 16F87XA comparators and voltage reference

The 16F87XA has two comparators, which share inputs with the ADC inputs. Comparator 1 has inputs on AN0 and AN3, and comparator 2 on AN1 and AN2. They are controlled by the **CMCON** register, seen at address $9C_H$ in Figure 7.6. A number of different configurations are possible. For example, comparator outputs can be routed externally, to pins RA4 and RA5, or they can simply appear as bits in the **CMCON** register. These can be easily looked up in the Microchip data [Ref. 7.1] and are not enumerated here. Importantly, the comparators also form an interrupt source. A change in state of either comparator will set the **CMIF** flag, seen in Figure 7.10.

The 16F873A also has a voltage reference module, under the control of the **CVRCON** register (address $9D_H$). This is a resistor ladder, connected at one end to the supply voltage (via a switching transistor), which allows different voltage values to be selected as reference. The voltage reference can be used as an input to one or both comparators, and can be output through pin 4 of the 16F873A. In this case, it can be used as a very crude digital-to-analog converter output.

11.8   Applying the Derbot circuit for measurement purposes

We have now reached a stage where we can acquire input analog voltages, process the data acquired and then display it. This is a very powerful position, as it is the basis of many a measurement system. This section describes how the Derbot PCB and hand controller can be used to create certain measurement tools, using programs already described. The hand controller connects to the Derbot 'bus' connector, which lies physically at the lower end of the microcontroller IC holder. It should be loaded with its standard program, i.e. that of Program Example 10.3.

These projects can be created as stand-alone devices or incorporated into the AGV. If stand-alone, then of course the end product is a little bulkier than one would wish. The option remains to redesign the hardware to create a more compact outcome. The circuit required, if stand-alone projects are being made, is shown in Figure 11.15. This shows components and connections needed for all three items. The sections below will indicate which ones are needed for any single build.

When programming, ensure as always that all unused port bits are set as outputs. Check, therefore, *your* build against the program initialisation in use and adjust if necessary.

11.8.1   The electronic tape measure

The ultrasound test program, shown in part in Program Example 9.7, forms the basis of an ultrasonic 'tape measure', displaying in centimetres the distance from the sensor to a reflecting surface. It processes data in a similar way to the light meter program of Program Example 11.1 and therefore shares a number of features with it.

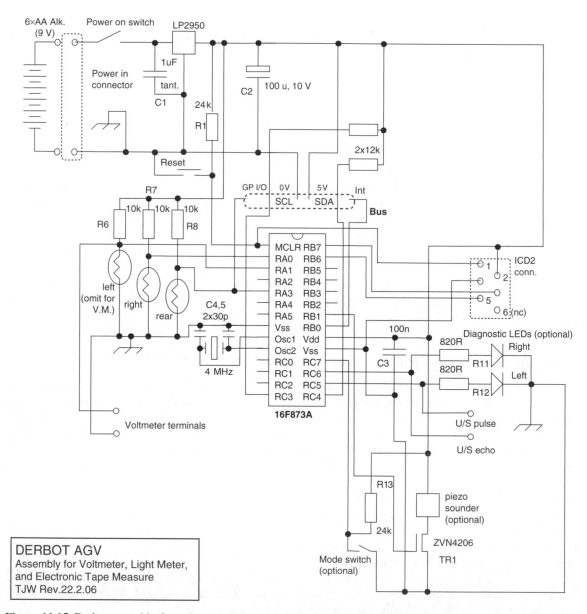

Figure 11.15 Derbot assembly for voltmeter, light meter and electronic tape measure

To build the tape measure as a stand-alone unit, the circuit of Figure 11.15 needs to be built on the Derbot PCB, but without the LDRs and their resistors. A Devantech SRF04 or SRF05 ultrasonic sensor should be used. A set of connection points for this is available just forward of the motor locations on the Derbot PCB. Terminal pins should be soldered in here. Applying the sensor data [Ref. 8.7], connect 0 V, 5 V, pulse and echo to the pins, and glue or attach the sensor to a place of your choosing on the PCB.

Complete the build as described and create an MPLAB® project around the **Dbt_US_Test** program, shown in part in Program Example 9.7. Adjust the initialisation to match your build, setting all unused port bits to output. Build the program and download to the microcontroller. Point the sensor at a reflective surface, and the distance between sensor and surface should be continuously displayed in centimetres on the hand controller display.

11.8.2 The light meter

To create the light meter, build the circuit of Figure 11.15 without the ultrasonic sensor. Download the **Dbt_light_meter** project from the book CD and create an MPLAB project around it. Adjust the program initialisation to match your build, setting all unused port bits to output. Build the program and download to the microcontroller. Run the program and see how all LDR readings are displayed. As discussed, each reading is in millivolts, but gives an indication of light intensity. Notice that the reading goes down as the light intensity increases.

11.8.3 The voltmeter

It is a simple matter to adapt the light meter program to a voltmeter. In this case, the LDRs should not be built onto the board. The voltmeter input is shown connected across the left LDR, although any other LDR connection points can be chosen. Input terminals for the voltmeter can be placed in the prototype area and connected appropriately.

Download the **Dbt_light_meter** project from the book CD and create an MPLAB project around it, giving it a name appropriate to this project. Adjust the program initialisation to match your build, setting all unused port bits to output (and noting that two ADC inputs are now no longer used). The program can otherwise be used as it is. It is better, however, to remove the two unused ADC readings and the two redundant displayed values, which will now have no meaning, unless a multi-input voltmeter is wanted. To test, connect a variable DC voltage input to your voltmeter input and connect a digital voltmeter across this. Check whether the two voltmeter readings agree.

This adaptation of the Derbot core design leads to a fairly inaccurate voltmeter, as the voltage reference is the power supply. The place of the rear LDR can, however, be taken by a voltage reference of appropriate value. It can be seen from Figures 11.6 or 11.8 that this input can be reconfigured as a reference input. The software scaling will need to change to reflect any change in reference value. By making these changes, a voltmeter of reasonable accuracy can be developed.

11.8.4 Other measurement systems

It is very easy to adapt the ideas already applied here to other measurement systems. This applies to any sensor that has a voltage output. There is, for example, a range of semiconductor temperature sensors that have a voltage output directly dependent on temperature. One of these could be used, either on a flying lead or soldered into the prototyping area. With the scaling adjusted, a digital thermometer can readily be developed.

11.9   Configuring the Derbot AGV as a light-seeking robot

To configure the Derbot as a light-seeking AGV, the LDRs and their resistors should be added to the board. Download the **Dbt_light_seek** project from the book CD and create an MPLAB project around it. Adjust the program initialisation to match your build, setting all unused port bits to output. Build the program and download to the microcontroller. The Derbot should be immediately ready to run in light-seeking mode. This works well in a space where there is a clear variation of light. The Derbot will struggle in a space where the light is mottled or patchy, and will sit completely still in a space that is uniformly lit!

Summary

- Most signals produced by transducers are analog in nature, while all processing done by a microcontroller is digital.
- Analog signals can be converted to digital form using an analog-to-digital converter (ADC). The ADC generally forms just one part of a larger *data acquisition system*. ADCs represent a fascinating interface between the analog and digital worlds.
- Considerable care needs to be taken in applying ADCs and data acquisition systems, using knowledge among other things of timing requirements, signal conditioning, grounding and the use of voltage references.
- The 16F873A has a 10-bit ADC module that contains the features of a data acquisition system. An understanding of such systems is essential in applying this module.
- Data values, once acquired, are likely to need further processing, including offsetting, scaling and code conversion. Standard algorithms exist for all of these, and Assembler libraries are published.
- A simple interface between the analog and digital world is the comparator, which is commonly used to classify an analog signal into one of two states.

References

11.1.   PICmicro® Mid-Range MCU Family Reference Manual (1997). Microchip Technology, Section 23, DS31023A.

11.2.   Fixed Point Routines (1996). Microchip Technology, Application Note 671, Document Number DS00617B.

11.3.   Watt-Hour Meter using PIC16C923 and CS5460. (2000). Microchip Technology, Application Note 671, Document Number DS00220A.

Section 4
Smarter Systems and
the PIC® 18FXX2

This section moves to more sophisticated processing power and matching programming techniques. The PIC 18 Series family is introduced and used as the underlying hardware. The section is, however, mainly concerned with software. The C programming language is introduced and applied. The issues of multi-tasking and real time, essential in the embedded world, are introduced, and the Salvo[TM] Real Time Operating System is applied.

12
Smarter systems and the PIC® 18FXX2

In this book so far we have studied the application of two microcontrollers from the PIC 16 Series – one small and one large. While these are good microcontrollers, certain tensions and limitations have emerged. Some aspects of the hardware are very limiting, for example the small size of the stack or the way all those interrupt sources have to share a single interrupt vector. As far as the instruction set is concerned, we recognise the value of the RISC approach. The absence of certain styles of instruction is, however, frustrating. Branch instructions, for example, have to be made out of a combination of **skip** and **goto**. Furthermore, the prospect of crafting major programs out of all those little Assembler instructions is becoming increasingly daunting.

Many of these limitations are experienced because the 16 Series has retained a core which was designed with only modest applications in mind. Now the number of peripherals has increased dramatically and memory sizes have soared. It is worth having a new beginning, addressing the issues that limit the 16 Series and rethinking the design of the microcontroller core. This is the basis of the PIC 18 Series.

When advancing from the 16F84A, we found that the 16F873A kept the core and enhanced the peripherals. We will now find that the 18 Series devices advance the core, while keeping peripherals reasonably constant. A very new style of microcontroller is produced as a result. The distinctive advantages of the PIC structure – RISC architecture, high speed and so on – is of course retained. New features appear, however, that allow the PIC microcontroller to move into a bigger arena – features which enhance real time operation, ease the use of high-level languages and allow interaction with much larger memories. Of the 18 Series the 18F242, along with its close relations in the 18FXX2 sub-family, are chosen for detailed study.

This chapter anticipates that the reader is upgrading from a 16 Series device. It makes comparison with the 16 Series, as it is interesting to see how design concepts have evolved. The description of the instruction set in particular is developed through such a comparison.

From this chapter onwards the book will use the C language for programming, rather than Assembler. It is therefore less important to know the microcontroller hardware in intimate detail. The C compiler looks after that side of things. This comes as quite a relief, as the 18 Series devices are not simple. For this reason, you may find that some topics in this chapter are introduced with a 'lighter touch' than equivalent sections in, say, Chapter 7.

After introducing the 18 Series and the 18F242, the chapter focuses on its central features, primarily core and memory. In particular, it introduces:

- The overall architecture of the 18FXX2 group of microcontrollers
- The structure and operation of the 18 Series microcontroller core

- The 18 Series instruction set
- The memory structure, and how it is accessed and addressed
- The 18FXX2 interrupt structure
- The 18FXX2 power supply, reset and oscillator.

12.1 The main idea – the PIC 18 Series and the 18FXX2

The PIC 18 Series microcontrollers dramatically enhance the PIC core, making it suitable for advanced embedded projects. Despite many features which are new, it has been designed to make upwards migration from a 16 Series device easy, so that the designer making this move will find many things which are familiar. The principal characteristics we shall see are:

Similar to 16 Series

- RISC (Reduced Instruction Set Computer), pipelined, 8-bit CPU, with single Working (W) and Status registers
- Many peripherals identical or very similar
- Similar packages and pinouts
- Many Special Function Register (SFR) and bit names unchanged
- All but one of the 16 Series instructions are part of the 18 Series instruction set
- Instruction cycle made up of four oscillator cycles.

New for 18 Series

- The number of instructions more than doubled, with 16-bit instruction word
- Enhanced Status register
- Hardware 8×8 multiply
- More external interrupts
- Two prioritised interrupt vectors
- Radically different approach to memory structures, with increased memory size
- Enhanced address generation for program and data memory
- Bigger Stack, with some user access and control
- Phase-locked loop (PLL) clock generator.

The 18FXX2 microcontrollers form a set of four closely related devices, whose main characteristics are represented in Table 12.1. In many ways this table is similar to Table 2.1 and the 16F87XA microcontrollers. All 18FXX2 devices have an instruction set of 75 instructions, with a clock oscillator that can run from DC to 40 MHz. There are also 'low-power' versions of each microcontroller available, coded 18LFXX2. The full manufacturer's data on this family is found in Ref. 12.1.

The pin connections of the 18FXX2 family, for the dual-in-line packages, are shown in Figure 12.1. The figure is very similar to Figure 7.1, with the ports, power supply, oscillator and reset lying in the same places. This of course allows upgrade with minimum change.

Table 12.1 The 18FXX2 sub-family

Device number	No. of pins*	Memory	Peripherals/special features
18F242	28	16 KB program memory (8K instructions**) 768 bytes RAM, 256 bytes EEPROM	3 parallel ports, 4 counter/timers, 2 capture/compare/PWM modules, 2 serial communication modules, 5 10-bit ADC channels
18F252	28	32 KB program memory (16K instructions**) 1536 bytes RAM, 256 bytes EEPROM	3 parallel ports, 4 counter/timers, 2 capture/compare/PWM modules, 2 serial communication modules, 5 10-bit ADC channels
18F442	40	16 KB program memory (8K instructions**) 768 bytes RAM, 256 bytes EEPROM	5 parallel ports, 4 counter/timers, 2 capture/compare/PWM modules, 2 serial communication modules, 8 10-bit ADC channels
18F452	40	32 KB program memory (16K instructions**) 1536 bytes RAM, 256 bytes EEPROM	5 parallel ports, 4 counter/timers, 2 capture/compare/PWM modules, 2 serial communication modules, 8 10-bit ADC channels

*For DIP package only.
**Single-word instructions, noting that some are two words.
ADC, analog-to-digital converter; PWM, pulse width modulation.

12.2   The 18F2X2 block diagram and Status register

The block diagram of the 18F2X2 microcontrollers, which are the 28-pin devices of Figure 12.1(a), is shown in Figure 12.2. Take some time to identify the principal features of this important diagram

Almost central to the diagram is the CPU (Central Processing Unit), containing the 8-bit ALU (Arithmetic Logic Unit), the Working register 'WREG' (sometimes called the accumulator) and an 8-bit × 8-bit hardware multiply unit. CPU action is determined by the instruction received from program memory, which is transferred through the Instruction register. This is seen above and to the left of the CPU block. An important element of the CPU, though not shown in this diagram, is the Status register.

Program memory is seen to the top left of the diagram. Its address bus enters the memory 'Address Latch'. With its 21 bits, it is possible to address 2^{21} locations, i.e. 2097152 (2 Mbyte) locations. Table 12.1 shows that the 18F242 only needs to address 16K locations, which require only 14 bits. The other lines are therefore redundant here. The 16-bit bus carrying the instruction word is seen leaving the 'program memory data latch'. This can be seen working its way down to, among other places, the Instruction register, already mentioned. To the right of the program memory is an area labelled 'program memory address generation'. Central to this of course is the Program Counter. Below the Program Counter is the Stack, which contains 31 locations. A table pointer provides a means of accessing tables or other data in program memory, under user program control.

The data memory is seen almost top centre of the diagram. Like the program memory, its address generation forms a significant block of the diagram overall, with a bank of File Select Registers (FSR0, for example)

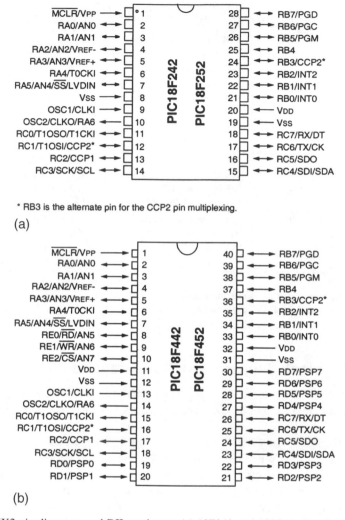

* RB3 is the alternate pin for the CCP2 pin multiplexing.

(a)

(b)

Figure 12.1 PIC 18FXX2 pin diagrams and DIL packages. (a) 18F242 and 18F252. (b) 18F442 and 18F452

and a Bank Select Register (BSR). The data memory address is 12 bits, which can address 4096 bytes. Again, Table 12.1 shows us that this address bus is not fully exploited. Data transfer to and from the data memory is through the main data bus.

With the separate address bus and data input/output for each of program memory and data memory, we can at this point confirm the underlying Harvard structure of the microcontroller.

Power supply connections for the microcontroller, V_{DD} and V_{SS}, appear towards the bottom left of the diagram. Looking back at Figure 12.1 shows that *two* pins are dedicated to the V_{SS} connection for either

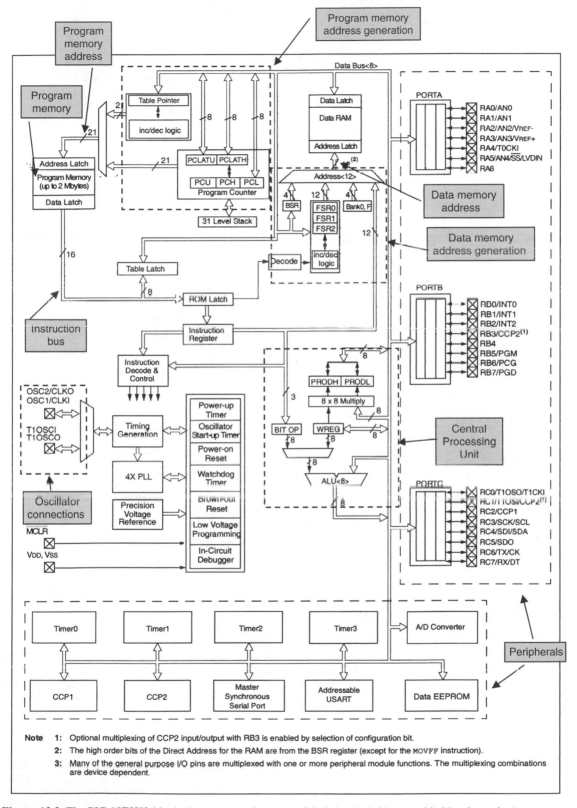

Figure 12.2 The PIC 18F2X2 block diagram (supplementary labels in shaded boxes added by the author)

of the two packages shown. This ensures a good 0 V connection. The smaller package has a single V_{DD} connection, while the larger has two. Associated with power supply is the Power-on Reset, which ensures reset when power is switched on; the Power-up Timer, which can be used to maintain the microcontroller in a state of reset for a fixed time after power-up; and the Brown-out Reset, which will force the microcontroller into reset if the power supply dips.

Oscillator inputs, both for the main oscillator and for Timer 1, are seen at the left of the diagram. Linked to the main oscillator is the Timing Generation circuit, which develops the various internal clock signals. Also associated with the main oscillator is the Oscillator Start-up Timer, which can be used to ensure that the oscillator is running reliably before the microcontroller itself starts to run. A new oscillator element is the phase-locked loop (PLL). This introduces the option of multiplying up the oscillator frequency, so that overall operation can be speeded up. This has an impact on operating flexibility and on power consumption.

In the lower left region of the diagram we also see the Watchdog Timer and in-circuit debugger. The first is a timing element that can be used to force a restart of microcontroller operation, if the program crashes in some way. The second provides a diagnostic capability that is actually designed into the microcontroller. Its use relies on connection of the microcontroller to a host computer. If used, program execution can be controlled from the computer. In this case it is also possible to return diagnostic data, including values of microcontroller registers, to the computer for display. This feature has already been described in Section 7.11 of Chapter 7.

Finally, the parallel ports, containing both their input/output function and all interconnection with the peripherals, are placed on the right-hand side. The other peripherals appear along the bottom of the diagram. Among these is another memory block, using EEPROM technology.

As an important part of the CPU, the Status register is shown in Figure 12.3. It contains a full 5 bits of status information on the operation most recently performed by the microcontroller. The limited number of Status bits (Figure 7.3) is arguably one of the weaknesses of the 16 Series microcontroller. The 18 Series adds 2 new bits. These are **OV** (bit 3), which signals an overflow of the 8-bit range, and **N**, which indicates that a twos complement number is negative. As the sign bit in a twos complement number is the MSB, the **N** bit is simply the MSB of the result. These extra bits allow improved program branching and better mathematical capability. The other information in the figure is self-explanatory.

12.3   The 18 Series instruction set

The full instruction set of the PIC 18 Series is shown in Appendix 5, with 75 distinct instructions! It is a super-set of the PIC 16 and 17 Series instruction sets, so any program which runs on one of those microcontrollers can be expected to run on the 18 Series. With this many instructions, the set begins to have a CISC (Complex Instruction Set Computer – see Chapter 1) feel to it, losing some of the simplicity we expect of RISC. Note that there is also an 'extended' 18 Series instruction set, which adds a number of instructions to this one. These are described in Chapter 13.

Looking at Table A5.1, it seems a daunting prospect to get to know all those instructions. The good news is: we don't need to! After this chapter, almost all the programming we do will be in C, so someone else

U-0	U-0	U-0	R/W-x	R/W-x	R/W-x	R/W-x	R/W-x
—	—	—	N	OV	Z	DC	C

bit 7   bit 0

bit 7-5   **Unimplemented:** Read as '0'

bit 4   **N:** Negative bit
This bit is used for signed arithmetic (2's complement). It indicates whether the result was negative (ALU MSB = 1).

1 = Result was negative
0 = Result was positive

bit 3   **OV:** Overflow bit
This bit is used for signed arithmetic (2's complement). It indicates an overflow of the 7-bit magnitude, which causes the sign bit (bit7) to change state.

1 = Overflow occurred for signed arithmetic (in this arithmetic operation)
0 = No overflow occurred

bit 2   **Z:** Zero bit

1 = The result of an arithmetic or logic operation is zero
0 = The result of an arithmetic or logic operation is not zero

bit 1   **DC:** Digit carry/borrow bit
For ADDWF, ADDLW, SUBLW, and SUBWF instructions

1 = A carry-out from the 4th low order bit of the result occurred
0 = No carry-out from the 4th low order bit of the result

Note: For borrow, the polarity is reversed. A subtraction is executed by adding the two's complement of the second operand. For rotate (RRF, RLF) instructions, this bit is loaded with either the bit 4 or bit 3 of the source register.

bit 0   **C:** Carry/borrow bit
For ADDWF, ADDLW, SUBLW, and SUBWF instructions

1 = A carry-out from the Most Significant bit of the result occurred
0 = No carry-out from the Most Significant bit of the result occurred

Note: For borrow, the polarity is reversed. A subtraction is executed by adding the two's complement of the second operand. For rotate (RRF, RLF) instructions, this bit is loaded with either the high or low order bit of the source register.

Figure 12.3 The PIC 18FXX2 Status register

can worry about Assembler. However, an overview is still useful – we do, after all, need Assembler inserts in C sometimes.

For those migrating to the 18 Series from the 16 Series, Table 12.2 should be interesting. This gives a brief comparison of the two instruction sets. The first column lists all 16 Series instructions, approximately in the order given in Appendix 1. The second column gives the 18 Series equivalent, where there is one, and also lists all instructions that are completely 'new'.

To explore the 18 Series instructions, let's start by looking at the first two columns of Table A5.1. These give the Assembler mnemonic of the instruction with its operands, followed by what the instruction does.

Table 12.2 Comparison of 16 Series and 18 Series instruction sets

16 Series instruction	18 Series equivalents	Description
Byte-oriented file register operations		
addwf f,d	addwf f,d,a	Add W and f
	addwfc f,d,a	Add W and f with Carry
andwf f,d	andwf f,d,a	And W and f
clrf f	clrf f,a	Clear f
clrw	–	Clear W
comf f,d	comf f,d,a	Complement f
–	cpfseq f,a	Compare f with W, skip if equal
–	cpfsgt f,a	Compare f with W, skip if greater than
–	cpfslt f,a	Compare f with W, skip if less than
decf f,d	decf f,d,a	Decrement f
decfsz f,d	decfsz f,d,a	Decrement f, skip if zero
–	decfsnz f,d,a	Decrement f, skip if not zero
incf f,d	incf f,d,a	Increment f
incfsz f,d	incfsz f,d,a	Increment f, skip if zero
	incfsnz f,d,a	Increment f, skip if not zero
iorwf f,d	iorwf f,d,a	Inclusive OR f with W
movf f,d	movf f,d,a	Move f
–	movff f_s,f_d	Move source file f_s to destination file f_d
movwf f	movwf f,a	Move W to f
nop	nop	No operation – an intentional instruction
	nop	The second word of a two-word instruction, which is encoded to execute as a **nop** if it is accidentally interpreted as an instruction
–	mulwf f,a	Multipy W and f
–	negf f,a	Negate f
rlf f,d	rlfc f,d,a	Rotate left through Carry
	rlnfc f,d,a	Rotate left, no Carry
rrf f,d	rrcf f,d,a	Rotate right through Carry
	rrncf f,d,a	Rotate right, no Carry
–	set f	Set f
subwf f,d	subwf f,d,a	Subtract W from f
	subwfb f,d,a	Subtract W from f with borrow
–	subfwb f,d,a	Subtract f from W with borrow
swapf f,d	swapf f,d,a	Swap nibbles in f
–	tstfsz f,a	Test f, skip if zero
xorwf f,d	xorwf f,d,a	Exclusive OR W with f
Bit-oriented file register operations		
bcf f,b	bcf f,b,a	Clear bit b in register f
bsf f,b	bsf f,b,a	Set bit b in register f
–	btg f,d,a	Toggle bit b in register f
btfsc f,b	btfsc f,b,a	Test bit b in f, skip if clear
btfss f,b	btfss f,b,a	Test bit b in f, skip if set

continued

Table 12.2 Continued

16 Series instruction	18 Series equivalents	Description
Literal operations		
addlw k	**addlw k**	Add literal to W
andlw k	**andlw k**	And literal with W
iorlw k	**iorlw k**	Inclusive OR literal with W
movlw k	**movlw k**	Move literal to W
–	**movlb**	Move literal to BSR
–	**lfsr f,k**	Load FSR **f** with 12-bit literal **k**
–	**mullw**	Multiply literal with W
sublw k	**sublw k**	Subtract W from literal
xorlw k	**xorlw k**	Exclusive OR literal with W
Control operations		
call k	**call n,s** **rcall n**	Call subroutine, with (**s** = 1) or without (**s** = 0) saving context to Stack Relative call to subroutine
clrwdt	**clrwdt**	Clear Watchdog Timer
–	**daw**	Decimal adjust W
goto k	**goto n**	Go to absolute address, where **k/n** address anywhere in program memory space
–	**pop**	Pop top of return stack (TOS)
–	**push**	Push top of return stack (TOS)
–	**reset**	Software reset
retfie	**retfie s**	Return from interrupt, with (**s** = 1) or without (**s** = 0) retrieving context from Stack
retlw k	**retlw k**	Return with literal in W
return	**return s**	Return from subroutine, with (**s** = 1) or without (**s** = 0) retrieving context from Stack
sleep	**sleep**	Go into standby mode
–	**bc, bn, bnc, bnn, bnov, bnz, bov, bz**	Eight conditional branch instructions, one for each state of Status register bits **N**, **OV**, **Z**, **C**, all with 8-bit twos complement relative address **n**
–	**bra n**	Branch unconditionally 8-bit twos complement relative address **n**
Program memory Table Read/Write operations		
–	**tblrd*, tblrd*+, tblrd*−, tblrd+***	Four Table Read instructions, with pointer change respectively no change, post-increment, post-decrement, pre-increment
–	**tblwt*, tblwt*+, tblwt*−, tblwt+***	Four Table Write instructions, with pointer change respectively: no change, post-increment, post-decrement, pre-increment

The symbols used for the operands, for example **a**, **d** or **f**, are explained in Table A5.2. While some of these are familiar from the 16 Series, the operand bit **a** is new and leads to the possibility of instructions with three operands. This defines the memory area called 'Access RAM' described later in this chapter. Through this bit, the programmer now has a choice of whether to use Access RAM or not.

The third column of Table A5.1 indicates how many instruction cycles the instruction takes to execute. As we expect with a RISC and pipelined structure, all normal instructions execute in a single cycle. Those which cause a program branch take two. Skip instructions take one cycle if there is no skip, two if followed by a single-word instruction and three if followed by a two-word instruction. A small number of complex instructions also take more than one cycle.

The next column (column 4) shows the actual coding of the instruction. Fortunately, it is the assembler or compiler that generates this code. Most instructions are contained in a single 16-bit word, while just four occupy two 16-bit words. These are **call**, **goto**, **movff** and **lfsr**. Take a look at the machine code itself of these instructions. It is interesting to see that while the second word of any of them carries useful information, if taken alone it encodes as a **nop** instruction. This is because the most significant 4 bits form the **nop** machine code, for which the less significant bits are 'don't care'. This arrangement allows program execution to realign itself, if at any time the microcontroller tries to interpret this second word as a stand-alone instruction.

The final column of Table A5.1, 'Status affected', indicates which bits of the Status register (Figure 12.3) are affected by the action of the instruction. It is interesting from this point of view to compare 'identical' instructions, as they appear in the 16 Series (Appendix 1) and 18 Series instructions sets. For example, **addwf** in the 16 Series only affects the **C**, **DC** and **Z** bits. In the 18 Series, the same bits are affected, as well as **OV** and **N**. While the instruction is unchanged in its function, it has become more powerful, by its effect on more Status bits.

In relation to the 16 Series, and looking at Table 12.2, it can be seen that the 18 Series instructions fall into the categories listed below.

12.3.1 Instructions which are unchanged

These are instructions whose function and form is identical to the 16 Series. Many examples lie in the literal instructions, for example **addlw k**, **andlw k**. The only difference is the effect on the increased number of Status bits.

12.3.2 Instructions which have been upgraded

These instructions are almost the same as their 16 Series predecessors, but have added functionality or flexibility, due in part to changes in architecture. The most widespread example of this is the ability to select Access RAM as the target memory area, as mentioned above. It can be seen that this change is implemented in a large number of the byte-oriented arithmetic and logical operations, for example **addwf f,d,a** and **andwf f,d,a**.

Another interesting development is the flexibility attached to **call**, **return** and **retfie** instructions. A new operand **s** allows a choice to be made over whether context is saved to the Stack, and then retrieved from it or not.

To ensure upward compatibility, all these instructions assemble to valid 18 Series code, even if presented in 16 Series format.

12.3.3 New, variant, instructions

Some of the 16 Series instructions felt limited, and for effective use in certain situations had to be used in direct association with other instructions. The 18 Series instruction set adds variants to many of these instructions to ease these limitations. For example, the simple add instruction **addwf** is now also available as add with carry, **addwfc**. This simplifies an enormous number of 16-bit or greater additions. Similarly, subtract with borrow is also available, as are rotates with or without Carry and **incfsnz** (increment f, skip if *not* zero).

12.3.4 New instructions

Finally, there are many instructions that are just plain new. These derive in many cases from enhanced hardware or memory addressing techniques. Significant among arithmetic instructions is the multiply, available as **mulwf** (multiply W and f) and **mullw** (multiply W and literal). These invoke the hardware multiplier, seen already in Figure 12.2. Multiplier and multiplicand are viewed as unsigned, and the result is placed in the registers **PRODH** and **PRODL**. It is worth noting that the multiply instructions cause no change to the Status flags, even though a zero result is possible.

Other important additions to the instruction set are a whole block of Table Read and Write instructions, data transfer to and from the Stack, and a good selection of conditional branch instructions, which build upon the increased number of condition flags in the Status register. There are also instructions that contribute to conditional branching. These include the group of compares, for example **cpfseq**, and the test instruction, **tstfsz**.

A useful new move instruction is **movff**, which gives a direct move from one memory location to another. It should be noted that this codes in two words and takes two cycles to execute. Therefore, its advantage over the two 16 Series instructions which it replaces may seem slight. It does, however, save the value of the W register from being overwritten.

Some of these new instructions will be explored in the program example and exercises of Section 12.10.

12.4 Data memory and Special Function Registers

Table 12.1 showed the different memory sizes in the 18FXX2 family. In the next sections, where we look at memory, we will use the 18F242 as the main example. The other devices in the family are similar, but all have larger memory capability in one way or another. Those upgrading from the PIC 16 Series will need to be ready for some new approaches to memory structure.

12.4.1 The data memory map

The general data memory map of the 18F242 is shown in Figure 12.4. Each memory location is 1 byte, while the address is 12-bit, with the capability to address up to 4096 locations. The structure of the memory is effectively made up of 16 banks, each of 256 bytes. A special register, the Bank Select Register (BSR), holds the bits that select the bank. These bits are seen in Figure 12.4 and form the highest 4 bits of the 12-bit memory address. Figure 12.4 shows that only the lowest three banks are implemented as 'general-purpose

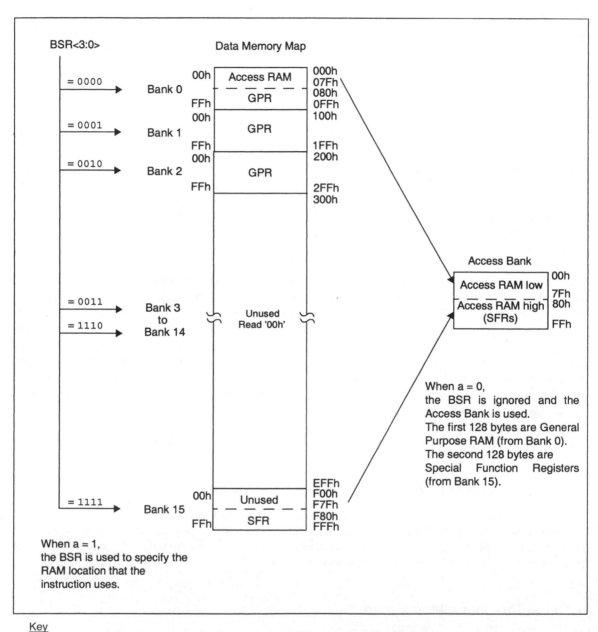

Key
BSR: Bank Select Register   GPR: General-purpose Register   SFR: Special Function Register

Figure 12.4 The PIC 18F242 data memory map

registers' (i.e. general-purpose RAM) in the 18F242. These make the 768 (3 × 256) bytes of data memory indicated in Table 12.1.

While the lower three banks of data memory are used for general-purpose RAM, the SFRs are contained in a block at the *top* end of memory, in the upper half of the top bank. They are shown in Figure 12.5. Note that the four columns shown are not themselves memory banks. In fact, taken together they only form part of the highest bank, as already seen in Figure 12.4.

12.4.2 Access RAM

We have already come across mention of 'Access RAM' while looking at the instruction set. Figure 12.4 shows two areas of memory that carry this name. These are the lowest and highest 128 bytes of memory. The lowest is general-purpose RAM, while the highest contains all the SFRs. The Access RAM concept provides a way of addressing a part of RAM quickly. While the two halves of Access RAM are at opposite ends of the memory map, when used as Access RAM they are treated as one continuous block of memory. The bits of the BSR are ignored and they then have just an 8-bit address, with the SFR addresses following directly on from the lower Access RAM addresses.

Access RAM is available to all instructions which have the **a** operand, as seen in Appendix 5 or Table 12.2. If **a** is set to 0, then only Access RAM is available and it is accessed as described above.

12.4.3 Indirect addressing and accessing tables in data memory

The concept of indirect addressing in the 16 Series was introduced in Chapter 5, Section 5.4.1. This described how a File Select Register (FSR) can be used to hold an address. If the program addresses the (non-existent) register **INDF**, then the address placed in the **FSR** is used as the address for that instruction.

This concept continues to apply in the 18 Series, but is extended to match the larger memory sizes involved, and with multiple **FSR** and **INDF** registers. To begin with, there are now three FSRs, which appear in Figure 12.2. Because of the much larger data memory size, each one needs to have 12 bits. They are therefore each made up of two memory locations and can be seen in the SFR map of Figure 12.5 as **FSR2H:FSR2L**, **FSR1H:FSR1L** and **FSR0H:FSR0L**.

To further complicate matters, there are *five* equivalents to the **INDF** register. These are shown in Table 12.3. They can also be found in the SFR map of Figure 12.5.

Now if an instruction addresses any of the locations shown in Table 12.3, then the 12-bit number held in the corresponding FSR is used as an indirect address, accessing a location in data memory upon which the instruction operates. During instruction execution the value of the FSR is modified as shown.

12.5 Program memory

As with the data memory, we use the smallest 18FXX2, the 18F242, as our example device when looking at program memory.

Address	Name	Address	Name	Address	Name	Address	Name
FFFh	TOSU	FDFh	INDF2[3]	FBFh	CCPR1H	F9Fh	IPR1
FFEh	TOSH	FDEh	POSTINC2[3]	FBEh	CCPR1L	F9Eh	PIR1
FFDh	TOSL	FDDh	POSTDEC2[3]	FBDh	CCP1CON	F9Dh	PIE1
FFCh	STKPTR	FDCh	PREINC2[3]	FBCh	CCPR2H	F9Ch	—
FFBh	PCLATU	FDBh	PLUSW2[3]	FBBh	CCPR2L	F9Bh	—
FFAh	PCLATH	FDAh	FSR2H	FBAh	CCP2CON	F9Ah	—
FF9h	PCL	FD9h	FSR2L	FB9h	—	F99h	—
FF8h	TBLPTRU	FD8h	STATUS	FB8h	—	F98h	—
FF7h	TBLPTRH	FD7h	TMR0H	FB7h	—	F97h	—
FF6h	TBLPTRL	FD6h	TMR0L	FB6h	—	F96h	TRISE[2]
FF5h	TABLAT	FD5h	T0CON	FB5h	—	F95h	TRISD[2]
FF4h	PRODH	FD4h	—	FB4h	—	F94h	TRISC
FF3h	PRODL	FD3h	OSCCON	FB3h	TMR3H	F93h	TRISB
FF2h	INTCON	FD2h	LVDCON	FB2h	TMR3L	F92h	TRISA
FF1h	INTCON2	FD1h	WDTCON	FB1h	T3CON	F91h	—
FF0h	INTCON3	FD0h	RCON	FB0h	—	F90h	—
FEFh	INDF0[3]	FCFh	TMR1H	FAFh	SPBRG	F8Fh	—
FEEh	POSTINC0[3]	FCEh	TMR1L	FAEh	RCREG	F8Eh	—
FEDh	POSTDEC0[3]	FCDh	T1CON	FADh	TXREG	F8Dh	LATE[2]
FECh	PREINC0[3]	FCCh	TMR2	FACh	TXSTA	F8Ch	LATD[2]
FEBh	PLUSW0[3]	FCBh	PR2	FABh	RCSTA	F8Bh	LATC
FEAh	FSR0H	FCAh	T2CON	FAAh	—	F8Ah	LATB
FE9h	FSR0L	FC9h	SSPBUF	FA9h	EEADR	F89h	LATA
FE8h	WREG	FC8h	SSPADD	FA8h	EEDATA	F88h	—
FE7h	INDF1[3]	FC7h	SSPSTAT	FA7h	EECON2	F87h	—
FE6h	POSTINC1[3]	FC6h	SSPCON1	FA6h	EECON1	F86h	—
FE5h	POSTDEC1[3]	FC5h	SSPCON2	FA5h	—	F85h	—
FE4h	PREINC1[3]	FC4h	ADRESH	FA4h	—	F84h	PORTE[2]
FE3h	PLUSW1[3]	FC3h	ADRESL	FA3h	—	F83h	PORTD[2]
FE2h	FSR1H	FC2h	ADCON0	FA2h	IPR2	F82h	PORTC
FE1h	FSR1L	FC1h	ADCON1	FA1h	PIR2	F81h	PORTB
FE0h	BSR	FC0h	—	FA0h	PIE2	F80h	PORTA

Note 1: Unimplemented registers are read as '0'.
     2: This register is not available on PIC18F2X2 devices.
     3: This is not a physical register.

Figure 12.5 The PIC 18F242 Special Function Registers

Table 12.3 'Virtual' registers used in indirect addressing

'Virtual' register addressed n = 0, 1 or 2	Action following instruction invoking FSR
INDF*n*	No change to FSR*n*
POSTINC*n*	The FSR is automatically incremented following access
POSTDEC*n*	The FSR is automatically decremented following access
PREINC*n*	The FSR is automatically incremented preceding access
PLUSW*n*	The value in WREG is added to FSR*n*, to form indirect address. Neither FSR nor WREG are changed

12.5.1 The program memory map

The program memory map is shown in Figure 12.6. It can be seen that it occupies only a modest proportion of the overall capacity offered by the 21 bit bus. Each memory location has a size of 1 byte. With the normal instruction word being 16 bits, each instruction therefore takes 2 or 4 bytes. The less significant byte is stored in a location with an even address. The reset vector, where the program starts, is shown as memory location 0000, and the two interrupt vectors at locations 0008_H and 0018_H. Interrupt Service Routines (ISRs) must be written to start at one of these locations.

12.5.2 The Program Counter

The Program Counter, seen at the top of Figure 12.6, is 21 bits wide. With each instruction stored as 2 or 4 bytes, the Program Counter increments twice, or four times, every instruction.

Access to the Program Counter is available through SFRs in the memory map. The least significant byte is called **PCL**. It is readable and writeable, and can be seen as memory location $FF9_H$ in Figure 12.5. The middle program counter byte, and the higher 5 bits, are not directly readable or writeable. Updates to these may, however, be made through the **PCLATH** and **PCLATU** registers, seen in both Figures 12.2 and 12.5. Contents of these locations are transferred to the Program Counter by any operation that writes to **PCL**. Similarly, any operation that reads **PCL** will cause the relevant higher bits of the Program Counter to be transferred to **PCLATH** and **PCLATU**.

12.5.3 Upgrading from the 16 Series and computed *goto* instructions

With the 16 Series, we used the computed **goto** as a means of retrieving data from a look-up table, for example Figure 5.5 or Program Example 5.4. An offset was added to the Program Counter to cause a jump to one of a list of **retlw** instructions. That instruction then caused a return to the main program, with the selected byte of data carried in the W register. This is still possible in the 18 Series, but it is important to remember that each instruction now takes 2 bytes in program memory. Therefore, any offset added to the Program Counter in a 16 Series program must be doubled to create the same effect in an 18 Series program. An example of how to do this is given in Ref. 12.2.

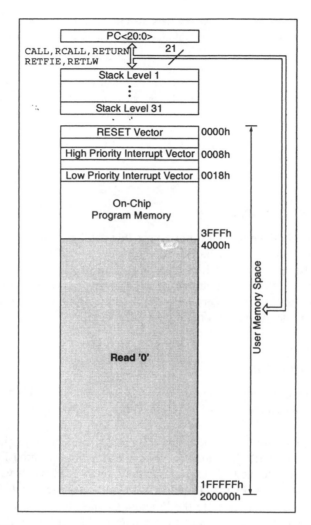

Figure 12.6 The PIC 18F242 program memory map and return address stack

12.5.4 The Configuration registers

Whereas the 16 Series microcontrollers have a single 14-bit Configuration Word, the 18 Series has a whole set of Configuration registers, reflecting the greater complexity and flexibility of the device. The registers are shown in Table 12.4 and a summary of the function of each bit in Table 12.5. It is easy to see that the usual features – oscillator settings, Watchdog Timer, power-up timing, Brown-out Reset and code protection – are there. Most of these have grown in the options they offer. There are now, for example, eight oscillator settings, an alternative Watchdog mode and more brown-out settings. More bits are therefore needed.

The Configuration registers are mapped in program memory starting at memory location 300000_H. Like program memory, they are 8-bit locations. The two device identification registers, **DEVID1** and

Table 12.4 18 Series Configuration Words: 18FXX2 Configuration registers and device identifications (Table provided by Microchip)

File name		Bit 7	Bit 6	Bit 5	Bit 4	Bit 3	Bit 2	Bit 1	Bit 0	Default/unprogrammed value
300001h	CONFIG1H	–	–	|OSCSEN	–	–	FOSC2	FOSC1	FOSC0	--1- -111
300002h	CONFIG2L	–	–	–	–	BORV1	BORV0	BOREN	|PWRTEN	---- 1111
300003h	CONFIG2H	–	–	–	–	WDTPS2	WDTPS1	WDTPS0	WDTEN	---- 1111
300005h	CONFIG3H	–	–	–	–	–	–	–	CCP2MX	---- ---1
300006h	CONFIG4L	|DEBUG	–	–	–	–	LVP	–	STVREN	1--- -1-1
300008h	CONFIG5L	–	–	–	–	CP3	CP2	CP1	CP0	---- 1111
300009h	CONFIG5H	CPD	CPB	–	–	–	–	–	–	11-- ----
30000Ah	CONFIG6L	–	–	–	–	WRT3	WRT2	WRT1	WRT0	---- 1111
30000Bh	CONFIG6H	WRTD	WRTB	WRTC	–	–	–	–	–	111- ----
30000Ch	CONFIG7L	–	–	–	–	EBTR3	EBTR2	EBTR1	EBTR0	---- 1111
30000Dh	CONFIG7H	–	EBTRB	–			–	–		-1-- ----
3FFFFEh	DEVID1	DEV2	DEV1	DEV0	REV4	REV3	REV2	REV1	REV0	**(1)**
3FFFFFh	DEVID2	DEV10	DEV9	DEV8	DEV7	DEV6	DEV5	DEV4	DEV3	0000 0100

Table 12.5 18 Series Configuration Words: 18FXX2 summary of configuration bits and device identifications

Configuration bit(s)	Summary of function (unprogrammed value is 1, active (enabled) value generally 0)
OSCSEN	Clock Source Switching Enable bit
FOSC2:FOSC0	Selects one of eight oscillator modes (see Table 12.6)
BORV1, BORV0	Selects Brown-out Reset voltage, to 2.5, 2.7, 4.2 or 4.5 V
BOREN	Brown-out Reset Enable bit
PWRTEN	Power-up Timer Enable bit
WDTPS2:WDTPS0	Watchdog Timer Postscale bits, eight values from 1:1 to 1:128
WDTEN	Watchdog Timer Enable bit
CCP2MX	CCP2 Multiplex, selects RC1 (1) or RB3 (0)
DEBUG	Background Debug Enable bit
LVP	Low-Voltage Program Enable bit
STVREN	Stack Full/Underflow Reset Enable bit
CP3:CP0	Code Protection bits
CPD	Data EEPROM Code Protection bit
CPB	Boot Block Code Protection bit
WRT3:WRT0	Program Memory Write Protection bits
WRTD	Data EEPROM Write Protection bit
WRTB	Boot Block Write Protection bit
WRTC	Configuration Register Write Protection bit
EBTR3:EBTR0	Table Read Protection bits
EBTRB	Boot Block Table Read Protection bit
DEV2:DEV0	Device ID bits: 000 = 18F252, 001 = 18F452, 100 = 18F242, 101 = 18F442
REV4:REV0	Revision ID bits
DEV10:DEV3	Further Device ID bits

DEVID2, are readable only and contain pre-programming information identifying the device and its revision number.

12.6 The Stacks

The Stack in an advanced microprocessor is a versatile memory block, which can be used both automatically – say, for saving a return address in a subroutine call – as well as by the programmer, for short-term data storage. Indeed, in many cases multiple Stacks are used. The 16 Series Stack seen earlier was, however, a small and mechanistic structure. With only eight levels, it is linked directly to the Program Counter, and just saves or returns its value under certain subroutine call or interrupt conditions.

The 18 Series Stack moves some way to taking on the features of the larger microprocessor. The automatic functions remain, for subroutine call and so on, but there is also some user access. There is also limited stacking, in another memory area, of key data registers.

12.6.1 Automatic Stack operations

The main Stack is called, in the 18 Series, the 'Return Address Stack'. This distinguishes it from the other smaller Stack locations that also exist. It is seen as part of Figure 12.6. It consists of 31 Read/Write memory locations, each of 21 bits. This figure also shows the Assembler mnemonics of those instructions that cause it automatic access. All of these relate to subroutine call or return, with the exception of **retfie**, the return from interrupt instruction.

12.6.2 Programmer access to the Stack

Not shown in Figure 12.6 is the Stack Pointer, which holds the current Stack address. It is set to zero on all resets and its value is changed by all automatic Stack accesses. Thus, it is incremented when a value is pushed onto the Stack and decremented when a value is popped off. It is *also* configured as part of an SFR, called **STKPTR**. As with all SFRs, this is readable and writeable by the programmer. **STKPTR** is seen as memory location FFC_H in Figure 12.5. The 5 lower bits of **STKPTR** form the Stack Pointer. The register also contains bits that flag Stack overflow or underflow – respectively **STKOVF** (also called **STKFUL**, bit 7) and **STKUNF** (bit 6).

The value held in the top location of the Stack is accessible to the programmer, and is both readable and writeable. Being 21 bits, it occupies three register locations, **TOSU**, **TOSH** and **TOSL**. These can again be seen in the register map of Figure 12.5, above **STKPTR**. The instructions **PUSH** and **POP** allow the programmer to push the current Program Counter value onto the Stack, or retrieve the top of the Stack and place it in the Program Counter.

12.6.3 The Fast Register Stack

The 18 Series structure provides not only a Stack for the Program Counter, but also (in primitive form) a separate 'Stack' for the **STATUS**, **WREG** and **BSR** registers. These are single memory locations, not directly accessible to the programmer, which together are called the *Fast Register Stack*. When an interrupt occurs, either high or low priority, the values of the three registers listed are saved. The stacked values are returned if a 'fast' return from interrupt is selected. In other words, the **retfie** instruction should be used with the **s** operand set to 1. Section 12.7.7 returns to this issue.

The Fast Register Stack can also be used with subroutine calls and returns. As Appendix 5 shows, both **call** and **return** instructions can be used with the **s** bit set, invoking use of the Fast Register Stack. It is of course only safe to do this if interrupts are not being used. Otherwise, an interrupt occurring during a subroutine will cause the Fast Register Stack to be overwritten.

12.7 The interrupts

The 18FXX2 microcontrollers offer a sophisticated interrupt structure that shows considerable advance over the 16 Series. Improvements include the introduction of a second interrupt vector, allowing a high priority and a low priority vector. This has already been seen in the program memory map of Figure 12.6. All interrupts but one can be allocated to high or low priority. There are more external interrupts, and automatic context saving, by use of the 'fast return' Stack (Section 12.5).

12.7.1 An interrupt structure overview

The interrupt structure is shown in Figure 12.7. It is fairly complex, but an understanding of it will lead to effective use of the microcontroller interrupts. Interrupt sources tend to appear from the left of the diagram. The three major outputs are to the right. Two of these lead to the interrupt vectors. Activation of either of these outputs causes a CPU interrupt, with Interrupt Service Routine (ISR) starting at one or other of the interrupt vectors seen in Figure 12.6. There is also a 'Wake-up' output, implemented if in Sleep mode.

As we explore the diagram, we will look for these features:

- The interrupt sources (noting, however, that not all are shown)
- The source enabling logic
- The source prioritisation logic
- The overall prioritisation enabling logic
- The overall (global) enabling logic.

12.7.2 The interrupt sources, their enabling and prioritisation

Let us start by identifying some of the interrupt sources. A useful first block to look at has been labelled 'External interrupt sources and Timer 0'. This block contains five AND gates, with all but one having three inputs. Look at the top one, with inputs labelled **TMR0IF**, **TMR0IE** and **TMR0IP**. This is the Timer 0 input and it displays a pattern that is repeated many times over. The input **…IF** is the Interrupt Flag bit, set if the Timer 0 interrupt has occurred; the input **…IE** is the Enable bit, a bit in an SFR which can be set or cleared by the program. The input **…IP** is the Priority bit, also in an SFR and set or cleared by the program. If *all* these inputs are high, i.e. the interrupt is enabled, it is selected as high priority, and the interrupt flag has been set, then the output of the AND gate goes high and the interrupt is routed through to the next stage of gating. Notice that the outputs of all AND gates in this block are ORed together.

The block just described is repeated towards the bottom of the diagram. Now, however, the third input to the Timer 0 AND gate is $\overline{\text{TMR0IP}}$. It is possible to see that *every* interrupt source (except one, which we will mention below) appears once in the top half of the diagram and once in the bottom. In the top half, it is enabled if its priority bit is high; in the bottom half, it is enabled if the bit is low. Again, the outputs of all AND gates in this block are ORed together.

The one source that is not prioritised is the external interrupt 0. This is always high priority and is labelled as such in the diagram.

The interrupts from the microcontroller peripherals, towards the left of the diagram, follow a similar pattern to the external interrupts. Again, they can be selected for high or low priority. They are not all shown, as there are so many. The generalised pattern for high priority sources appears towards the top left of the diagram, repeated for low priority towards the bottom left. As before, the logic ensures that for any one interrupt source, *either* high priority *or* low priority can be selected. As an example, the input for Timer 1 is given. Again, the outputs of all AND gates in each block are ORed together.

12.7.3   Overall interrupt prioritisation enabling

While we have seen that it is possible to prioritise individual interrupt sources, we may not want to use this facility. Therefore, it is possible to enable or disable the whole prioritisation process. This is done by **IPEN**, the Interrupt Priority Enable bit. **IPEN** is the MSB in the **RCON** register, which appears in Figure 12.14.

Look now at the block towards the centre of Figure 12.7, labelled 'priority steering logic'. The line **IPEN** appears three times in this block (once incorrectly labelled as **IPE**). If it is *low*, then interrupts from the lower half of the diagram, coming either from peripheral sources or from external sources, are routed up to the upper half of the diagram through the two AND gates in the block. All interrupts in this case are routed through to the high priority vector and there is no effective prioritisation. There is one more piece of enabling in this state, that we return to soon. When **IPEN** is low, as just described, the interrupt system is compatible with the 16 Series.

If **IPEN** is *high*, then high priority and low priority interrupts are each routed towards their own vector. Interrupt prioritisation is enabled and individual interrupt sources can be placed in the low or high priority domain, with their individual priority control bits.

12.7.4   Global enabling

There are two levels of global interrupt enabling, controlled by bits **GIE/GIEH** and **PEIE/GIEL**. These have a somewhat dual function, depending on the state of **IPEN** – hence the dual nature of their name.

It can be seen that **GIE/GIEH** controls both of the AND gates leading to the interrupt vectors. Through this it plays its 'global interrupt enable' role. If **GIE/GIEH** is low, there will be no interrupt, whatever the state of **IPEN**. When **IPEN** is low, all interrupts are routed towards the high interrupt vector, so it genuinely acts as a 'global enable'. When **IPEN** is high, it still has the power to disable both high and low priority interrupt. However, it cannot on its own enable the low priority one, as **PEIE/GIEL** is also involved. Therefore, it acts as *enable* for only the high priority path.

When **IPEN** is low, **PEIE/GIEL** acts as an enable line for all unmasked peripheral interrupts. It performs this function through the OR gate at the centre of the 'priority steering logic'. When **IPEN** is high, it acts as 'global enable' for the lower priority inputs, through its connection to the output AND gate for the lower priority vector.

12.7.5 Other aspects of the interrupt logic

Two final elements in this circuit are the line going from the high priority interrupt output down to the lower priority control logic. We can see that the action of this is that if a high priority interrupt is asserted, then the low priority path is blocked. The reverse is not true, however, and a high priority interrupt *can* interrupt a low priority interrupt. There are also lines from either interrupt path up to an OR gate which leads to Wake-up from Sleep. The action of these lines is independent of the state of the two 'global enable' lines.

12.7.6 The Interrupt registers

With the formidable number of bits seen in Figure 12.7, it is clear that a good number of registers will be needed to hold them all. Each interrupt source but one now needs a Flag bit, an Enable bit and a Priority bit, and all the control bits are needed as well. Beyond this, there is further control over some inputs, for example setting the active edge on external interrupts.

The design approach in creating the Interrupt registers has been to retain as far as possible those registers that are used in the 16 Series. This allows a comparatively easy upgrade path for the system designer. Therefore, the 16F87XA **INTCON** register, appearing in Figure 7.11, is almost exactly replicated in the 18 Series **INTCON** register, seen in Figure 12.8. Some small retitling of bits has taken place, and the functionality of **GIE** and **PEIE** has been extended, as just discussed.

To the **INTCON** register are added two further Interrupt control registers, **INTCON2** and **INTCON3**. These are shown in Figures 12.9 and 12.10 respectively. They contain the control bits for the interrupts that appear in the **INTCON** register. The bits are self-explanatory. It can be seen that the 'odd one out' is **RBPU**, which is not an interrupt bit at all, but controls the Port B pull-ups. In the 16 Series it was placed in the Option register, which does not exist in the 18 Series.

Enable bits, Flag (or 'Request') bits and Priority bits for the peripheral interrupt sources are placed in the **PIE1**, **PIE2**, **PIR1**, **PIR2**, **IPR1** and **IPR2** registers. These are summarised in Figures 12.11 and 12.12. Again, to improve upward compatability, they are very similar to the 16 Series registers of the same name, excluding of course the priority registers. It is interesting to compare them, by looking back at Figures 7.12 and 7.13. By doing this, you can see which interrupts have been added (Timer 3 and Low Voltage Detect),

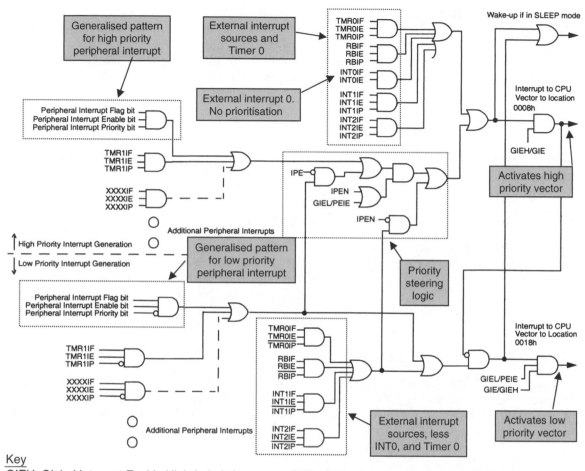

Figure 12.7 The PIC 18F242 interrupt logic (supplementary labels in shaded boxes added by the author)

and which have been removed (Comparator). Further details on the operation of some individual peripheral interrupts are given in the section on that peripheral, mainly in Chapter 13.

12.7.7   Context saving with interrupts

With the Fast Register Stack, described in Section 12.6.3 of this chapter, context saving can in some circumstances be delightfully easy. The programmer must decide first if the three registers saved on this stack, **WREG**, **STATUS** and **BSR**, are adequate for the purpose. If not, or if the fast return from interrupt is not used, then the programmer will need to write code to save all necessary registers at the start of the ISR and retrieve them at the end. It is important also to remember that a high priority interrupt can interrupt one of lower priority. In so doing, the interrupt of high priority would overwrite the contents of the Fast

R/W-0	R/W-0	R/W-0	R/W-0	R/W-0	R/W-0	R/W-0	R/W-x
GIE/GIEH	PEIE/GIEL	TMR0IE	INT0IE	RBIE	TMR0IF	INT0IF	RBIF

bit 7   bit 0

bit 7   **GIE/GIEH:** Global Interrupt Enable bit

When IPEN = 0:
1 = Enables all unmasked interrupts
0 = Disables all interrupts

When IPEN = 1:
1 = Enables all high priority interrupts
0 = Disables all interrupts

bit 6   **PEIE/GIEL:** Peripheral Interrupt Enable bit

When IPEN = 0:
1 = Enables all unmasked peripheral interrupts
0 = Disables all peripheral interrupts

When IPEN = 1:
1 = Enables all low priority peripheral interrupts
0 = Disables all low priority peripheral interrupts

bit 5   **TMR0IE:** TMR0 Overflow Interrupt Enable bit
1 = Enables the TMR0 overflow interrupt
0 = Disables the TMR0 overflow interrupt

bit 4   **INT0IE:** INT0 External Interrupt Enable bit
1 = Enables the INT0 external interrupt
0 = Disables the INT0 external interrupt

bit 3   **RBIE:** RB Port Change Interrupt Enable bit
1 = Enables the RB port change interrupt
0 = Disables the RB port change interrupt

bit 2   **TMR0IF:** TMR0 Overflow Interrupt Flag bit
1 = TMR0 register has overflowed (must be cleared in software)
0 = TMR0 register did not overflow

bit 1   **INT0IF:** INT0 External Interrupt Flag bit
1 = The INT0 external interrupt occurred (must be cleared in software)
0 = The INT0 external interrupt did not occur

bit 0   **RBIF:** RB Port Change Interrupt Flag bit
1 = At least one of the RB7:RB4 pins changed state (must be cleared in software)
0 = None of the RB7:RB4 pins have changed state

         Note: A mismatch condition will continue to set this bit. Reading PORTB will end the mismatch condition and allow the bit to be cleared.

Figure 12.8 The PIC 18FXX2 **INTCON** register

R/W-1	R/W-1	R/W-1	R/W-1	U-0	R/W-1	U-0	R/W-1
RBPU	INTEDG0	INTEDG1	INTEDG2	—	TMR0IP	—	RBIP

bit 7   bit 0

bit 7   **RBPU**: PORTB Pull-up Enable bit
          1 = All PORTB pull-ups are disabled
          0 = PORTB pull-ups are enabled by individual port latch values

bit 6   **INTEDG0**:External Interrupt0 Edge Select bit
          1 = Interrupt on rising edge
          0 = Interrupt on falling edge

bit 5   **INTEDG1**: External Interrupt1 Edge Select bit
          1 = Interrupt on rising edge
          0 = Interrupt on falling edge

bit 4   **INTEDG2**: External Interrupt2 Edge Select bit
          1 = Interrupt on rising edge
          0 = Interrupt on falling edge

bit 3   **Unimplemented:** Read as '0'

bit 2   **TMR0IP**: TMR0 Overflow Interrupt Priority bit
          1 = High priority
          0 = Low priority

bit 1   **Unimplemented:** Read as '0'

bit 0   **RBIP**: RB Port Change Interrupt Priority bit
          1 = High priority
          0 = Low priority

Figure 12.9 The PIC 18FXX2 **INTCON2** register

Register Stack, and the low priority interrupt loses its context! In such cases it is not safe to use the Fast Register Stack for low priority interrupts. Context for these should be saved in software.

12.8   Power supply and reset

12.8.1   Power supply

The supply voltage requirements of the 18LFXX2 and the 18FXX2 are shown in Figure 12.13. This shows that the 18LFXX2 devices can operate with a supply from 2.0 to 5.5 V, and the 18FXX2 from 4.2 to 5.5 V. The low-power device cannot, however, run at full speed at the lower voltage. The data [Ref. 12.1] shows that its maximum clock frequency at minimum supply voltage is 4 MHz. This rises to 40 MHz at 4.2 V.

12.8.2   Power-up and Reset

In Section 2.8 of Chapter 2 we explored the Reset circuitry of a simple PIC microcontroller, the 16F84A. The 18FXX2 controllers have a reset structure built directly on the model of Figure 2.11, with just a few

R/W-1	R/W-1	U-0	R/W-0	R/W-0	U-0	R/W-0	R/W-0
INT2IP	INT1IP	—	INT2IE	INT1IE	—	INT2IF	INT1IF

bit 7   bit 0

bit 7 **INT2IP:** INT2 External Interrupt Priority bit
1 = High priority
0 = Low priority

bit 6 **INT1IP:** INT1 External Interrupt Priority bit
1 = High priority
0 = Low priority

bit 5 **Unimplemented:** Read as '0'

bit 4 **INT2IE:** INT2 External Interrupt Enable bit
1 = Enables the INT2 external interrupt
0 = Disables the INT2 external interrupt

bit 3 **INT1IE:** INT1 External Interrupt Enable bit
1 = Enables the INT1 external interrupt
0 = Disables the INT1 external interrupt

bit 2 **Unimplemented:** Read as '0'

bit 1 **INT2IF:** INT2 External Interrupt Flag bit
1 = The INT2 external interrupt occurred (must be cleared in software)
0 = The INT2 external interrupt did not occur

bit 0 **INT1IF:** INT1 External Interrupt Flag bit
1 = The INT1 external interrupt occurred (must be cleared in software)
0 = The INT1 external interrupt did not occur

Figure 12.10 The PIC 18FXX2 **INTCON3** register

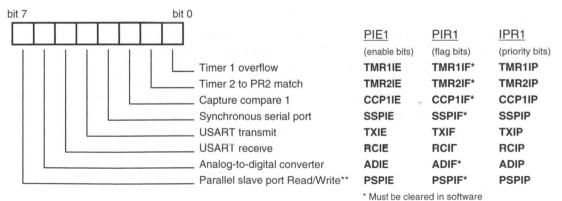

	PIE1	PIR1	IPR1
	(enable bits)	(flag bits)	(priority bits)
Timer 1 overflow	TMR1IE	TMR1IF*	TMR1IP
Timer 2 to PR2 match	TMR2IE	TMR2IF*	TMR2IP
Capture compare 1	CCP1IE	CCP1IF*	CCP1IP
Synchronous serial port	SSPIE	SSPIF*	SSPIP
USART transmit	TXIE	TXIF	TXIP
USART receive	RCIE	RCIF	RCIP
Analog-to-digital converter	ADIE	ADIF*	ADIP
Parallel slave port Read/Write**	PSPIE	PSPIF*	PSPIP

* Must be cleared in software
** 18F4X2 only. All bits reserved in 18F2X2.

Figure 12.11 18FXX2 **PIE1/PIR1/IPR1** (Peripheral Interrupt Enable/Peripheral Interrupt Request/Peripheral Interrupt Priority) registers

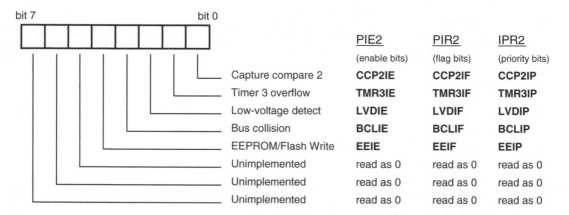

All flag bits must be cleared in software

Figure 12.12 18FXX2 **PIE2/PIR2/IPR2** (Peripheral Interrupt Enable/Peripheral Interrupt Request/Peripheral Interrupt Priority) registers

extra sources of reset. These are Stack over- or under-flow, Brown-out (already seen in the 16F873A) and Software Reset.

Besides adding further sources of reset, the 18 Series goes beyond this in an interesting way, by providing some history of what the source of reset was. Therefore, coming out of the Reset condition is not the completely fresh start that it is with simple microcontrollers. Now we can find out why we were forced into reset. In certain circumstances this can be very valuable, say if the Watchdog Timer has timed out. This information is provided through the **RCON** register, whose bits indicate what type of reset has occurred most recently. We have already met **RCON**, as its MSB is the interrupt **IPEN** bit.

A listing of all 18XX2 resets is given in Figure 12.14. This also shows the value of the Program Counter, the **RCON** register bits and the two Stack overflow bits (described in Section 12.5 of this chapter), after the reset has occurred. Now at the restart of a program it is possible to test the state of **RCON**, with the chance of introducing customised action if a particular type of reset has occurred.

The Software Reset indicated in Figure 12.14 is caused by execution of the instruction **reset** (Table A5.1). This replicates an external reset caused by a Logic 0 on input $\overline{\text{MCLR}}$. The conditions to ensure Power-on Reset are given in Figure 12.13, along with the different possible settings for the Brown-out Reset.

12.9   The oscillator sources

So far in this chapter we have seen that where 18 Series hardware features have been compared with 16 Series, the theme has been: evolve from the 16 Series, but give better performance and more flexibility. We will find similar developments as we look at the 18FXX2 clock sources.

PIC18LFXX2 (Industrial)			Standard Operating Conditions (unless otherwise stated) Operating temperature -40°C ≤ TA ≤ +85°C for industrial					
PIC18FXX2 (Industrial, Extended)			Standard Operating Conditions (unless otherwise stated) Operating temperature -40°C ≤ TA ≤ +85°C for industrial -40°C ≤ TA ≤ +125°C for extended					
Param No.	Symbol	Characteristic	Min	Typ	Max	Units	Conditions	
D001	VDD	**Supply Voltage**						
D001		PIC18LFXX2	2.0	—	5.5	V	HS, XT, RC and LP Osc mode	
D001		PIC18FXX2	4.2	—	5.5	V		
D002	VDR	**RAM Data Retention Voltage**[1]	1.5	—	—	V		
D003	VPOR	**VDD Start Voltage** to ensure internal Power-on Reset signal	—	—	0.7	V	See Section 3.1 (Power-on Reset) for details	
D004	SVDD	**VDD Rise Rate** to ensure internal Power-on Reset signal	0.05	—	—	V/ms	See Section 3.1 (Power-on Reset) for details	
D005	VBOR	**Brown-out Reset Voltage**						
		PIC18LFXX2						
		BORV1:BORV0 = 11	1.98	—	2.14	V	85°C ≥ T ≥ 25°C	
		BORV1:BORV0 = 10	2.67	—	2.89	V		
		BORV1:BORV0 = 01	4.16	—	4.5	V		
		BORV1:BORV0 = 00	4.45	—	4.83	V		
D005		PIC18FXX2						
		BORV1:BORV0 = 1x	N.A.	—	N.A.	V	Not in operating voltage range of device	
		BORV1:BORV0 = 01	4.16	—	4.5	V		
		BORV1:BORV0 = 00	4.45	—	4.83	V		

Legend: Shading of rows is to assist in readability of the table.
Note 1: This is the limit to which VDD can be lowered in SLEEP mode, or during a device RESET, without losing RAM data.

Figure 12.13 The PIC 18FXX2 power supply parameters

As with the 16F87XA group, the 18FXX2 devices have internal oscillator driver circuits, along with two external pins, **OSC1** and **OSC2**, for connection of any necessary external components. These pins can be seen in Figure 12.1. Interestingly, they share with other functions. The peripheral Timer 1 also has an oscillator source, with external connections labelled **T1OS0** and **T1OS1** (alongside other shared functions).

The 18FXX2 offers all four clock oscillator modes of the 16F87XA, and to these adds four more. These are all seen in Table 12.6. The relevant configuration bits are found in Configuration Word 1 (Table 12.4).

The oscillator modes have the characteristics described below.

Condition	Program Counter	RCON Register	Bits 4, 3, 2, 1, 0 resp. of **RCON** Register					Bits 7 and 6 resp. of **STKPTR** Register	
			$\overline{RI}$	$\overline{TO}$	$\overline{PD}$	$\overline{POR}$	$\overline{BOR}$	STKFUL	STKUNF
Power-on Reset	0000h	0--1 1100	1	1	1	0	0	u	u
$\overline{MCLR}$ Reset during normal operation	0000h	0--u uuuu	u	u	u	u	u	u	u
Software Reset during normal operation	0000h	0--0 uuuu	0	u	u	u	u	u	u
Stack Full Reset during normal operation	0000h	0--u uu11	u	u	u	u	u	u	1
Stack Underflow Reset during normal operation	0000h	0--u uu11	u	u	u	u	u	1	u
$\overline{MCLR}$ Reset during SLEEP	0000h	0--u 10uu	u	1	0	u	u	u	u
WDT Reset	0000h	0--u 01uu	1	0	1	u	u	u	u
WDT Wake-up	PC + 2	u--u 00uu	u	0	0	u	u	u	u
Brown-out Reset	0000h	0--1 11u0	1	1	1	1	0	u	u
Interrupt wake-up from SLEEP	PC + 2[1]	u--u 00uu	u	1	0	u	u	u	u

Legend: u = unchanged, x = unknown, - = unimplemented bit, read as '0'

Figure 12.14 The PIC 18FXX2 sources of reset, and subsequent Program Counter and Flag values (supplementary labels in shaded boxes added by the author)

Table 12.6 Oscillator modes

Mode	Description	Config. bits FOSC2:FOSC0
LP	Low power	000
XT	Crystal/resonator	001
HS	High-speed crystal/resonator	010
RC	External resistor/capacitor	011
EC	External clock	100
ECIO	External clock with **OSC2** configured as RA6	101
HS + PLL	High-speed crystal/resonator with phase-locked loop	110
RCIO	External resistor/capacitor with **OSC2** configured as RA6	111

12.9.1   *LP, XT, HS and RC oscillator modes*

These modes of operation are the same as the modes of the same name as used in the 16 Series, and described in Section 3.5.3 of Chapter 3. That section can be read again as a review.

12.9.2   *EC, ECIO and RCIO oscillator modes*

In the EC (external clock) mode of oscillator operation, an external clock source is connected to the **OSC1** pin. This is not a new concept to PIC microcontrollers and was identified as a possible mode of operation with the 16 Series. The difference in the mode discussed here is that the feedback device in the internal drive circuit is switched off, thereby saving a little current.

The EC mode of operation uses only one pin as input, potentially leaving the **OSC2** pin unused. The ECIO mode grabs this pin and creates of it an extra bit for Port A, bit RA6. A similar approach is taken with the RCIO mode – the RC clock oscillator only uses one pin, leaving pin **OSC2** again available to be used as a port pin.

12.9.3   *HS + PLL oscillator mode*

A phase-locked loop (PLL) is a clever piece of analog and digital circuitry that can be used, among other things, to multiply by an integer number the frequency of a signal. PLLs are finding increasing usage in microcontrollers to manipulate the frequency of clock signals. This can allow certain sections of the microcontroller to run faster than others, or to run the microcontroller at a clock frequency faster than the oscillator itself. The 18FXX2 PLL can be enabled if the microcontroller is set in HS + PLL mode, and then multiplies the oscillator signal by a factor of 4. Therefore, for example, the oscillator can run at 10 MHz, but with the PLL running, the internal clock frequency will be 40 MHz. This can have the effect of reducing external electromagnetic interference.

Because, just like a crystal oscillator, a PLL takes finite time to settle into stable operation, a timer is included to delay CPU start-up until the PLL has settled. This actually forms part of the Power-up Timer (PWRT), pictured in Figure 2.11 in a 16 Series version. The added delay is symbolised as T_{PLL} and is set to be 2 ms.

12.9.4   *Clock source switching*

If a microcontroller is designed into a power-conscious application, and particularly if that application runs over a long period of time, it can be beneficial to be able to switch the clock speed. Thus, it could run fast for periods of intensive activity and slower for less activity. In a way this is a less extreme variation on using the Sleep mode.

The ability to switch clock sources is controlled by the $\overline{\text{OSCSEN}}$ configuration bit (Tables 12.4 and 12.5). If enabled, the clock source can be selected from one of just two sources – the main clock oscillator (in whatever mode it has been configured) and the Timer 1 oscillator. The selection is controlled by the **SCS** bit, which is the least significant (and only) bit in the **OSCCON** register. When it is low the main oscillator is selected, when high the Timer 1 oscillator is selected.

It is of course important when switching between two unsynchronised oscillators, of different frequency, to ensure that unwanted glitches do not occur in the clock signal. Therefore, the 18FXX2 contains circuitry to ensure error-free switching. When the **SCS** bit changes state, program execution freezes at the start of the next instruction cycle. Eight cycles of the oscillator signal that it is switching to are counted before the switch is implemented. The CPU can then proceed. The precise details of the switch, which depends also on the type of oscillator in use, can be found in the full data [Ref. 12.1].

12.10   Introductory programming with the 18F242

While most programming in this section of the book is to be done with the C language, it is worth simulating some trial programs in Assembler, in order to gain an initial familiarity with the instruction set. If you have followed this book from the beginning, you will be familiar with the MPLAB® development environment and the simulator MPSIM™. Let us use MPLAB to develop some simple programs. Review Section 4.6 of Chapter 4 for a refresher if needed.

12.10.1   Using the MPLAB IDE for the 18 Series

Try opening MPLAB and, using Configure > Select Device, select the 18F242. See that all familiar development tools remain available. Then, using Configure > Configuration Bits, see the considerably increased number of bits that are available. These were seen in Tables 12.4 and 12.5, and in MPLAB form are shown in Figure 12.15.

Configuration Bits			
Address	**Value**	**Category**	**Setting**
300001	27	Oscillator	RC-OSC2 as RA6
		Osc. Switch Enable	Disabled
300002	0F	Power Up Timer	Disabled
		Brown Out Detect	Enabled
		Brown Out Voltage	2.5V
300003	0F	Watchdog Timer	Enabled
		Watchdog Postscaler	1:128
300005	01	CCP2 Mux	RC1
300006	85	Stack Overflow Reset	Enabled
		Low Voltage Program	Enabled
300008	0F	Code Protect 00200-01FFF	Disabled
		Code Protect 02000-03FFF	Disabled
300009	C0	Data EE Read Protect	Disabled
		Code Protect Boot	Disabled
30000A	0F	Table Write Protect 00200-01FFF	Disabled
		Table Write Protect 02000-03FFF	Disabled
30000B	E0	Data EE Write Protect	Disabled
		Table Write Protect Boot	Disabled
		Config. Write Protect	Disabled
30000C	0F	Table Read Protect 00200-01FFF	Disabled
		Table Read Protect 02000-03FFF	Disabled
30000D	40	Table Read Protect Boot	Disabled

Figure 12.15 Setting the 18F242 Configuration Word in MPLAB – all values shown are default

12.10.2 The Fibonacci program

Program Example 5.6 is a program that calculates a Fibonacci series, using the 16 Series CPU.

> **Programming Exercise 12.1**
>
> Create a project in MPLAB called **Fibonacci-18**. Copy from the book CD the source file of Program Example 5.6 into it. Using Configure > Select Device, set the chosen microcontroller to be the 18F242. Introduce any changes essential to allow the program to run correctly. By doing this, you should be able to illustrate that the 18 Series instruction set is a super-set of the 16 Series, and that the changes needed are minimal. Notice that in doing this we are invoking Assembler default values. For example, a direct code substitution like this will not specify the value of the instruction operand 'a' (Appendix 5). Therefore, the default value of 1 is applied.

Programming Exercise 12.1 shows that an 18 Series device can run using just 16 Series instructions. However, this wastes the powerful new features of the 18 Series CPU. Program Example 12.1 adapts the Fibonacci program in a modest way to the 18 Series, with the changes highlighted in bold. This illustrates use of some of the new instructions.

```
;*******************************************************************
;Fibo_18
;In a Fibonacci series each number is the sum of the two
;previous ones, e.g. 0,1,1,2,3,5,8,13,21....
;This program calculates Fibonacci numbers within an 8-bit range,
;first going up and then down.
;Program intended for simulation only, hence no input/output.
;The program demonstrates addition, subtraction, compare, and conditional
;branching.
;TJW 10.10.05                               Tested 10.10.05
;*******************************************************************
;Configuration bits need not be set

        list p=18F242
        #include P18F242.inc
;no i/o ports used

;these memory locations hold the Fibonnaci series.
fib0    equ   10  ;lowest number (oldest when going up, newest when reversing
                                  ;down)
fib1    equ   11 ;middle number
fib2    equ   12 ;highest number
fibtemp equ   13 ;temporary location for newest number
counter equ   14 ;indicates which value we have reached, opening value is 3

        org 00
```

Program Example 12.1 The Fibonacci series generator, adapted for the 18F242

```
;Initialse BSR
        movlb 00 ;clear BSR
;preload initial values
        movlw 0
        movwf fib0
        movlw 1
        movwf fib1
        movwf fib2
        movlw 3
        movwf counter ;have preloaded the first three numbers, so start at 3
;
forward movf  fib1,0
        addwf fib2,0
        bc      reverse ;reverse down the series if we have overflowed
        movwf fibtemp ;latest number now placed in fibtemp
        incf counter,1
;now shuffle numbers held, discarding the oldest
        movff fib1,fib0
        movff fib2,fib1
        movff fibtemp,fib2
        goto forward
;when reversing down, we will subtract fib0 from fib1 to form new fib0
reverse movf  fib0,0
        subwf fib1,0
        movwf fibtemp       ;latest number now placed in fibtemp
        decf  counter,1
;now shuffle numbers held, discarding the oldest
        movff fib1,fib2
        movff fib0,fib1
        movff fibtemp,fib0
;test if counter has reached 3, in which case return to forward
        movlw 3
        cpfseq counter
        goto reverse
        goto  forward
        end
```

Program Example 12.1 Continued

Programming Exercise 12.2

From the book CD take the source file of Program Example 12.1. Create a project around it and build it. Simulate the program, and use View > Special Function Registers and View > File Registers to view these two areas of memory. Notice how their structure differs from the 16 Series. Scroll the Special Function Registers window to see **PCL**. Step through the program and see how this increments twice for every instruction, as described in Section 12.5.2. Also see the Fibonacci numbers appearing in the data registers.

Summary

- The 18 Series microcontrollers represent a very clear step forward in the PIC design strategy. The CPU and memory structure are radically redeveloped, while many peripherals are retained.
- The instruction set is increased to 75 distinct instructions, with big new capability in arithmetic, program branching, table access and memory usage.
- Data memory is structured to give a much greater RAM capacity and a separate grouping of Special Function Registers.
- Program memory has greatly increased capacity, with larger address bus, and the 16-bit instructions are now split into 2 bytes for storage. The Stack is deeper and more flexible.

References

12.1. PIC 18FXX2 Data Sheet. (2002). Microchip Technology Inc., Document no. DS39564B; www.microchip.com

12.2. Migrating Designs from PIC16C74A/74B to PIC18C442 (1999). Microchip Technology Inc., Application Note AN716, Document no. DS00716A.

13
The PIC® 18FXX2 peripherals

The purpose of this chapter is to introduce the PIC 18FXX2 peripherals. For a microcontroller family with many peripherals, the chapter is surprisingly short. There are two reasons for this. First, as already stated in Chapter 12, many of the 18 Series peripherals are similar or identical to those used in the 16 Series. We will build on this knowledge and not repeat material already in the book. Second, because we are moving to program in C, we will benefit from the support that the C compiler gives. We will find C library functions to undertake all interaction with the peripherals, and in normal usage we just won't need an intimate knowledge of the peripherals. Energies previously spent (while programming in Assembler) in learning every fine detail of a peripheral can therefore be diverted to the creative task of writing working programs. This is a liberating step.

The descriptions that follow make repeated reference to 16 Series peripherals in earlier chapters. If you know the peripherals well, then this will act as a useful refresher and show the small changes that have been introduced in the 18 Series. If you don't know the 16 Series peripherals, you will learn them (and hence those of the 18 Series) in this way.

So that we do not view the 18FXX2 as representing all there is to know on the 18 Series, the chapter closes with a final section on another member of the family, the 18F2420. This has some interesting characteristics that we need to know about, notably the extended instruction set and nanowatt technology.

This chapter aims to:

- Review all the PIC 18FXX2 peripherals, in most cases drawing on knowledge of their 16 Series equivalents
- Glimpse some further enhancements which are available in the 18 Series world.

The chapter can be read through. Alternatively, it can be skipped and simply used as a reference from later chapters when needed.

13.1   The main idea – the 18FXX2 peripherals

The peripherals of the 18FXX2 microcontrollers are summarised in Table 12.1 and seen for the 18F2X2 pair in Figure 12.2. When comparing them to the peripherals of the 16F87XA, in Figure 7.2, it can be seen that, in almost every case, the same peripherals appear, with the same name and in the same pattern. Differences to note are that the 18FXX2 claims an extra timer, Timer 3, but does not have the comparator module of the 16F87XA.

13.2 The parallel ports

The parallel ports of the 18FXX2 are very similar to those of the 16 Series, both in structure and interfacing. They each have a **PORTX** register for data transfer and a **TRISX** register to set data direction. There is just one significant difference we need to take note of, which will be described shortly.

When working with a port we may need to do four things: set the data direction, read an input value, set an output value and read back an output value previously written. The 16 Series designs had one weakness in all of this – it was not good at doing the fourth in this list. Suppose a port bit such as the one in Figure 7.15(a) was set to output and a bit value written to it. If the port was then read by the CPU, it was impossible to be certain that the value read was equal to the value previously written. This is because the reading is of the actual port bit pin value. This could be the value output by the bit circuitry or it could be a different value forced by an external device connected to the pin.

To get round this small problem, the 18 Series ports introduce an interesting development. Each port has a third register, **LATA** for Port A, **LATB** for Port B and so on, which holds the value of the latched output port bit. This can be read by the program, and the programmer can have complete confidence that he/she is reading the value previously stored at the port and nothing else.

We will now survey the 18XX2 ports, introducing this and other changes.

13.2.1   The 18FXX2 Port A

The position of the Port A pins can be seen in the pin layout diagram of Figure 12.1. Unsurprisingly, the pattern is almost the same as the 16F873A, with the digital I/O features of the port being shared with the ADC (analog-to-digital converter) inputs. The Timer 0 input is shared with bit 4 of the Port, which has an Open Drain output. There is the possibility of an extra port bit, RA6, if ECIO or RCIO oscillator modes are used, as described in Section 12.9.2 of Chapter 12.

The three registers associated with Port A – **PORTA**, **TRISA** and **LATA** – can be seen in the fourth column of the register map (Figure 12.5). The working of the extra data latch register **LATA** is illustrated in Figure 13.1. This diagram is very similar to its 16 Series equivalent (Figure 7.15(a)). With the new register **LATA** introduced, one expects to see an extra bistable in the circuit somewhere. Interestingly, this is not the case. A quick look at the diagram shows that **LATA** and **PORTA** share the same data latch; a write to one is equivalent to a write to the other. The only real difference in the circuit is the **LATA** buffer appearing at the top of the diagram. A read to **LATA** activates this buffer and the value held on the **PORTA/LATA** data latch is transferred to the data bus. Therefore, the **LATA** is not really a different register at all – but addressing it allows a direct reading of the output of the **PORTA** data latch.

13.2.2   The 18FXX2 Port B

The position of the Port B pins can be seen in the pin layout diagram of Figure 12.1 and its three primary registers – **PORTB**, **TRISB** and **LATB** – in Figure 12.5. **LATB** plays an identical function to **LATA**, discussed above.

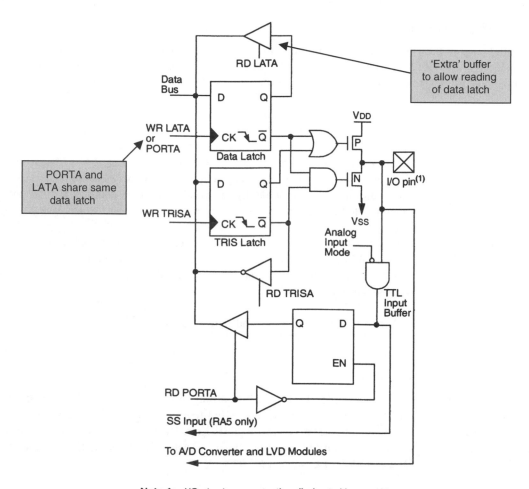

Note 1: I/O pins have protection diodes to VDD and VSS.

Figure 13.1 18FXX2 RA0–RA3 and RA5 pins (supplementary labels in shaded boxes added by the author)

The port keeps most of the characteristics of the 16 Series, being comparatively straightforward in design. There is some increase in the number of shared functions, especially with the introduction of more external interrupt sources. It can be seen that bits 5–7 share with the in-circuit debug functions of **PGD**, **PGC** and **PGM**, the last of these having moved from its 16 Series position. The three external interrupt inputs share with the lower 3 bits of the port. An interesting addition is the introduction of an alternative CCP2 connection, on bit 3. This relieves pressure on the very 'busy' pin of Port C bit 1, and allows the Timer 1 external oscillator to be used at the same time as CCP2.

Internal pull-ups are still available on all pins, with the controlling bit, $\overline{\text{RBPU}}$, now placed in register **INTCON2** (Figure 12.9). The interrupt on change function, whereby a change on any of pins 4–7 causes an interrupt flag to be set, is also in place. The Enable and Flag bits, **RBIE** and **RBIF**, are in the **INTCON** register (Figure 12.8).

13.2.3 The 18FXX2 Port C

As with the other ports, the position of the Port C pins can be seen in the pin layout diagram of Figure 12.1 and its three primary registers – **PORTC**, **TRISC** and **LATC** – in Figure 12.5. As with the 16F87XA, this port shares its pins with the serial ports and the CCP functions, while bit 0 is shared with the Timer 1 input. This bit and bit 1 can also be used for an external oscillator input for Timer 1. Because of these shared functions, the port pin driver circuits are comparatively complex. The pin function can, moreover, be taken over by peripheral functions – care is therefore needed. This override does not, however, include the reading of the **LATC** register. This can be read whatever mode a port pin is operating in.

13.2.4 The parallel slave port

This is implemented on the 40-pin devices and is the same as described in Section 7.12 of Chapter 7. The parallel port itself is Port D, with 3 bits of Port E being available for handshaking. The mode of operation of the port is controlled primarily by the **TRISE** register, which is the same as in Figure 7.26. The only addition to the port is the inclusion of an **LATD** register, functioning as described for Port A above.

13.3 The timers

All versions of the 18FXX2 have no less than four programmable timers, as well as a Watchdog Timer. We will survey the timers in turn, making extensive use of information already provided for their 16 Series equivalents.

13.3.1 Timer 0

The 18FXX2 Timer 0 draws its roots clearly from the 16 Series Timer 0. It can, however, operate either in 8-bit or in 16-bit mode. Its control register, **T0CON**, is shown in Figure 13.2. The lower 6 bits of this have similar function to the 16 Series **OPTION** register of Figure 6.9, which it replaces. The overall operating mode is selected by bit 6, which determines whether it is in 8-bit or 16-bit mode.

In 8-bit mode the action of Timer 0 is effectively the same as the 16 Series Timer 0 of Figure 6.8. Unlike the 16 Series Timer 0, which shares its prescaler with the WDT, the prescaler is entirely assigned to the timer. Therefore, the **PSA** bit of **T0CON** simply determines whether or not the prescaler is in circuit, it does not assign it to the Watchdog Timer.

In 16-bit mode the action of the timer is as shown in Figure 13.3. The lower byte of the counter itself is called **TMR0L**, while the higher byte is simply called **TMR0**, as it is the same register location as the 8-bit version.

R/W-1	R/W-1	R/W-1	R/W-1	R/W-1	R/W-1	R/W-1	R/W-1
TMR0ON	T08BIT	T0CS	T0SE	PSA	T0PS2	T0PS1	T0PS0

bit 7   bit 0

bit 7   **TMR0ON:** Timer0 On/Off Control bit
         1 = Enables Timer0
         0 = Stops Timer0

bit 6   **T08BIT:** Timer0 8-bit/16-bit Control bit
         1 = Timer0 is configured as an 8-bit timer/counter
         0 = Timer0 is configured as a 16-bit timer/counter

bit 5   **T0CS:** Timer0 Clock Source Select bit
         1 = Transition on T0CKI pin
         0 = Internal instruction cycle clock (CLKO)

bit 4   **T0SE:** Timer0 Source Edge Select bit
         1 = Increment on high-to-low transition on T0CKI pin
         0 = Increment on low-to-high transition on T0CKI pin

bit 3   **PSA:** Timer0 Prescaler Assignment bit
         1 = TImer0 prescaler is NOT assigned. Timer0 clock input bypasses prescaler.
         0 = Timer0 prescaler is assigned. Timer0 clock input comes from prescaler output.

bit 2-0 **T0PS2:T0PS0:** Timer0 Prescaler Select bits
         111 = 1:256 prescale value
         110 = 1:128 prescale value
         101 = 1:64 prescale value
         100 = 1:32 prescale value
         011 = 1:16 prescale value
         010 = 1:8 prescale value
         001 = 1:4 prescale value
         000 = 1:2 prescale value

Figure 13.2 The Timer 0 control register, **T0CON**

An interesting problem occurs in 16-bit timers when operating in the 8-bit environment. Suppose the program needs to read the value held in the timer. The 2 bytes are read in turn. It is possible, however, that an increment occurs after the first byte has been read, maybe causing an overflow from lower to higher byte. The value of the 2 bytes that have been read can therefore be seriously in error.

The solution to this problem is seen in Figure 13.3. A buffer, **TMR0H**, is included alongside the timer higher byte, **TMR0**. It is impossible to access the higher byte directly. Whenever the lower byte is read, however, the value of **TMR0** is simultaneously transferred to **TMR0H**. This can be read in a later instruction. Its value is guaranteed to correspond exactly to the value of the lower byte, when it was read. Similarly, if the programmer wishes to write to the timer, the program should *first* write the required higher byte to **TMR0H**. When the lower byte is written to **TMR0L**, then the value stored in **TMR0H** is transferred simultaneously to **TMR0**. Again, the 16-bit value being loaded into the timer is guaranteed to be correct, and uncorrupted by increments occurring within a 2-byte transfer.

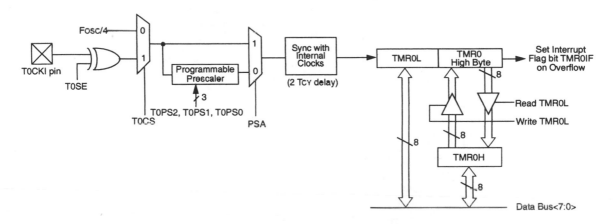

Figure 13.3 Timer 0 operating in 16-bit mode

In either mode of Timer 0, an interrupt is generated when the counter overflows from its maximum value (FF_H for the 8-bit and $FFFF_H$ for the 16-bit). The same bit, **TMR0IF**, seen in Figure 12.7, is used for both interrupts.

Timer 0, in 16-bit mode and with interrupt, is used in a number of the C example programs in later chapters.

13.3.2   Timer 1

The 18FXX2 Timer 1 in its basic form is nearly identical to the 16 Series Timer 1, as seen in Figure 9.1. It has a control register **T1CON**. This is the same as Figure 9.2, except that bit 7 (not used in the 16 Series) is called **RD16**. When set to 1 this bit enables the '16-bit Read/Write' mode. This is just as described for Timer 0 and when enabled operates as shown in Figure 13.4. There is now a buffer register **TMR1H** for the higher byte, allowing synchronised data transfer to or from the timer. A small further difference with the 16 Series Timer 1 is the ability to clear the timer through a 'special event' from the CCP module. This is also seen in Figure 13.4 and is described in Section 13.4.

13.3.3   Timer 2

The 18F242 Timer 2 is the same as the 16 Series Timer 2, shown in Figure 9.4. Similarly, it has a control register **T2CON**, the same as in Figure 9.5.

13.3.4   Timer 3

Timer 3 is structurally the same as Timer 1, so presents no new problems of understanding. It is shown in its basic form in Figure 13.5, with its control register, **T3CON**, shown in Figure 13.6. As with Timer 1, this timer can be operated as a timer with internal clock source input, as a counter with external input or as a timer with external oscillator input. For the last two, inputs are shared with Timer 1. Thus, if external

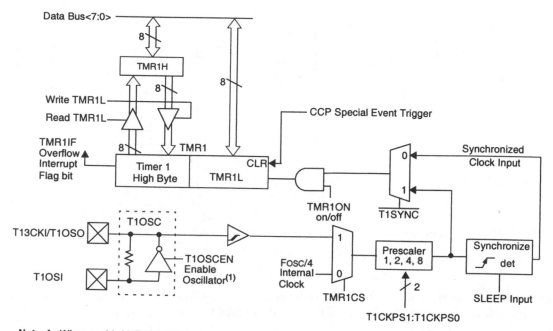

Note 1: When enable bit T1OSCEN is cleared, the inverter and feedback resistor are turned off. This eliminates power drain.

Figure 13.4 Timer 1 operating in 16-bit Read/Write mode

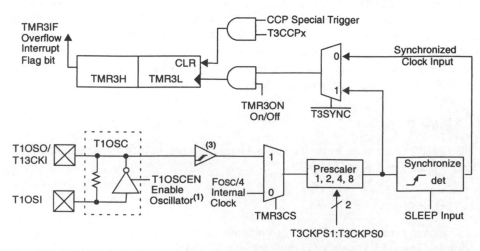

Note 1: When enable bit T1OSCEN is cleared, the inverter and feedback resistor are turned off. This eliminates power drain.

Figure 13.5 The Timer 3 block diagram

R/W-0	R/W-0	R/W-0	R/W-0	R/W-0	R/W-0	R/W-0	R/W-0
RD16	T3CCP2	T3CKPS1	T3CKPS0	T3CCP1	T3SYNC	TMR3CS	TMR3ON

bit 7   bit 0

bit 7 **RD16:** 16-bit Read/Write Mode Enable bit

1 = Enables register Read/Write of Timer3 in one 16-bit operation
0 = Enables register Read/Write of Timer3 in two 8-bit operations

bit 6-3 **T3CCP2:T3CCP1:** Timer3 and Timer1 to CCPx Enable bits

1x = Timer3 is the clock source for compare/capture CCP modules
01 = Timer3 is the clock source for compare/capture of CCP2,
        Timer1 is the clock source for compare/capture of CCP1
00 = Timer1 is the clock source for compare/capture CCP modules

bit 5-4 **T3CKPS1:T3CKPS0:** Timer3 Input Clock Prescale Select bits

11 = 1:8 Prescale value
10 = 1:4 Prescale value
01 = 1:2 Prescale value
00 = 1:1 Prescale value

bit 2 **T3SYNC:** Timer3 External Clock Input Synchronization Control bit
(Not usable if the system clock comes from Timer1/Timer3)
When TMR3CS = 1:
1 = Do not synchronize external clock input
0 = Synchronize external clock input
When TMR3CS = 0:
This bit is ignored. Timer3 uses the internal clock when TMR3CS = 0.

bit 1 **TMR3CS:** Timer3 Clock Source Select bit

1 = External clock input from Timer1 oscillator or T1CKI
        (on the rising edge after the first falling edge)
0 = Internal clock (Fosc/4)

bit 0 **TMR3ON:** Timer3 On bit

1 = Enables Timer3
0 = Stops Timer3

Figure 13.6 The Timer 3 control register, **T3CON**

oscillator is used, it is that of Timer 1, which must be enabled with the **T1OSCEN** bit in **T1CON**. If external input is used, it is the same input as Timer 1, i.e. pin 11 in the case of the 18F2X2.

A study of Figure 13.6 shows that bits 0, 1, 2, 4, 5 and 7 have the same function as those of **T1CON** (Figure 9.2). Bit 7 shows that the '16-bit Read/Write' mode is again available as described for Timers 0 and 1. If not selected, the timer operates as shown in Figure 13.5. If selected, the timer's interface with the data bus takes the '16-bit Read/Write' form shown in Figure 13.4. Reading or writing to the higher byte is then buffered, with the buffer being memory mapped as **TMR3H**.

It is evident in bits 6 and 3 of **T3CON** that the timer can be linked to a CCP (capture/compare/PWM) module, and used for capture and compare instead of, or alongside, Timer 1. The 'CCP special trigger' is seen coming in to Figure 13.5, being gated by one of these control bits. If either bit 3 or 6 is high, this path is enabled. These functions are described further in Section 13.4.

13.3.5   The Watchdog Timer

The Watchdog Timer (WDT) is identical in concept to the 16 Series WDT, as described in Section 6.5 of Chapter 6. It is a free-running down counter which, if enabled and allowed to time out, will cause the microcontroller to be reset. It is enabled by the **WDTEN** bit of Configuration register **CONFIG2H** (Table 12.4). It has its own dedicated postscaler (unlike the 16 Series, where it is shared with Timer 0), whose setting is determined by bits **WDTPS2** to **WDTPS0** in the same Configuration register. Reference 12.1 indicates that the time-out period is typically 18 ms, with minimum of 7 ms and maximum of 33 ms, for a postscaler value of 1. If the postscaler is set to 128, the typical time-out value rises to 2.3 s. Both WDT and postscaler are cleared by execution of the **clrwdt** instruction.

An important development in design strategy for the WDT is the inclusion of a software WDT enable bit, **SWDTEN**. This is the LSB, and the only active bit, in the register **WDTCON**, seen at memory location FD1$_H$ in Figure 12.5. If the WDT has been *disabled* in the Configuration register, then it can be *enabled* by setting the **SWDTEN** bit. If the WDT is enabled in the Configuration register, then the **SWDTEN** bit has no effect. The ability to switch the WDT on and off goes rather against the whole concept of the WDT – how wise is it to have a safety feature that can be switched off? However, it allows the WDT to be enabled for certain modes of operation and disabled for others. As with all safety or reliability features, it should be used with caution.

13.4   The capture/compare/PWM (CCP) modules

The 18FXX2 has two CCP modules, as seen in Figure 12.2. These are very similar to the 16 Series modules, applying exactly the principles described in Section 9.4.1 of Chapter 9. The CCP modules work with Timers 1, 2 and 3 to provide capture, compare and PWM operation. An important difference with the 16 Series, and a big step forward in terms of flexibility of operation, lies with the addition of Timer 3 and its interlinking with the CCP modules.

13.4.1   The control registers

The CCP modules are controlled primarily by the **CCPxCON** registers, seen in Figure 13.7. These are very similar to the 16 Series **CCPxCON** registers of Figure 9.7. There is, however, a little difference in detail in the modes which are available, as a study of the lower 4 significant bits will show. A new mode, when bits 3–0 are set to 0010, is, for example, available in the 18FXX2. It can be seen that there is a name change, but not a function change, in bits 4 and 5 of these registers.

13.4.2   Capture mode

The CCP configured for Capture mode is shown in Figure 13.8. This is a direct equivalent of Figure 9.8 and applies the principles described in Section 9.4.2 of Chapter 9. Both inputs are, however, drawn out here.

U-0	U-0	R/W-0	R/W-0	R/W-0	R/W-0	R/W-0	R/W-0
—	—	DCxB1	DCxB0	CCPxM3	CCPxM2	CCPxM1	CCPxM0

bit 7   bit 0

bit 7-6   **Unimplemented:** Read as '0'

bit 5-4   **DCxB1:DCxB0**: PWM Duty Cycle bit1 and bit0

<u>Capture mode:</u>
Unused

<u>Compare mode:</u>
Unused

<u>PWM mode:</u>
These bits are the two LSbs (bit1 and bit0) of the 10-bit PWM duty cycle. The upper eight bits (DCx9:DCx2) of the duty cycle are found in CCPRxL.

bit 3-0   **CCPxM3:CCPxM0**: CCPx Mode Select bits

0000 = Capture/Compare/PWM disabled (resets CCPx module)
0001 = Reserved
0010 = Compare mode, toggle output on match (CCPxIF bit is set)
0011 = Reserved
0100 = Capture mode, every falling edge
0101 = Capture mode, every rising edge
0110 = Capture mode, every 4th rising edge
0111 = Capture mode, every 16th rising edge
1000 = Compare mode,
       Initialize CCP pin Low, on compare match force CCP pin High (CCPIF bit is set)
1001 = Compare mode,
       Initialize CCP pin High, on compare match force CCP pin Low (CCPIF bit is set)
1010 = Compare mode,
       Generate software interrupt on compare match (CCPIF bit is set, CCP pin is unaffected)
1011 = Compare mode,
       Trigger special event (CCPIF bit is set)
11xx = PWM mode

Figure 13.7 The **CCP1CON** and **CCP2CON** registers (addresses FBD$_H$ and FBA$_H$)

The major structural difference is that both Timers 1 and 3 are available to be used. This allows two input signals of rather different time characteristics to be tested, as Timers 1 and 3 can now both be used, and both optimised, in terms of frequency and prescaler setting, for the signal they are 'capturing'. Selection of the timer to be used is determined by the values of the **T3CCPx** bits, already seen in the **T3CON** register in Figure 13.6. Once this selection is made, capture operation is the same as for the 16 Series.

13.4.3 Compare mode

The CCP configured for Compare mode is shown in Figure 13.9. This is a direct equivalent of Figure 9.9 and applies the principles described in Section 9.4.3 of Chapter 9. As with the Capture mode above, both inputs are drawn out here. The major structural difference, of both Timers 1 and 3 being available, is shown.

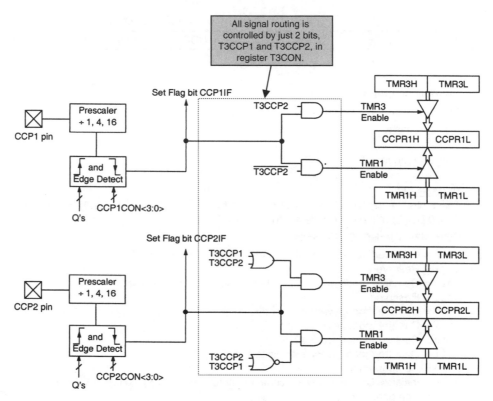

Figure 13.8 Capture mode operation (supplementary label in shaded box added by the author)

Selection of the timer to be used is again determined by the values of the **T3CCPx** bits, as indicated in Figure 13.6.

The 'special event' action is summarised at the top left of the diagram. It is selected through setting the right combination of the lower 4 bits in the **CCP1CON** or **CCP2CON** registers (Figure 13.7).

13.4.4 Pulse width modulation

The CCP modules configured for PWM follow the principles described in Section 9.5.1 of Chapter 9. They behave exactly like the 16 Series module, hence acting exactly as described in Section 9.5.2 of that chapter, and as seen in Figure 9.11.

13.5 The serial ports

The 18FXX2 has two major serial modules, the Master Synchronous Serial Port (MSSP) and the Address-able Universal Synchronous Asynchronous Receiver Transmitter (USART), as seen in Figure 12.2. The MSSP is identical to the 16 Series peripheral of the same name. Thus, it can be configured either in

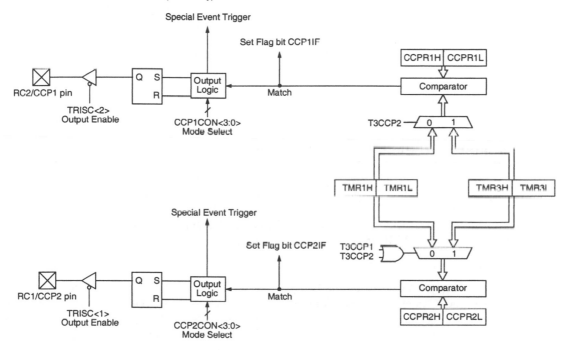

Figure 13.9 Compare mode operation

SPI (Serial Peripheral Interface) mode or in I²C mode. The action of both ports is summarised briefly in the sections that follow.

13.5.1 The MSSP in SPI mode

In this mode the MSSP is configured as in Figure 10.7. Data is transferred into or out of the serial port through the **SSPBUF** register, seen in Figure 12.6 at address FC9$_H$. The port is controlled by two registers, **SSPCON1** and **SSPSTAT**. These are seen in Figures 10.8 and 10.9 respectively, while their position on the 18 Series memory map can be seen in Figure 12.5 (addresses FC6$_H$ and FC7$_H$ respectively). The behaviour of the peripheral in this mode is exactly as described in Section 10.3 of Chapter 10. As interrupt prioritisation is introduced in the 18 Series (see Section 12.7 of Chapter 12), the MSSP interrupt (**SSPIF** in **PIR1**, Figure 12.11) can also be prioritised. This is controlled by the **SSPIP** bit in register **IPR1** (also Figure 12.11).

13.5.2 The MSSP in I²C mode

The MSSP in I²C mode behaves the same as the 16 Series MSSP. Thus, it applies all the principles of I²C, as described in Section 10.6 of Chapter 10. It uses the same control registers, **SSPCON1**, **SSPCON2**,

SSPSTAT and **SSPADD**. In Slave mode it configures as shown in Figure 10.17, and in Master mode as in Figure 10.19. Overall, it behaves exactly as described in Section 10.7 of Chapter 10.

13.5.3 The USART

This module is the same as the 16 Series peripheral of the same name, and thus behaves as described in Section 10.10 of Chapter 10. With the higher maximum oscillator frequency, it has the potential to run at higher baud rates. Its interrupts (bits **RCIF** and **TXIF** in **PIR1**, Figure 12.11) can be prioritised and are subject to the interrupt enabling mechanism of the 18 Series.

13.6 The analog-to-digital converter (ADC)

The 18FXX2 ADC module is effectively the same as that of the 16F873A, described in Section 11.3 of Chapter 11. It has the block diagram of Figure 11.6, with eight inputs available in the 18F442 and 452, and five on the 18F242 and 252. These are seen in Figure 12.1. The ADC is controlled by the **ADCON0** and **ADCON1** registers, now mapped in locations $FC2_H$ and $FC1_H$ (Figure 12.5). The conversion result is placed in **ADRESL** and **ADRESH** ($FC3_H$ and $FC4_H$, Figure 12.5). The analog input model is as shown in Figure 11.10.

Figure 13.9 shows that the CCP module in Compare mode can be set to trigger an ADC conversion. For this to occur the ADC module must be switched on, by the **ADON** bit in **ADCON0** (Figure 11.7). The input multiplexer must also have been preset, with sufficient time allowed for acquisition, as described in Section 11.3.4 of Chapter 11. If the timer causing the special event trigger (Timer 1 or 3) is left running continuously, this provides a useful means of performing a periodic analog-to-digital conversion.

13.7 Low-voltage detect

The ability to detect a falling power supply is very valuable in an embedded system. In a battery-powered product this can be used to detect a failing battery. It can also be used to detect power being switched off. In either case, the microcontroller may wish to activate a warning signal or exercise an orderly shutdown, perhaps saving key operating variables to EEPROM. In the case of complete power loss, residual charge on the system reservoir capacitor still provides some time to do this.

Low-voltage detect is available in many voltage regulator ICs, particularly those intended for battery supply. These will generate an output if the low voltage is detected, which can be used as an interrupt to a microcontroller. The 18FXX2, however, has its own low-voltage detect, with its own interrupt source. The general concept is shown in Figure 13.10. The power supply voltage V_{DD} is applied to the top of a resistor chain. The other end of this can be connected to ground through the switching transistor, whose gate voltage is labelled **LVDEN**. This can be seen to be bit 4 of the **LVDCON** register (Figure 13.11).

If **LVDEN** is set high, then the resistor chain acts as a potential divider, the different nodes of which are taken to a multiplexer. One of these inputs is selected as the multiplexer output, controlled by the setting of the lower 4 bits of **LVDCON**. This multiplexer output value is compared with an internally generated

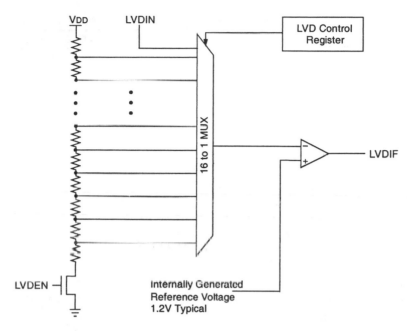

Figure 13.10 Low-voltage detect circuit

reference voltage of 1.2 V. If the multiplexer output voltage falls below the reference voltage, then the comparator output, **LVDIF**, goes high. This is the low-voltage detect interrupt flag, which can be seen in Figure 12.12. If enabled, an interrupt is generated and the microcontroller can make an appropriate response to its failing power supply.

It is interesting to look at the 'trip' voltages which can be selected, as seen in Figure 13.11, and compare these both with the legal operating range of the device and the possible brown-out settings, as seen in Figure 12.13. Clearly, it would not be useful to set the low-voltage detect below the brown-out reset, as the device would always reset before the low-voltage detect could kick in! A designer may therefore choose not to use brown-out reset, but to place all power-line monitoring with the low-voltage detect.

An alternative to using a voltage derived from the internal resistor divider chain is to use an external input. This can be routed through the low-voltage detect input pin, **LVDIN** (pin 7 on either package, Figure 12.1). The signal path is seen in Figure 13.10 and the necessary settings in Figure 13.11. If selected, this external voltage is compared directly with the 1.2 V internal voltage.

As the low-voltage detect when enabled consumes some current (in the region of 30 μA), it is possible to save power by enabling it only periodically. In this case the voltage reference requires some settling time. The state of the reference is indicated by the **IRVST** bit in the **LVDCON** register. The interrupt should only be enabled when this bit has been set.

U-0	U-0	R-0	R/W-0	R/W-0	R/W-1	R/W-0	R/W-1
—	—	IRVST	LVDEN	LVDL3	LVDL2	LVDL1	LVDL0

bit 7   bit 0

bit 7-6   **Unimplemented:** Read as '0'

bit 5   **IRVST:** Internal Reference Voltage Stable Flag bit
1 = Indicates that the Low Voltage Detect logic will generate the interrupt flag at the specified voltage range
0 = Indicates that the Low Voltage Detect logic will not generate the interrupt flag at the specified voltage range and the LVD interrupt should not be enabled

bit 4   **LVDEN:** Low Voltage Detect Power Enable bit
1 = Enables LVD, powers up LVD circuit
0 = Disables LVD, powers down LVD circuit

bit 3-0   **LVDL3:LVDL0:** Low Voltage Detection Limit bits
1111 = External analog input is used (input comes from the LVDIN pin)
1110 = 4.5V - 4.77V
1101 = 4.2V - 4.45V
1100 = 4.0V - 4.24V
1011 = 3.8V - 4.03V
1010 = 3.6V - 3.82V
1001 = 3.5V - 3.71V
1000 = 3.3V - 3.50V
0111 = 3.0V - 3.18V
0110 = 2.8V - 2.97V
0101 = 2.7V - 2.86V
0100 = 2.5V - 2.65V
0011 = 2.4V - 2.54V
0010 = 2.2V - 2.33V
0001 = 2.0V - 2.12V
0000 = Reserved

Note: LVDL3:LVDL0 modes which result in a trip point below the valid operating voltage of the device are not tested.

Figure 13.11 The **LVDCON** register

It should be noted that the effect of any reset is to clear the **LVDEN** bit and hence disable the low-voltage detect.

13.8   Applying the 18 Series in the Derbot-18

Due to the pin compatibility between the 16F873A and the 18F242, the Derbot is immediately compatible with the 18 Series device – no changes are needed whatsoever! This includes all parallel I/O, ADC, PWM, counters and timers, and I^{2}C. The Derbot, loaded with a 18F242 microcontroller, is used in the chapters that follow for all program examples. To indicate this change of microcontroller (but no other change), it is called 'Derbot-18'.

From a hardware point of view there is little that is done with the 18F242 that could not have been done with the 16F873A. The big step forward lies in the use of C, which works much better with the 18 Series. This in its turn leads to the use of the Real Time Operating System that we meet towards the end of the book.

13.9   The 18F2420 and the extended instruction set

The 18FXX2 microcontrollers are a successful and powerful group, yet they represent only one point of achievement for the 18 Series microcontroller. There is an interesting 18 Series family which is very closely related to the 18FXX2 family. It is made up of a set of devices which are numbered with an extra 0 on the identifier. Thus, the 18F242 has an equivalent 18F2420, the 18F442 an equivalent 18F4420, and so on. Each has the same memory capability, a similar set of peripherals and the same basic pin pattern as its equivalent.

We will just summarise the important developments in this sub-family, as it shows some of the other characteristics that can be found in the 18 Series. Full details can be found in Ref. 13.1.

13.9.1   Nanowatt technology

Many embedded systems are battery powered and it is essential for them to consume as little power as possible. We do not give in this book a detailed discussion of how power consumption can be reduced. This can be found in Ref. 1.1. As explained there, three simple ways of reducing power consumption, for a given circuit and technology, are:

- Reducing the supply voltage
- Reducing the clock frequency
- Switching off unused circuit sections.

The Sleep mode has been available in all microcontrollers we have looked at so far, as a means of reducing power when the device was not in action. There are many situations, however, where the microcontroller needs to keep running, but the power consumption must be minimised. The nanowatt Microchip approach makes this possible. It includes the features summarised below.

Alternate run modes

Instead of running from the main clock oscillator, the microcontroller can be switched to run from the Timer 1 oscillator. Therefore, when the CPU needs to run at full speed it can do this from the main oscillator. For periods of time when there is little activity it only needs to run slowly. It can then switch to the Timer 1 oscillator, running at a lower frequency. This concept has already been discussed in Section 12.9.4 of Chapter 12. It can result in considerable power saving. There is also in this microcontroller group an internal oscillator block, which provides a further set of clock sources. Changes are under the control of the program code, so all switches can be invoked as the program runs.

Multiple idle modes

It is not difficult to imagine situations when it is necessary to keep peripherals going, but allow the CPU to stop running. This can happen in only a very limited way in Sleep mode, but is the basis of the idle modes. An idle mode is entered by presetting an Idle Enable SFR bit called **IDLEN** and then executing the **sleep** instruction. In this case an idle mode is entered and the clock source which was in use before executing **sleep** is retained. The different idle modes correspond to the different clock sources available. Running in idle mode from a slow clock source results in a very striking reduction in power consumption.

13.9.2   The extended instruction set

The 18F2420 has the regular 18 Series instruction set, with its 75 instructions, which we met in Chapter 12. It has also an optional set of eight extra instructions. These must be enabled by setting an **XINST** bit in the configuration setting. When using these instructions, the microcontroller is said to be operating in *extended mode*.

The extra instructions are intended to enhance the efficiency of a C compiler. Therefore, it is unlikely that a programmer would use them in an Assembler program, and we will not even list them here. They all relate to the ability of the microcontroller to undertake indirect and indexed addressing, and hence to improve the capability of the compiler in working with the software stack and other features.

It is important to recognise the existence of the extended instruction set, as the C18 C compiler, which we are about to use, makes reference to it.

13.9.3   Enhanced peripherals

A number of peripherals in this sub-family are available in 'enhanced' form. These are summarised below.

Enhanced CCP modules

These modules are available only on the larger 18F4420 and 18F4520 devices. They carry the basic CCP features described in Section 13.4, but include four PWM outputs and the ability to 'invert' the output polarity by having it active low instead of active high.

Enhanced Addressable USART

This serial communications module provides a number of important upgrades, which include automatic baud rate detection, an automatic Wake-up mode and other features that make it suitable for the Local Interconnect Network (LIN) bus.

ADC module

The ADC module now has an amazing 10 possible inputs for 28-pin versions and 13 for 40- or 44-pin versions! With changes to the CCP modules, it is possible to set up a repetitive acquisition sequence with minimal program interaction.

Summary

- 18 Series microcontrollers adopt the peripherals of the 16 Series, with some upgrades.
- This adoption is always within the context of the 18 Series architecture, most notably a revised register map and interrupt structure.
- Aside from this, some peripherals are adopted wholesale and can be applied exactly like their 16 Series counterpart.
- Other peripherals have been upgraded in the transfer to the 18 Series. A designer migrating from the 16 Series needs to check what these are and what their impact is.
- Generally, peripherals are designed to power up in the mode closest to the 16 Series version.
- A set of 18 Series enhancements is represented in the 18F2420 and its close relations. Significant among these are its 'nanowatt' capability and the extended instruction set.

Reference

13.1. PIC18F2420/2520/4420/4520 Data Sheet. (2004). Microchip Technology Inc., Document no. DS39631A.

14
Introducing C

In recent chapters we have developed increasingly complicated programs; in doing this it has become more and more difficult to apply Assembler programming. It was difficult to manage the complexity of the program, program errors were hard to find, program flow was difficult to control and even quite simple mathematical tasks (like the scaling in the light meter program) were laborious to implement.

The alternative is to change the programming strategy. Figure 4.1 showed the programmer's dilemma, and three alternative approaches were described. We initially adopted the third of these approaches, Assembler, as it represented a way of writing programs which very directly controlled the system hardware. Because of the problems just described, however, it is now sensible to explore another option. High-level languages were invented to cope with program complexity and simplify debugging. Can we apply one to the embedded environment? The answer is yes – for now we choose C, a language with outstanding credentials for our purpose.

This chapter, and the three which follow, aim to introduce C from first principles, leading to a working knowledge of its key features, as they apply to the embedded environment. Not every aspect of C is covered, particularly not its more advanced features.

The Microchip C18 compiler, which is designed to work within the MPLAB® environment, is used as a vehicle for this study. This can be purchased or a free student version downloaded from the Microchip website. It is also on the book CD. It is assumed that readers have access to this.

Our study will operate at several levels. First of all, the language C itself is introduced. This knowledge is fairly independent of embedded systems and can be applied at the desktop computer or elsewhere. In parallel with this we explore how C is applied to the embedded system and particularly in the PIC® environment. To allow these two strands to move forward in parallel, the C18 compiler is introduced at an early stage in the chapter. The practice of writing C programs is then developed through graded examples, applied in part to the Derbot-18 AGV. It does not take long, in this introduction to C, to begin to recognise some of its advantages over Assembler.

At the end of this chapter you should have developed an understanding of:

- The core features of the programming language C
- The core features of the MPLAB C18 C compiler, including its libraries.

Readers with experience of C may wish to skip the first three sections of the chapter and will find some of the later sections easy going.

14.1 The main idea – why C?

Section 4.1 of Chapter 4 briefly introduced the concept of the high-level language (HLL). An HLL is a programming language that makes use of language and structures that are easy for us human beings to understand. At the same time it has clearly defined rules, so that a program can be written which has the precision to be converted to machine code.

Whatever the HLL used, an intermediary computer program is required to convert the program as written into the computer machine code. If the program is converted *before* program execution, then the converting program is called a *compiler*. C is an example of a language using a compiler. If it is converted as the program executes, it is called an *interpreter*. Basic is an example of a language using an interpreter.

One great advantage of the HLL is that it allows *portability*. The program that the programmer writes does not depend on the computer that it is going to run on. It is the compiler or interpreter that is machine dependent and creates the actual machine code. Thus, the same source code can (again theoretically – we must stress this) be used to run on entirely different computers. This is, of course, quite unlike the Assembler approach, which is entirely dependent on the target computer.

When adopting an HLL for embedded applications, we want the benefits that it can bring, of simpler and more reliable coding. At the same time, however, we want to retain as far as possible the benefits of working close to the hardware, which Assembler programming allowed us. C was written for the much simpler computers of the 1970s. While it has in many cases been overtaken by other languages for applications on desktops and workstations, it remains a very powerful language for working close to the computer hardware, as we do in embedded systems. It shows the features that we want of an HLL, but still allows hardware elements to be accessed.

14.2 An introduction to C

Although a comparatively simple language in today's terms, C can lead to complex and sophisticated programs. This section aims to provide the very basics of C, using a simple example program. The intention is that the reader can then move as soon as possible to using the MPLAB C18 compiler and start writing simple programs for PIC 18 Series microcontrollers.

14.2.1 A little history

C was developed in the late 1970s at the Bell Labs, in New Jersey, USA. Its first publicly available documentation was in a publication by Brian Kernighan and Dennis Ritchie in 1978. This became so well known that it is often called the 'K&R' version. In 1989 a version of C was adopted by the American National Standards Institute (ANSI), as standard X3.159-1989. It is important to recognise this standard, as it is the one that the Microchip C18 compiler is based upon. This standard became very widely recognised and adopted, and one finds many references to 'ANSI C'. In 1990, the International Organisation for Standards (ISO) adopted the same version as an International Standard, with amendments made in 1995 and 1999. The 1999 version contains extensions which are not implemented in many compilers targeted at embedded systems.

14.2.2   A first program

A tradition has developed that books about C should start with an example program which outputs the words 'Hello world' to a display. In many embedded systems we won't have such a display to write to. As an alternative, an early need in almost any embedded system is to be able to output data to a port. Therefore, our first example program will just increment a number and output it to Port B of an 18 Series microcontroller. This program is seen in Program Example 14.1.

```
/************************************************************************
Example1.
Introductory Example of C Programming, with PIC 18 Series Microcontroller.
8-bit value output by Port B is continuously incremented.
Files c018i.o and p18f242.lib are included by the Linker Script.
TJW 21.10.05                                          Tested 23.10.05
*************************************************************************/

//Include 18F242 header file, for all processor-specific declarations
#include <p18f242.h>

unsigned char counter;       //specify counter as unsigned character

void main (void)             //main function starts here
{
  TRISB = 0;                 // initialise all bits of PORTB as output
  counter = 1;               //counter value is initialised to 1

  while (1)
    {
    PORTB = counter;         // Move 'counter' value to Port B
    counter = counter + 1;   //Increment counter
    }
}
```

Program Example 14.1 Incrementing the output value of Port B

14.2.3   Laying out the program – declarations, statements, comments and space

C is a so-called free-form programming language. That means that there is not a strict layout format to which programs must adhere. This is unlike Assembler, where the position of a word on the line can be crucial.

Crudely speaking, a C program is made up of: *declarations*, which set the scene and initialise things; *statements*, where the programming action takes place; *comments*, which provide a commentary to the human reader on what is going on; and *space*, which provides essential gaps between the words and symbols used, and is also used to improve clarity through the way the code is laid out. Let us look at each of these in turn.

Comments

Comments start with the combination /* and end with */. In this form they can run over more than one line, and can follow or precede statements or declarations.

It can be seen that the first seven lines of the program example are made up of comments, even though the comment delineators merge somewhat with the lines of asterisks. Following the title block there is a blank line. This is not a problem with C – such lines can be used to improve the intelligibility of the program.

An alternative comment format is to precede the comment with a double slash, //. Such a comment lasts only for the remainder of the line in which it appears and needs no terminating symbol. It is convenient to use both comment styles, the first for major blocks of comment, the second for single lines. This is the practice adopted in this book; it can be seen in the example program.

As with Assembler programming, it is good to make liberal use of comments, indicating clearly but briefly what is meant to be going on in the program.

Declarations

Declarations are used in a number of ways to create program elements, like variables and functions, and to indicate their properties. This is important, as all variables and functions in C must be declared before they can be applied. Characteristics specified include the type of data element (for example, whether fixed or floating point), the allocation of memory storage or the characteristics of a function. A declaration is terminated with a semicolon.

In the example program, the line

```
unsigned char counter;     //specify counter as unsigned character
```

is a declaration, whose meaning will shortly be explained.

In simple programs, declarations tend to appear as one of the first things in the program, which seems to make sense. As we meet more complex programs, however, we will see that declarations can occur within the program, with significance attached to the location of the declaration.

Statements

Statements are where the action of the program takes place. They perform mathematical or logical operations, and establish program flow. Every statement which is not a block (see below) ends with a semicolon.

There are a number of different categories of statement. The commonest is the *expression* statement, which includes mathematical manipulations. Example expression statements from Program Example 14.1 are:

```
TRISB = 0;       //initialise all bits of PORTB as output
counter = 1;     //counter value is initialised to 1
```

Statements are executed in the sequence they appear in the program, except where program branches take place.

Code blocks

Declarations and statements can be grouped together into *blocks*. A block is contained within braces (curly brackets). An example block from Program Example 14.1 is seen here:

```
while (1)
  {
  PORTB = counter;       //Move 'counter' value to Port B
  counter = counter + 1; //Increment counter
  }
```

Blocks can be written within other blocks, each within its own pair of braces. Keeping track of these pairs of braces is an important pastime in C programming, as in a complex piece of software there can be numerous ones nested within each other. It is common, and very good, practice to indent them such that matching pairs fall directly below each other on the page, with each nested pair being indented deeper into the page. In this way it is possible to keep track of brace pairs.

Space

The judicious use of space in a C program can make a big difference to its clarity. A space is required to separate words which would otherwise merge into one, for example in the example declaration quoted above. Further space, including blank lines, is ignored by the compiler and is used by the programmer to optimise the program layout. This applies both to blank lines and to indents within lines. For example, in the example block seen directly above, the braces are separated out onto their own lines and placed directly above each other. The program would compile the same if it were written:

```
while (1){PORTB = counter; // Move 'counter' value to Port B
counter = counter + 1;}    //Increment counter
```

It would, however, be less clear to read, in particular when looking for where the code block ended. The need for good layout increases dramatically as program complexity increases.

14.2.4   C keywords

C has a set of just 32 *keywords*. These are shown in Tables A6.1–A6.3, each with summary description. It can be seen that a good number relate to data type. In our example program, the declaration

```
unsigned char counter;    //specify counter as unsigned character
```

declares a variable called **counter** and uses the keywords **unsigned char** to specify its type as an unsigned character.

Other keywords (Table A6.2) relate to program flow. In this example the **while** keyword sets up a continuous loop, as described in Section 14.2.8.

Keywords are recognised by the compiler, which expects them to be applied within a defined context. They cannot be used for any other purpose, for example as the name of a variable.

14.2.5   The C function

C programs are structured from *functions*. Every program must have at least one function, called *main*. Program execution starts with this function and the program is contained within it.

Apart from the main function, functions are in some ways similar to Assembler subroutines. They are used in a similar way, generally to contain an identifiable program action. Good program structures tend to have much of the program contained within functions, with the main function calling subsidiary ones. Any function may call another.

What distinguishes a C function from an Assembler subroutine is the control exercised in how data is passed between calling program and function. Data elements, called *arguments*, can be passed to a function. They must, however, be of a type which is declared in advance. Only one return variable is allowed, whose data type must also be declared. The data passed to the variable is a *copy* of the original. Therefore, the function does not itself modify the value of the variable named. The impact of the function should thus be predictable and controlled. The terminology *parameter* is often used in place of *argument*. Distinction between the two terms is made in detailed specifications of the C language. In these chapters we will, however, use them interchangeably.

A function is defined in a program by a block of code having particular characteristics. Its first line forms the function header. The function header from the example program, shown here, illustrates the general format:

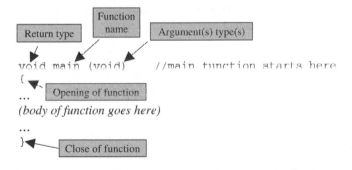

The return type is given first. In this example, the keyword **void** is used to indicate that there is no return value expected. This is common practice for the **main** function – after all, to where or what would it return a value? After the function name, in brackets, one or more data types may be listed, which identify the arguments which must be passed *to* the function. In this case (again as may be expected with **main**) there are no arguments transferred and the keyword **void** is again used to indicate this.

Following the function header, a pair of braces encloses the code which makes up the function itself. This could be anything from a single line to many pages. The final statement of the function may be a **return**, which will specify the value returned to the calling program. This is not essential if no return value is required.

It can be seen that, for clarity, the program is laid out so that the braces which enclose the **main** function are aligned fully left and that the braces containing the **while** statement are indented.

In Program Example 14.1 the **main** function is the only one. A number of further issues arise when multiple functions are used. These are introduced in later chapters.

14.2.6   Data type and storage

Variables within a C program have four attributes: their name, type, value and storage location. It has already been said that the type (for example, whether signed or unsigned, fixed or floating point) must be declared before use. Once the type has been specified, the compiler can determine the amount of memory needed to store the variable. Its value can then, if needed, be initialised.

The words used to define data type are shown in Table A6.1. The actual memory size applied to each data type can vary between compilers. The data types available in the MPLAB C18 compiler, along with their storage size, are shown in Table A6.4. For example, in Program Example 14.1 the variable **counter** is defined as being an unsigned character. Table A6.4 shows that the C18 compiler will assign it an 8-bit memory location. Although the name implies it must be a character, in fact it can be used for any 8-bit number. This is a useful data type for the many single-byte variables that we use in the PIC environment. We will see later that **PORTB** and **TRISB** are also defined as unsigned characters.

Data names must start with a letter. When writing complex programs it is common practice to start the name with a letter or letters which identify the type of the variable, for example the name **counter** in the example program could be changed to **uicounter**, to remind the programmer that it is an unsigned integer. This helps the programmer remember its type and reduces programming errors. This practice is not, however, adopted in the example programs in this book.

When writing numbers in a program, the default radix for integers is decimal, with no leading 0 (zero). Octal numbers are identified with a leading 0. Hexadecimal numbers are prefixed with 0x.

14.2.7   C operators

C recognises a diverse set of operators, which are shown in Table A6.5. The symbols used are familiar, but their application is *not* always the same as in conventional algebra. For example, a single 'equals' symbol, '=', is used to assign a value to a variable. A double equals sign, '==', is used to represent the conventional 'equal to'. Thus, in Program Example 14.1, the line

```
TRISB = 0;        // initialise all bits of PORTB as output
```

means that the variable called **TRISB** is assigned the value 0. This can be read as 'the variable **TRISB** takes the value 0'. In the 18F242 header file **TRISB** has been defined as an unsigned character, hence an 8-bit number. Due to the port action, the program line quoted above causes all bits of Port B to be set as outputs.

Operators have a certain order of precedence, shown in the table. The compiler applies this order when it evaluates a statement. If more than one operator at the same level of precedence occurs in a statement, then those operators are evaluated in turn, either left to right or right to left, as shown in Table A6.5.

As a very simple example, the line in the example program

```
counter = counter + 1;
```

contains two operators. Table A6.5 shows that the addition operator has precedence level 4, while the assign operator has precedence 14. The addition is therefore evaluated first, followed by the assign. The outcome is that the variable **counter** is incremented by 1.

14.2.8 Control of program flow, and the **while** keyword

All the keywords of Table A6.2 are associated with program flow, for example looping and branching. As a first example, the **while** keyword allows a statement, or block of statements, to be executed repeatedly, as long as a particular condition holds true. This is the first of a number of branching and looping structures that we will meet.

The general **while** structure is:

```
while (conditional expression)
statement;
```

This will cause the **statement** to be executed repeatedly, as long as the conditional expression evaluates 'true' (i.e. non-zero). If it is no longer true, then program execution after the loop proceeds.

If more than one statement needs to be associated with the **while**, then a series of statements can be enclosed in curly brackets, as shown:

```
while (conditional expression)
{statement 1;
statement 2;
statement 3;
}
```

Note that the **while** condition is evaluated at the *beginning* of loop execution. If it holds true, then the whole loop is executed, even if the condition changes as the loop executes.

In this example, a continuous loop is forced by putting '1' for the conditional expression.

```
while (1)
   {
   PORTB = counter;        // Move 'counter' value to Port B
   counter = counter + 1;
   }
```

The two statements which fall within the **while** braces thus repeat indefinitely.

14.2.9   The C preprocessor and its directives

The process of compiling is made up of a number of distinct stages. The first of these is undertaken by the *preprocessor*. This responds to any preprocessor *directives* which it finds. These act in a way similar to Assembler directives, giving instructions to the compiler itself. Example preprocessor directives appear in Table A6.6. The format of the preprocessor directive requires that each directive occupies a line to itself. It is not terminated with a semicolon.

The line in the example program

```
#include <p18f242.h>      //for all 18F242 declarations
```

uses the **#include** directive to include a processor-specific header file. This file, specific to the C18 compiler, contains the declarations necessary for this particular 18 Series processor and saves having to spell them out in the source code. It contains declarations for all the SFRs (Special Function Registers), including those for Port B used in this program.

14.2.10   Use of libraries, and the Standard Library

Because C is a simple language, much of its functionality derives from standard functions and macros which are available in the libraries accompanying any compiler. A C library is a set of precompiled functions, in the form of object files, which can be linked in to the application. The contents of the *Standard Library* are defined in the ANSI standard. It includes functions for input and output, a range of mathematical functions (for example, all trigonometric functions) and other data handling functions.

In addition to the Standard Library, as we shall see, a compiler may have its own library of functions, intended specifically for its target environment.

14.3   Compiling the C program

When the C source code is complete, it is *compiled*. This process leads eventually to the production of a file containing the machine code equivalent of the source, which the target computer can execute.

We have already met the concept of header files and library files. These are extensively used, with the result that hardly any C source files are stand-alone. Once the 'extra' files are incorporated, the final executable program is built up from a number of contributing files, often in quite a complex way. In turn, the process of compiling a C program creates a range of output files.

The process of compiling, along with the files used and generated, is illustrated in a general way in Figure 14.1. The main program, the C source file, is written in the C language, in a file with the extension **.c**. This is very likely to include (using the preprocessor **#include** directive) other standard files, for example the processor-specific header file we have already seen. A source file, together with any included header files, is known as a *translation unit*.

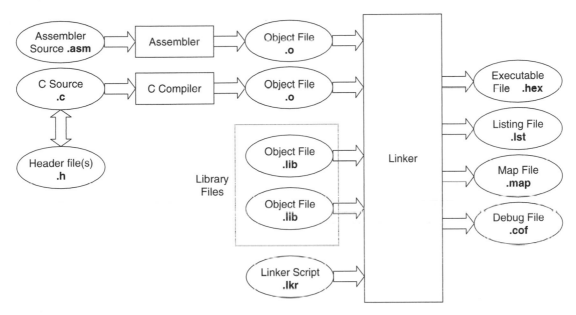

Figure 14.1 A C file structure, with C18 file extensions

When the source program is complete, it is compiled using the C compiler to produce an *object file*. This consists of relocatable code, not yet fully mapped to the processor memory map. Other files can also be developed, including those in Assembler, and all compiled or assembled in a similar way, leading to object files containing (mainly) relocatable code.

At this stage, in all but the simplest programs, it is likely to be combined with other files, which exist already at the object file stage. These may be from the libraries which accompany the compiler, or they may be previously developed files of the programmer or company. It is the task of the *Linker* to combine these different files together, to create one single executable file. In doing this, it is guided by the *Linker Script*, a file which defines the processor memory map and provides other information, including the possibility of drawing in further precompiled files. The opening title block in Program Example 14.1 makes reference to two object files, **c018i.0** and **p18f242.lib**, which are linked in to the program by the Linker. How this is done, and the role they play, is explained in Chapter 17.

After the linking process is complete, and assuming no errors in the process, a series of output files is produced. Those for the Microchip C18 compiler are detailed in Section 14.5.4.

14.4   The MPLAB C18 compiler

The MPLAB C18 compiler is Microchip's own C compiler, written especially for the PIC 18 Series microcontroller. It follows the ANSI X3.159-1989 standard, except that it also contains a number of extensions, designed to optimise its use with the PIC microcontroller.

The C18 compiler operates within the main MPLAB IDE environment, working alongside its Assembler, Linker and Librarian. Once installed, it can be linked in to this tool suite. Unlike MPLAB, the compiler must be purchased, at a cost in the region of $150. At the time of writing, a student version is available free, as a download from the Microchip website. There is also a copy on the book CD.

The compiler is reasonably well documented, primarily through Refs 14.1 and 14.2. These are available from the Microchip website. In the sections which follow, the compiler is introduced to a level adequate to run all the example programs given. Further details can then be found from this reference information.

14.4.1   Specification of radix

MPLAB C18 recognises the C radix specifications mentioned in Section 14.2.6. It also introduces the very useful '0b' prefix for binary numbers. For example:

```
TRISA = 0b'10000110';      // initialise PORTA
TRISB = 0x86;              // initialise PORTB
TRISC = 134;               // initialise PORTC
```

puts the same values into each TRIS register.

14.4.2   Arithmetic operations

The ISO/ANSI C standard requires that all arithmetic operations are done at **int** precision (i.e. 16 bits) or higher. The C18 compiler, reflecting the 8-bit world it serves, steps outside this requirement and undertakes arithmetic with **char** data types.

In the sections which follow it is assumed that the reader has installed MPLAB IDE and the MPLAB C18 compiler, following the simple procedure given in Ref. 14.1. Version 3.00 of the C18 compiler is used.

14.5   A C18 tutorial

Having described a simple yet plausible C program, let us try to compile it and simulate. In MPLAB open a new project with suitable name; the images which follow use **example1** as the entirely unoriginal name. Open a new file and enter into it the code of Program Example 14.1. Save this with a file name, for example **example1.c**, and add it to the project, using Project > Add Files to Project.

14.5.1   The Linker and Linker Scripts

As the compile process automatically uses the Linker, it is essential to include a Linker Script. We examine the form of this in Chapter 17, for now we just need to specify it. On the project window, as seen in Figure 14.2, right-click on the **Linker Scripts** line. An **Add File** prompt will appear. Click on this, and searching through the **Add Files to Project** window, find the folder **mcc18/lkr/**. Within this, click on the **18f242.lkr** file. The project window should now appear as in Figure 14.2.

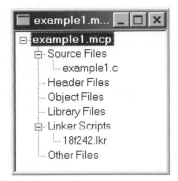

Figure 14.2 Incorporating the Linker Script

Note that we are invoking the Linker Script here in its original form and location. For more advanced projects it is better to make a copy of it and place it in the project directory. This allows the possibility of modifying the file and avoids the risk of corrupting the original.

14.5.2   Linking header and library files

Our example program itself invokes a header file, through the line

```
#include <p18f242.h>
```

It is therefore important to ensure that the search path for this is recognised. The Linker Script calls up other library files, which must also be found. Check that these paths are set up correctly, using **Project > Build Options > Project > General**. The window of Figure 14.3 should then appear. If your compiler is installed to its default location, it should be possible just to click on **Suite Defaults**. This selects the paths shown. Then click on OK.

14.5.3   Building the project

It should now be possible to build the project. As with Assembler, click on **Project > Build All**. If all is correctly entered and linked, the **Build Succeeded** message should ultimately appear in the Output window, as seen in Figure 14.4. Notice that the build process takes considerably longer than Assembler, even for a tiny program like this, and that there are a number of distinct stages. These are shown in Figure 14.4. At the compile stage the window shows that a number of compile options, indicated by **-Ou- -Ot- -Ob-** and so on, are being implemented. These are the compiler defaults and are not of immediate interest at this early stage of compiler use.

If the project did not build successfully, it will be necessary to explore the resulting error messages carefully. The compiler first checks that the program has correct syntax, i.e. it is written to obey the formatting rules of C. Even if you did not get any errors, try removing a semicolon or inserting some other small error, and check the response of the compiler on build. If there is a syntax error, the compiler reports this in the

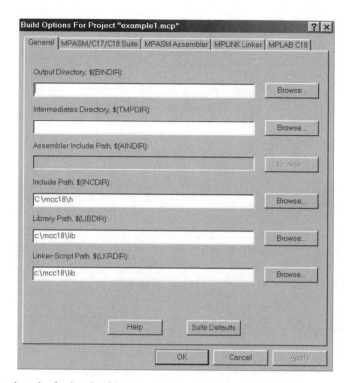

Figure 14.3 Setting search paths for header files

Output window, with a line number. Line numbers may be switched on using **Edit > Properties > Editor**, in the MPLAB window, and clicking the **Line Numbers** box. It is worth noting that the line number where the compiler perceives the error may not be the exact one where correction is needed. As a simple example, try 'commenting out' the opening brace of **main**. You will see that the syntax error is reported as being in the line which follows – which on its own is a perfectly good piece of C code!

When all syntax errors are removed, the compiler checks for other errors and reports an error reference number. The error numbers, with brief descriptions of the error and possible solutions, are listed in Ref. 14.2.

14.5.4   Project files

A successfully built project will yield the files shown in Figure 14.5. Many can also be seen in the diagram of Figure 14.1.

The input and output files within the project window are as follows:

- *Source file* (**.c**). This is the original source file, written in C.
- *Symbol and debug file* (**.cod**). This a file for debug purposes, used with MPLAB IDE version 5.xx and earlier.

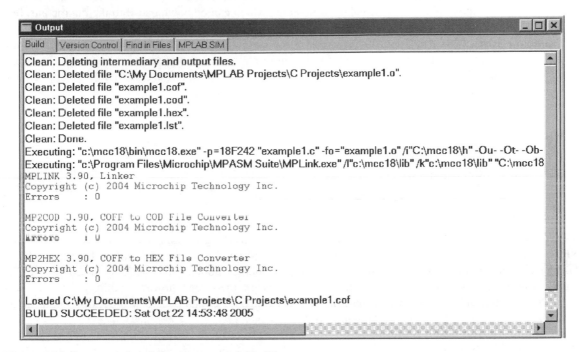

Figure 14.4 Output window following successful build

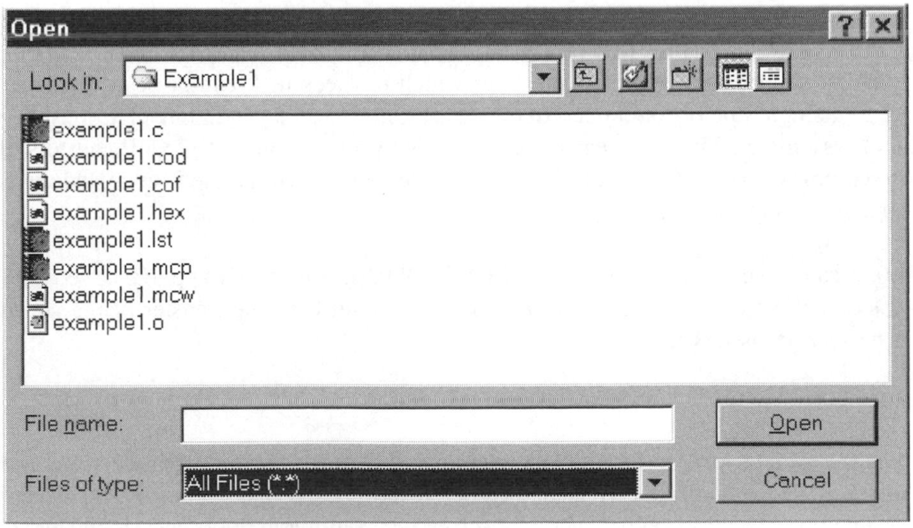

Figure 14.5 Files associated with a simple C project

- *COFF object module file* (**.cof**). This file provides debugging information for MPLAB IDE v6.xx or later.
- *Executable file* (**.hex**). This is the actual program code, which can be downloaded to the microcontroller, or used for simulation or emulation purposes.
- *Listing file* (**.lst**). This file shows the original source code alongside the object code. Symbol values, memory usage information, and error and warning information are also provided.
- *Object file* (**.o**). This file contains the relocatable code. It is the output of the compiler or Assembler, and forms the input to the Linker. Object files are also found in the library.

14.6   Simulating a C program

Explore now the simulation of Example 1. While the program itself is exceedingly simple, the simulation will reveal a number of interesting things about how the overall C program is built, and will bring us back again to Figure 14.1.

As in Section 4.7 of Chapter 4, select MPLAB's simulator by invoking **Debugger > Select Tool > MPLAB SIM**. Using **View > Watch**, open a Watch window, and display the values of **counter**, **PortB** and **PCL**. Using the debugger toolbar (Figure 4.9), reset the program counter, noticing that the Program Counter (**PCL**) clears to zero, as expected. Another program window opens on the screen, however, entirely unasked! This is **c018i.c**, a start-up program which the Linker has invoked. It contains certain processor initialisation essential for correct operation of C. We return to it in Chapter 17.

Single-step through the program (or use Animate) and you will find that execution transfers ultimately to the source program. It is then possible to see in the Watch window the values of **counter** and **PORTB** being incremented. From here, program execution stays indefinitely in the **while** loop.

An interesting way of seeing how the C program has been created is by viewing the *Disassembly Listing*, by selecting **View > Disassembly Listing**. In this window (as seen in Figure 14.6) it is possible to see both the original C source code and the Assembler which replaces it. It can be seen that some lines of C code translate to a single line of Assembler. In most cases, however, a single line of C must be replaced by several lines of Assembler. This is an early indication that programming in C will lead to simpler source files. As more complex lines of C are introduced, the scale factor between the C code and the Disassembly Listing will become more marked.

In Figure 14.6, it can be seen that the address of **TRISB**, $0F93_H$, is correctly applied, as seen in Figure 12.5. The memory location $8A_H$ in Bank 0 has been allocated to **counter**. The transfer of the **counter** value to **PORTB** can be seen in the instruction

```
MOVFF 0x8a,0xf81
```

where 0xf81 is the address of Port B. The incrementing of **counter** can similarly be seen. Notice that the Assembler instructions, as indicated by their memory locations, are placed out of sequence. The branch instruction, at memory location $00F2_H$, actually forms the continuous loop and should lie at the end of the listing.

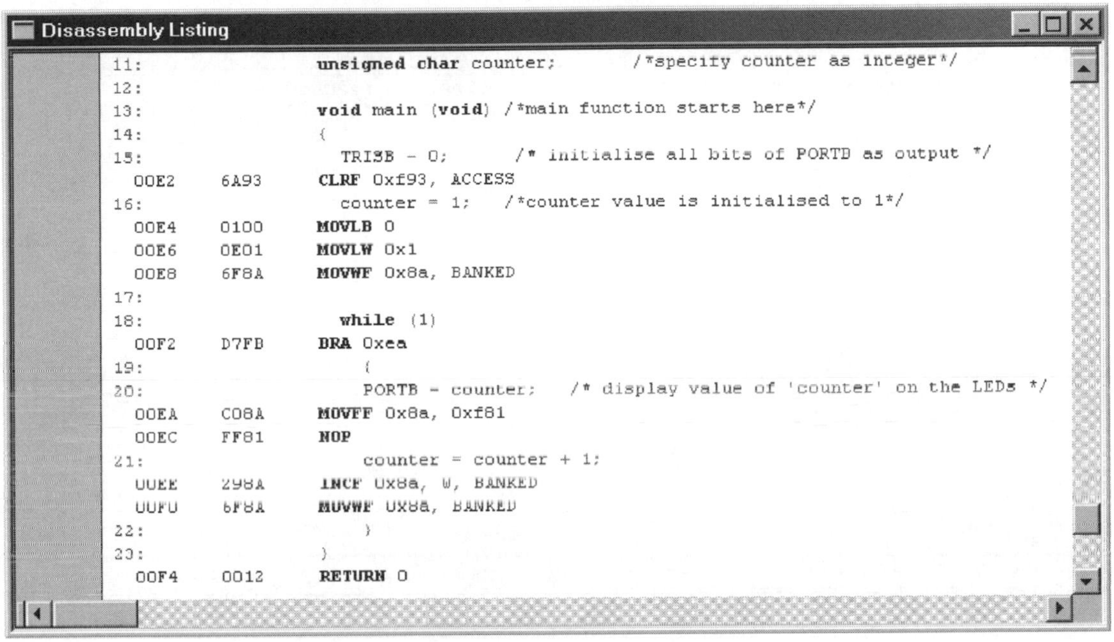

Figure 14.6 The Disassembly Listing for part of Program Example 14.1

14.7 A second C example – the Fibonacci program

Let's now take our C programming a step further, with a program of slightly increased complexity. Program Example 14.2 provides a C version of a program already familiar to us, the Fibonacci series generator. This calculates a Fibonacci series, first by going up to a certain level and then working backwards again. This action is repeated indefinitely.

```
/**********************************************************************
Fibonacci
In a Fibonacci series each number is the sum of the two previous numbers.
This program calculates Fibonacci numbers within an 8-bit range,
Files c018i.o and p18f242.lib are included by the Linker Script.
Program intended for simulation only, hence no input/output.
TJW 21.10.05                                    Tested 23.10.05
;*********************************************************************/

#include <p18f242.h>

//these memory locations hold the Fibonacci series
unsigned char fib0; //lowest number
                //(oldest when going up, newest when reversing down)
```

Program Example 14.2 The Fibonacci series generator

```
unsigned char fib1; //middle number
unsigned char fib2; //highest number
unsigned char fibtemp;  //temporary location for newest number
unsigned char counter;  //indicates which value series has reached

void main (void)
{
 fib0 = 0;
 fib1 = 1;
 fib2 = 1;
 counter = 3; //have preloaded the first three numbers, so start at 3
loop:
        do
        {
        fibtemp = fib1 + fib2;
        counter = counter + 1;
//now shuffle numbers held, discarding the oldest
        fib0 = fib1; //first move middle number, to overwrite oldest
        fib1 = fib2;
        fib2 = fibtemp;
        }
        while (counter<12);
//when reversing down, we will subtract fib0 from fib1 to form new fib0
        do
        {
        fibtemp = fib1 - fib0; //latest number now placed in fibtemp
        counter = counter - 1;
//now shuffle numbers held, discarding the oldest
        fib2 = fib1; //first move middle number, to overwrite oldest
        fib1 = fib0;
        fib0 = fibtemp;
        }
        while (fib0>0);
        goto loop;
}
```

Program Example 14.2 Continued

14.7.1   *Program preliminaries – more on declaring variables*

Following the opening comments, the early program lines declare five unsigned character variables, while the first three lines within **main** initialise three of these to certain values. There are certain ways in which this process can be shortened. Firstly, a declaration of one data type need not be confined to a single variable. Therefore, the five variables *could* all be declared together, as

```
unsigned char fib0, fib1, fib2, fibtemp, counter;
```

It would be up to the programmer to decide whether this format was preferable. One disadvantage is that it is less easy to add a comment per variable, should this be desired.

It is also possible to initialise a variable at the time of declaration. The following would be a possible format:

```
unsigned char fib0 = 0;    //lowest number
unsigned char fib1 = 1;    //middle number
unsigned char fib2 = 1;    //highest number
```

Again, it would be up to the programmer to decide whether this format gave any advantage.

14.7.2 The *do–while* construct

There are two loops in the Fibonacci program. These are constructed here using the **do** and **while** keywords. The block of code following the **do** word is executed as long as the **while** condition (**counter** < **12** for the first loop) is 'true', i.e. it has a non-zero value. In such a loop, the **do** code is always executed at least once before the **while** condition is tested. This is different from a **while** loop, which will not execute even once if its condition is not true on entry.

When reversing down the series, the loop is repeated until **fib0** has been found to have reached a value of 0, using another test in a **while** statement.

14.7.3 Labels and the *goto* keyword

Immediately before the first **do** loop there appears the expression **loop:**. This is a label, identified by its terminating colon. The last line of the program returns execution to **loop:**, using the keyword **goto**. The only application of such labels is as targets for **goto** instructions; they are not used for any other purpose. A **goto** causes unconditional branching within a given function; it cannot branch to another function. The **goto** branch is not much loved in C circles, as its uncontrolled use leads to unstructured programs.

14.7.4 Simulating the Fibonacci program

As always, it is worth simulating this program. Copy the program from the book CD into an appropriate project and simulate with MPLAB SIM. Open a Watch window, with **PCL** and all variables displayed. Single-step through the program and, once in **main**, observe how the looping is controlled. Notice that when **counter** reaches 12, the **do** code block is completed before execution moves to the lower loop. An alternative structure for this loop is explored in Chapter 15 and seen in Program Example 15.2.

14.8 The MPLAB C18 libraries

We close this chapter with a survey of the library functions available in the C18 compiler. The compiler has an extensive set of libraries, which reflects both the C Standard Library, as well as many functions specific to the PIC environment. Most programs written using the compiler will almost inevitably use at least one of these functions, if not many. Certainly, *all* example programs from this point on use library functions.

Each library function has its own entry in the libraries reference manual [Ref. 14.3]. This indicates the action of the function, the arguments to be passed, the return type and any header file which must be included. The functions fall into the categories described in the following sections.

14.8.1 Hardware peripheral functions

These are a set of functions which relate to the microcontroller peripherals. There are functions for enabling and configuring a peripheral, changing its mode of operation, reading it and disabling it. For the ADC, for example, the functions shown in Table 14.1 are available. It can be seen (by reference to Section 11.3 of Chapter 11 if necessary) that these provide all the functionality which would normally be required when using the ADC. The actual setting is conveyed in the function arguments, which are detailed in the libraries manual [Ref. 14.3]. A number of these functions are used in program examples in Chapter 16.

The peripherals can, of course, still be manipulated directly through their SFRs, but usually the library function makes programming easier, more visible and more reliable. Knowledge of the detail of the SFRs remains important and in some cases still essential. This helps to interpret how the function parameters should be set.

These functions are unique to the C18 compiler and do not exist in the C Standard Library.

14.8.2 The software peripheral library

This library provides drive functions for a number of external devices which can be included in a system. These include the Hitachi HD44780 LCD driver (see Chapter 8), the MCP2510 CAN (Controller Area Network – described in Chapter 20) interface, as well as functions for generating serial interchange in software.

Like the hardware peripheral library, these functions are unique to the C18 compiler and do not exist in the C Standard Library.

Table 14.1 C18 library ADC functions

Function	Action
OpenADC( )	Configures the ADC
SetChanADC( )	Selects the channel to be used
ConvertADC( )	Starts an ADC conversion
BusyADC( )	Tests whether ADC is currently busy
ReadADC( )	Reads the result of an ADC conversion
CloseADC( )	Disables the ADC

14.8.3 The general software library

The functions in this library are a mix of functions from the C Standard Library and ones specific to Microchip. They fall into the following categories:

Character classification

These functions match the requirements of the Standard C **ctype** Library and provide tests for characters to determine their nature. Examples are shown in Table 14.2.

Data conversion

These functions provide conversion between one form of data representation and another. This can be extremely useful in the embedded environment, as we convert data from binary form to a string of characters for display, or alternatively read a string in from a keypad and convert to binary. Examples are shown in Table 14.3. They represent a mix of standard C functions and C18 'specials'.

Memory and string manipulation

This set of functions allows manipulation of memory. Most are drawn from the C Standard Library, although there are some small differences. Example functions are shown in Table 14.4.

Delays

All the delay functions available are listed in Table 14.5. The first is a fixed single-cycle delay, compiling just as a **nop** (no operation) instruction. All the others allow a programmable delay, based on

Table 14.2 Example character classification functions

Function	Action
isalnum( )	Determines if character is alphanumeric
isalpha( )	Determines if character is alphabetic
iscntrl( )	Determines if character is control
isdigit( )	Determines if character is decimal digit

Table 14.3 Example string/character conversion functions

Function	Action
atop( )	Converts string to signed byte
atof( )	Converts string to floating-point value
atoi( )	Converts string to 16-bit signed integer
atol( )	Converts string to long integer
itoa( )	Converts 16-bit signed integer to string

Table 14.4 C18 general software library delay functions

Function	Action
memchr( )	Searches for a value in specified memory area
memcmp( )	Compares the contents of two arrays
memcpy( )	Copies a buffer from data or program memory to data
memset( )	Initialises an array with repeated memory value

Table 14.5 C18 general software library delay functions

Function	Action
Delay1TCY( )	Delay in one instruction cycle
Delay10TCYx( )	Delay in multiples of 10 instruction cycles
Delay100TCYx( )	Delay in multiples of 100 instruction cycles
Delay1KTCYx( )	Delay in multiples of 1000 instruction cycles
Delay10KTCYx( )	Delay in multiples of 10 000 instruction cycles

Table 14.6 Example C18 general software library reset functions

Function	Action
isBOR( )	Determines if brown-out was the cause of reset
isLVD( )	Determines if low voltage was the cause of reset
isMCLR( )	Determines if master clear was the cause of reset
isPOR( )	Determines if power-on was the cause of reset

the instruction cycle time of the microcontroller. This provides a very useful facility in the embedded environment.

Reset

These functions allow the programmer to test a source of reset, from information provided in the 18 Series **RCON** register (Figure 12.14). Example functions are shown in Table 14.6.

14.8.4 The maths library

This library provides the mathematical functions required in the C Standard Library. Variables are generally floating point. Examples are shown in Table 14.7.

14.9 Further reading

When studying C, it is worth noting the different types of references that are available, which can be used to broaden one's knowledge of the language. Some, like Ref. 14.4, give a general introduction to C.

Table 14.7 Example C18 maths library functions

Function	Action
sin( )	Computes the sine
cos( )	Computes the cosine
tan( )	Computes the tangent
sqrt( )	Computes the square root
log10( )	Computes the log to the base 10
pow( )	Computes the exponential x^y

Often, these are very much in the context of programming with a desktop computer, with emphasis on data input from keyboard and output to computer screen. There are also a number of useful reference books, like Ref. 14.5. These give complete reference information on the language, without attempting to structure the material as a teaching text. It can be very useful to have easy access to one of these, to check details of syntax and so on. There are also a number of books targeted at the embedded environment, such as Refs 14.6 and 14.7. These get much closer to our interest here. They do, however, tend to be microcontroller or compiler specific, and in their detail are somewhat less useful when applied to a different processor.

Summary

- Although it is a high-level language, C contains features that allow it be extremely effective at the embedded system level. It remains the high-level language of choice for many embedded applications.
- The core features of C are comparatively simple and logical, and can be learned without too much difficulty.
- The MPLAB C18 C compiler is a powerful software tool, drawing on the strengths of the C language, but optimised for the PIC 18 Series.
- The MPLAB C18 C compiler can be used as a tool within the MPLAB IDE. Thus, all the expertise that a developer has with this environment can immediately be applied to the development of C programs.
- Writing in C requires a combination of knowledge of the language itself, along with the library functions that are available for the compiler and processor that are in use. There is a rich collection of library functions in the C18 compiler, providing general utilities, peripheral control, data manipulation and mathematical functions.

References

14.1. MPLAB C18 C Compiler Getting Started (2003). Microchip, Document No. DS51295B.
14.2. MPLAB C18 C Compiler User's Guide (2005). Microchip, Document No. DS51288J.
14.3. MPLAB C18 C Compiler Libraries (2005). Microchip, Document No. DS51297F.

14.4. Austin, M. and Chancogne, D. (1999). *Engineering Programming in C, Matlab and Java*. Wiley, New York. ISBN0-471-00116-3.

14.5. Prinz, P. and Kirch-Prinz, U. (2003). *C Pocket Reference*. O'Reilly. ISBN 0-596-00436-2.

14.6. Pont, M. J. (2002). *Embedded C*. Addison-Wesley, Great Britain. ISBN 0 201 79523 X.

14.7. van Sickle, E. (2003). *Programming Microcontrollers in C*, 2nd edn. Elsevier, USA. ISBN 1-878707-57-4.

15
C and the embedded environment

The C programming language was introduced, in overview, in Chapter 14. This chapter now aims to start applying that skill to writing real embedded C programs. It begins with that most essential of embedded requirements, the manipulation of individual bits, and then moves on to interaction with peripherals. While doing this, many more details of C are introduced, as they come up in example programs.

Examples in this chapter are mostly applied to the Derbot-18 AGV. They can all be simulated, so it does not matter too much whether or not you have the hardware. Most are unashamed reworkings of the Assembler programs which have appeared earlier in the book. By adapting Assembler programs, a comparison can be made between the different programming languages, and the reader who is working through the book is on familiar territory as far as the target hardware is concerned.

By working through this chapter, you should develop a good understanding of:

- How to access and manipulate single bits
- How to write simple functions and call them
- How to invoke library functions, including those for control of the microcontroller peripherals
- How to give some structure to C programs, making appropriate use of functions, and looping and branching constructs.

15.1   The main idea – adapting C to the embedded environment

Now we have adopted a high-level language (HLL), it is as if we are seeking the best of both worlds. We want the benefits of the HLL, but we retain our determination to work close to the hardware – setting up peripherals, setting or clearing individual port bits, and ultimately setting up and responding to multiple interrupts. This tension is resolved in interesting ways, which we now begin to explore.

15.2   Controlling and branching on bit values

Program Example 14.1 illustrated in a very simple way the use of the microcontroller ports. This relied on the port SFRs (Special Function Registers) being declared in the header file, the data direction being set up by writing to the TRIS register, and then moving data to the port by writing a whole byte.

Fundamental to developing programs for embedded systems is, of course, the ability to read and set single bits. Program Example 15.1, which moves the state of the microswitches on the Derbot to the LEDs, demonstrates how this is done in C. Essentially, it rewrites Program Example 7.1. As it must, the program contains a **main** function, as well as two user-defined functions, for initialisation **initialise( )** and for diagnostics **diagnostic( )**. Locate each of these in the program listing. In implementing the diagnostic flashing of the LEDs, the program also makes use of a C18 library function. The way the functions are used will be discussed in Section 15.3.

```
/*********************************************************************
Sw_to_led_18C                                Uses PIC 18F242
Runs on Derbot-18. Moves state of front microswitches to leds
Files c018i.o and p18f242.lib are included by the Linker Script.
TJW 22.10.05                                 Tested 24.10.05
*********************************************************************
Clock is 4MHz
Configuration Word all default, except: crystal oscillator (HS),
power-up timer on, brown-out detect off, WDT off, LV Program disabled*/

        #include <p18F242.h>
        #include <delays.h>          //header file for delays

//Function Prototypes (Library prototypes are in Header files)
        void initialise (void);
        void diagnostic (void);

void main (void)
{
        initialise();          //call initialise function
        diagnostic();          //call diagnostic function

//move microswitch states to diag leds
loop:
        if (PORTBbits.RB4 == 0)
        PORTCbits.RC6 = 0;
        else PORTCbits.RC6 = 1;

        if (PORTBbits.RB5 == 0)
        PORTCbits.RC5 = 0;
        else PORTCbits.RC5 = 1;

        goto   loop;
}

//Initialises SFRs, and sets initial outputs.
//Assumes hardware is "Build Stage 1". All unused port bits set to output.
void initialise (void)
{       TRISA = 0b00000000; //All bits output (none used in this program)
        TRISB = 0b00110000; //Bits 5 and 4 (microswitches) only are input
        TRISC = 0b10000000; //All bits output, except bit 7 (mode switch)
//Switch all outputs off
        PORTA = 0;
        PORTB = 0;
        PORTC = 0;
}
```

Program Example 15.1 Derbot – moving microswitch states to LEDs

```
//Diagnostic: switches leds on for 1s (Tcy = 1us)
void diagnostic (void)
{       PORTCbits.RC6 = 1;
        PORTCbits.RC5 = 1;
        Delay10KTCYx (100);
        PORTCbits.RC6 = 0;
        PORTCbits.RC5 = 0;
        Delay10KTCYx (100);
}
```

Program Example 15.1 Continued

15.2.1 *Controlling individual bits*

The bits of each port are defined in the microcontroller header file, using a C structure that is described in Section 17.9 of Chapter 17. For the purposes of this program, it is enough to recognise that a port bit can be specified by the format **PORTxbits.Rxy**, where *x* indicates the port and *y* the bit in that port. As an example, the diagnostic function lies at the end of the program listing. In it, we see bits 5 and 6 of Port C being set to Logic 1 in the lines:

```
PORTCbits.RC6 = 1;
PORTCbits.RC5 = 1;
```

With this simple step we now have the ability to set or clear individual bits in a register, as long as they have been previously declared. As all microcontroller SFRs and their bits are declared in the header file; this represents a great step forward.

15.2.2 *The **if** and **if–else** conditional branch structures*

The action in this program is built around a conditional **if–else** branching structure. This allows a program to contain a choice between two separate paths of action. An example of the structure appears in the **main** function, as quoted here:

```
if (PORTBbits.RB4 == 0)
PORTCbits.RC6 = 0;
else PORTCbits.RC6 = 1;
```

This can be interpreted as: *if bit 4 of Port B is at Logic 0, then set bit 6 of Port C to 0; otherwise (else) set it to 1*. A block of code, rather than just a single line, can also be associated with either the **if** and/or the **else**, in which case it must be enclosed in curly brackets. For example:

```
if (PORTBbits.RB4 == 0)
{PORTCbits.RC6 = 0;
PORTCbits.RC0 = 1;
}
else PORTCbits.RC6 = 1;
```

This would cause two Port C bits to change, if Port B bit 4 was found to be zero.

It is also possible to use the **if** structure on its own. In this case there is no alternative action if the condition tested is not true. An example is:

```
if (PORTBbits.RB4 == 0)
PORTCbits.RC6 = 0;
if (PORTBbits.RB5 == 0)
PORTCbits.RC5 = 0;
```

In this case, Port C bit 6 is set to 0 if Port B bit 4 is at 0, but no action is taken if Port B bit 4 is at Logic 1. The same can be seen to happen in the lines which follow, with bit 5 of each port.

Notice in these examples how the assignment operator '=' and the equal to operator '==' are used. As we have seen, the first is used to assign a value to a variable. The second is used, within the **if** construct, to test whether a variable is equal to a particular value.

15.2.3   Setting the configuration bits

Settings for the configuration bits are indicated in the program listing. For the time being, these should be set in the MPLAB **Configuration Bits** window, as seen in Figure 12.15. This is not, of course, a very satisfactory procedure and in Chapter 17 we will see a way of embedding the settings in the program. Note that the settings in the window can be returned to their default condition by right-clicking in the window, and clicking **Reset to Defaults** in the dialogue box.

15.2.4   Simulating and running the example program

It is interesting to simulate this program in the MPLAB® simulator. Having created and built the project, set up the simulation with the following steps.

- In MPLAB, select the simulator with **Debugger > Select Tool > MPLAB SIM.**
- Open a Watch window, and select Port B and Port C as variables for display.
- Set up a stimulus controller and simulate inputs for the microswitch inputs, RB5 and RB4, with **Toggle** as the **Action.**
- Place breakpoints in the **diagnostic( )** function, as shown in Figure 15.1(a).
- Use **Debugger > Settings > Osc/Trace**, to set the Processor Frequency to 4 MHz.
- Use **Debugger > Stopwatch** to display the Stopwatch window, as shown in Figure 15.1(b).

Now reset and run the program to the first breakpoint, which will be the first in Figure 15.1(a). At this point zero the Stopwatch and then run to the next breakpoint. This is just the line below. To get there, however, the program has to execute the **Delay10KTCYx( )** function. The Stopwatch should now be exactly as shown in Figure 15.1(b). Exactly one second of simulated time has elapsed in execution of the function. This is a satisfying confirmation of the accuracy of this function.

Now set one further breakpoint in the line below the label **loop:**. Run the program to here and then single-step through the loop. Using the stimulus controller, change the values of the microswitch inputs (Port B bits 4 and 5) and observe how the program loop responds.

(a)

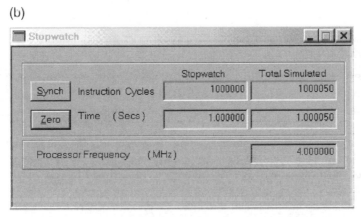

```
C:\...\sw_to_led_18.c

//Diagnostic: switches leds on for 1s (Tcy = 1us)
void diagnostic (void)
{    PORTCbits.RC6 = 1;
     PORTCbits.RC5 = 1;
     Delay10KTCYx (100);
     PORTCbits.RC6 = 0;
     PORTCbits.RC5 = 0;
     Delay10KTCYx (100);
}
```

(b)

```
Stopwatch

                              Stopwatch      Total Simulated
  Synch   Instruction  Cycles    1000000          1000050

  Zero    Time   (Secs)         1.000000         1.000050

  Processor Frequency   (MHz)                    4.000000
```

Figure 15.1 Simulation settings for Program Example 15.1. (a) Suggested breakpoints in diagnostic function. (b) Stopwatch, after completion of delay function

If you have the Derbot-18 hardware, download the program in the usual way. The program function is simple; the satisfaction comes from seeing your first C program running in hardware!

15.3   More on functions

Now that we have a program example with several functions, it is useful to pause to explore further the way functions are used. In particular, we need to know how a function is written, how it is called, how data is passed to it and how it is returned.

15.3.1   The function prototype

When **main** is not the only function in the program, it is necessary to include for every function a *function prototype*. This is a declaration which informs the compiler the type of the function's argument(s), if any, and its return type. It is similar to the function header seen earlier and has the same general format. For example, the **Delay10KTCYx( )** library function, described below, has the prototype:

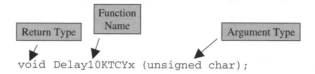

```
void Delay10KTCYx (unsigned char);
```

It can be seen that one argument is sent, an unsigned character, while no return value is expected.

Prototypes for library functions appear in the library header files and do not need to be repeated in the user code. In this case, the prototype just discussed can be found in the **delays.h** header file. Prototypes for the two user-defined functions in the program example can be seen towards the top of the program. This shows that neither expects a return value and neither carries any arguments.

15.3.2   The function definition

The actual code of a function is called the *function definition*. It follows the format described in Section 14.2.5 of Chapter 14. The definition for the **Delay10KTCYx( )** function is contained within the general software library (Table 14.5) and is merged with the main program at the time of linking.

The definitions for the two user-defined functions can be seen towards the end of the program listing. They are placed here for clarity. They can, however, be placed anywhere in the program listing, as long as they are not inside another function definition. These definitions are easy to follow.

The **initialise** function sets up the SFRs and initialises the ports to 0. Strictly speaking initialising variables to 0 should be unnecessary, as the ANSI standard requires it anyway. With C18 this is only done if the **c018iz.o** start-up utility is used (described later, in Section 17.7.1 of Chapter 17). We do *not* use this in the example programs in this book and variables are hence not initialised to zero as a matter of course. The **diagnostic** function sets bits 5 and 6 of Port C (the two LED output bits) to 1 and calls the delay function. It then clears the same bits to zero, before calling the same delay again.

15.3.3   Function calls and data passing

A function is called by quoting its name and placing the necessary arguments in the brackets which must follow. Where there are no arguments, the brackets must still be there, but can be left empty. The function calls in Program Example 15.1 are simple and self-explanatory. Notice that the parameter 100_D is passed to the delay function.

It is important to be aware of a number of features of function calls which are not evident from this particular example. Importantly, *a function call has the type and value of its return type*. This is an assertion of some significance! It means that a call can be inserted into an expression; the function is evaluated and its return value acts in its place in the expression. This is done several times in Program Example 16.1 and the programs which follow. An example from there is as follows:

```
ldr_rt = ReadADC()&0x03FF;   // read it, AND out unwanted bits
```

Here the function call **ReadADC( )** is placed within a statement. It is evaluated first and its return value then takes its place in the statement.

Remember also that any parameters passed to the function are *copied* to it, with the original value (if a declared variable) retained. It uses these in its internal execution, to generate the return value. It does not in itself modify the value of the variable.

15.3.4   Library delay functions, and *Delay10KTCYx( )*

All the delay functions available in the C18 general software library are shown in Table 14.5. These all have a function prototype of the form seen already for **Delay10KTCYx( )**, with an unsigned character (i.e. 8-bit word) acting as the multiplier to set the actual delay.

The **Delay10KTCYx( )** function used here allows the longest of the delays. It introduces a software delay in multiples of 10 000 instruction cycles, with a maximum of 255×10^4 cycles. Thus, with a clock running at 4 MHz and the variable expressed as 100 (as in this example), the total delay is $10\,000 \times 100 \times 1\,\mu s$, i.e. 1 second.

The **delays.h** header file, required for this function, can be seen included in the early stages of this program.

15.4   More branching and looping

15.4.1   Using the *break* keyword

Now that we have a mechanism in C which gives access to individual register bits, we can make use of this in further bit testing and setting operations.

Returning to the Fibonacci program of Program Example 14.2, let us explore another way of constructing the first loop. The purpose of this loop is to generate the Fibonacci series, within the limits of the 8-bit number used (specified as type **unsigned char**). In Program Example 14.2 we did this in a rather artificial way, by limiting the number of values calculated to a known 'safe' maximum.

Try now replacing the first loop with the program section in Program Example 15.2. This establishes what appears to be a continuous loop, using a **while (1)** construct. Within the loop, however, is a possible exit strategy, based on the **break** keyword. The statement that it is part of tests the value of the Carry bit and causes an exit if it is high. If you simulate with this revised version, you will see that an exit from the loop is forced immediately following this statement, once the Carry bit goes high.

```
while (1)
{fibtemp = fib1 + fib2;
if (STATUSbits.C == 1) break; //exit loop if Carry bit set
 counter = counter + 1;
//now shuffle numbers held, discarding the oldest
fib0 = fib1; //first move middle number, to overwrite oldest
fib1 = fib2;
fib2 = fibtemp;
}
```

Program Example 15.2 Alternative first loop for the Fibonacci program

Given a way of testing the Carry bit, one might ask – why not make this the **while** condition? The loop could then be started:

```
while (STATUSbits.C != 1)   //loop while the Carry bit is not 1
{fibtemp = fib1 + fib2;
...
```

The problem here is that the condition is only tested at the end of the loop. By this time the addition has overflowed and incorrect numbers will have been loaded into at least one location in the Fibonacci series. The loop could, of course, be restructured so that the addition was at the bottom of the loop and the test of the Carry bit then took immediate effect.

15.4.2   Using the *for* keyword

The **for** keyword provides another means of 'packaging' conditions for a loop. It has the general format:

```
for (initialisation; condition; modification)
statement, or statements in braces
```

The three expressions within the **for** brackets, called here **initialisation**, **condition** and **modification**, are all defined by the programmer. Moving immediately to an example, the first loop in Program Example 14.2 could be rewritten as shown here:

```
for (counter = 0;counter<12;counter = counter + 1)
{fibtemp = fib1 + fib2;
//now shuffle numbers held, discarding the oldest
        fib0 = fib1;  //first move middle number, to overwrite oldest
        fib1 = fib2;
        fib2 = fibtemp;
    }
```

In the first expression, **counter** is initialised to 0. This occurs only once, when the loop is entered. The condition tested is whether **counter** is less than 12, and the modification caused is an increment to the value of **counter**. This does not occur on the first loop iteration. When the program runs, the loop is repeatedly executed, with **counter** being incremented each time. When it is incremented to 12, this is immediately detected by the condition expression and no further loop iterations occur.

It is interesting to modify the Fibonacci program to contain this code, and simulate.

Any of the three expressions associated with **for** can be omitted. If the condition is left out, then there is no test and the loop is continuous. Initialisation and modifications can still, however, apply. A simple way of creating a continuous loop is by entering no expressions at all, giving:

```
for(;;)
{...
```

This is a direct alternative to:

```
while(1)
{...
```

15.5   Using the timer and PWM peripherals

We turn now to controlling microcontroller peripherals through the use of library functions. To do this we will use the Derbot 'blind navigation' program, first introduced as Program Example 8.4. It is shown in a C version in Program Example 15.3. The program makes use of library functions for Timer 2 and PWM, and writes a number of functions of its own. These tend to replicate the subroutines of the original program.

The program simply causes the Derbot to run forward until it hits an obstacle, detected by a microswitch. It then reverses and turns, turning right if the left microswitch was hit and vice versa for the right microswitch. After this it returns to running forward. These simple moves require the use of the microcontroller PWM facility, which in turn requires the setting of Timer 2.

```
/***********************************************************************
Dbt_blind_Nav_PWM_C
Derbot moves by "blind" navigation.
Moves forward, and reverses and turns on bump.
Files c018i.o and p18f242.lib are included by the Linker Script.
Fixed rate PWM applied to set reasonable speeds.
TJW 3.11.05                                        Tested 7.11.05
************************************************************************
Clock is 4MHz
Configuration Word all default, except: crystal oscillator (HS),
power-up timer on, brown-out detect off, WDT off, LV Program disabled*/

#include <p18F242.h>
#include <delays.h>          //header file for delays
```

Program Example 15.3 Derbot 'blind navigation' program

```
#include <timers.h>          //header file for Timers
#include <pwm.h>             //header file for PWM

/*function prototypes, reproduced from Header Files for information
      void OpenPWM1 (char);
      void OpenPWM2 (char);
      void OpenTimer2 (unsigned char);
      void Delay10KTCYx (unsigned char); */

//User-defined function prototypes
      void diagnostic (void);
      void leftmot_fwd (void);
      void rtmot_fwd (void);
      void rev_left (void);
      void rev_rt (void);

void main (void)
{
/*Initialises SFRs, and sets initial outputs. Assumes hardware is "Build
Stage 2". All unused port bits set to output. Used bits are identified.*/
      TRISA = 0b00000000;   //All bits output, 2 & 5 used for motor enables.
      TRISB = 0b00110000;   //Bits 5 and 4 (microswitches) only are input,
      TRISC = 0b10000000;   //All bits output except 7 (mode switch),
                            //1 & 2 used for PWM
      ADCON1 = 0b00000110;  //Set Port A for digital i/o

//Switch all outputs off
      PORTA = 0;
      PORTB = 0;
      PORTC = 0;
//call diagnostic function
      diagnostic();
//Enable PWM
      OpenTimer2 (TIMER_INT_OFF & T2_PS_1_1 & T2_POST_1_1);
      OpenPWM1 (0xFF); //Enable PWM1 and set period
      OpenPWM2 (0xFF); //Enable PWM2 and set period

      while (1)
      {
      //start motors
      leftmot_fwd ();
      rtmot_fwd ();

//test for bumps - reverse and turn if either microswitch closes
      if (PORTBbits.RB4 == 0) //Test right uswitch
      rev_left ();
      if (PORTBbits.RB5 == 0) //Test left uswitch
      rev_rt ();
      Delay10KTCYx (10);
      }
}
```

Program Example 15.3 Continued

```
/******************************************************************
Motor Drive Functions
******************************************************************/
      void leftmot_fwd (void)      //sets left motor running forward
{     CCPR2L = 196;
      PORTAbits.RA5 = 1;           //enable motor
}


      void rtmot_fwd (void)        //sets right motor running forward
{     CCPR1L = 196;
      PORTAbits.RA2 = 1;           //enable motor
}


      void leftmot_rev (void)      //sets left motor running in reverse
{     CCPR2L = 60;
      PORTAbits.RA5 = 1;           //enable motor
}


      void rtmot_rev (void)        //sets right motor running in reverse
{     CCPR1L = 60;
      PORTAbits.RA2 = 1;           //enable motor
}


      void rev_rt (void)           //reverses and then turns to right
{     PORTCbits.RC6 = 1;           //set right led
      PORTAbits.RA5 = 0;           //stop motors
      PORTAbits.RA2 = 0;
      PORTBbits.RB1 = 1;           //small bleep from sounder
      Delay10KTCYx (50);
      PORTBbits.RB1 = 0;           //clear sounder
      leftmot_rev ();              //reverse both motors
      rtmot_rev ();
      Delay10KTCYx (200);
      leftmot_fwd ();              //left motor forward to turn
      Delay10KTCYx (100);
      PORTCbits.RC6 = 0;           //clear led
}


      void rev_left (void)         //reverses and then turns to left
{     PORTCbits.RC5 = 1;           //set left led
      PORTAbits.RA5 = 0;           //stop motors
      PORTAbits.RA2 = 0;
      PORTBbits.RB1 = 1;           //small bleep from sounder
      Delay10KTCYx (50);
      PORTBbits.RB1 = 0;
      leftmot_rev ();              //reverse both motors
      rtmot_rev ();
      Delay10KTCYx (200);
      rtmot_fwd ();                //right motor forward to turn
```

Program Example 15.3 Continued

```
          Delay10KTCYx (100);
          PORTCbits.RC5 = 0;              //clear led
}
```

```
...
```
diagnostic function same as Program Example 15.1.
```
...
```

Program Example 15.3 Continued

15.5.1 Using the timer peripherals

There are four 18FXX2 timers and for each of these there are four library functions. These are shown in Table 15.1, where *x* can be 0, 1, 2 or 3. Full details of their associated arguments are given in Ref. 14.3.

This program uses just one timer-related function, **OpenTimer2( )**. It represents a style of peripheral drive library function that we shall see again. In this, the argument is made up of a bit mask, created by performing a logical AND of a number of settings. These are specified in the library reference [Ref. 14.3] and for this function are reproduced in Table 15.2. In this case three such settings must be chosen, for interrupt enable, prescale and postscale. These can be seen applied in the function call, as quoted here:

```
          OpenTimer2 (TIMER_INT_OFF & T2_PS_1_1 & T2_POST_1_1);
```

Table 15.1 Timer library functions

Function	Action
OpenTimer*x*( )	Configures Timer *x*
ReadTimer*x*( )	Reads Timer *x*
WriteTimer*x*( )	Writes to Timer *x*
CloseTimer*x*( )	Closes Timer *x*

Table 15.2 Settings options for **OpenTimer2( )**

Value	Effect
Interrupt	
TIMER_INT_ON	Interrupt enabled
TIMER_INT_OFF	Interrupt disabled
Prescaler	
T2_PS_1_1	1:1 prescale
T2_PS_1_4	1:4 prescale
T2_PS_1_16	1:16 prescale
Postscaler	
T2_POST_1_1	1:1 postscale
T2_POST_1_2	1:2 postscale
...	...
T2_POST_1_16	1:16 postscale

This enables the timer, disables its interrupt and sets pre- and postscaler to divide-by-one. It effectively replaces the two Assembler lines shown below, quoted from Program Example 9.2:

```
movlw B'00000100'  ;switch on Timer2, no pre or postscale
movwf t2con
```

In terms of code lines saved, the C version gives little advantage. The benefit lies elsewhere, however. Using library functions like these, the programmer no longer needs to get into the detail of the peripheral structure or of its SFR bits. Once the requirements of the function are understood, then a peripheral can be applied with limited understanding of its internal working.

15.5.2 Using PWM

The concept of PWM, and use of the peripheral, was described in Section 9.5 of Chapter 9. The hardware is built around Timer 2 and can initially be difficult to understand. However, it ends up being easy to use. Table 15.3 shows the PWM library functions that are available for the 18FXX2 microcontroller, where x can take the value 1 or 2.

The **OpenPWMx( )** function is used in Program Example 15.3, in the two lines copied here:

```
OpenPWM1 (0xFF);  //Enable PWM1 and set period
OpenPWM2 (0xFF);  //Enable PWM2 and set period
```

The function enables the CCP module in PWM mode and loads the **PR2** register, seen in Figure 9.11. Of course, the **PR2** register is shared between the CCP modules and can only be set to one value. Therefore, it is to be expected that the argument to both function calls is the same. To set the repetition rate, equation (9.2) is applied. In this case, with no prescale on Timer 2 and a function argument of $0xFF_H$, a PWM frequency of 3.906 kHz results.

In this program example the **SetDCPWMx( )** function is not used to set and change speed. As only 8-bit resolution is applied, the **CCPR1L** and **CCPR2L** registers are written to directly.

15.5.3 The main program loop

The main program loop is formed again with a **while (1)** construct. The microswitches are tested in turn with an **if** statement, as shown here:

```
if (PORTBbits.RB4 == 0)  //Test right uswitch
rev_left ();
```

Table 15.3 PWM library functions

Function	Action
OpenPWMx( )	Configures period and timebase of PWM x
SetDCPWMx( )	Writes a 10-bit duty cycle value to PWM x
ClosePWMx( )	Disables PWM x

```
if (PORTBbits.RB5 == 0)   //Test left uswitch
rev_rt ();
```

If either microswitch is activated, its associated input value is 0. Then either **rev_left** or **rev_rt** is called. It should not be too difficult to follow either of these functions through. The AGV pauses and both motors are then set in reverse for a fixed period. Then one or other is set forward (while the other continues in reverse) to cause a turn. Program execution then returns to the main loop and the AGV moves forward again. This loop makes use of the motor drive functions, as well as the **Delay10KTCYx( )** function.

Summary

This chapter has begun to show how C can, in a practical way, be applied to the embedded environment and the PIC® 18 Series microcontroller.

- Individual bits in memory registers can easily be accessed and manipulated.
- There are a variety of branching and looping constructs which allow clearly defined program flow.
- It is easy to identify and use library functions; these greatly simplify interaction with the microcontroller peripherals.
- It is not difficult to write and use functions; a well-structured program will locate distinct tasks in functions, with the main program showing a high number of function calls.

16
Acquiring and using data with C

Having established a grounding in the world of embedded C, it is now appropriate to explore how C can be applied to the acquisition – and then the manipulation – of data.

The first point of interest will be to interface to the ADC peripheral with C. We will find that there are a number of useful library functions that ease this task. Having acquired some data, we will need to consider how it can be stored and manipulated. This will lead us into the field of C arrays and strings. As we will need to move data around, it will be appropriate to look at how to use the I^{2}C peripheral. As overall this can become a complex field of study, and we are only working at an introductory level, only integer data manipulation will be considered.

As in the previous chapter, examples will mostly be applied to the Derbot-18 AGV, but all can be simulated to good effect.

At the end of the chapter, you should have a good understanding of:

- How to invoke library functions for use with the 18FXX2 ADC and I^{2}C peripherals
- How to work with arrays, strings and pointers
- How to invoke library functions which facilitate string manipulation.

16.1 The main idea – using C for data manipulation

One of the strengths of C lies in its ability to work with data. It defines data types, controls data movement and protects data from unwanted changes. In its use in the desktop computer environment, it has many library functions that are designed to make it easy to move blocks of data around. We will see a few of these capabilities in this chapter, applied to the embedded environment. While it has already been mentioned that we will be using only integer data in this chapter for reasons of simplicity, it is also worth stating that floating-point routines on an 8-bit microcontroller like the 18 Series are rather time-consuming in execution. This gives another reason to avoid them, except in situations where it is absolutely necessary.

16.2 Using the 18FXX2 ADC

Program Example 16.1 provides a more substantial example of C programming, again for the Derbot-18 AGV. It is the light-seeking program, first introduced in Assembler as Program Example 11.3. With its three light-dependent resistors (LDRs), the Derbot seeks light, coming to a halt when all sensors are at a similar light level. The program provides useful further examples of conditional branching and introduces use of the ADC. Note that some line numbers are embedded within the comments.

```
/****************************************************************************
Dbt_light_seek_c
Derbot seeks light. PWM applied. Speed is dependent on light difference
(front to back), so Derbot comes to a halt when light difference is minimal.
Microswitches used for bump detection.
Files c018i.o and p18f242.lib are included by the Linker Script.
TJW 12.11.05                                             Tested 27.11.05
****************************************************************************
Clock is 4MHz.
Configuration Word all default, except: crystal oscillator (HS),
power-up timer on, brown-out detect off, WDT off, LV Program disabled*/

#include <p18F242.h>
#include <adc.h>
#include <timers.h>
#include <pwm.h>
#include <delays.h>

/*function prototypes, reproduced from Header Files for information only
        void OpenPWM1 (char);
        void OpenPWM2 (char);
        void OpenTimer2 (unsigned char);
        void Delay10KTCYx (unsigned char);
        void Delay10TCYx (unsigned char);
        void OpenADC (unsigned char, unsigned char);
        void SetChanADC (unsigned char);
        void ConvertADC(void);
        char BusyADC(void);
        int ReadADC(void);                      */

//User-defined function prototypes
        void leftmot_fwd (void);
        void rtmot_fwd (void);
        void leftmot_rev (void);
        void rtmot_rev (void);
        void rev_left (void);
        void rev_rt (void);
        void fwd_left (void);
        void fwd_rt (void);
        void rotate_rt (void);
        void rotate_left (void);
        void diagnostic (void);

//Declare Variables
        int ldr_rt;          //right ldr value
        int ldr_left;        //left ldr value
        int ldr_rear;        //rear ldr value
        int ldr_ave;         //computed average of front ldrs
        int ldr_diff;        //difference between front ldrs, left - right
        int ldr_fwd;         //ave fwd speed required
        int fwd_dr_left;     //offset added to left PWM for fwd motion
        int fwd_dr_rt;       //offset added to right PWM for fwd motion
```

Program Example 16.1 Derbot 'light-seeking' program

```
//Main Program
     void main (void)
{
//Line 57 Initialise. Active bits identified. Unused bits set as outputs.
     TRISA = 0b00001011;  //ADC channels set as inputs,
                          //bits 2&5 are motor enables
     TRISB = 0b00110000;  //bits 4 & 5 are uswitch inputs
     TRISC = 0b10000000;  //bit 7 is mode switch, 1 & 2 are PWM
//Enable Timer 2, with pre- and post-scalers divide-by-1
     OpenTimer2  (TIMER_INT_OFF & T2_PS_1_1 & T2_POST_1_1);
     OpenPWM1 (0xFF);      //Enable PWM1 and set period
     OpenPWM2 (0xFF);      //Enable PWM2 and set period
//Enable ADC. Port A bits 0,1,3 are analog input, internal reference,
       //right justify result
OpenADC(ADC_FOSC_8 & ADC_RIGHT_JUST & ADC_3ANA_0REF,ADC_CH0 & ADC_INT_OFF);
//Switch all outputs off
     PORTA = PORTB = PORTC = 0;
//call diagnostic function
     diagnostic();
//enable motors at idle speed
     CCPR1L = CCPR2L = 0x80;
     PORTAbits.RA5 = 1;
     PORTAbits.RA2 = 1;


//*********************************************************************
//this is main loop.
//*********************************************************************
while (1)
   {
//Line 83 check first for collisions
     if (PORTBbits.RB4 == 0) //Test right uswitch
     rev_left ();
     if (PORTBbits.RB5 == 0) //Test left uswitch
     rev_rt ();
//Read and store all ldr values
     //left channel
     SetChanADC (ADC_CH0);
     Delay10TCYx (2);             //delay for 20us approx acquisition time
     ConvertADC();
     while (BusyADC());           //wait for conversion to complete
     ldr_left = ReadADC()&0x03FF;  // read it, AND out unwanted bits
     ldr_left = 1024 - ldr_left;   //reverse polarity
     //Line 96  right channel
     SetChanADC (ADC_CH1);
     Delay10KTCYx (2);            //delay for 20us approx acquisition time
     ConvertADC();
     while (BusyADC());
     ldr_rt = ReadADC()&0x03FF;   // read it, AND out unwanted bits
     ldr_rt = 1024 - ldr_rt;      //reverse polarity
     //rear channel
     SetChanADC (ADC_CH3);
     Delay10KTCYx (2);            //delay for 20us approx acquisition time
     ConvertADC();
```

Program Example 16.1 Continued

```
      while (BusyADC());
      ldr_rear = ReadADC()&0x03FF;   // read it, AND out unwanted bits
      ldr_rear = 1024 - ldr_rear;    //reverse polarity
//Line 110 Compute some intermediate variables
      ldr_diff = (ldr_left - ldr_rt);  //difference between ldrs
      ldr_ave = (ldr_left + ldr_rt);   //average front two ldrs
      ldr_ave = (ldr_ave>>1);          //divide this +ve no. by 2
      ldr_fwd = ldr_ave - ldr_rear;  //fr. to back difference, for fwd speed
      if (ldr_fwd < 0) ldr_fwd = 0;    //set minimum value
//Line 116 determine action, by comparing LDR readings
      if (ldr_left > ldr_rt)
      {if (ldr_left > ldr_rear)
            fwd_left();              //ldr_left is brightest, go forward left
        else rotate_left ();        //rear is brightest, rotate towards light
      }
      else
      {if (ldr_rt > ldr_rear)
            fwd_rt();               //ldr_rt is brightest, go forward left
            else rotate_rt ();
      }
      Delay10KTCYx (10);
   }            //end of while
 }              //end of main


/****************************************************************
Movement Functions
One of these four functions selected every loop iteration
****************************************************************/
/*light is front left, hence move forward left. Algorithm is:
fwd drive left = ldr_fwd - ldr_diff, fwd drive right = ldr_diff + ldr_fwd*/
      void fwd_left (void)
      {
      fwd_dr_left = ldr_fwd - ldr_diff;
      if (fwd_dr_left < 0) fwd_dr_left = 0;  //set to zero if -ve
      fwd_dr_left = fwd_dr_left>>1;    //rotate right to scale down value
      if (fwd_dr_left > 127) fwd_dr_left = 127;  //limit maximum value
      fwd_dr_rt = ldr_fwd + ldr_diff;
      if (fwd_dr_rt < 0) fwd_dr_rt = 0;        //set to zero if -ve
      fwd_dr_rt = fwd_dr_rt>>1;         //rotate right to scale down value
      if (fwd_dr_rt >127) fwd_dr_rt = 127;    //limit maximum value
      CCPR1L = 0x80 + fwd_dr_rt;         //set right motor, which is greater
      CCPR2L = 0x80 + fwd_dr_left;       //set left motor, which is less
      }
/*light is front right, hence move forward right. Algorithm is:
fwd drive right = ldr_fwd + ldr_diff, fwd drive left = ldr_fwd - ldr_diff
(noting "polarity" of ldr_diff*/
      void fwd_rt (void)
      {
      fwd_dr_rt = ldr_fwd + ldr_diff;
      if (fwd_dr_rt < 0) fwd_dr_rt = 0;        //set to zero if -ve
      fwd_dr_rt = fwd_dr_rt>>1;         //rotate right to scale down value
      if (fwd_dr_rt > 127) fwd_dr_rt = 127;    //limit maximum value
      fwd_dr_left = ldr_fwd - ldr_diff;
```

Program Example 16.1 Continued

```
      if (fwd_dr_left < 0) fwd_dr_left = 0;    //set to zero if -ve
      fwd_dr_left = fwd_dr_left>>1;        //rotate right to scale down value
      if (fwd_dr_left >127) fwd_dr_left = 127;   //limit maximum value
      CCPR1L = 0x80 + fwd_dr_rt;        //set right motor, which is less
      CCPR2L = 0x80 + fwd_dr_left;      //set left motor, which is greater
      }

//fixed speed left rotation (light is at rear left)
      void rotate_left (void)
      {rtmot_twd ();
      leftmot_rev ();
      }

//fixed speed right rotation (light is at rear right)
      void rotate_rt (void)
      {leftmot_fwd ();
      rtmot_rev ();
      }
/*****************************************************************
Motor Drive Functions
*****************************************************************/
...
(same functions as Program Example 15.3 - blind navigation program)
...
```

Program Example 16.1 Continued

With this increase in program complexity, it is useful to pause to look at the program layout. There are varied practices adopted for this; the goal of each is program clarity. The practice adopted here, where there are nested code blocks, is for the opening brace of each nested block to be indented one step to the right. Notice first the opening brace of the **main** function, which is fully left. Then note the position of the opening brace of the major **while** loop starting at line 83. This is indented one tab to the right. Further code blocks within this are indented further right again, with matching braces always lying directly below each other. The end of the **while** loop and end of **main** are both commented, and in this layout fairly easy to see. The position of a 'fully left' pair of braces in this book is generally reserved for the **main** function. Therefore, the braces for the smaller functions are indented, but still paired vertically. Major comments are given a full line or lines, while smaller comments are just placed to the right of the code.

16.2.1   The light-seeking program structure

This program, with around 20 functions, illustrates how functions proliferate as complexity increases. Around half are from the libraries, while the rest are user-defined. Function prototypes for the library functions appear in the associated header files and are thus not needed in the source file. They are, however, copied in as comments here, just for information. Prototypes for the others are included for real.

The **main** function starts with initialisation followed by the diagnostic function, in the usual way. The program then enters a continuous loop, making use of the **while** keyword. Within this it first tests the front microswitches, responding if necessary in a way seen in Program Example 15.3. It then follows broadly the flow diagram of Figure 11.13.

16.2.2 Use of the ADC

Chapter 11, Section 11.3, describes the ADC structure for both the 16F873A and the 18F242. Table 14.1 shows all six of the functions associated with the ADC. This program uses all but one of them. The most complicated of these is **OpenADC**, whose function prototype is quoted early in the program and repeated here:

```
void OpenADC (unsigned char, unsigned char);
```

The two unsigned characters required as arguments are made up of bit masks, in a way similar to that described for the **OpenTimer2( )** function in Section 15.5.1 of Chapter 15. The options for the ADC are more complex than for Timer 2. They are not reproduced here, but can easily be looked up in Ref. 14.3. The function is used here as follows:

```
//Enable ADC. Port A bits 0,1,3 are analog input, internal reference,
     //right justify result
OpenADC(ADC_FOSC_8 & ADC_RIGHT_JUST & ADC_3ANA_0REF,ADC_CH0 & ADC_INT_OFF);
```

As the comment indicates, this function makes the following settings:

- The ADC conversion speed is set, determined by the oscillator which drives it. With a minimum conversion clock cycle time (T_{AD}) of 1.6 μs, the 4 MHz internal oscillator is divided by 8 to give a T_{AD} of 2.0 μs.
- The result is right justified, as seen in Figure 11.9.
- Three channels are used for input, and the voltage reference is internal. This sets up the lower 4 bits in register **ADCON1** to be 0100, as seen in Figure 11.8.
- Channel 0 is currently selected. This has no impact on the program that follows.
- The interrupt is turned off.

The use of this function effectively replaces these lines in Assembler:

```
        bsf     status,rp0
...
        movlw   B'10000100'   ;select port A bits 0,1,3 for analog input
        movwf   adcon1            ;right justify result
        bcf     status,rp0
        movlw   B'01000001'   ;set up ADC: clock Fosc/8, switch ADC on but not
                                                  ;converting,
        movwf   adcon0        ;input channel selection currently irrelevant
```

The conversion process which follows reflects the data acquisition flow diagram of Figure 11.5. It is repeated for each LDR and appears as reproduced here, in this case for channel 0:

```
        SetChanADC (ADC_CH0);
        Delay10TCYx (2);               //delay for 20us approx acquisition time
```

```
ConvertADC();
while (BusyADC());          //wait for conversion to complete
ldr_left = ReadADC()&0x03FF;  // read it, AND out unwanted bits
ldr_left = 1024 - ldr_left;   //reverse polarity
```

Here the functions used are fairly straightforward: **SetchanADC( )** selects the input channel, after which $20\,\mu s$ ensures adequate acquisition time. The conversion is initiated with **ConvertADC( )**. The **BusyADC( )** function tests whether the conversion is still ongoing and returns a 1 if it is. Recall from Section 15.3.3 of Chapter 15 that a function call acts as its return value. Here the function call is embedded within the **while** construct. Its use causes the program to loop at that point until the conversion is complete.

The result of the conversion is then read with the **ReadADC( )** function. The function is embedded within an expression, acting as its return value, which is the ADC result. This is a 10-bit value, with consequent maximum possible value of 1023_D. The result is ANDed with $03FF_H$ to ensure that no higher bits are present in the reading. While this is probably an unnecessary move, it helps to illustrate how the function can be placed within an expression. Due to the hardware configuration of the LDRs, the magnitude of the result *decreases* with increasing light intensity. To make the subsequent arithmetic simple, the result is then subtracted from its 10-bit maximum, 1024_D, so that the value of **ldr_left** increases with increasing light.

16.2.3   Further use of *if–else*

Following the data conversions, and having undertaken some intermediate calculations, the program then determines, from line 116, which (out of four) possible paths of action to take. In doing this, it is following the flow diagram of Figure 11.13. If one of the front LDRs is brightest, it will veer in that direction. If the rear LDR is brightest, it will rotate, either clockwise or anticlockwise. These decisions are made with three **if–else** tests, repeated below. It can be seen that the first **if** has an **if–else** nested within it, as does the corresponding **else**.

```
//determine action, by comparing LDR readings
     if (ldr_left > ldr_rt)
     {if (ldr_left > ldr_rear)
          fwd_left();          //ldr_left is brightest, go forward left
       else rotate_left ();     //rear is brightest, rotate towards light
     }
     else
     {if (ldr_rt > ldr_rear)
          fwd_rt();            //ldr_rt is brightest, go forward right
          else rotate_rt ();
     }
```

16.2.4   Simulating the light-seeking program

It is interesting to simulate this program with the MPLAB® simulator, whether or not you have a Derbot. This allows examination of the branching that is used and the operations that are applied. Set up the simulation with the following steps.

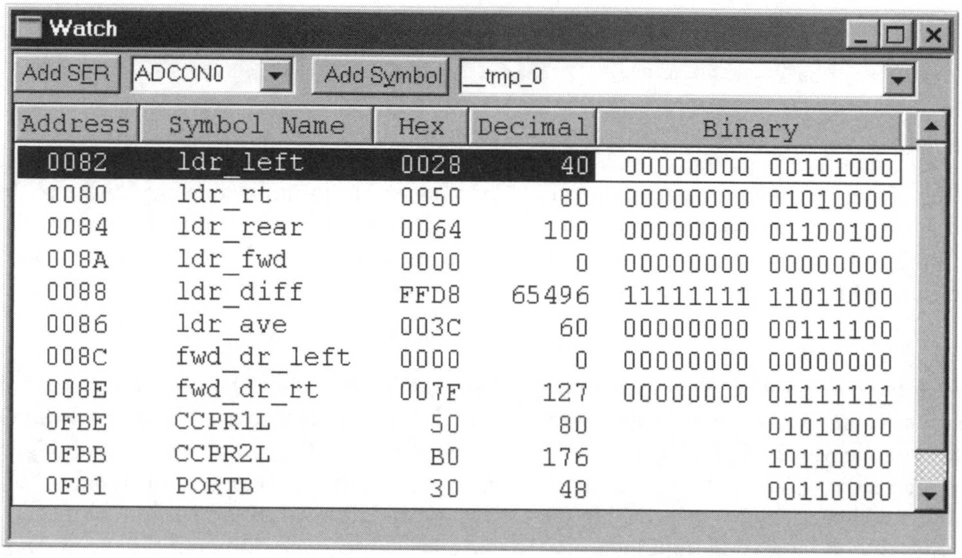

Figure 16.1 Watch window for 'light-seeking' simulation

- In MPLAB, select the simulator with **Debugger > Select Tool > MPLAB SIM.**
- Open a Watch window and select the variables shown in Figure 16.1 for display.
- Set up a stimulus controller and simulate inputs for the microswitch inputs, RB5 and RB4.
- Using **Debugger > Settings > Osc/Trace**, set the Processor Frequency to 4 MHz. This is only important here because if set too high, the simulator is clever enough to recognise and flag that the converter cycle time, T_{AD} (Chapter 11, Section 11.3.2), is too short.
- Insert breakpoints at lines 84 and 111.

Run the program to the first breakpoint. With the stimulus controller, 'fire' bits 4 and 5 of Port B to 1, to simulate the microswitches being inactive. Now run the program again to the breakpoint at line 111. The Output window will warn 'No stimulus file attached to ADRESL for A/D'. Don't worry about this.

Now enter in the Watch window the first set of trial values, for **ldr_left**, **ldr_right** and **ldr_rear**, from Table 16.1. (Enter the decimal value for each and the other columns will follow.) From here, single step through the program, using the 'Step Into' Debugger button. See how the correct function is chosen, **fwd_left**, in this first case, and in the Watch window see the outcome of the calculations made. The results for each set of input figures are shown in the table. When you reach the first set of results, run the program back to the first breakpoint and then down to the second, and enter another set of results. Continue to loop in this way, entering different trial results and observing the program response in terms of its calculations and looping.

Table 16.1 Trial values in the 'light-seeking' simulation

Condition	Input values from ADC (decimal)			Resulting action and values (decimal)		
	ldr_left	ldr_rt	ldr_rear	Function selected	fwd_dr_left	fwd_dr_rt
left>right>rear	0100	0080	0040	fwd_left	015	035
right>left>rear	0080	0100	0040	fwd_rt	035	015
left≫right>rear	0200	0040	0020	fwd_left	00	127
rear>left>right	0080	0040	0100	rotate_left	–	–
rear>right>left	0040	0080	0100	rotate_right	–	–

If you have a Derbot AGV, this is an entertaining program to run. It is enhanced later in this chapter by the addition of a display capability.

16.3   Pointers, arrays and strings

We saw in Chapter 14 how data elements were declared and used. Much data exists, however, in the form of *sets* of variables, for example in a string of data being prepared to send to a display. In this section we look therefore at how sets of data can be defined and used, introducing *arrays* and *strings*, and the *pointers* which access them.

16.3.1   Pointers

Instead of specifying a variable by name, we can specify its address. In C terminology such an address is called a *pointer*. A pointer can be loaded with the address of a variable by using the unary operator '&', like this:

```
my_pointer = &fred;
```

This loads the variable **my_pointer** with the *address* of the variable **fred**; **my_pointer** is then said to *point* to **fred**.

Doing things the other way round, the value of the variable pointed to by a pointer can be specified by prefixing the pointer with the '*' operator. For example, *__my_pointer__ can be read as 'the value pointed to by **my_pointer**'. The * operator, used in this way, is sometimes called the *dereferencing* or *indirection* operator. The indirect value of a pointer, for example *__my_pointer__, can be used in an expression just like any other variable.

A pointer is declared by the data type it points to. Thus,

```
int *my_pointer;
```

indicates that **my_pointer** points to a variable of type **int**.

Table 16.2 C18 pointer sizes

Pointer type	Pointer size
Data memory	16-bit
Near program memory	16-bit
Far program memory	24-bit

Reflecting the Harvard memory structure of the PIC® microcontroller, the C18 compiler allows pointers to be set up both for program memory and for data memory. The resulting size of the pointer is shown in Table 16.2. The meaning of *near* and *far* in this context is described in Section 17.6.5 of Chapter 17.

16.3.2   Arrays

An *array* in C is defined as a set of data elements, each of which has the same type. Any data type can be used. Array elements are stored in consecutive memory locations. An array is declared with its name, data type and (optionally) the number of elements in the array. For example, the declaration

```
unsigned char message1[8];
```

defines an array called **message1**, containing eight characters. It is easy to recognise arrays by the use of the square brackets which follow the name.

Elements within an array can be accessed with an index, starting with value 0. Therefore, for the above array, **message[0]** selects the first element and **message[7]** the last. The index can be replaced by any variable which represents the required value.

Importantly, *the name of an array is set equal to the address of the initial element*. Therefore, when an array name is passed in a function, what is passed is this address.

16.3.3   Using pointers with arrays

Pointers can be set up to point at arrays using the operators just introduced. For example, the statement

```
ADC_val_ptr = &ADC_val_BCD[0];
```

assigns the address of the first element of the array **ADC_val_BCD** to the pointer **ADC_val_ptr**. This assignment, if desired, can be combined with the pointer declaration itself, as follows:

```
int *ADC_val_ptr = &ADC_val_BCD[0];
```

Following this assignment, the value of **ADC_val_BCD[0]** (the first element in the array) is equal to ***ADC_val_ptr** (the value pointed to by **ADC_val_ptr**). It follows that the values of **ADC_val_BCD[1]** and ***(ADC_val_ptr + 1)** are also equal, as are **ADC_val_BCD[2]** and ***(ADC_val_ptr + 2)**, and so on.

Because the name of the array is also its first address, the above assignment could be written as

$$\texttt{ADC\_val\_ptr = ADC\_val\_BCD;}$$

It follows further that **ADC_val_BCD[i]** is equivalent to ***(ADC_val_BCD + i)**, the name of the array effectively now being used as the pointer.

A further notational development is that the pointer can be used with an index. Thus, **ADC_val_ptr[i]** is the same as ***(ADC_val_ptr + i)**. It seems like the pointer and the array name are almost interchangeable, and that the pointer may be barely necessary. Remember, however, that the array name is a constant – the array is fixed in memory, whereas the pointer is a genuine variable, with all the properties of the variable.

16.3.4   Strings

A *string* is a particular type of array, made of characters of type **char**, ending with the null character '\0'. The size of the string array must therefore be at least 1 byte more than the string itself, to accommodate this final character.

16.3.5   An example program: using pointers, arrays and strings

The slightly improbable Program Example 16.2 demonstrates some of the ways in which arrays, strings and pointers are used. It has nothing to do with embedded systems and is for simulation only. An array of characters, called **list**, is declared, with each value preset to zero. A string, called **item1**, is declared, containing the word 'Apple'. A similar one is declared, with the word 'Pear'. A *single* character is declared, with the name **plural**. Notice that strings are contained within double inverted commas and characters within a single. Two pointers are defined. The first, **pntr1**, is set to the address of the first element of the **item1** string. The second, **pntr2**, is set to the address of the variable **number**.

```
/*********************************************************************
Strings&chars_c
This program, for simulation only, explores characters, strings and pointers.
Files c018i.o and p18f242.lib are included by the Linker Script.
TJW 29.11.05                                          Tested 30.11.05
*********************************************************************/
//Configuration bits need not be set
#include <p18F242.h>

//data type definitions
      char counter;                    //index for list
      char list[8] = {0,0,0,0,0,0,0,0};
      char number = 0;
      char item1[] = "Apple";
      char item2[] = "Pear";
      char plural = 's';
      char *pntr1 = &item1[0];   //Pointer to "Apple" string
      char *pntr2 = &number;
```

Program Example 16.2 Working with pointers, arrays and strings

```
// Main function
      void main (void)
{
loop:
//Do apples
      counter = 0;                    //set index to list
      list[counter] = number;        //indicate number of items
      while (item1[counter] != 0) //indicate type of item
      {list[counter+1] = *(pntr1+counter);
      counter = counter + 1;
      }
      if (number >1)list[counter+1] = plural;  //set the item to plural
      else list[counter+1] = 0x20;  //return item to single, with ASCII space

//Do pears
      counter = 0;                    //set index to list
      list[counter] = number;  //indicate number of items
      while (item2[counter] != 0)  //indicate type of item
      {list[counter+1] = *(item2+counter);
      counter = counter + 1;
      }
      if (number >1)list[counter+1] = plural;  //set the item to plural
      else list[counter+1] = 0x20;  //return item to single, with ASCII space
      counter = counter + 1;
      list[counter+1] = 0x20;         //insert ASCII space
      number = *pntr2 + 1;
      if (number > 9)number = 0;
      goto loop;
}
```

Program Example 16.2 Continued

16.3.6   A word on evaluating the **while** condition

The **while** keyword was introduced in Section 14.2.8 of Chapter 14. While we have used it a number of times for endless loops, this is the first time that we use it in a conditional sense, in the line:

```
      while (item1[counter] != 0)  //indicate type of item
```

There is a similar usage a few lines later. Here we have spelled out the condition for looping, that the array element identified by the value of **counter** must not be equal to zero. We can in fact simplify this, as was explained in Chapter 14. The loop will continue executing as long as the conditional expression is 'true' (i.e. non-zero) and will be left when the expression is zero. Therefore, a simpler way of writing the **while** statement is:

```
      while (item1[counter])  //indicate type of item
```

This will have an effect in the program identical to the original version of the line.

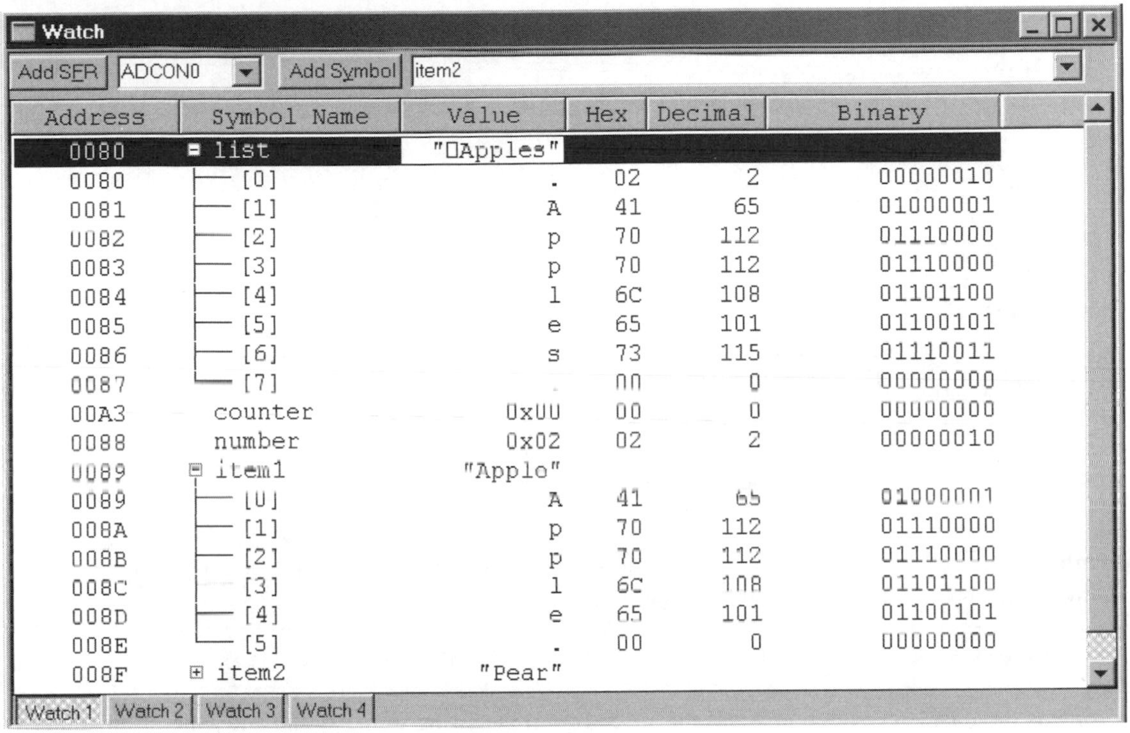

Address	Symbol Name	Value	Hex	Decimal	Binary
0080	list	"□Apples"			
0080	[0]	.	02	2	00000010
0081	[1]	A	41	65	01000001
0082	[2]	p	70	112	01110000
0083	[3]	p	70	112	01110000
0084	[4]	l	6C	108	01101100
0085	[5]	e	65	101	01100101
0086	[6]	s	73	115	01110011
0087	[7]	.	00	0	00000000
00A3	counter	0x00	00	0	00000000
0088	number	0x02	02	2	00000010
0089	item1	"Apple"			
0089	[0]	A	41	65	01000001
008A	[1]	p	70	112	01110000
008B	[2]	p	70	112	01110000
008C	[3]	l	6C	108	01101100
008D	[4]	e	65	101	01100101
008E	[5]	.	00	0	00000000
008F	item2	"Pear"			

Figure 16.2 Watch window for simulation of Program Example 16.2

16.3.7 Simulating the program example

If you have the C18 compiler, create a project round Program Example 16.2 (the source code is on the book CD), build it and simulate with the MPLAB simulator.

Open a Watch window, showing the variables seen in Figure 16.2. This gives an opportunity to see how the simulator handles the display of arrays. The figure shows that they can be displayed by name only, as with **item2**, or they can be expanded to a full listing, as with **list** and **item1**. Notice that **item1** has been created with six elements, the last one being a null character.

Set breakpoints at the locations shown in Figure 16.3 and run the program down to the first. Then single step carefully through the program, noting the action of each line. There are two program sections, identified by the comments 'Do apples' and 'Do pears'. The action of each section is essentially to populate the array **list** with a number and the type of commodity, i.e. apple or pear.

After initialising **counter**, the number is transferred to the first location in the array, with the line

```
list[counter] = number;                //indicate number of items
```

This uses the simplest form of array accessing, where the number contained in the square brackets indicates the array element. In this case, the first element of the array is assigned the value **number**.

```
//Do apples
    counter = 0;                        //set index to list
    list[counter] = number;             //indicate number of items
    while (item1[counter] != 0)         //indicate type of item
    {list[counter+1] = *(pntr1+counter);
    counter = counter + 1;
    }
    if (number >1)list[counter+1] = plural; //set the item to plural
    else list[counter+1] = 0;           //return item to single

//Do pears
    counter = 0;                        //set index to list
    list[counter] = number;             //indicate number of items
    while (item2[counter] != 0)         //indicate type of item
    {list[counter+1] = *(item2+counter);
    counter = counter + 1;
```

Figure 16.3 Suggested breakpoint locations for Program Example 16.2

A **while** loop is now set up. A string is known to terminate with a null character, so a test is made for this. The 'not equal to' operator is used in this line:

```
while (item1[counter] != 0)                 //indicate type of item
```

As long as the character is not null, it is transferred from the string **item1** to the array **list**, in this line:

```
{list[counter+1] = *(pntr1+counter);
```

The list element is again determined with an index in the squared bracket. Now the value of **counter** is offset by 1, as its first element is already populated. The value assigned to it is the indirect value of **pntr1**, offset by the value of **counter**.

A test follows for whether the number of apples is greater than 1. If so, an 's' must be added to the commodity type. Otherwise, a space is inserted.

```
if (number >1)list[counter+1] = plural;  //set the item to plural
else list[counter+1] = 0x20;  //return item to single, with ASCII space
```

Notice that in the 'Pears' section of the program, the string is transferred in a different way. Now the array element in the **item2** string is determined using the array name as the base address, offset in turn by the value of **counter**. It is the indirect value of this calculated address which is transferred to the list.

```
{list[counter+1] = *(item2+counter);
```

Towards the end of the loop the value of **number** is incremented. On the right-hand side of the assignment, **number** is accessed by using the indirect value of its pointer, i.e. **number** itself.

```
number = *pntr2 + 1;
```

Table 16.3 Example I²C library functions

Function	Action
OpenI2C( )	Configures the SSP module for I²C
StartI2C( )	Generates an I²C Start condition
WriteI2C( )	Writes a single byte to the I²C
ReadI2C( )	Reads a single byte from the I²C
StopI2C( )	Generates an I²C Stop condition

Single step through this program until you understand what each line of code is doing. Then run it from breakpoint to breakpoint, and see how the contents of the **list** array update in each loop iteration.

One thing may strike you as a little odd – that the strings we have defined have found their way into data memory, as the Watch window indicates. Normally, we expect to find such strings in program memory. We return to this issue in the next chapter, when we explore how it is possible to control in which memory type strings and other constants are placed.

16.4   Using the I²C peripheral

We saw in Chapter 8 that I²C is a useful standard for serial communication, but with some complexity in use. The C18 compiler library provides us with some very useful functions, which allow most I²C functionality to be implemented in a simple and reliable way. Reference 14.3 lists no less than 15 functions for use with I²C, although some are synonyms of others. Examples are shown in Table 16.3, in the order in which they might be used.

16.4.1   An example I²C program

A simple program which uses the I²C microcontroller capability, and applies the functions of Table 16.3, is shown in Program Example 16.3. The program is intended for the Derbot AGV, and sends a single character, followed by a string, to the hand controller.

```
/*********************************************************************
Dbt_I2C_test_c
Sends single character, and then string, periodically on I2C.
Set up as Master.
Files c018i.o and p18f242.lib are included by the Linker Script.
TJW 12.11.05                                      Tested 4.12.05
*********************************************************************
Clock is 4MHz.
Configuration Word all default, except: crystal oscillator (HS),
power-up timer on, brown-out detect off, WDT off, LV Program disabled*/

        #include <p18F242.h>
        #include <i2c.h>              //header file for I2C
```

Program Example 16.3 Using I²C to send a character and string to the Derbot hand controller

```
        #include <delays.h>          //header file for delays
        #define slave_addr1 0xA4   //Adress of Derbot Hand Controller I2C node.

//Function Prototypes. Library function prototypes are found in Header file
        void diagnostic (void);

//constants & variables
        unsigned char message[] = " Derbot";
        unsigned char *i = &message[0];   //pointer to message[]
        char loop_cntr = 0;
/*********************************************************************
This is main function.
*********************************************************************/
        void main (void)
{
//Initialise Ports
        TRISA = 0b00001011;          //ADC channels set as inputs
        TRISB = 0b11001000;
        TRISC = 0b10011000;          //I2C bits are both set as ip
//Switch all outputs off
        PORTA = PORTB = PORTC = 0;
//call diagnostic function (flash leds)
        diagnostic();
//Initialise I2C
        OpenI2C (MASTER, SLEW_OFF);
        SSPADD = 0x07;               //set up 125kHz baud rate
loop:
//Send single character
        i = &message[0];
        StartI2C();                  //send start condition
        WriteI2C (slave_addr1);      //send address word, function waits until
                                     //write is complete
        loop_cntr = loop_cntr| 0x30;  //convert counter to ASCII
        WriteI2C (loop_cntr);
        loop_cntr = loop_cntr&0x0F;  //retrieve counter from ASCII
        StopI2C();                   //send stop condition
        Delay10KTCYx (100);
//Send a string
        StartI2C();                  //send start condition
        WriteI2C (slave_addr1);      //send address word, function waits until
                                     //write is complete
        while (*i)                   //Test for null character
        {
        WriteI2C (*i);
        i++;
        Delay1KTCYx (5);             //delay needed for hand controller
        }
        StopI2C();                   //send stop condition
        Delay10KTCYx (100);
        loop_cntr++;
        if      (loop_cntr == 10) loop_cntr = 0;
        goto loop;
}                                    //end of main
```

Program Example 16.3 Continued

```
//Diagnostic: switches leds on for 1s (Tcy = 1us)
void diagnostic (void)
...
(same as diagnostic function, Program Example 15.1)
...
```

Program Example 16.3 Continued

The opening lines of the program indicate which header files are to be included and defines A4$_H$ as being the address of the hand controller slave node, as described in Section 10.8 of Chapter 10. They also declare the character string which forms the message that is to be sent and declare a pointer for it.

The I^{2}C port is partially initialised with the **OpenI2C( )** function. This has two arguments, detailed in Ref. 14.3, which determine the operating mode (whether Slave or Master) and select the slew rate. Note that it is still necessary to set the baud rate by writing to the **SSPADD** register. An I^{2}C message is initiated with the **StartI2C( )** function, which puts an I^{2}C Start condition on the serial link. This is followed by the address byte being sent, being passed as an argument to the **WriteI2C( )** function. This sets the **R/W̄** bit in the transmitted word low, as seen in Figure 10.13. The ASCII version of the loop counter value is formed, which is then sent through another use of the **WriteI2C( )** function. The message is terminated with a **StopI2C( )** function.

The string is sent by techniques which are mainly familiar, although certain developments of these are applied. A new I^{2}C message is initiated with the **StartI2C( )** function, followed by another sending of the slave address. A **while** loop is then set up, with the condition

```
while (*i)
```

With **i** as the pointer to the string, *****i** indicates an element of the string. Remembering that the last string element is always a null character, the loop will be repeated until the end is reached, whereupon it will be exited.

16.4.2   Use of ++ and −− operators

Towards the end of the program example we see the ++ operator applied to both the index **i** and to **loop_cntr**. The operator, which causes an increment to the variable to which it is applied, can be seen in Table A6.5. Therefore,

i++;   is apparently the same as   i = i + 1;

There are, however, some important differences. This operator can be placed *before* the variable, in which case it indicates *pre-increment*. If it is placed after the variable, it indicates *post-increment*. The difference is not of significance in this program. It can, however, be understood by looking at these two examples:

```
index = 4;                          index = 4;
new_val = index++;                  new_val = ++index;
```

In the example on the left, **new_val** takes the value 4 and **index** is then incremented. In the example on the right, **index** is (pre-)incremented and **new_val** then takes the value 5.

The decrement operator, $--$, is applied in exactly the same way.

16.5   Formatting data for display

We have seen now how characters, and character strings, can be sent over an I^2C link to an LCD display. What we have not yet done is generate some meaningful data to be displayed. This section develops the Derbot 'light-seeking' program (Program Example 16.1), so that the values read by the light-dependent resistors are shown on the hand controller display.

16.5.1   Overview of example program

In Program Example 16.1, values are read from the ADC as 10-bit numbers. To convert this to a character string we need convert the value to BCD and then to ASCII. We did this in Assembler, in Program Example 11.2, and very laborious it proved to be. Can C work better for us?

The answer to this question is a resounding 'yes'. C has many functions that are designed to convert data from one format to another. A few examples are shown in Table 14.3. What we need for this application is a function that will take our 10-bit ADC output and convert it into a character string. This is very conveniently provided by the function **itoa** (read this as i-to-a – integer to ASCII), seen in the table.

Program Example 16.4 shows sections of the program **light_seek_&_disp**, which appears in full on the book CD. The sections shown are extensions to Program Example 16.1. Two new functions have been written. One, **disp_int( )**, formats the ADC output value and sends the resulting character string on the I^2C link. The other, **send_space( )**, simply sends a series of spaces to the display, to optimise information layout on the LCD.

The data is displayed only once every 10 iterations of the main loop. To do it any faster makes the data flicker in an annoying way. A loop counter, **loop_cntr**, has been inserted in the main loop and is incremented for every loop iteration. When data is to be displayed, it can be seen that **disp_int( )** is called three times, once for each of the LDRs. These two results will be placed on the first line of the two-line display. Spaces are sent before and after the display of the rear LDR, to centre it on the second display line.

```
...
    #include <stdlib.h>        //for itoa function
    #include <string.h>        //for strlen function
...
```
(this program section is placed within the main program loop, after the ADC values have been read)

Program Example 16.4 Formatting data for LCD display

```
//Display ldr values every 10 loops
        if (loop_cntr == 10)
        {
        disp_int (ldr_left);        //display left ldr value on lcd
        disp_int (ldr_rt);          //display right ldr value on lcd
        send_space (2);             //centre rear display
        disp_int (ldr_rear);        //display rear ldr value on lcd
        send_space (2);             //fills second line, forcing line feed
        loop_cntr = 0;
        }
...
...
/**********************************************
Display Functions
**********************************************/
//Converts an integer to string, and sends to lcd via I2C, filling with spaces to
ensure 4 digits are always sent
        void disp_int (int op_int)
{
        char disp_val[5];           //will hold the string representation of
                                        //any ldr value
        char *disp_val_ptr;         //pointer to disp_val[]
        char space_no;              //number of spaces to be inserted
        disp_val_ptr = &disp_val[0];
        itoa (op_int, disp_val_ptr);//first convert to a BCD string
        space_no = 4 - strlen(disp_val_ptr); //find how many spaces needed
//Now send the message
        StartI2C();                 //send start condition
        WriteI2C (slave_addr1);     //send address word
        while (space_no)            //fill up with leading spaces
        {WriteI2C (' ');
        Delay1KTCYx(5);             //little delay needed by hand controller
        space_no--;
        }
//send the string
        while (*disp_val_ptr)
        {
        WriteI2C (*disp_val_ptr);
        disp_val_ptr++;
        Delay1KTCYx(5);             //delay needed for hand controller
        }
        StopI2C();                  //send stop condition
}

//Sends space to lcd via I2C
        void send_space (char space_no)
{
        StartI2C();                 //send start condition
        WriteI2C (slave_addr1);     //send address word,
//send space
        while (space_no)
        {
        WriteI2C (' ');
```

Program Example 16.4 Continued

```
      Delay1KTCYx(5);              //delay needed for hand controller
      space_no--;
      }
      StopI2C();                   //send stop condition
}
```

Program Example 16.4 Continued

16.5.2   Using library functions for data formatting

Let's take a close look at the function **disp_int( )**, in Program Example 16.4, as it includes some important features. Notice first that an array, a pointer and a character variable are declared at the beginning. These will exist only for the duration of the function. The array is set for five locations, as the maximum digit count from a 10-bit number will be 4 (1023_D) and a fifth byte is needed for the terminating null character. The argument transferred to the function is labelled **op_int**. The pointer is initialised to point to the start of the array.

The **itoa( )** function is then called. This has the function prototype:

```
      char * itoa (int value, char * string);
```

where **value** is the integer to be converted, **string** is the string where the result is to be placed and the return value is the pointer to the string. Its implementation is in the line:

```
      itoa (op_int, disp_val_ptr);//first convert to a BCD string
```

It can be seen that **op_int** is the variable to be converted and that the resulting string is to be placed, by use of its previously declared pointer **disp_val_ptr**, in the array **disp_val**.

It would seem that this character string could be sent straight away to the display. It may, however, be of any length from one to four digits. To ensure that it always occupies the same location on the display, it is necessary first to find out how long it is. This is done with the **strlen( )** function, found in the general software library. The function measures the length of a string and returns its value. Its prototype is:

```
      size_t strlen(const char *string);
```

Here **string** is the string to be measured and the return value, **size_t**, contains the string length. The function is used to compute a value for **space_no**, which holds the number of spaces which must be sent to make up the value to four digits. These are sent before the data itself. The string is then sent and the function terminates.

16.5.3   Program evaluation

This program combines the function of the earlier light meter (Program Example 11.2) and light-seeking programs in one. It can be simulated in a similar way to that described in Section 16.2.4, inserting trial values into **ldr_left**, **ldr_rt** and **ldr_rear**, observing their conversion to a string, and the string length being tested.

It is also a very satisfying program to run on the Derbot, as both its action and the data displayed provide a very explicit demonstration of what the program is doing.

Summary

This chapter aimed to show how C can be applied in acquiring and using integer data in embedded systems. The main points were:

- The 18FXX2 ADC and the I^{2}C serial port can be driven in a straightforward way using library functions.
- Arrays and strings, with their associated pointers, provide powerful ways of dealing with sets of data. Some care is needed in understanding the way C deals with these.
- There are some useful library functions for manipulating data strings. These are particularly useful for formatting data in readiness for display.

17
More C and the wider C environment

We have now reached a stage where we should have some level of confidence in writing simple C programs for the PIC® microcontroller. There still remain gaps in our knowledge, however. One of these is the use of interrupts. An exploration of this takes us to the very boundary of how C is used and actually makes us step outside the language altogether. Furthermore, as our programs become bigger, it is useful to know about the wider environment in which we are programming, for example those other files which we link in or include. We have used these so far with little knowledge of how they work.

The aim of this chapter is therefore twofold. One is to develop knowledge of language aspects which enable closer working with the hardware. This includes use of assembler inserts and interrupts. The other is to expand our knowledge of the wider context within which C operates. To allow this to happen, we will need to develop our knowledge of certain other aspects of C itself.

On completion of this chapter you should have developed a good understanding of:

- The use of assembler inserts
- The use of interrupts
- Further aspects of data definition and storage, and how memory usage can be controlled
- The files usually linked in to an application, including header and start-up files
- The Linker and Linker Script.

As in other chapters, examples are used as widely as possible.

17.1 The main idea – more C and the wider C environment

While this chapter is meant to develop your expertise further in C, the one thing it will probably not do much is develop your skill in writing C code itself. Instead, we meet C at its limits, confronting the need to work very close to the hardware through interrupts or out-of-the-ordinary memory maps. We will also look at those files outside the source code, as well as the Linker, which pulls them all together.

To help relate these different elements, a number of examples will be taken from the file which starts all the C program examples in this book, the **c018i.c** start-up file. This contains a number of interesting programming features. Through studying these, we will at the same time learn something about this important program component. It is not simple code, however, and you don't need to understand it all. It is suggested that you print out the source version of it and have it to hand as you work through the chapter. It is a comparatively short piece of code – a print-out will require about two and a half pages. To find it, simulate any C file in this book with MPLAB® and it automatically pops up on the screen as the simulation starts. Otherwise, find it in the **mcc18\src\traditional\startup** folder of the C18 installation.

17.2   Assembler inserts

Despite the usefulness of C in the embedded environment, there remain times when it is still better to use Assembler. These times include:

- When certain instructions cause very processor-specific actions, for which C simply has no equivalent – in the PIC world these include the instructions **SLEEP** or **CLRWDT**
- Where the timing requirements of program execution are very specific and the programmer needs to have direct control over how a certain program section is written
- Where a section of program must execute very fast and the programmer wishes to write it in the most efficient way possible.

It is therefore useful to be able to switch to assembler programming, when necessary, within a C program. This is called *in-line assembler*. It presents an opportunity distinct from writing a whole program section in Assembler and linking it into the main program through the build process, as illustrated in Figure 14.1.

The MPLAB C18 compiler allows assembler inserts to be placed in a C program. These can range from a single instruction to a whole block. One might expect the C compiler to refer blocks of in-line assembler code back to MPASM™, the regular MPLAB assembler. However, it doesn't. The C18 compiler has its own internal assembler, which is applied to these sections of in-line code.

The C18 in-line assembler differs from the regular MPLAB Assembler in a number of significant ways. The major differences are:

- The assembler section must be contained between the identifiers **_asm** and **_endasm**
- Assembler directives may not be used
- Comments must be in C or C++ format
- No operand defaults are applied; operands must be fully specified
- Full text versions of Table Read/Write instructions (as seen at the end of Table A5.1) must be used
- The default radix is decimal
- Literals are specified using C radix notation
- Labels must end with a colon.

An example of in-line assembler code is given in Program Example 17.1. While at first glance it looks like a regular piece of code, there are in fact no less than six of the characteristics of in-line assembler coding applied. All are labelled and relate directly to the list above. The example is taken from the start-up file **c018i.c**.

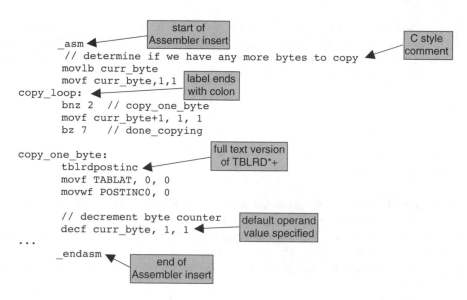

Program Example 17.1 A fragment of the start-up code **c018i.c**

In-line assembler should be used with some care, especially if the block of Assembler is any length. As we know, Assembler imposes less discipline than C. It is easy to step outside this discipline, even unknowingly, while writing a section in Assembler, and in so doing corrupt the structure of the host program. As a beginner it is unwise to write inserts that impact on variables or functions that are declared in the C file. If a big block of Assembler is to be written, it is better to write it as a separate file, assemble it with the MPASM assembler and link it into the main program. This will help to ensure that order is maintained in memory mapping, usage of variables and calling of functions.

17.3   Controlling memory allocation

One of the benefits of working with a high-level language should be that the programmer does not need to worry about the memory map or how memory is allocated. The C compiler will, if we want it to, look after almost all this. Somewhere, of course, the compiler needs to be given information about the memory of the computer for which it is compiling. This is hidden in the Linker Script, which is described in Section 17.10.

There are situations in the embedded environment, however, when we want to take back control of memory allocation. These include those very hardware-specific activities, like dealing with interrupts or configuration bits, as well as the broader issues of optimising use of the memory map.

The techniques that are available in the C18 compiler for control of memory allocation are varied. Some are complex and should only be used by experienced programmers. A few need to be recognised by all. These are now explored.

17.3.1   Memory allocation pragmas

We have already (in Chapter 14, Section 14.2) come across the concept of the preprocessor directive. A special type of directive is the **#pragma**. This allows C to be customised by a specific compiler – every time **#pragma** is used, the statement which follows it is specific for that compiler.

The C18 compiler has four pragmas to control memory allocation. These change the *section* (i.e. a specifically identified memory block) into which the compiler puts data. They are shown in Table 17.1. The full format of each allows for a number of different options, given in full in Ref. 14.2. These can be quite complex, so we only go into limited detail here.

The **#pragma** met most often is the first in the table, which has the general format of:

#pragma code (*section name*) (*=address*)

This acts in a way like the **org** directive in Assembler, specifying – if it is needed – where program code should be placed in memory. Both the terms in brackets are optional. It is one of these pragmas which starts every C18 program we write, coming at the beginning of the **c018i.c** start-up file. The opening lines are shown here:

```
#pragma code _entry_scn=0x00
void _entry (void)
{
_asm goto _startup _endasm
}
#pragma code _startup_scn
void
_startup (void)
{
...
```

Here the specified format for **#pragma code** is applied with a section name of **_entry_scn** and the address of 0x00. A few lines down the pragma is used again, with a section name of **_startup_scn**, although this time an address is not specified. This allows the Linker to set the address. Both **_entry_scn** and **_startup_scn** are names reserved by C18, the former to locate the reset vector and the latter to hold the start-up code.

17.3.2   Setting the Configuration Words

With the large number of configuration bits in the PIC 18 Series microcontrollers, it is attractive to set them within the program. Version 3.0 of the C18 compiler allows Configuration Words to be set simply,

Table 17.1 Pragmas for memory allocation

Pragma	Effect
# pragma code ...	Locates program code in program memory
# pragma romdata ...	Locates data in program memory
# pragma udata ...	Locates uninitialised user variables in data memory
# pragma idata ...	Locates initialised user variables in data memory

using the **#pragma config** directive. The actual settings to be used are processor specific. The options, even for one microcontroller, are surprisingly extensive. Therefore they are not reproduced here, but can be found for all 18 Series PIC microcontrollers in Ref. 17.1.

Program Example 17.2 illustrates settings appropriate for the Derbot-18. It is useful to look back at Table 12.4 to check the configuration options. One **#pragma config** line is used per Configuration Word, with the format of the directive drawn from Ref. 17.1. Any configuration bits not defined are left at their default value.

```
#pragma config OSC = HS, OSCS = OFF   //oscillator type is HS, oscillator switch is off
#pragma config PWRT = ON, BOR = OFF   //power-up timer is on, brown-out detect is off
#pragma config WDT = OFF              //watchdog timer is off
#pragma config STVR = ON, LVP = OFF   //Stack overflow reset enable is on,
                                      //low voltage programming is off
```

Program Example 17.2 Setting configuration bits for the Derbot

Once a configuration bit is set with a pragma directive, it overrides any setting in the MPLAB IDE **Configuration Bits** window. During a project build, the compiler sets the bits according to their setting in the program. It is interesting to check this in practice by setting some configuration bits 'wrong' in the Configuration Bits window and then building the program. Note, however, that if a bit is not specified in the source code, but is set in a particular way in the **Configuration Bits** window, then the build process does *not* force the bits back to their default value. As in Section 7.11.3 of Chapter 7, click **Configure > Settings > Program Loading** in MPLAB and check the **Clear configuration bits upon loading the program** box. This ensures that on program download the window setting is cleared, and that it is only the configuration bits defined in the program which are downloaded.

17.4   Interrupts

Interrupts present a number of challenges in the C environment. When working with interrupts we are working very close to the hardware, yet a high-level language tends to distance us from it. A number of distinct and important actions must be taken in order to allow the 18 Series interrupts to work successfully. The interrupt must be enabled and allocated to the desired priority. The ISR must be located in program memory at the right start address (noting that there are two interrupt vectors in the 18 Series structure) and context saving must be managed. Check Figure 12.7 and the accompanying description for a reminder of these points if needed.

17.4.1   The Interrupt Service Routine

Interrupt Service Routines in C are similar to C functions, except of course they are called by occurrence of an interrupt and terminate with a return from interrupt instruction. They can have local variables (i.e. variables declared within the ISR) and access global ones. Global variables which are accessed by an ISR should, however, be designated **volatile**. This indicates that the variable value can be changed outside normal program operation. As the ISR can be called anywhere, it is not allowed to transfer any parameters or return values.

17.4.2   Locating and identifying the ISR

The C18 compiler uses several pragmas, first to locate the start of the ISR at the reset vector and then to distinguish the ISR from a regular function.

Like the reset vector, the C18 compiler does not automatically start the ISR at the high or low interrupt vector in program memory. This requires the use of the **#pragma code**, already described. This is used to locate the start of the ISR correctly.

The ISR is identified in the program through use of a pragma. Two are available for ISR definition:

- **#pragma interrupt** *function_name* (**save** − *save_list*). This pragma declares *function_name* to be a high priority ISR. The Fast Register Stack (Section 12.6.3 of Chapter 12) is used to save the minimum context − the **STATUS**, **WREG** and **BSR** registers. The interrupt is ended with a fast return from interrupt.
- **#pragma interruptlow** *function_name* (**save** = *save_list*). This pragma declares *function_name* to be a low priority ISR. The software stack is used to save the minimum context. This slows response to the interrupt. The interrupt is ended with a normal return from interrupt.

Further context saving, beyond the minimum, can be achieved in either interrupt type by specifying register(s) to be saved in the optional **save** section of the pragma.

17.5   Example with interrupt on overflow – flashing LEDs on the Derbot

Program Example 17.3 uses the 18F242 Timer 0 interrupt on overflow to flash the LEDs on the Derbot. This seemingly simple application allows us to take some useful further steps in C programming.

```
/****************************************************************
Flashing LEDS
Flashes Derbot LEDs, driven by Timer 0 interrupt on overflow.
Demonstrates: Use of Timer 0 peripheral, Interrupts, and inline assembly.
TJW 30.10.05
****************************************************************/

#include <p18F242.h>
#include <timers.h>

#pragma config OSC = HS, OSCS = OFF  // HS oscillator, oscillator switch off
#pragma config PWRT = ON, BOR = OFF  //power-up timer on, brown-out detect off
#pragma config WDT = OFF             //watchdog timer off
#pragma config STVR = ON, LVP = OFF  //Stack overflow reset enable on,
                                     //low voltage programming off
//function prototype, repeated for info
void timer0_isr (void);

unsigned char counter = 0;
```

Program Example 17.3 'Flashing LEDs' program, using Timer 0 interrupts

```
//Define the high interrupt vector to be at 0008h
#pragma code high_vector=0x08
void interrupt (void)
{
  _asm GOTO timer0_isr _endasm   //jump to ISR
}

#pragma code   //Return to default code section

//Function timer0_isr specified as high-priority ISR
#pragma interrupt timer0_isr

//timer0_isr function. No transfer of parameters, as required by ISRs
void timer0_isr (void)
{
  counter = counter + 1;
  PORTC = counter<<5;         //Shift Counter left, and move to PORT C
  INTCONbits.TMR0IF = 0;      //Clear TMR0 interrupt flag
}
void main (void)
{
  //Initialise
      TRISC = 0b10000000;
      PORTC = 0;              //Switch outputs off

/*Initialise TMR0: interrupt enabled,16-bit operation, internal clock,
prescaler divide by 4, hence (with 4MHz clock)period of 1usx64kx4 = 262ms*/

  OpenTimer0 (TIMER_INT_ON & T0_SOURCE_INT & T0_16BIT & T0_PS_1_4);
  INTCONbits.GIE = 1;        //Enable global interrupt

  while (1)    //Await interrupts
    {
    }
}
```

Program Example 17.3 Continued

It can be seen that the **main** function of this program is very short – once initialisation is complete all significant action is contained within the ISR. All the **main** function does is to initialise Port C and set its value to zero, initialise Timer 0 (see below), set the Global Interrupt Enable and finally enter an endless **while** loop, waiting for interrupts to occur.

17.5.1   Using Timer 0

We have already met the library functions available for the 18 Series timers in Section 15.5.1 of Chapter 15. Timer 0 itself is described in Section 13.3.1 of Chapter 13, with its various modes of operation evident from Figure 13.2. The timer's header file **timers.h** must be included in any program that will use it, as is seen here.

It can be seen that this program uses the function **OpenTimer0** in the initialisation section of **main**, as shown below:

```
OpenTimer0 (TIMER_INT_ON & T0_SOURCE_INT & T0_16BIT & T0_PS_1_4);
```

Timer set-up data is specified in the argument of this function. Information on the full range of options for this can be found in Ref. 14.3. This implementation enables the timer interrupt on overflow, sets the clock source to be the internal oscillator, selects 16-bit operation (as opposed to 8-bit) and sets the prescaler to be divide-by-4. Once this is set, the timer free runs and generates a series of interrupts to which the microcontroller can respond. With these settings, the counter requires 65 536 cycles to count through its range and each input cycle has a duration of 4 µs. The interval between interrupts is therefore 65 536 × 4 µs, or 262.144 ms.

17.5.2   Using interrupts, and the ISR action

This program example uses a single interrupt, from Timer 0. Prioritisation has not been enabled, so by default the high priority vector is used. As mentioned, the interrupt vectors need to be specified in the program listing, making use of the **#pragma code** option. To set the high priority vector (Figure 12.6), the following is used:

```
#pragma code high_vector = 0x08
```

This specifies that the code section which follows is to be placed in code memory starting at memory location 08_H. It has been named **high_vector** by the programmer. This is not a reserved name and another could be chosen. Alternatively, for the low priority vector:

```
#pragma code low_vector=0x18
```

specifies that code which follows is to be located at code memory location 18_H. It has been named **low_vector** by the programmer.

With the vector correctly placed, there follows in the program a single line of in-line assembler code, forcing a jump to the main body of the ISR:

```
_asm GOTO timer0_isr _endasm //jump to ISR
```

To 'undo' the action of the earlier pragmas, and allow the compiler to control code location again, the pragma:

```
#pragma code
```

is applied. This returns code location to the default.

The ISR, **timer0_ISR**, appears as a function prototype near the start of the program. The action of the ISR is simple. The value of **counter** is first incremented. It is then shifted left five times and assigned to Port C.

The effect of this is to transfer the least significant 2 bits of **counter** to the LEDs, which are located at bits 5 and 6 of Port C. The Timer 0 interrupt flag is then cleared.

This simple interrupt example uses just a single interrupt and does not enable the 18 Series interrupt prioritisation. The use of two, prioritised, interrupts is illustrated in Program Example 19.6.

17.5.3 Simulating the flashing LEDs program

It is interesting to simulate Program Example 17.3, whether or not it is downloaded to hardware. Create a project round the program using the source code on the book CD. Build it and simulate with the MPLAB simulator. Open a Watch window, and display **PCL**, **counter**, **PORTC** and **INTCON**, as seen in Figure 17.1(a). Step quickly through the early stages of the program, noting how the **OpenTimer0** function appears. Once **main** is entered and the user initialisation is complete, the program enters the continuous **while** loop, waiting for the interrupt.

Now force an interrupt by setting high the Timer 0 interrupt flag (bit 2 of the **INTCON** register) in the Watch window. With further single stepping, program execution jumps to the ISR, via the assembler insert at the high priority interrupt vector. The actions of the ISR can be observed by continued single stepping.

Let us now examine the action of the timer interrupt. First, place a breakpoint at the very start of the interrupt routine, as shown in Figure 17.1(b). If run, the program will now stop whenever it reaches this point, i.e. every time the timer interrupt occurs. From the toolbar, select **Debugger > Settings > Osc/Trace**

(a)

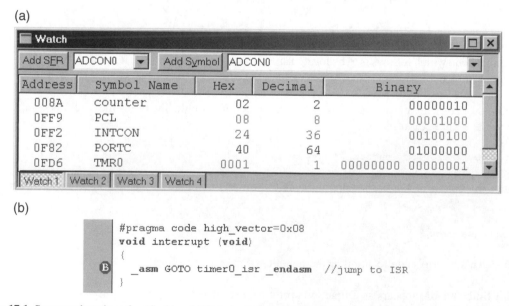

(b)

Figure 17.1 Suggested settings for 'flashing LEDs' simulation. (a) Watch window. (b) Breakpoint location

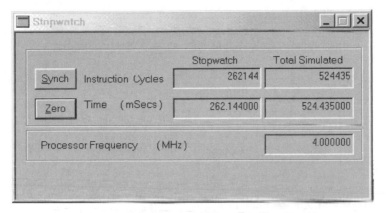

Figure 17.2 Stopwatch for 'flashing LEDs' simulation

and set the oscillator frequency to 4 MHz. Open the Stopwatch window, as seen in Figure 17.2, from the Debugger pull-down menu.

From wherever you are in the program, run to the breakpoint. Zero the Stopwatch and run again. After a moment, program execution should again stop at the breakpoint. The Stopwatch time displayed should be 262.144 ms, exactly as shown in Figure 17.2. This confirms the calculated time between overflows. At the same time, the Watch window values should appear similar to Figure 17.1(a). Timer 0 has just overflowed and is beginning to step up from zero again; the lower byte of the Program Counter (**PCL**) has been set to the high priority reset vector, 08_H, and the timer interrupt flag, bit 2 of **INTCON**, is set.

If you have the Derbot-18 hardware, then you can use this program to provide a pleasing display of flashing LEDs.

17.6   Storage classes and their application

The rest of this chapter aims to look at the wider environment within which a C source file is placed. Notable among these are three files used by just about every program: the microcontroller header file, the start-up file and the Linker Script. To understand these, it is necessary to explore a few more aspects of C. We start that process here, with storage classes.

17.6.1   Storage classes

As we have begun to see, the C language controls the use of data carefully, in terms of how it is declared and how it can be used. This is necessary partly because C programs can be complex things, made up of different files and functions, written at different times by different people, and saved in different stages of compilation. Functions and data may, for example, be declared in one file, but need to be accessible to others.

A characteristic of data used in C is therefore its *storage class*. This defines its status within and across blocks of code, functions and files. Table A6.3 shows the four C keywords which are used in this connection.

Table 17.2 Storage classes: effect of specifier and position

Storage class specifier	Declared outside all functions	Declared inside a function
none	File scope Static duration External linkage	Block scope Automatic duration No linkage
auto	–	Block scope Automatic duration No linkage
static	File scope Static duration Internal linkage	Block scope Static duration No linkage
extern	File scope Static duration External linkage	Block scope Static duration External linkage

The C18 compiler uses only three of these, **auto**, **static** and **extern**. These are somewhat peculiar terms and don't seem to make much sense, even when we realise that **auto** is short for automatic and **extern** for external. It is useful to know that the use of the word *automatic* was borrowed from other computer languages, and implies a variable which comes into existence for a particular purpose within a certain function, but does not exist at other times. Static, on the other hand, implies a variable which has some form of continuous existence. External implies a variable which has continuous existence and that can be accessed by any function.

Returning to the storage class, it determines three things: the *scope*, *duration* and *linkage* of the data. Let us examine each of these in turn. Table 17.2 summarises the points made, for all possible combinations of application. Some of these are, of course, more widely used than others and we only make use of a selection in the example programs of this book.

17.6.2 Scope

The scope of a variable determines the part of the program in which it can be used. Two possible scopes are:

- *Block scope*. The variable can only be used in the block of code within which it is declared, starting with the point of declaration. Variables of this type are called *local* variables. The same name for local variables can be used in different blocks of the program and they will be unrelated.
- *File scope*. The variable can only be used in the file within which it is declared, starting with the point of declaration. The variable must be declared outside all blocks.

17.6.3 Duration

The duration of a variable can be one of two:

- *Automatic storage duration*. An automatic variable is declared within a block of code and is recreated every time program execution enters that block. When the block ends, the variable ceases to exist and the memory occupied by it is freed. The C keyword **auto** defines this storage duration. As it is the default duration when a variable is defined within a block, the keyword is not often used.

- *Static storage duration*. A static variable exists throughout program execution and is identified by the keyword **static**. Static variables may still be local to a block, but retain their existence outside that block. Variables declared outside all blocks, whether or not the keyword is explicitly used, are interpreted as static.

17.6.4 Linkage

Both within files, and in the case where several files are used, it is important to consider how variable names can be recognised as referring to the same variable. There are thus three possible types of *linkage*:

- *External linkage*. If a variable or function is externally linked, it is recognised throughout the program, wherever it is declared. The name is recognised by the Linker. A variable has external linkage if it is declared with the storage class specifier **extern**, or if it is declared outside all functions, with no storage class specified.
- *Internal linkage*. A variable has internal linkage if it is declared **static**, outside all functions. The variable remains internal to the translation unit, but is recognised throughout it. The Linker has no 'knowledge' of it.
- *No linkage*. This is the state of all other variables, for example those with automatic duration.

17.6.5 Working with 18 Series memory

A further complication – or opportunity – arises with the specification of storage in the C18 compiler. Due to its Harvard memory structure and flexible use of memory, the PIC microcontroller presents a challenge to how C treats memory allocation. The C18 compiler therefore introduces the storage qualifiers **far** and **near**. These act as keywords and are effectively C18-specific extensions to C. They indicate the microcontroller memory size or the way memory should be used. Each can be applied to two more keyword extensions, **rom** and **ram**. These are used if the type of memory must be specified in the declaration of a variable or constant. When data is declared without storage qualification, the default is **ram** and **far**. The action of all of these is summarised in Table A6.8, along with a brief description.

Two memory models can also be specified, *small* and *large*. The properties of each of these is summarised in Table A6.9. They are selected by command line options, with small being the default. The only difference is in pointer size needed. Only the small model, which is the default, is needed for programs of introductory or medium length.

17.6.6 Storage class examples

In the first of two code fragments from the **c018i.c** file, taken from towards the top, we see variables being declared. Here the word **extern** is explicitly being used, indicating variables which have external linkage. All three variables also appear in the **18f242.h** header file. The use of the qualifier **near** indicates that they are to be placed in Access RAM. The data types, some of which we have not met before, can be checked by reference to Table A6.4.

```
extern volatile near unsigned long short TBLPTR;
extern near unsigned FSR0;
extern near char FPFLAGS;
```

The second example, below, shows the start of the **_do_cinit**( ) function, with comments removed. Four variables are declared at the head of the function. Each is **static**, so it will have continuous existence. Being declared within a block, i.e. the function, they will be local to that function.

```
void _do_cinit (void)
{
  static short long prom;
  static unsigned short curr_byte;
  static unsigned short curr_entry;
  static short long data_ptr;
...
```

The implications of the way these variables are declared is explored further when we come to simulate the **c018i.c** file, in Section 17.7.3.

17.7   Start-up code: c018i.c

We come now at last to viewing the **c018i.c** file in its entirety. It comes as a surprise when simulating a first C program that the simulator doesn't go straight to **main**. Surely this is what all the textbooks tell us, and shouldn't **main** just start at the reset vector? There are, however, things which need to be set up for the C program to run correctly, even before it starts. These include anything needed for correct operation of the C program, for example software stacks for transfer of data, and the initialisation of all variables and constants. This may be due to a requirement of C itself, or because values have been initialised in the program itself.

Initialisation routines that perform these functions are included with every compiler and may be nearly invisible to the programmer. If we depend on something, however, it is worth developing at least some sort of acquaintance with it.

17.7.1   The C18 start-up files

The C18 compiler provides three start-up program files, at varying levels of complexity. These are available as pre-compiled object files, which are linked to the main program at the build stage. They are normally linked in to the user application with the Linker Script. A source file version is also available. The start-up routine is the program element that is placed at the reset vector, so it is the very first thing that the CPU executes. It initialises the software stack and initialises all data which has a defined starting value. It then jumps to the user's **main** function.

The programs illustrated so far have all made use of the **c018i.c** file. When you simulate any of the example C programs with MPSIM^{TM}, this is what you see if you start to single step through the program. It is intended for programs with the processor operating in non-extended mode. The **c018i_e.c** version is for extended mode operation.

A simpler version of the start-up file is **c018.c**. This simply sets up the software stack and jumps to main, without any data memory initialisation. A more complex version is **c018iz.c**. This does the same as **c018i.c**, but also sets all uninitialised variables to zero, as required in strict ANSI C. These two versions are for non-extended mode operation. Their extended mode equivalents are **c018_e.c** and **c018iz_e.c**.

17.7.2 The *c018i.c* structure

The opening of the **c018i.c** program section has already been quoted in Section 17.3.1.

Its first action is to initialise **FSR1** and **FSR2**, which are used for the software stack (and hence not available to the programmer). It then calls a function called **_do_cinit**. This initialises variables in RAM, if there are any that need it. At the end of this function, the **main** function is called, in the lines shown below.

```
loop:
  // Call the user's main routine
  main ();
  goto loop;
```

It is interesting to see from this that if **main** ever executes a return, then it is immediately called again.

17.7.3 Simulating *c018i.c*

You will have passed through **c018i.c** many times if you are simulating the programs in this book. Let us now, however, step through it with a little more interest as to what is going on. We can also use it to check up on the storage class of certain variables that we have already seen in examples.

Open the project you made for Program Example 14.1 and enable the MPLAB simulator. Open the Watch window and select the variables shown in Figure 17.3. Reset the simulator. It is interesting to observe that the variables declared in the source code appear to be valid, as does **TBLPTR**, declared towards the beginning of the program. The other variables in the Watch window, those declared *inside* the **_do_cinit** function, are specified as being 'out of scope', as seen in the figure. This is in accordance with the description of Section 17.6.2. Single step through the program until the **_do_cinit** function is entered. Notice how these three variables suddenly come into scope and take on a (zero) value. If you continue to single step through the program, you will see that execution returns from the function at an early stage. This simple program requires no initialisation. The **main** function is then called.

Now, in the source code, move the declaration of **counter**, in the line

```
unsigned char counter;     //specify counter as unsigned character
```

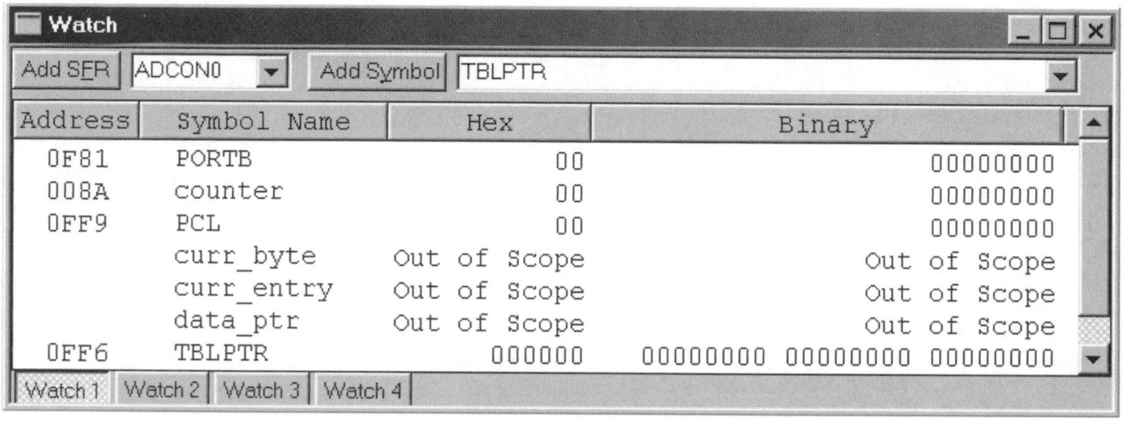

Figure 17.3 Watch window for **c018i.o** execution, Program Example 14.1

to just *inside* the **main** function. Build the program again and simulate. Notice now that, at the beginning of program execution, **counter** is 'out of scope'. By declaring it within a function, it has lost its external linkage, as shown in Table 17.2. If you single step through the program, you will find it becomes in scope once **main** is entered.

Open now the project of Program Example 16.2, which has plenty of lists to initialise. Set up a Watch window with the variables shown in Figure 17.4. The lowest three variables will again initially be out of scope.

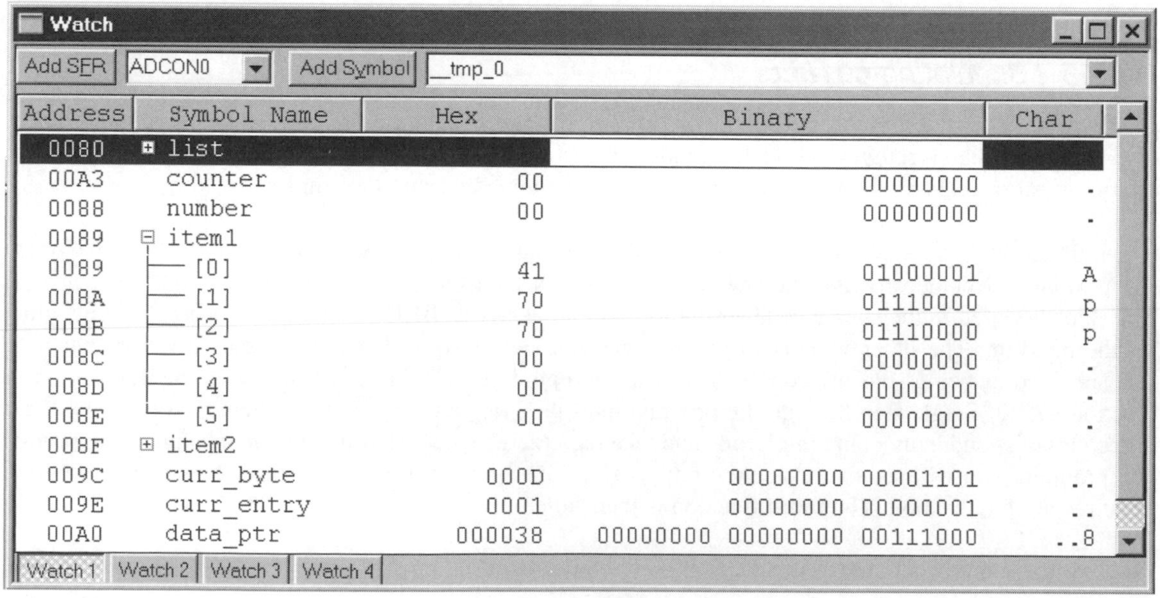

Figure 17.4 Watch window for **c018i.o** execution, Program Example 16.2

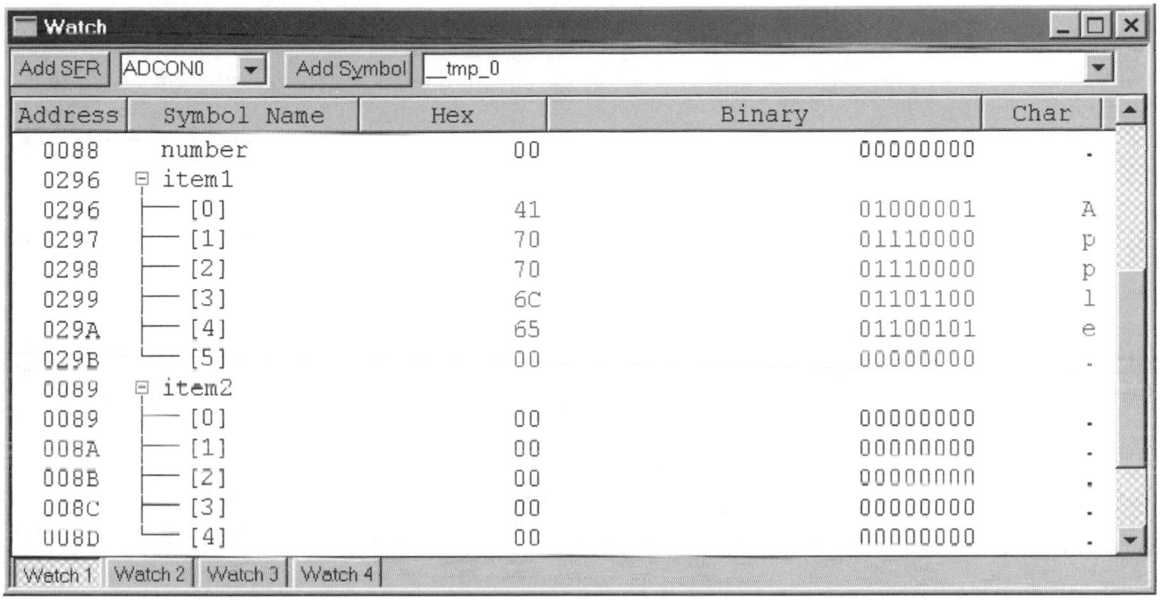

```
┌─────────────────────────────────────────────────────────────────────────────────┐
│ ■ Watch                                                              _ □ x        │
├─────────────────────────────────────────────────────────────────────────────────┤
│ [Add SFR] [ADCON0  ▼] [Add Symbol] [__tmp_0                              ▼]        │
├──────────┬───────────────┬──────────────┬────────────────────────┬────────┬──────┤
│ Address  │  Symbol Name  │     Hex      │         Binary         │  Char  │  ▲   │
├──────────┼───────────────┼──────────────┼────────────────────────┼────────┼──────┤
│   0088   │    number     │      00      │       00000000         │   .    │      │
│   0296   │ ⊟ item1       │              │                        │        │      │
│   0296   │   ├── [0]     │      41      │       01000001         │   A    │      │
│   0297   │   ├── [1]     │      70      │       01110000         │   p    │      │
│   0298   │   ├── [2]     │      70      │       01110000         │   p    │      │
│   0299   │   ├── [3]     │      6C      │       01101100         │   l    │      │
│   029A   │   ├── [4]     │      65      │       01100101         │   e    │      │
│   029B   │   └── [5]     │      00      │       00000000         │   .    │      │
│   0089   │ ⊟ item2       │              │                        │        │      │
│   0089   │   ├── [0]     │      00      │       00000000         │   .    │      │
│   008A   │   ├── [1]     │      00      │       00000000         │   .    │      │
│   008B   │   ├── [2]     │      00      │       00000000         │   .    │      │
│   008C   │   ├── [3]     │      00      │       00000000         │   .    │      │
│   008D   │   └── [4]     │      00      │       00000000         │   .    │  ▼   │
├──────────┴───────────────┴──────────────┴────────────────────────┴────────┴──────┤
│ │Watch 1│ Watch 2 │ Watch 3 │ Watch 4 │                                           │
└─────────────────────────────────────────────────────────────────────────────────┘
```

Figure 17.5 Watch window for Program Example 16.2 – use of **rom** keyword

Single step through the program, into the **_do_cinit** function. As before, the lower three variables take on numerical values. Continue single stepping and notice that program execution enters a major loop within the function. Ultimately, you will see the character strings being populated as data is moved over into data memory from program memory. The figure shows the 'Apple' string partially completed.

The character strings we have been looking at are only placed in data memory because that is the default location, as Section 17.6.5 has indicated. Let us explore using the **rom** extension keyword to put (or in fact leave) one of them in program memory, which is much more sensible. In the declaration of the 'Apple' string in the source file, insert the word **rom** as follows:

```
rom char item1[] = "Apple";
```

Rebuild the program and reset the simulator. Notice now (from the addresses used) that the 'Apple' string has been placed into program memory, as seen in Figure 17.5. It is immediately available and no initialisation of memory is needed for it.

17.8   Structures, unions and bit-fields

In the section that follows this, we will be looking at microcontroller header files. A large part of these apply certain data types which we have yet to meet. These are therefore now introduced.

Structures and *unions* are both sets of related variables, defined through the C keywords **struct** and **union**. In a way they are like arrays, but in both cases they can be of data elements of *different* types.

Structure elements, called *members*, are arranged sequentially, with the members occupying successive locations in memory. A structure is declared by invoking the **struct** keyword, followed by an optional name (called the structure *tag*), followed by a list of the structure members, each of these itself forming a declaration. For example:

```
struct resistor {int val; char pow; char tol;};
```

declares a structure with tag **resistor**, which holds the value (**val**), power rating (**pow**) and tolerance (**tol**) of a resistor. The tag may come before or after the braces holding the list of structure members.

Structure elements are identified by specifying the name of the variable and the name of the member, separated by a full stop (period). Therefore, **resistor.val** identifies the first member of the example structure above.

Like a structure, a union can hold different types of data. Unlike the structure, union elements all begin at the same address. Hence the union can represent only one of its members at any one time, and the size of the union is the size of the largest element. It is up to the programmer to track which type is currently stored! Unions are declared in a format similar to that of structures.

Unions, structures and arrays can occur within each other. We will see an example of this in the following section.

In embedded systems we are very interested in identifying and accessing individual bits. The bit-field capability of C assists with just that, where a bit-field is a set of adjacent bits within a single word. Bit-fields can only be declared as members of a structure or union. The format for declaring the bit-field is:

```
type [name]:width
```

Here *type* is either a signed or unsigned integer, *width* is the number of bits and *name* is an optional name.

17.9   Processor-specific header files

The processor-specific header files are very important in embedded C. They include definitions for all the Special Function Registers (SFRs) and their bits, as well as some useful extras, for example extra features for working with Assembler. It is instructive to look further at one of the processor header files. We will do this with the **18f242.h** file. You can find it in the **mcc18\h** folder of the C18 software.

17.9.1   SFR definitions

An excerpt of the 18F242 header file is shown as Program Example 17.4. In this we see Port B and its bits being declared. The first line of the excerpt uses no less than four of the C keywords to define the **PORTB** type. Use of **unsigned char** specifies it as single byte, while **volatile** indicates that it can be changed outside program control. Finally, **extern** indicates that the variable can be accessed outside this file. The use of **near** indicates that it is placed within Access RAM.

By looking carefully at this program example it is possible to see that the declaration of port bits is arranged as a union named **PORTBbits**, containing two structures. Like Port B the union is specified as **extern volatile near**. It can be seen that the first structure within the union is a list of the conventional names of the port bits, each declared as a single-bit bit-field.

The second structure lists the alternative uses of the port bits, each again being a bit-field. As both these structures belong to a union, they effectively occupy the same memory space and can be used as alternatives to each other.

```
extern volatile near unsigned char PORTB;
extern volatile near union {
  struct {
    unsigned RB0:1;
    unsigned RB1:1;
    unsigned RB2:1;
    unsigned RB3:1;
    unsigned RB4:1;
    unsigned RB5:1;
    unsigned RB6:1;
    unsigned RB7:1;
  };
  struct {
    unsigned INT0:1;
    unsigned INT1:1;
    unsigned INT2:1;
    unsigned CCP2:1;
    unsigned :1;
    unsigned PGM:1;
    unsigned PGC:1;
    unsigned PGD:1;
  };
} PORTBbits;
```

Program Example 17.4 Port B declaration – part of the **18F242.h** file, version 1.6

We are now at last in a position to understand the format we have used for several chapters to identify port and other SFR bits. When we write, for example

```
PORTBbits.RB7 = 1;
```

we now know that this invokes the member **RB7** (a bit-field) of a structure, which in turn is a member of the union **PORTBbits**.

17.9.2 *Assembler utilities in the header file*

Program Example 17.5 shows the **#define** preprocessor directive being used to define certain 18 Series Assembler instructions. By doing this they can be used in a C program, without even invoking the usual in-line assembler procedure. To the casual observer their use will appear as functions. We find this technique applied again in Chapter 19, with the Salvo Real Time Operating System.

```
...
#define Nop()     {_asm nop _endasm}
#define ClrWdt()  {_asm clrwdt _endasm}
#define Sleep()   {_asm sleep _endasm}
#define Reset()   {_asm reset _endasm}
...
```

Program Example 17.5 Part of the **18F242.h** file – Assembler utilities

17.10   Taking things further – the MPLAB Linker and the .map file

Figure 14.1 shows the central part that the Linker plays in any build process. For straightforward applications it is not necessary to understand how the Linker works. An approximate understanding of the Linker is, however, useful even in simple applications to appreciate how a program is put together, particularly in terms of finding and understanding those 'hidden' files which are linked in. For more advanced applications the programmer may want to modify or rewrite the Linker file provided. This section introduces the MPLAB Linker, MPLINK™. Reference information on this can be found in Ref. 17.2.

17.10.1   What the Linker does

As the build process of Figure 14.1 shows, the Linker takes object files as its input and combines these together to create executable code, which can be downloaded to the microcontroller. It also provides background information on how it has allocated memory, which can be used for debug purposes. The object files may be application code, generated first as C or Assembler. Alternatively, they may be general-purpose library files. In either case, the code they contain is largely *relocatable*. That means that addresses in memory, whether data or program, have not yet been assigned. It is the function of the Linker to locate all the object files into memory and ensure that they link across to each other correctly. It can also control allocation of the software stack. It is guided in all this by the *Linker Script*, which contains essential information about the memory map of the microcontroller that is to be used. In undertaking its task, the Linker may well uncover programming faults, for example addresses which clash, or inadequate information.

17.10.2   The Linker Script

The Linker Script is a text file made up of a series of Linker directives, which tell the Linker where the available memory is and how it should be used. Thus, they reflect exactly the memory resources and memory map of the target microcontroller. Standard Linker Scripts are provided in MPLAB for all available PIC 18 Series microcontrollers. In a standard C18 installation they can be found in the **mcc\lkr** folder. The script for the 18F242, the **18f242.lkr** file, is used in every C program example in this book. It is reproduced as Program Example 17.6. An alternative but very similar version, not specific to a C18 implementation, can also be found in any MPLAB installation.

```
// $Id: 18f242.lkr,v 1.1 2003/12/16 14:53:08 GrosbaJ Exp $
// File: 18f242.lkr
// Sample linker script for the PIC18F242 processor

LIBPATH .

FILES c018i.o
FILES clib.lib
FILES p18f242.lib

CODEPAGE    NAME=vectors    START=0x0         END=0x29          PROTECTED
CODEPAGE    NAME=page       START=0x2A        END=0x3FFF
CODEPAGE    NAME=idlocs     START=0x200000    END=0x200007      PROTECTED
CODEPAGE    NAME=config     START=0x300000    END=0x30000D      PROTECTED
CODEPAGE    NAME=devid      START=0x3FFFFE    END=0x3FFFFF      PROTECTED
CODEPAGE    NAME=eedata     START=0xF00000    END=0xF000FF      PROTECTED

ACCESSBANK  NAME=accessram  START=0x0         END=0x7F
DATABANK    NAME=gpr0       START=0x80        END=0xFF
DATABANK    NAME=gpr1       START=0x100       END=0x1FF
DATABANK    NAME=gpr2       START=0x200       END=0x2FF
ACCESSBANK  NAME=accesssfr  START=0xF80       END=0xFFF         PROTECTED

SECTION     NAME=CONFIG     ROM=config

STACK SIZE=0x100 RAM=gpr2
```

Program Example 17.6 Linker Script for 18F242

Let us now explore this example Linker Script. Our goal at this stage is to appreciate what it is saying, not to write a new file. Therefore, we will not worry about exact formats.

- Linker comments. All comments are preceded by //. All text following this on a line is ignored by the Linker. As may be expected, the comments seen here provide title and version information.
- Directive **LIBPATH**. This provides an optional search path for files to be included. It is not used in this example.
- Directive **FILES**. This directive specifies object files for linking. Three files are specified here:

  – **c018i.o**. This is the object code version of the start-up file, already described in this chapter
  – **clib.lib**. This contains the standard C library supported by the C18 compiler
  – **p18f242.lib**. This file contains processor-specific information and effectively works alongside the processor-specific header file.

- Directive **CODEPAGE**. This directive is used to allocate program memory. It is used no less than six times in this example, with the primary purpose of conveying the microcontroller memory maps. The main block of memory, to which the Linker can allocate program code, is located from address $02A_H$ to $03FF_H$. This accords with the memory map of Figure 12.6. A space above this is reserved for vectors. Blocks for configuration data and device identification are also reserved, corresponding to the locations shown in Table 12.4. Further space is reserved for EEPROM and identification.
- Directive **ACCESSBANK**. This directive, used twice in this example, allocates access data memory. The first time it is used, the RAM located in access memory is labelled **accessram** and is correctly located in

the address range 0 to 7F$_H$. In the second, the SFR memory block is identified, located in the memory map and labelled **accesssfr**. This memory block is specified as being protected, which stops the Linker allocating it for general-purpose usage. The absolute memory allocations made elsewhere for the SFRs are thus preserved.

* Directive **DATABANK**. This directive is similar to **ACCESSBANK** and uses the same format. It is used to specify banked RAM. Its implementation in this example can be seen to follow exactly the data memory map seen in Figure 12.4. Each block is available to be used by the Linker, so none is protected.
* Directive **SECTION**. This directive allows a name identified in the source code with a **#pragma** directive to be linked across to a block of memory identified in the Linker Script. In this case the connection is being made for configuration memory, so that data generated by use of **#pragma config** (as illustrated in Program Example 17.2) is placed in the right location.
* Directive **STACK SIZE**. This directive allows the software stack location and size to be specified. In this example it can be seen that a stack of size 100$_H$ is specified, located in RAM Bank 2.

17.10.3   The .map file

The result of the Linker's action is that all code is mapped correctly into the different categories of memory. How can this be checked? The answer lies in the **.map** file, an optional file which we can ask the compiler to generate. This file shows all memory allocation. For a given project, it can be generated by clicking **Project > Build Options > Project > MPLINK Linker > Generate map File.**

Following a successful project build, the **.map** file can be found along with other output files, with name *project_name*.map. The **.map** file is not a pretty sight for the casual observer, as it contains *all* the address locations of *all* symbols used, as well as the memory mapping derived from the Linker Script. However, it can be useful as a diagnostic tool if one is having trouble working out how a variable has been treated, or what has happened to a memory location or block of memory. Another useful feature of the **.map** file is that it shows the proportion of memory used. This, of course, becomes very important as programs grow.

Fragments of the **.map** file for Program Example 14.1 are shown in Program Example 17.7. It shows where some of the main program sections are placed and goes on to indicate that program memory usage is a modest 1 per cent!

```
                         Section Info
            Section      Type      Address    Location  Size(Bytes)
          ---------    ---------  ---------   --------- ---------
          _entry_scn      code    0x000000    program   0x000006
              .cinit   romdata    0x00002a    program   0x000002
          _cinit_scn      code    0x00002c    program   0x00009e
    .code_example1.o      code    0x0000ca    program   0x000024
        _startup_scn      code    0x0000ee    program   0x000018

    ...
```

Program Example 17.7 Fragments of **.map** file for Program Example 14.1

```
            Program Memory Usage
              Start          End
            ---------     ---------
            0x000000      0x000005
            0x00002a      0x000105
226 out of 16664 program addresses used, program memory utilization is 1%
```

Program Example 17.7 Continued

Summary

- It may still be necessary from time to time to step outside the strict confines of C to make use of Assembler.
- It is not difficult to use interrupts in C, but an understanding is needed of how the interrupt vectors are defined and how the service routine is constructed.
- To work with larger programs, it is useful to develop further knowledge of different data types and storage classes.
- The development of a program in C involves far more than simply writing source code. A wide selection of other files can (and effectively must) be used. It is useful to have some understanding of what these are, how they work and how they relate to each other.
- The Linker brings the various contributing files together. Knowledge of the Linker at an appreciation level is useful for simple programs. A detailed knowledge becomes important when writing major pieces of software.

At the end of these four chapters on C, we have reached an introductory but useful understanding of the C language, as applied to embedded systems. This should allow you to go on to write increasingly complex C programs, as indeed is done in Chapter 19. While the basics of C have been introduced, there are plenty of features of C which haven't. A deeper and wider knowledge of C can be gained by more programming experience, studying good example programs and reading from the various specialist C books that are available.

References

17.1. PIC18 Configuration Settings Addendum (2005). Microchip Technology Inc., Document No. DS51537C.
17.2. MPASM™ Assembler, MPLINK™ Object Linker, MPLIB™ Object Librarian User's Guide (2005). Microchip Technology Inc., Document No. DS33014J; www.microchip.com

18
Multi-tasking and the Real Time Operating System

Almost every embedded system has more than one activity that it needs to perform. A program for the Derbot AGV, for example, may need to sense its environment through bump and light sensors, measure the distance it has moved, and hence calculate and implement drive values for its motors. As a system becomes more complicated, it becomes increasingly difficult to balance the needs of the different things it does. Each will compete for CPU time and may therefore cause delays in other areas of the system. The program needs a way of dividing its time 'fairly' between the different demands laid upon it.

An important parallel aspect of the need to time-share the CPU is the need to ensure that things are happening in time. This is very important in almost every embedded system, and the problem just gets worse when there are multiple activities competing for CPU attention.

This chapter explores the demands of systems which have many things on their minds. It investigates the underlying challenges and comes up with a strategy for dealing with them – the Real Time Operating System. This leads to a completely new approach to programming – it's no longer the program sequence which determines what happens next, but the operating system that is controlling it!

Having worked through the chapter, you should have a good understanding of:

- The challenges posed by multi-tasking
- The meaning and implication of 'real time'
- How simple multi-tasking can be achieved through sequential programming
- The principles of the Real Time Operating System.

As this chapter forms a broad introduction to real time programming, there are no actual program examples in it. However, the chapter acts as preparation for Chapter 19, which has a number of significant examples.

18.1 The main ideas – the challenge of multi-tasking and real time

Many of us in this busy modern world feel we spend our lives multi-tasking. A parent may need to get two or three children ready for school – one has lost a sock, one feels sick and the other has spilt the milk. And the dog needs feeding, the saucepan is boiling over, the mailman is at the door and the phone is ringing. Many things need to be done, but we can only do one thing at a time. The microcontroller in an embedded system can feel similarly harassed. It can be surrounded by many things, each demanding its attention. It will need to decide what to do first and what can be left till later.

Let us start this chapter by exploring the nature both of multi-tasking and of real time.

18.1.1   Multi-tasking – tasks, priorities and deadlines

Figure 18.1 shows a simplified flow diagram of a program we were looking at in Chapter 16, the Derbot light-seeking program. This program was used at the time primarily to introduce certain concepts in C, and we did not spend much time thinking of its structure.

The program is made up of a number of distinct activities, each of which appears in the figure. Let us immediately adopt the practice of calling such activities *tasks*. A task is *a program strand or section which has a clear and distinct purpose and outcome*. Multi-tasking simply describes a situation where there are many tasks which need to be performed, ideally simultaneously. Four tasks are identified in this example.

The program is structured so that each task is placed in a super loop and each is executed in turn. One task, the display (where data is sent to the hand controller for display on the LCD), executes only once every 10 cycles of the loop. The loop is ended by a delay, which determines the loop repetition rate and hence influences the program timing characteristics.

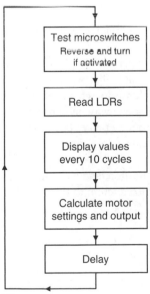

Key: LDR = Light-dependent resistor

Figure 18.1 Simplified flow diagram of the Derbot light-seeking program

This is a very simple multi-tasking example. The program has a number of tasks. It performs them strictly in rotation, taking a little rest before it starts over again.

In reality, of course, the tasks are not all of equal importance. Going back to the harassed parent: if the phone rings, the dog probably has to wait for its dinner. Therefore, we recognise that different tasks have different *priorities*. A high priority task should have the right to execute before a low priority task. Each task also has a *deadline*, or could have one estimated for it. The phone *has* to be answered within 30 seconds, otherwise the caller rings off; the children *have* to be ready to leave home by 8.30 a.m., otherwise they miss the bus, and so on. The concept of priorities is linked to that of deadlines. Generally, a task with a tight deadline requires a high priority – but more of that later.

The tasks of this example program are shown in Table 18.1, together with some preliminary classification. Three levels of priority are used, with three estimated deadlines. In assigning these priorities, a microswitch being pressed is viewed as an emergency; the AGV has collided with something and the motor has probably stalled. It is therefore of high priority. In contrast, the human user will barely notice if the display function is delayed by a second or so. It can therefore be of lower priority.

18.1.2   So what is 'real time'?

How does the Derbot light-seeking program, or indeed the harassed parent, relate to the concept of *real time*? The concept is widely talked about, but seems often to be surrounded by a certain mystique.

A simple but completely effective definition of operating in real time, already adopted in Ref. 1.1, is as follows:

> *A system operating in real time must be able to provide the correct results at the required time deadlines.*

Notice that this definition carries no implication that working in real time implies high speed, although this can often help. It simply states that what is needed must be ready at the time it is needed. Therefore, if the parent gets all the children to school on time, feeds the dog before it starts to howl, opens the door to the postman before he goes away and answers the phone before the caller hangs up, then the real time requirements of this environment have been met. Similarly, if the Derbot meets all deadlines quantified in Table 18.1, it too is operating in real time.

Underlying this simple definition, and all it implies, lies a multitude of program design challenges. The rest of this chapter forms an introduction to these, and to how they can be resolved.

Table 18.1 Tasks in the Derbot light-seeking program

Task	Priority	Deadline (ms)
Respond to microswitches	1	20
Read LDRs	2	50
Calculate and set motor speed	2	50
Display	3	500

18.2   Achieving multi-tasking with sequential programming

The type of programming we have engaged in to date, whether in Assembler or C, is sometimes called *sequential programming*. This simply implies that the program executes in the normal way, each instruction or statement following the one before, unless program branches, or subroutine or function calls, take place. Later in this chapter we will make a departure from this type of program. For now, let us explore how we can optimise it for multi-tasking applications, addressing some of the weaknesses of the program structure of Figure 18.1.

18.2.1   Evaluating the super loop

This program appears to run quite well, but that is mainly because it is not very demanding in its needs. Let us consider some of its drawbacks, all related to each other:

- *Loop execution time is not constant.* The time it takes to run through the loop once is equal to the sum of the times taken by each task, plus the delay time. Clearly, this could vary. If, for example, a microswitch is activated, the Derbot reverses and then turns. This results in a particularly long loop execution, which may upset some other activity in the loop.
- *Tasks interfere with each other.* Once a task gets the chance to execute, it will keep the CPU busy until it has completed what it needs to do. Agreed, each is written so that it shouldn't take too long. Suppose, however, that the display task starts to execute and the AGV suddenly hits a wall. The processor will continue sending data to the display and will then complete the delay routine before it finds out that an emergency has occurred.
- *High priority tasks don't get the attention they need.* We have already recognised that some tasks are more important than others. They should therefore have higher priority, a concept we have already met in the context of interrupts. In this continuous loop structure, every task has the same priority.

We need to find a way of structuring programs to recognise the nature and needs of the tasks they contain, and to meet their real time demands.

18.2.2   Time-triggered and event-triggered tasks

It is easy to recognise that some tasks are *time triggered* and others *event triggered*. A time-triggered task occurs on completion of a certain period of time and is usually periodic. An example of this is the reading of the LDRs in this program. Event-triggered tasks occur when a certain event takes place. In this case the pressing of a microswitch is a good example.

18.2.3   Using interrupts for prioritisation – the foreground/ background structure

To address the problem of lack of prioritisation among the tasks in the light-seeking program, it would be possible to transfer high priority task(s) to interrupts. These would then gain CPU attention as soon as it was needed, particularly if only one interrupt was used. The program structure would then appear

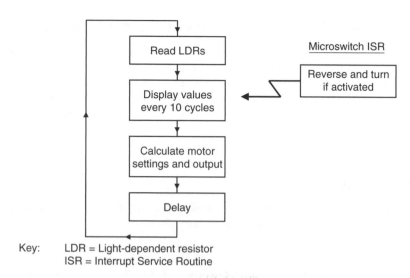

Key: LDR = Light-dependent resistor
        ISR = Interrupt Service Routine

Figure 18.2 Using an interrupt for prioritisation in the Derbot light-seeking program

as in Figure 18.2. It is quite likely that tasks in the loop would be time triggered, where they would be able to use the repetition rate of the loop as a time base. Tasks driven by interrupts are likely to be event triggered.

With this simple program structure we have achieved a reliable repetition rate for tasks in the loop, and prioritisation for tasks that need it. This is sometimes called a *foreground/background program structure*. Tasks with higher priority and driven by interrupts are in the foreground (when they need to be), while the lower priority tasks in the loop can run almost continuously in the background.

18.2.4 Introducing a 'clock tick' to synchronise program activity

To minimise the impact of the variable task execution times on the overall loop execution time, it is possible to trigger the whole loop from a timed interrupt, say from a timer overflow. The program would then have the structure of Figure 18.3. The main loop is now triggered by a timer overflow, so occurs at a fixed and reliable rate. Time-triggered tasks can base their own activity on this repetition rate. Event-triggered tasks, through the interrupts, can occur when needed. Task timings, of course, have to be calculated and controlled, so that the loop has adequate time to execute within the time allowed and the event-triggered tasks do not disturb too much the repetitive nature of the loop timing.

The idea of a regular timer interrupt used in this way to synchronise program activity was introduced in Chapter 9 and illustrated in Figure 9.3. As mentioned there, it is usually called a 'clock tick'. The idea is simple, but it becomes fundamental to many program structures we are about to consider. The clock tick should not be confused with the clock oscillator itself, even though the tick is usually derived from the oscillator.

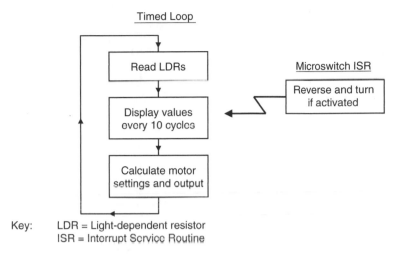

Figure 18.3 Using a timed interrupt in the Derbot light-seeking program

18.2.5 A general-purpose 'operating system'

The structure which emerges in Figure 18.3 can be abstracted into a general-purpose 'operating system', as shown to the right in Figure 18.4. Here the main loop contains a series of low or medium priority tasks. It is driven by a 'clock tick'. The general structure of each task is shown on the left. As needed, each task has an enable flag (a bit in a memory location) and each has a task counter. Tasks which need to execute every clock tick will do so. Many will only need to execute less frequently, at an interval set by the value of their task counter. Tasks can be enabled or disabled by the setting or clearing of their enable flag, by each other or by the ISRs.

This general-purpose operating system structure can be adapted to form the framework of a multi-tasking program. Its general concepts are applied in practical designs in Refs 18.1 and 18.2. The former describes the complete design of a multi-tasking metronome based on the PIC^{TM} 16F84.

If several tasks are allocated to interrupts, then interrupt latency obviously suffers, as one ISR has to wait for another to complete. This must be analysed carefully in a very time-sensitive system. Reference 1.1 explores this topic in some depth.

18.2.6 The limits of sequential programming when multi-tasking

The approach to programming for multi-tasking just described will generally be acceptable, as long as:

- There are not too many tasks
- Task priorities can be accommodated in the structure
- Tasks are moderately well behaved, for example their requirement for CPU time is always reasonable, and interrupt-driven tasks don't occur too often.

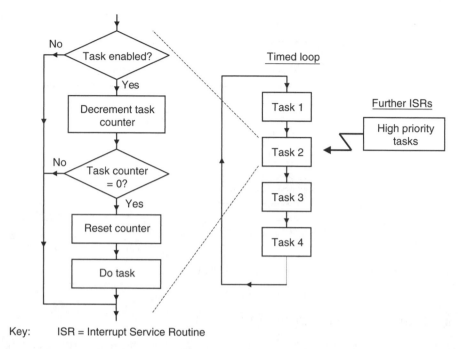

Key: ISR = Interrupt Service Routine

Figure 18.4 A general-purpose 'operating system' structure, using sequential programming

If these conditions are not met, it is necessary to consider a radically different programming strategy. The natural candidate is the Real Time Operating System.

18.3   The Real Time Operating System (RTOS)

When using a true operating system, we move away from the assumptions of normal sequential programming, as outlined in Section 18.2 above. Instead, we hand over control of the CPU and all system resources to the operating system. It is the operating system which now determines which section of the program is to run and for how long, and how it accesses system resources. The application program itself is subservient to the operating system and is written in a way that recognises the requirements of the operating system. Because we are concerned to meet real time needs, we make use of a particular type of operating system which meets this requirement, the *Real Time Operating System* (RTOS).

A program written for an RTOS is structured into tasks, usually (but not always) prioritised, which are controlled by the operating system. The RTOS performs three main functions:

- It decides which task should run and for how long
- It provides communication and synchronisation between tasks
- It controls the use of resources shared between the tasks, for example memory and hardware peripherals.

An RTOS itself is a general-purpose program utility. It is adapted for a particular application by writing tasks for it and by customising it in other ways. While you can write your own RTOS, it is pretty much a specialist activity and generally done by specialists. There are a number of companies which develop and supply such operating systems, usually targeted towards one particular type of market and scale of processor. If you buy one of these, then you are buying a 'COTS RTOS' – a commercial off-the-shelf real time operating system! We explore using such an RTOS in Chapter 19.

18.4   Scheduling and the scheduler

A central part of the RTOS is the *scheduler*. This determines which task is allowed to run at any particular moment. Among other things, the scheduler must be aware of what tasks are ready to run and their priorities (if any). There are a number of fundamentally different scheduling strategies, which we consider now.

18.4.1   Cyclic scheduling

Cyclic scheduling is simple. Each task is allowed to run to completion before it hands over to the next. A task cannot be discontinued as it runs. This is almost like the super loop operation we have seen earlier in this chapter.

A diagrammatic example of cyclic scheduling is shown in Figure 18.5. Here the horizontal band represents CPU activity and the numbered blocks the tasks as they execute. Tasks are seen executing in turn, with Task 3 initially the longest and 2 the shortest. In the third iteration, however, Task 1 takes longer and the overall loop time is longer. Cyclic scheduling carries the disadvantages of sequential programming in a loop, as outlined above. At least it is simple.

18.4.2   Round robin scheduling and context switching

In round robin scheduling the operating system is driven by a regular interrupt (the 'clock tick'). Tasks are selected in a fixed sequence for execution. On each clock tick, the current task is discontinued and the next is allowed to start execution. All tasks are treated as being of equal importance and wait in turn for their slot of CPU time. Tasks are *not* allowed to run to completion, but are *pre-empted*, i.e. their execution is discontinued mid-flight. This is an example of a *pre-emptive* scheduler.

The implications of this pre-emptive task switching, and its overheads, are not insignificant and must be taken into account. When the task is allowed to run again, it must be able to pick up operation seamlessly, with no side-effect from the pre-emption. Therefore, complete context saving (all flags, registers and other

Figure 18.5 Cyclic scheduling – Tasks 1, 2 and 3 execute in turn

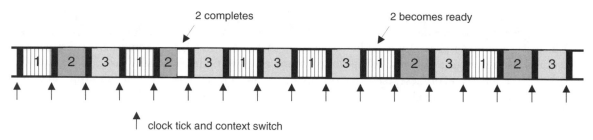

Figure 18.6 Round robin scheduling

memory locations) must be undertaken as the task switches. Time-critical program elements should not be interrupted, however, and this requirement will need to be written into the program.

A diagrammatic example of round robin scheduling is shown in Figure 18.6. The numbered blocks once more represent the tasks as they execute, but there is a major difference from Figure 18.5. Now each task gets a slot of CPU time, which has a fixed length. The clock tick, which causes this task switch, is represented in the diagram by an arrow. When that time is up, the next task takes over, whether the current one has completed or not. At one stage Task 2 completes and does not need CPU time for several time slices. It then becomes ready for action again and takes its turn in the cycle.

As the task and context are switched, there is an inevitable time overhead, which is represented by the black bars. This is the time taken serving the requirements of the RTOS, which is lost to the application program.

18.4.3   Task states

It is worth pausing at this moment to consider what is happening to the tasks now they are being controlled by a scheduler. Clearly, only one task is running at any one time. Others may need to run, but at any one instant do not have the chance. Others may just need to respond to a particular set of circumstances and hence only be active at certain times during program execution.

It is important, therefore, to recognise that tasks can move between different states. A possible state diagram for this is shown in Figure 18.7. The states are described below. Note, however, that the terminology used and the way the state is effected vary to some extent from one RTOS to another. Therefore, in some cases several terms are used to describe a certain state.

- *Ready (or eligible)*. The task is ready to run and will do so as soon as it is allocated CPU time. The task leaves this state and enters the active state when it is started by the scheduler.
- *Running*. The task has been allocated CPU time and is executing. A number of things can cause the task to leave this state. Maybe it simply completes and no longer needs CPU time. Alternatively, the scheduler may pre-empt it, so that another task can run. Finally, it may enter a blocked or waiting state for one of the reasons described below.
- *Blocked/waiting/delayed*. This state represents a task which is ready to run, but for one reason or another is not allowed to. There are a number of distinct reasons why this may be the case, and indeed this single

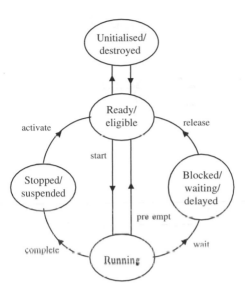

Figure 18.7 Task states

state on the diagram could be replaced by several, if greater detail was wanted. The task could be waiting for some data to arrive or for a resource that it needs, which is currently being used by another task, or it could be waiting for a period of time to be up. The state is left when the task is released from the condition which is holding it there.

- *Stopped/suspended/dormant.* The task does not at present need CPU time. A task leaves this state and enters the ready state when it is activated again, for whatever reason.
- *Unitialised/destroyed.* In this state the task no longer exists as far as the RTOS is concerned. An implication of this is that a task does not need to have continuous existence throughout the course of program execution. Generally, they have to be created or initialised in the program before they can run. If necessary they can later be destroyed and possibly another created instead. Removing unneeded tasks from the task list simplifies scheduler operation and reduces demands on memory.

18.4.4   Prioritised pre-emptive scheduling

We return now to our survey of scheduling strategies, armed with a greater understanding of the lifestyle of tasks. In round robin scheduling tasks become subservient to a higher power – the operating system – as we have seen. Yet all tasks are of equal priority, so an unimportant task gets just as much access to the CPU as one of tip-top priority. We can change this by prioritising tasks.

In the *prioritised pre-emptive scheduler*, tasks are given priorities. High priority tasks are now allowed to complete before any time whatsoever is given to tasks of lower priority. The scheduler is still run by a clock tick. On every tick it checks which ready task has the highest priority. Whichever that is gets access to the CPU. An executing task which still needs CPU time, and is highest priority, keeps the CPU. A low priority

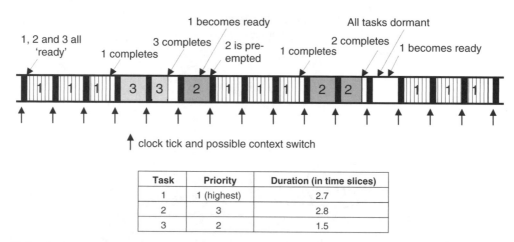

Figure 18.8 Prioritised pre-emptive scheduling

Task	Priority	Duration (in time slices)
1	1 (highest)	2.7
2	3	2.8
3	2	1.5

task which is executing is replaced by one of higher priority, if that has become ready. The high priority task becomes the 'bully in the playground'. In almost every case it gets its way.

The way this scheduling strategy works is illustrated in the example of Figure 18.8. This contains a number of the key concepts of the RTOS and is worth understanding well. The diagram shows three tasks, each of different priority and different execution duration. At the beginning, all are ready to run. Because Task 1 has the highest priority, the scheduler selects it to run. At the next clock tick, the scheduler recognises that Task 1 still needs to run, so it is allowed to continue. The same happens at the next clock tick and the task completes during the following time slice. Task 1 does not now need CPU time and becomes suspended. At the next clock tick the scheduler therefore selects the ready task which has the highest priority, which is now Task 3. This also runs to completion.

At last Task 2 gets a chance to run! Unfortunately for it, however, during its first time slice Task 1 becomes ready again. At the next clock tick the scheduler therefore selects Task 1 to run again. Once more, this is allowed to run to completion. When it has, and only because no other task is ready, Task 2 can re-enter the arena and finally complete. Following this, for one time slice, there is no active task and hence no CPU activity. Task 1 then becomes ready one more time and starts to run again to completion.

18.4.5   Cooperative scheduling

The scheduling strategy just discussed, prioritised pre-emptive scheduling, represents classic RTOS action. It is not without disadvantage, however. The scheduler must hold all context information for all tasks that it pre-empts. This is generally done in one stack per task and is memory intensive. The context switching can also be time-consuming. Moreover, tasks must be written in such a way that they can be switched at any time during their operation.

An alternative to pre-emptive scheduling is *cooperative* scheduling. Now each task must relinquish, of its own accord, its CPU access at some time in its operation. This sounds like we're blocking out the operating system, but if each task is written correctly this need not be. The advantage is that the task relinquishes control at a moment of its choosing, so it can control its context saving and the central overhead is not required.

Cooperative scheduling is unlikely to be quite as responsive to tight deadlines as pre-emptive scheduling. It does, however, need less memory and can switch tasks quicker. This is very important in the small system, such as one based on a PIC microcontroller.

18.4.6   The role of interrupts in scheduling

So far, we have not mentioned interrupts in connection with the RTOS. Should ISRs themselves form tasks, as was done in structures like that of Figure 18.4? The answer is no. The first use of interrupts is almost always to provide the clock tick, through a timer interrupt on overflow. Beyond this, ISRs are usually used to supply urgent information to the tasks or scheduler. The interrupt could, for example, be set to signal that a certain event has occurred, thereby releasing a task from a blocked state (Figure 18.7). The ISRs themselves are not normally used as tasks.

18.5   Developing tasks

Having established the concept of the task and how they are scheduled, let us now consider how they are written.

18.5.1   Defining tasks

It is an interesting early requirement of the programmer to actually choose which activities of the system will be defined as tasks. The number of tasks created should not be too many. More tasks generally imply more programming complexity, and for every task switch there is a time and memory overhead.

A useful starting point is to consider what the deadlines are and then to allocate one task per deadline. A set of activities which are closely related in time are likely to serve a single deadline and should therefore be grouped together into a single task. A set of activities which are closely related in function and interchange a large amount of data should also be grouped into a single task.

For example, in the Derbot light-seeking program of Figure 18.1, the super loop at one stage reads the three LDRs, then makes some calculations, then sets the motor speed. It also periodically sends data to the display. At any time, the microswitches may be activated. How many RTOS tasks should there be? The central activities are closely related in time and in function, and do share data. Writing to the display could be set as a distinct task – it occurs less often than the others and is of comparatively low priority. As the reading of the LDRs supplies data directly to the motor calculations and the motor control, all these activities could be grouped into a single task. Alternatively, the LDR reading could be separated into its own task. Finally, the microswitch response could be allocated as a further task.

18.5.2   Writing tasks and setting priority

Tasks should be written as if they are to run continuously, as self-contained and semi-autonomous programs, even though they may be discontinued by the scheduler. They cannot call on a section of another's code, but can access common code, for example C libraries. They may depend on services provided by each other and may need to be synchronised with each other. In either case, the RTOS will have special services to allow this to happen.

In all cases but the most simple, the RTOS allows the programmer to set task priorities. In the case of *static* priority, priorities are fixed. In the case of *dynamic* priority, priorities may be changed as the program runs. One way of looking at priority is to consider how important a task is to the operation and well-being of the system, its user and environment. Priority can then be allocated:

- Highest priority: tasks essential for system survival
- Middle priority: tasks essential for correct system operation
- Low priority: tasks needed for adequate system operation – these tasks might occasionally be expendable or a delay in their completion might be acceptable.

Priorities can also be considered by evaluating the task deadlines. In this case high priority is given to tasks which have very tight time deadlines. If, however, a task has a demanding deadline, but just isn't very important in the overall scheme of things, then it may still end up with a low priority.

18.6   Data and resource protection – the semaphore

Several tasks may need to access the same item of shared resource. This could be hardware (including memory or peripheral) or a common software module. This requires some care. A method for dealing with this is by the *semaphore*. A semaphore is allocated to each shared resource, which is used to indicate if it is in use.

In a *binary semaphore*, the first task needing to use the resource will find the semaphore in a GO state and will change it to WAIT before starting to use the resource. Any other task in the mean time needing to use the resource will have to enter the blocked state (Figure 18.7). When the first task has completed accessing the resource, it changes the semaphore back to GO. This leads to the concept of *mutual exclusion*; when one task is accessing the resource, all others are excluded.

The *counting semaphore* is used for a set of identical resources, for example a group of printers. Now the semaphore is initially set to the number of units of resource. As any task uses one of the units, it decrements the semaphore by one, incrementing it again on completion of use. Thus, the counting semaphore holds the number of units that are available for use.

As an effect of setting a semaphore to the WAIT state is that another task becomes blocked, they can be used as a means of providing time synchronisation and signalling between different tasks. One task can block another by setting a semaphore and can release it at a time of its choosing by clearing the semaphore.

Remember the 'bully in the playground', the high priority task mentioned in Section 18.4.4? By using a semaphore, a low priority task can turn the tables on the bully! If a low priority task sets a semaphore for a resource that the high priority task needs, it can block that task. This leads to a dangerous condition known as *priority inversion*. This is beyond the scope of this book, but is also illustrative of the many finer details in the world of real time programming, which are well worth further exploration. Reference 18.3 is a useful starting point for more detailed reading.

18.7   Where do we go from here?

The theory of the RTOS goes well beyond what has been discussed in this chapter and becomes specialised to different types of application. As with so many things, the theory takes on meaning when applied to a real situation. This is what happens in the chapter that follows.

Summary

- The requirement of multi-tasking, common to almost every embedded system, carries with it some valuable concepts – of tasks, deadlines and priorities.
- A system operating in real time is one that is able to meet its deadlines.
- Simple multi-tasking real time systems can be achieved using conventional sequential programming.
- More sophisticated multi-tasking real time systems require the use of a Real Time Operating System.
- Use of a Real Time Operating System requires that programs are structured in a different way, with the programmer clearly understanding the underlying principles of the operating system.

References

18.1. Wilmshurst, T. (2002). Exploring real-time programming. *Electronics World*, pp. 54–60, January; http://www.softcopy.co.uk/electronicsworld/
18.2. A Real-Time Operating System for PICmicro™ Microcontrollers (1997). Microchip Technology Inc., Application Note 585.
18.3. Simon, D. E. (1999). *An Embedded Software Primer*. Addison-Wesley. ISBN 0-201-61569-X.

19
The Salvo^{TM} Real Time Operating System

All the concepts introduced about the RTOS (Real Time Operating System) in Chapter 18 are a theoretical nicety unless we can apply them to a working system. Preferably, this should be one that can be run on a PIC®-based embedded system. Our chance to do this comes with Salvo^{TM} – 'the RTOS that runs in tiny places^{TM}'. Salvo is a commercially available RTOS specially intended for the small embedded system, with a version which works with the Microchip C18 compiler. Best of all, there is a free Salvo LITE version! This gives us the opportunity to enter the exciting yet challenging world of the RTOS, and without too much trouble to write simple, illustrative and working programs.

The main aim of this chapter is to provide an introduction to Salvo, to a level at which effective yet simple programs can be written. The purpose of this is to see a real RTOS applied to practical situations, rather than to become an expert user of Salvo. Therefore, the deeper detail of using Salvo is left to the User's Guide.

The chapter is essentially built around three example programs, which progress through applications of key RTOS concepts. Having reached the end of the chapter, you should have a good understanding of:

- The basics of Salvo, an example RTOS
- The operation of a real RTOS from a practical point of view
- The advantages and disadvantages of working with an RTOS.

19.1   The main idea – Salvo, an example RTOS

Salvo was originally developed, in Assembler, for the data acquisition system of a racing car. Once its wider applicability was recognised, it was rewritten in C and adapted for general use. Its target application is the small embedded system. Salvo now needs to run with a C compiler, it no longer works with Assembler. Versions of it are available for many of the major embedded system compilers. Salvo is supplied by Pumpkin Real Time Software Inc.

Selected information on Salvo is given in the sections that follow, intended to be enough to write some interesting introductory programs. For full details on Salvo, you should refer to the supplier's reference information, notably Refs 19.1–19.3.

19.1.1   Basic Salvo features

Salvo can run multiple prioritised tasks and works with a cooperative scheduler. This is one of the keys to its low memory demands – as discussed in Section 18.4.5 of Chapter 18, cooperative scheduling is less demanding of memory. The number of tasks (in the fully featured version) is limited only by available RAM, while 16 priority levels are available. Tasks can also share priority levels. Salvo supports a range of different 'events', including binary and counting semaphores, messages and message queues.

Salvo is supplied as a very extensive set of files – source, header, library and others. These effectively act as extra services that are added to the host compiler. The programmer works with the host compiler in the usual way, but incorporates the Salvo files as needed. In so doing, he/she must of course follow the requirements of the Salvo RTOS. The program as developed by the programmer is finally a combination of original source code, Salvo and compiler header and source files, and Salvo and compiler library files. The build process and the main contributory files are summarised in Figure 19.1. The output is an executable file, which can be downloaded to program memory.

19.1.2   Salvo versions and references

There are a number of versions of Salvo available. At the de luxe end is Salvo Pro, a highly configurable, fully featured version. The freeware version of Salvo, called Salvo Lite, contains a subset of Salvo functionality. At the time of writing this may be downloaded from the Pumpkin website: http://www.pumpkininc.com/ – a copy is also available on the book CD. The version which matches the compiler in use, in our case the Microchip C18 compiler, must be chosen. Salvo Lite permits three tasks and five events. This sounds like a modest limit. In fact, it allows surprisingly useful and advanced programs to be developed.

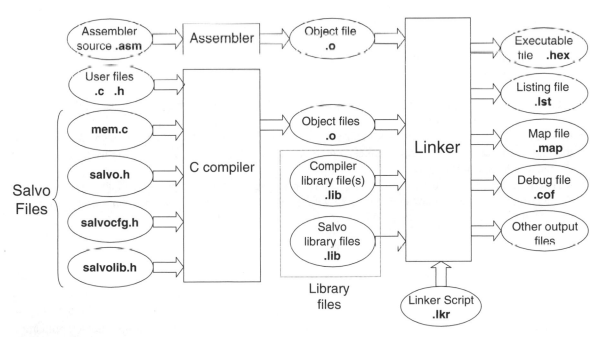

Figure 19.1 Salvo build process

Salvo comes with a big but readable user manual [Ref. 19.1]. This is written to support use of all versions of Salvo, right up to Salvo Pro. For the beginner, aiming to use Salvo Lite, some of it can be daunting. Nevertheless, it has good and informative introductory chapters. It contains all general information for the RTOS, but nothing that is compiler specific. This is contained within further reference manuals. For the Microchip C compiler the important ones are (at the time of writing) Refs 19.2 and 19.3.

19.2   Configuring the Salvo application

One of Salvo's principal features is that it is highly reconfigurable. It is therefore important early on to have a picture of how this configuration takes place. This section introduces some of the essential features of Salvo files and configuration, and will lead to a better understanding of the build process shown in Figure 19.1.

19.2.1   Building Salvo applications – the library build

There are two main ways to build a Salvo application: *library build* and *source-code build*. The latter requires manipulation of the Salvo source code and is for advanced players only. Anyway, it can only be undertaken with Salvo Pro, so we will not consider it here. That leaves us with library build – building the application using the library features available.

Different libraries are available for use, each one with a different set of features. For a given application, the most suitable library must be chosen. Once this is done, the library services are fixed. The user source code then makes calls to the Salvo functions contained in the chosen library.

Apart from the library files, certain other source and header files must be incorporated, as seen in Figure 19.1. The most important ones are as follows.

- **salvo.h**. This is Salvo's main header file and must be included in any source file that uses Salvo services. It should not be modified by the user. This file in turn includes **salvocfg.h**.
- **mem.c**. This is a major file, supplied with Salvo. It holds global objects, which define characteristics for the features used, like tasks, semaphores and so on. It should not be modified by the user, although the contents of **salvocfg.h** impact upon it.
- **salvocfg.h**. This file, written by the user, determines much of the configuration of the system for the application. It sets certain key elements, like which library is to be used, and how many tasks and events there will be. Further details are given in Section 19.4.4.

19.2.2   Salvo libraries

Salvo has a large set of standard libraries, which contain much of the RTOS functionality. There are different library sets for each compiler and different versions within compilers. These support different memory models and different combinations of features. One of the skills in configuring a Salvo application lies with selecting the library which has the features needed and nothing more.

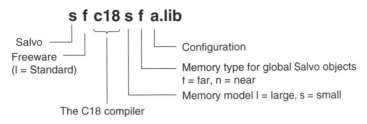

Figure 19.2 Salvo library naming, for C18 compiler

Table 19.1 Library services available for each library configuration

	Library configuration				
	a	**d**	**e**	**m**	**t**
Multi tasking	+	+	+	+	+
Delays	+	+	−	−	+
Events	+	−	+	−	+
Idling	+	+	+	−	+
Task priorities	+	+	+	−	+
Time-outs	−	−	−		+

+ = enabled; − = not enabled.

Salvo Lite is provided with a set of freeware libraries. These are like the standard libraries, but have limited capability. The coding used for the C18 compiler is shown in Figure 19.2. The final letter of the library name indicates its 'configuration', which is defined in Table 19.1. The **sfc18sfa.lib** library, for example, has all features except time-outs. This makes it a comparatively large library. If only multi-tasking was needed in an application, it would be better to use a library with 'm' configuration. This would lead to more efficient coding and less memory utilisation.

19.2.3 Using Salvo with C18

When using Salvo, it is essential to ensure that a compatible combination of MPLAB® IDE, C18 compiler and Salvo is used. This book uses MPLAB version 7.22, with C18 version 3.00 and Salvo version 3.2.3.c. To install Salvo, it should be downloaded from the Salvo website and the usual installation procedure followed. This is straightforward, but is detailed in Ref. 19.1. The installation will result in the set of folders shown in Figure 19.3.

19.3 Writing Salvo programs

This section gives an introduction to programming with the Salvo RTOS, before we look at a first program example in the section which follows.

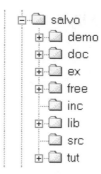

Figure 19.3 Salvo Lite folders

19.3.1   Initialisation and scheduling

Many of the features of the Salvo RTOS are contained within its C functions, found in the Salvo libraries. Part of the skill of programming with Salvo lies in knowing these functions and understanding what they do. Some examples, exactly the ones that will be used in our first program example, are given in Table 19.2. This gives the name of the function or service, a summary of its action and the parameters (if any) that it takes. The contents of the table will be referred to and explained repeatedly over the next few pages – do not worry if all its details are not immediately clear.

Table 19.2 Example core Salvo services

Function/service	Action and parameter(s)
OSInit( )	Initialises the operating system, including data structures, pointers and counters. Must be called before any other Salvo function. No parameters.
OSSched( )	The Salvo scheduler. On every call it chooses – from those which are eligible – the next task to run. Multi-tasking can occur when this is called repeatedly. No parameters.
OS_Yield( )*	Unconditional return to scheduler (1) Context switch label, often defined through use of **_OSLabel( )**.
OSCreateTask(, ,)	Creates a task *and* starts it (i.e. makes it eligible). (1) Pointer to task starting address – usually the task's name. (2) Pointer to task TCB (task control block). (3) Priority – a number from 0 (highest) to 15 (lowest).
OSStartTask( )	Makes a stopped task eligible. (1) Pointer to the task TCB (task control block).
_OSLabel( )	Defines a unique label required for each context switch. (1) Label name.
OSTCBP( )	Defines a pointer to specified control block, in this case to the task control block. (1) An integer from 1 to **OSTASKS**, where **OSTASKS** appears in the **salvocfg.h** file and specifies the number of tasks.

*Can cause a context switch.

Some of the functions shown can cause a context switch; by this means the cooperative scheduling is implemented. All Salvo functions that the user can call are prefixed with 'OS' or 'OS_'. The latter is used if the service contains a conditional or unconditional context switch.

To enable and initialise the RTOS the function **OSInit( )** must be called before any other Salvo function. Tasks can then be created with a call to **OSCreateTask( )**, ensuring that all arguments are properly specified.

The scheduler is contained within the **OSSched( )** function. On every call, this does three things, as listed below:

- It selects and runs the most eligible (i.e. highest priority) task. Tasks with the same priority are run on a round robin basis.
- It processes the event queue, if events are being used. Remember, events include semaphores and messages. As a result of this, certain tasks may become eligible to run.
- It processes the delay queue, where delays in Salvo are an important means of controlling when a task executes. Again, a task may become eligible to run.

19.3.2   Writing Salvo tasks

Each Salvo task is written as a C function and follows the general pattern for the writing of tasks described in Section 18.5 of Chapter 18. There are further important requirements, specific to Salvo. These are summarised below:

- All tasks are initially 'destroyed'. Tasks must be created using the Salvo **OSCreateTask( )** function. They can be created anywhere in the program. In practice, many are created early in the **main** function.
- Tasks are generally made up of an optional initialisation, followed by an infinite loop, which must contain at least one context switch.
- The context switch *can* be provided by a call to the function **OS_Yield( )**, although there are other functions which also cause a switch. With this (or equivalent) call, the task relinquishes access to the CPU and hands control back to the scheduler. This is the basis of the cooperative scheduling used by Salvo.
- Tasks cannot take any parameters.
- Tasks use **static** variables, therefore task data is unchanged when the task is not running. Variables of type **auto** *can* be used if the data does not need to be retained following a context switch.

The operating characteristics of a task are contained within its task control block (TCB). This is a block of memory allocated uniquely to the task, which contains (among other things) the task's start address, state and priority.

Tasks can, in general, follow the state diagram of Figure 18.7, with Salvo-specific interpretations of each state. These are introduced in the following pages.

19.4   A first Salvo example

A first example of a Salvo-based program is shown in Program Example 19.1. It uses exactly the Salvo services summarised in Table 19.1. It contains just two tasks, of equal priority. One, **Count_Task**, increments a counter. The other, **Display_Task**, displays two bits of the counter value on two bits of Port C. As the Derbot hardware can be used, the two least significant bits of the counter are shifted over to the Derbot LEDs, which are on bits 5 and 6 of Port C.

```
/***********************************************************************
rtos_ex1                                An introductory Salvo example.

There are two tasks, of equal priority. One counts, the other displays the count.
Salvo Lite RTOS with sfc18sfm.lib library used.
Mainly for simulation but can run on Derbot.
TJW 21.12.05                                            Tested 21.12.05
***********************************************************************/

#include <salvo.h>
#undef OSC     //necessary for this Salvo version, as it also defines this name
#include  <p18f242.h>

#pragma config OSC = HS, OSCS = OFF  //HS oscillator, oscillator switch off
#pragma config PWRT = ON, BOR = OFF  //pwr-up timer on, brown-out detect off
#pragma config WDT = OFF             //watchdog timer off
#pragma config STVR = ON, LVP = OFF  //Stack overflow reset enable on,
                                     //low voltage programming off

//function prototypes. These functions are tasks.
void Count_Task( void );
void Display_Task( void );

//Define labels for context switches
_OSLabel(Count_Task1)
_OSLabel(Display_Task1)

//Define and initialise variable
unsigned char counter = 0;

/***********************************************************************
Task Definitions (configured as functions)
***********************************************************************/
void Count_Task( void )
{
    for (;;)                   //infinite loop
    {
    counter++;
    OS_Yield(Count_Task1); //context switch
    }
}
//
void Display_Task( void )
{
    for (;;)
    {
    PORTC = counter<<5; //Shift Counter left, and move to PORT C
```

Program Example 19.1 A first Salvo RTOS application

```
        OS_Yield(Display_Task1);
        }
}

/**********************************************************************
Main
**********************************************************************/
void main( void )
{
//Initialise
        TRISC = 0b10000000; //Setall Port C bits to output, except bit 7.
        PORTC = 0;          //Setall Port C outputs low
//Initialise the RTOS
    OSInit();
//Create Tasks
    OSCreateTask(Count_Task, OSTCBP(1), 10);
    OSCreateTask(Display_Task, OSTCBP(2), 10);
 //Set up continuous loop, within which scheduling will take place.
    for (;;)
    OSSched();
}
```

Program Example 19.1 Continued

19.4.1 *Program overview and the **main** function*

Looking down the program listing, this example looks at first like a regular small C program. The first indication that it is a Salvo program comes in the inclusion of **salvo.h**. A few lines lower, the **_OSLabel** macro is applied twice. This provides a means of defining labels that are used in the task context switches. There are only two context switches in this program, one in each task. We will see that the labels chosen, **CountTask1** and **DisplayTask1**, are applied for this purpose within the tasks.

The **main** function starts conventionally enough, with a little initialisation for Port C, the only port to be used. It then calls the **OSInit( )** function, which initialises the operating system and sets the scene for all RTOS action to come. The tasks themselves are then created, with two calls to **OSCreateTask( )**. The format of this is summarised in Table 19.1, where we see that *three* arguments must be provided. The task start address is identified simply by the task name, from its function prototype. The TCB start address is defined using the **OSTCBP( )** macro. Numbering the argument to this from 1 upwards allocates a TCB block to each task. Both tasks are then set to the same priority. Any value from 1 to 16 could be chosen; arbitrarily the value 10 is used. A continuous loop is then established, causing a repeated call to the scheduler **OSSched( )**. The action of this will be to activate the most eligible task.

Configuration bits can be set in the program in the way described in Chapter 17. The version of Salvo used leads to a small clash, however, as the name **OSC** is defined for two different purposes in **salvo.h** and **p18F242.h**. Pumpkin Inc. advise that it is not needed for the C18 implementation. Therefore, it is undefined, using the **#undef** preprocessor directive in the line before **p18F242.h** is included.

19.4.2   Tasks and scheduling

It can be seen that the tasks themselves are written as functions. Each is a continuous loop, with each loop containing an **OS_Yield( )** function call. The argument to this function call is the context switch label, already defined at the top of the program. Every time the task is activated, it will execute until it reaches the **OS_Yield( )** call. Here it returns control back to the scheduler, and the task moves from the 'running' state to the 'ready' or 'eligible' state (Figure 18.7). When the scheduler activates the task again, it picks up execution at the line immediately following the **OS_Yield( )** call and returns to the 'running' state.

19.4.3   Creating a Salvo/C18 project

Creating a project with Salvo is initially just like creating any other C18 project. The steps outlined here are similar to those described in Ref. 19.3, where some troubleshooting advice is given as well. This is an application note for MPLAB Version 6, but is reasonably applicable to Version 7.22. At the time of writing, there is not an Application Note for this later version.

If you wish to build this project for yourself, and you are strongly encouraged to do so, you should start by creating an MPLAB project in the normal way. The name **rtos_ex1** was used for this project. Create a folder just for this project (there will be a number of files in it), and copy to it the source file and **salvocfg.h** file from the folder on the book CD. From the C18 Directory, add the 18F242 Linker Script, and from the Salvo directory, add the **sfc18sfm.lib** library. Also from the Salvo directory, add the **mem.c** file. Your MPLAB project window should then appear as shown in Figure 19.4.

Look back at Figure 19.2 and Table 19.1 to determine the characteristics of the **sfc18sfm.lib** library. It should not be difficult to work out that this is a freeware library for the C18 compiler, applying the small memory model, with capability for multi-tasking only. This is an appropriate choice for this very simple program.

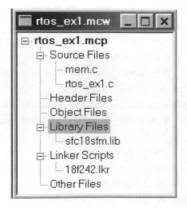

Figure 19.4 Files applied in Salvo_Ex1

19.4.4   Setting the configuration file

The Salvo RTOS is configured for a particular application by the settings in the **salvocfg.h** file, written by the programmer. The file is made up of a series of C define statements. A limited set of these is available in Salvo Lite, while many more are available in the fully featured versions. There are default values for every configuration option, so some (or all) of these can be adopted. The more tightly the configuration file matches the actual application, however, the more efficient is the final coding likely to be. There is, for example, no point in having memory set up for six tasks when only three are required.

The **salvocfg.h** file used for the **rtos_ex1** project is shown in Program Example 19.2. Comments are written to explain each line. It can be seen that the settings in this simple file relate either to library configuration or to program features, the latter including tasks, events and messages. The first line selects the fundamental option that pre-compiled libraries are to be used. It is important that the further library configuration options that follow actually match the library that has been selected. In this case, the **sfc18sfm.lib** library already selected is matched by the corresponding configuration setting, using the **OSM** code.

```
/******************************************************************
salvocfg.h file for rtos_ex1

TJW 8.1.06
******************************************************************/
//Library configuration
#define OSUSE_LIBRARY      TRUE  //Use precompiled Salvo library
#define OSLIBRARY_TYPE     OSF   //use freeware library
                                 //(OSL is standard library)
#define OSLIBRARY_GLOBALS  OSF   //Salvo objects far, in banked RAM
#define OSLIBRARY_CONFIG   OSM   //Set library configuration,
                                 //OSM = support multitasking only
#define OSLIBRARY_VARIANT OSNONE //No library variant

//Tasks, Events and Messages Configuration
#define OSEVENTS             0   //define maximum number of events
#define OSEVENT_FLAGS        0   //define maximum number of event flags
#define OSMESSAGE_QUEUES     0   //define maximum number of message queues
#define OSTASKS              2   //define maximum number of tasks
```

Program Example 19.2 The salvocfg.h file for the rtos_ex1 program

19.4.5   Building the Salvo example

A build using Salvo follows the same process as a normal C18 build. There are, however, more things which must be right for the program to build. Alongside all the possible errors of writing and linking a C program are the further requirements of a Salvo configuration.

The files shown in Figure 19.4 should all be selected. The correct search path for Salvo Include Files should be specified in the Build Options dialogue box, as shown in Figure 19.5(a). The options specified in Figure 19.5(b) should also be entered. The **SYSE** option is used by Salvo to identify the C18 compiler, within certain general-purpose files written for multiple compilers.

(a)   (b)

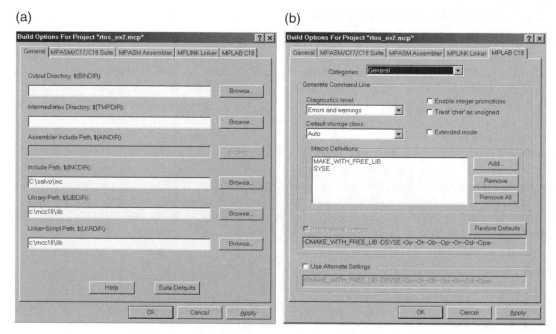

Figure 19.5 Setting build options for Salvo. (a) General. (b) MPLAB C18

If the files provided on the book CD are applied, then a build should proceed without problem. If errors persist, it may be necessary to refer to Salvo reference material or the Salvo User Forum (at the Pumpkin website). Check also the book website (www.embedded-knowhow.co.uk).

19.4.6   Simulating the Salvo program

This first example RTOS program can be simulated just like any other. Having successfully built the program, select MPLAB SIM.

Open a Watch window, and select **counter** and **PortC** for display. Insert the breakpoints shown in Figure 19.6. These are placed at the key points in this simple RTOS program. Then *run* the simulator to the first breakpoint, which is just before the RTOS initialises. Run it again, which takes you to the **OSSched( )** call. A further run will move to **Count_Task**, which runs first as it was created first. If you single step from here, you will see that execution remains sequentially within the task, until the **OS_Yield( )** function call is reached.

Further presses of run cause execution to alternate between tasks, visiting **OSSched( )** in between each task call. Every time execution returns to a task, it picks up exactly where it left off. As the tasks are of equal priority, they are being executed in round robin scheduling.

```
void Count_Task( void )
{
    for (;;) {                    //infinite loop
        counter++;
        OS_Yield(Count_Task1); //context switch
    }
}
//
void Display_Task( void )
{
    for (;;) {
    PORTC = counter<<5;           //Shift Counter left, and move to PORT C

        OS_Yield(Display_Task1);
    }
}
/*******************************************************************
Main
*******************************************************************/
void main( void )
{
//Initialise
    TRISC = 0b10000000; //Set all Port C bits to output, except bit 7.
    PORTC = 0;              //Set all Port C outputs low
//Initialise the RTOS
    OSInit();
//Create Tasks
    OSCreateTask(Count_Task, OSTCBP(1), 10);
    OSCreateTask(Display_Task, OSTCBP(2), 10);
 //Set up continuous loop, within which scheduling will take place.
    for (;;)
        OSSched();
}
```

Figure 19.6 Setting breakpoints for the RTOS simulation

Every time **Count_Task** runs, the value of **counter** in the Watch window increments, and every time **Display_Task** runs, the value of Port C is updated. In terms of functionality, we have achieved nothing startling. In terms of the way the program executes, it is a major new departure.

The program can also be run on the Derbot hardware. It is not particularly interesting to run it this way, however. As it is running at effectively uncontrolled speed, the LEDs just appear to be continuously on.

19.5   Using interrupts, delays and semaphores with Salvo

A key feature of any *Real Time* Operating System is, of course, its ability to manage real time activity. To do this, it is more or less essential to set up a continuous 'clock tick', at a fixed and reliable frequency, which can be used as the time base against which other things can happen. Once the clock tick is there, many Salvo features become possible.

Table 19.3 Example Salvo functions and services used with interrupts, timers and delays

Function/service	Action and parameter(s)
OSTimer( )	Checks to see if any delayed or waiting tasks have timed out. If yes, they are rendered eligible. Must be called at the desired tick rate if delay, time-out or elapsed time services are required. Often placed within timer interrupt ISR. No parameters.
OS_Delay(,)*	Causes current task to return to scheduler and delay by amount specified. Requires **OSTimer( )** to be in use. (1) Integer giving desired delay in system ticks, 8-bit value only. (2) Context switch label, often defined through use of **_OSLabel( )**.
OSEi( )	Enables interrupts (sets GEI and PEIE in **INTCON**; see Figure 12.8)
OSDi( )	Disables interrupts.

*Can cause a context switch.

To establish the clock tick we need to use a timer interrupt, and we then need services to count and respond to tick-based durations. Table 19.3 gives examples of Salvo services related to interrupts and timing, which we will be using.

19.5.1   An example program using an interrupt-based clock tick

Program Example 19.3, called **rtos_ex2**, is a development of our first RTOS example. It keeps the same two tasks, still of equal priority, but introduces an interrupt-driven clock tick, a delay and a binary semaphore. Each of these is discussed in detail in the sections that follow. The ISR for the clock tick appears in the program example that follows.

The program is structured in a way that should be recognisable, even though the new features are added. The **main** function initialises the microcontroller through a call to the function **Micro_Init( )**. This is followed by the RTOS initialisation, through a call to **OSInit( )**. After creating tasks and semaphore, the program enters the expected scheduling loop.

```
/****************************************************************************
rtos_ex2                                      A further Salvo example.

Applies Timer interrupt, clock tick, delays, and binary semaphore.
There are two tasks, of equal priority. One counts, the other displays the count.
Salvo Lite RTOS, with Library sfc18sfa used.
Can be simulated, or run on Derbot.
TJW 28.12.05                                   Tested 30.12.05
****************************************************************************/

#include <salvo.h>
#undef OSC    //necessary for this Salvo version, as it also defines this name
#include <p18f242.h>
#include <timers.h>
```

Program Example 19.3 Applying delays and semaphores

```
#pragma config OSC = HS, OSCS = OFF   //HS oscillator, oscillator switch off
#pragma config PWRT = ON, BOR = OFF   //pwr-up timer on, brown-out detect off
#pragma config WDT = OFF              //watchdog timer off
#pragma config STVR = ON, LVP = OFF   //Stack overflow reset enable on,
                                      //low voltage programming off

#define BINSEM_Display OSECBP(1)

//function prototypes.
void Micro_Init(void);

//These functions are tasks.
void Count_Task( void );
void Display_Task( void );

//Define and initialise variable
unsigned char counter;

//Define labels for context switches
_OSLabel(Count_Task1)
_OSLabel(Display_Task1)

/*********************************************************************
User-defined Functions, including RTOS Tasks.
*********************************************************************/
void Count_Task(void)
{
    for (;;) {                    //infinite loop
      counter++;
      OSSignalBinSem(BINSEM_Display);
      OS_Delay(20,Count_Task1);   //Task switch, delay for 20x10ms, (200ms)
                                  //Use smaller delay for simulation
    }
}

void Display_Task(void)
{
    for (;;) {                    //infinite loop
    OS_WaitBinSem(BINSEM_Display, 100, Display_Task1);
    PORTC = counter<<5;           //Shift Counter left, and move to PORT C
    OS_Yield(Display_Task1);
    }
}
void Micro_Init(void)
{
//Initialise Port C
    TRISC = 0b10000000;
    PORTC = 0;                    //Switch outputs off
/*Initialise TMR0: interrupt enabled,16-bit operation, internal clock,
prescaler divide by 16, hence (with 4MHz clock) input cycle period of 16us*/
    OpenTimer0 (TIMER_INT_ON & T0_SOURCE_INT & T0_16BIT & T0_PS_1_16);
    counter = 0;
}
```

Program Example 19.3 Continued

```
/*************************************************************************
Main
*************************************************************************/
void main( void )
{
//Initialise Microcontroller
      Micro_Init();
//Initialise RTOS
    OSInit();
    OSCreateTask(Count_Task, OSTCBP(1), 10);    //Create the Count_Task Task
    OSCreateTask(Display_Task, OSTCBP(2), 10);  //Create the Display_Task Task
    OSCreateBinSem(BINSEM_Display, 0);          //Create the Binary Semaphore
//Enable interrupts
    OSEi();
//Scheduling Loop
    for (;;)
        OSSched();
}
```

Program Example 19.3 Continued

19.5.2   Selecting the library and configuration

In the **rtos_ex1** example, the **sfc18sfa.lib** library was used. Now, however, we want to introduce delays and semaphores, the latter a type of 'event'. Looking at Table 19.1 tells us that the 'm' suffix library is no longer adequate. The best we can do is to go to an 'a' suffix library, even though this gives idling and priority capability, which we do not yet need.

With a different library, and new features, a new **salvocfg.h** file is needed for this project. This is the same as the one for **rtos_ex1**, except that two differences, the changed library and the inclusion of a semaphore (an 'event'), must be identified. The lines shown below are therefore inserted in place of the equivalent lines in the previous file. The full file is available on the book CD.

```
#define OSLIBRARY_CONFIG OSA   //Set library configuration,
                               //OSA = support multitasking, delays & events
...
#define OSEVENTS         1     //define maximum number of events
```

19.5.3   Using interrupts and establishing the clock tick

Interrupts can be introduced to a Salvo-based program by adding **USE_INTERRUPTS** in the Macro Definitions window (Figure 19.5(b)). This identifies to Salvo that a file called **isr.c** acts as the system ISR. Interrupts are then written as normal, as described in Section 17.4 of Chapter 17.

Program Example 19.4 shows the ISR for **rtos_ex2**. It follows the pattern of Program Example 17.3, except that it includes the all-important call to **OSTimer( )**. This function, seen in Table 19.3, allows the time-based features of Salvo to work. A variable **tick_counter** is also incremented on every ISR iteration. This is used in the simulation.

A clock tick period of 10 ms was chosen (somewhat arbitrarily) for this example program. It is established using the Timer 0 interrupt on overflow. This is enabled and configured through the **OpenTimer0( )** call, with settings as shown in the comments. With a 4 MHz clock oscillator and the prescaler set to divide-by-16, the clock input to the timer has a period of 16 µs. Theoretically, it therefore needs 625 cycles to produce the 10 ms period. The value reloaded into the timer must therefore be 65 536−625, or 64 911. After the program was simulated and measurements made using the Stopwatch feature, this number was adjusted upwards to take into account the not inconsiderable interrupt latency that was observed.

The timer interrupt is enabled within the call to **OpenTimer0( )**. The Global Interrupt Enable is set within the call to **OSEI( )**, in the **main** function. There is nothing Salvo-specific within this macro, it is just used here for convenience.

```
/******************************************************************
ISR for rtos_ex2
Timer 0 interrupt is high priority source.
Reloads Timer, and calls OSTimer()
TJW 30.12.05                                    Tested 30.12.05
******************************************************************/

#include <salvo.h>
#include <p18F242.h>
#include <timers.h>

//function prototype(s)
void timer0_isr (void);

static unsigned int tick_counter = 0;

//Define the high priority interrupt vector to be at 0008h
#pragma code high_vector=0x08

void interrupt (void)
{
  _asm GOTO timer0_isr _endasm //jump to ISR
}

#pragma code //Return to default code section

//Function timer0_isr specified as high-priority ISR
#pragma interrupt timer0_isr

//timer0_isr function.
void timer0_isr (void)
{
      WriteTimer0 (64918); //Reload value gives 625 cycles to overflow,
                           //less compensation for interrupt latency
      OSTimer();
      tick_counter++;      //increment tick counter, (for simulation)
      INTCONbits.TMR0IF = 0; //Clear TMR0 interrupt flag
}
```

Program Example 19.4 ISR for 'clock tick'

While we have now established a useful, if not essential, clock tick feature, we need to remember that in this form it will only be approximate. The RTOS does routinely disable interrupts, for example during **OSSched( )**. This worsens the interrupt latency and will cause delay in the response to the timer ISR. A timer with hardware auto-reload would give a more accurate time base. In either case, the time base formed is an approximation as, due to the cooperative scheduling, tasks can only respond to the clock tick when allowed by the action of other tasks. We will be able to assess the extent of this approximation in the simulation that is coming up.

19.5.4   Using delays

Now that we have a clock tick, we can synchronise tasks to this. One way of doing this is by using the **OS_Delay( )** function, with parameters as shown in Table 19.3. This function forces a context switch when called and can thus replace **OS_Yield( )**. It also introduces a delay before the next time the task can run, determined by the setting of the first parameter.

The **OS_Delay( )** function call is seen in the **Count_Task** function in Program Example 19.3, used in the following way:

```
OS_Delay (20,Count_Task1); //Task switch, delay for 20x10ms, (200ms)
                           //Use smaller delay for simulation
```

As the comment indicates, use of this delay causes the **Count_Task** function to occur periodically, every 200 ms.

19.5.5   Using a binary semaphore

The concept of the semaphore was introduced in Section 18.6 of Chapter 18. It can be used for resource protection or simply for signalling between tasks. It is a powerful feature, as it is effectively the simplest way of establishing communication between tasks. Program Example 19.3 uses a binary semaphore.

Salvo semaphores, like tasks, must first be created in the program. This allocates them an *event control block* (ECB) in memory, similar to the TCB for tasks. Here, essential information on the semaphore is stored. Semaphores can then be written to by one task and waited on by another. The functions that do these three things are shown in Table 19.4.

Our program example uses a single semaphore. As it is the only event in the program, it is allocated the ECB pointer **OSECB(1)**. The ECB pointer is used more than the TCB pointer in function calls, so it is worth allocating it a name. It is therefore given the name **BINSEM_Display** in this program line:

```
#define BINSEM_Display OSECBP(1)
```

The semaphore is created in the **main( )** function immediately after the tasks are created, in the line:

```
OSCreateBinSem(BINSEM_Display, 0);
```

Table 19.4 Example Salvo functions and services for binary semaphores

Function	Action and parameter(s)
OSCreateBinSem(,)	Creates binary semaphore. (1) Pointer to ECB (event control block). (2) Initial value (0 or 1).
OSSignalBinSem( )	Signals a binary semaphore. If no task is waiting it increments. If one or more tasks are waiting, then the one with highest priority is made eligible. (1) Pointer to semaphore ECB.
OS_WaitBinSem(, ,)*	Task waits (in 'wait' state) until binary semaphore is signalled. Wait state is exited when semaphore signals and task is highest priority, or if time-out expires. Semaphore is then automatically cleared. Time-out must be specified. (1) Pointer to ECB. (2) Time-out value (in system ticks). (3) Context switch label.
OSECB( )	Defines pointer to specified control block, in this case to the event control block. Similar to **OSTCB( )** in Table 18.2. An integer from 1 to **OSEVENTS**, where **OSEVENTS** defines the number of events in the **salvocfg.h** file.
OSNO_TIMEOUT	Used for time-out value if there is to be no time-out.

*Can cause a context switch.

This uses the ECB pointer name previously defined. The semaphore is initially set to 0.

The actual mechanics of this simple use of the binary semaphore can be understood by looking at the two tasks. Remember first of all that **Count_Task** only runs every 20 clock ticks, due to its use of **OS_Delay( )**. When it runs, it signals to the semaphore through its call to **OSSignalBinSem( )**. The action of signalling to the semaphore sets its value high. If a task is waiting for it, then that task becomes ready to run and the semaphore is cleared.

Meanwhile, **Display_Task** is waiting for the semaphore to go high, with the line:

```
OS_WaitBinSem(BINSEM_Display, 100, Display_Task1);
```

This indicates the name of the semaphore awaited, the time-out value and the context switch label. A time-out value must be indicated, even if it is not implemented (as in this case). The overall effect is that as soon as the count is updated in **Count_Task**, it is then displayed by **Display_Task**.

With the steps just taken, we have established simple inter-task communication and synchronisation. This is a great step forward, but there is a further advantage. A task that is waiting for a semaphore is not activated in any way by the scheduler. Hence CPU usage is made much more efficient.

19.5.6 Simulating the program

It is very interesting to simulate this program, and see how clock tick, tasks and the semaphore interact. Copy the source and **salvocfg.h** files from the Book CD, create it as a project and build.

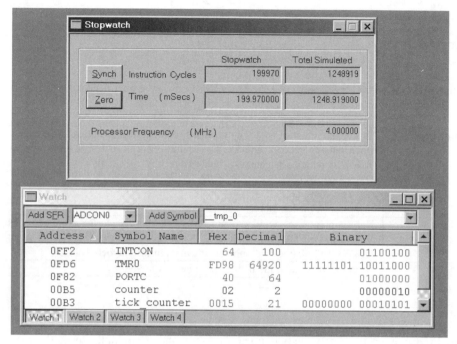

Figure 19.7 Watch window and Stopwatch used to track RTOS clock ticks

The following simulation settings are then suggested. Select MPLAB SIM as the Debugger and insert three breakpoints, one within each task and one in the ISR. Open a Watch window and select for display the variables shown in Figure 19.7. From the toolbar, select **Debugger > Settings > Osc/Trace** and set the oscillator frequency to 4 MHz. Open the Stopwatch window, as seen in the upper half of Figure 19.7, from the Debugger pull-down menu.

Using the simulator controls, run the program. It will halt at the first breakpoint it encounters. Assuming the timer interrupt has not yet occurred, this will be the breakpoint within **Count_Task**. This is the task created first, and the delay does not take effect until the first iteration of **OSDelay( )**. The value of **counter** is incremented to 1. The semaphore is then set and a further run will reach the breakpoint in **Display_Task**. The value of Port C here is set by the program to 20_H. The next run will find the breakpoint in the timer ISR.

At this point, Zero the time in the Stopwatch window. Further runs will repeatedly return to the timer ISR. These are the clock ticks. It can be seen that the Stopwatch time increments by almost exactly 10 ms every time. It is interesting to note that the error is not constant, but depends on other activities of the program, which are asynchronous with the timer. This reflects the approximation discussed in Section 19.5.3. After 20 clock ticks, the **Count_Task** delay is up and the task is revisited. In turn it sets its semaphore and **Display_Task** follows. It can be seen from the Stopwatch that both of these occur comfortably within one time slice.

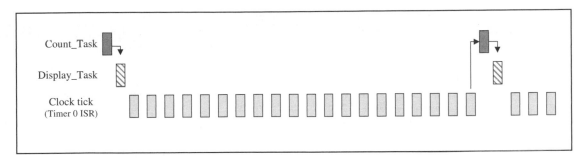

Figure 19.8 Breakpoint occurrences in program simulation

This general pattern of behaviour, as just described, is illustrated in Figure 19.8. This shows the broad overall sequencing of activity within the program; a precise horizontal time axis is not implied.

Tutorial task

The two screen shots in Figure 19.7 were taken from within a program simulation. From the information given, deduce in as much detail as possible where program execution has halted.

19.5.7 Running the program

If you have a Derbot, download the program and run it. The LEDs will flash in a pleasing manner. As you watch this, remember: each LED change is due to an accumulation of clock ticks taking place, leading to a task being released, leading to a counter being incremented, leading to a semaphore being switched, leading to an LED display being changed. We have achieved a very simple outcome, but the elegance and power of the underlying process is extremely satisfying. This elegance and power is also available for much more challenging applications.

19.6 Using Salvo messages and increasing RTOS complexity

The next step we take involves learning about just one new Salvo resource, the message. This leads us, however, into a program of significantly greater complexity than the earlier ones, in which we can exercise the RTOS features in a more practical and realistic way.

Messages provide a convenient way both of transferring data between tasks and of synchronising activity between them. They have some characteristics similar to semaphores – they need to be created and can then take a piece of data from one point in the program and transfer it to another, possibly releasing a task at the same time.

Terminology with messages can be a little confusing. We can create a 'message' (i.e. a Salvo data-carrying structure), which can then carry any number of messages (i.e. pieces of data moved from one part of the

Table 19.5 Example Salvo functions and services for messages

Function/service	Action and parameter(s)
OSCreateMsg(,)	Creates message, with ECB (event control block) pointer and initial value. (1) Pointer to message ECB. (2) Pointer to message.
OSSignalMsg(,)	Signals a message, i.e. attaches a data element to it. If one or more tasks are waiting, then the one with highest priority is made eligible. (1) Pointer to message ECB. (2) Pointer to message.
OS_WaitMsg(, , ,)*	Task waits (in 'wait' state) until message is signalled. Then makes message pointer (param. 2) point to it, waiting task then continues with execution. Task also continues if timed out. Time-out must be specified. (1) Pointer to message ECB. (2) Pointer for message. (3) Time-out value (in system ticks). (4) Label.
OSECB( )	As in Table 19.4
OSNO_TIMEOUT	As in Table 19.4.
OStypeMsgP	Define data type as message pointer. One of a number of predefined Salvo data types, which must be used as appropriate.

*Can cause a context switch.

program to another). Therefore, we call the actual Salvo service the message and then talk about 'signalling' the message when a piece of data is actually attached to it.

In Salvo the message signal itself can be of any data type, from character to array. The information that is actually passed is the *pointer* to the signal. It is up to the programmer to ensure that this is pointing to the data required when it is signalled and that it is dereferenced properly at the receiving end. Note that a message can be sent from anywhere in the program. Thus, for example, an interrupt can signal a message, which is then received by a task.

Sample Salvo message functions and services are given in Table 19.5. These will all be applied in Program Example 19.5. As a Salvo message is a type of 'event', an ECB is set up for each message, just as it was for a semaphore.

19.7 A program example with messages

Program Example 19.5 gives the listing of a fairly substantial Salvo-based program. It builds on the Derbot 'blind navigation' program of Program Example 15.3, but includes the use of an ultrasound sensor, of the type described in Section 18.6.5 of Chapter 8. This was mounted pointing upwards, as shown in Figure 19.9. Note that the sensor shares port bits with the LEDs, so these are not available in this program. The program

Figure 19.9 Derbot with upward facing ultrasound sensor

acts as the 'blind navigation' program, except that when the Derbot runs under an overhanging object, for example under a chair, it detects this, rotates and moves away.

The program has two tasks, now of different priorities, and two interrupts, also of different priorities. Unlike earlier examples, each task forms a substantial block of code and each contains more than one context switch. A single 'message' is defined; this is used to carry signals to the motor control task from both an interrupt and from the other task. A number of delays are also used.

The **main** function is very similar to that of the previous program example, except that a message is created instead of a semaphore.

```
/*********************************************************************
rtos_ex3
Implements Derbot Blind Navigation, with upward-looking US sensor to
detect if AGV is going under an overhang.
Tasks are: Ultrasound Sensor (higher priority), Motor_set (lower priority).
Interrupts are: Microswitch (Low priority) & Timer 0 (High, for clock tick).
One message and numerous delays are also used.
Applies Salvo LITE with Library sfc18sfa. Can run on Derbot, or be simulated.
TJW 3.1.06                                            Tested 4.1.06
*********************************************************************/
//Clock is 4MHz

#include <salvo.h>
#undef OSC    //necessary for this Salvo version, as it also defines this name
#include <p18f242.h>
```

Program Example 19.5 Derbot 'blind navigation' with ultrasound overhang detector

```
#include <timers.h>        //header file for delays
#include <delays.h>        //header file for delays
#include <pwm.h>           //header file for PWM

#pragma config OSC = HS, OSCS = OFF  //HS oscillator, oscillator switch off
#pragma config PWRT = ON, BOR = OFF  //pwr-up timer on, brown-out detect off
#pragma config WDT = OFF             //watchdog timer off
#pragma config STVR = ON, LVP = OFF  //Stack overflow reset enable on,
                                     //low voltage programming off


//User-defined function prototypes
      void Micro_Init(void);
      void leftmot_fwd (void);
      void rtmot_fwd (void);
      void leftmot_rev (void);
      void rtmot_rev (void);
```
 Line 32
```
//These functions are tasks.
       void Motor_Task( void );   //Sets motor according to messages received
       void USnd_Task( void );    //Fires Ultrasound Sensor periodically


//Define labels for context switches
_OSLabel(Motor_Task1)
_OSLabel(Motor_Task2)
_OSLabel(Motor_Task3)
_OSLabel(Motor_Task4)
_OSLabel(Motor_Task5)
_OSLabel(USnd_Task1)
_OSLabel(USnd_Task2)
_OSLabel(USnd_Task3)
_OSLabel(LED_Task1)
_OSLabel(LED_Task2)


//Carries messages from microswitch and ultrasound
#define Msg_to_Motor OSECBP(1)


      char Hole = 0x18;    //This value used, but never tested
```
 Line 53
```
/*****************************************************************************
User-defined Functions, including RTOS Tasks.
*****************************************************************************/
//This task controls motor action, determined by messages recd from elsewhere
void Motor_Task( void )
{     static char msge;    //hold message once recd
      OStypeMsgP msgP;     //Declare msgP as special Salvo pointer type
      for (;;)             //set up the infinite Task loop
      {
      rtmot_fwd ();        //set motors running forward. This is status quo
      leftmot_fwd ();                //until message arrives
      //Wait for message
      OS_WaitMsg(Msg_to_Motor,&msgP,OSNO_TIMEOUT,Motor_Task1);
```

Program Example 19.5 Continued

```
// Line 67 Proceed when message arrives
      msge = *(char*)msgP;
      PORTAbits.RA5 = 0;   //stop motors for 500ms
      PORTAbits.RA2 = 0;
      OS_Delay (50,Motor_Task1);
      if( (msge == 0x80)|| (msge == 0x01) ) //was it a microswitch?
            {
            rtmot_rev ();                   //Yes, so both motors reverse
            leftmot_rev ();
            OS_Delay (100,Motor_Task2);
            if(msge == 0x80)              //was left uswitch hit?
              {leftmot_fwd ();   //Yes, so turn right
              OS_Delay (80,Motor_Task3);
              }
              else              //right uswitch was hit
              {rtmot_fwd ();     //so turn left
              OS_Delay (80,Motor_Task4);
              }
            }
            else          //We're under an overhang, hence turn on spot
            {rtmot_rev ();
            leftmot_fwd ();
            OS_Delay (200,Motor_Task5);
            }
      }                        //end of "for" loop
}
```

Line 86

Line 93
```
/*Task periodically pulses Ultrasound, and sends a message if an overhang
detected. In this case, it suspends pulsing, to allow Derbot to exit*/
void USnd_Task(void)
{      int echo_time = 0;   //counts ultrasound distance measurement
   for (;;)           //set up the infinite Task loop
      {
      OS_Delay (20,USnd_Task1); //Task switch, and delay for 20x10ms, (200ms)
      OSDi();          //disable interrupts, this measurement is time sensitive
      echo_time = 0;
      PORTCbits.RC5 = 1; //output us pulse.
      Delay10TCYx(2);    //20us delay approx, gives pulse width
      PORTCbits.RC5 = 0;
      Delay10TCYx(30);   //pause for op to be set high; ie blank for 5cm
```

Line 107
```
//Values in this loop are adjusted experimentally to give detection threshold
//of 30cm approx
      while (echo_time < 50)   //limit the measurement to close objects
         {Delay10TCYx(1);    //10us delay
         echo_time++;         //increment the counter
         if(PORTCbits.RC6 == 0) //send message if target detected
           {OSSignalMsg(Msg_to_Motor,(OStypeMsgP)&Hole);
           OSEi();          //enable interrupts before delay
           OS_Delay (250,USnd_Task2); //Suspend the USnd,
                              //to allow Derbot to exit "hole"
```

Program Example 19.5 Continued

```
               OS_Delay (250,USnd_Task3);
               break;
               }
            }
       }
   OSEi();                     //enable interrupts
   OS_Yield(USnd_Task2);
   }                           //end of "for" loop
}
```

// **Line** 126 This function initialises the Microcontroller peripherals

```
void Micro_Init(void)
{
//Initialise Ports
       TRISA = 0b00000000; //All bits output, 2 & 5 used for motor enables.
       TRISB = 0b00110000; //Bits 5 and 4 (microswitches) only are input,
       TRISC = 0b11000000; //All bits output except 7 (mode switch) & 6
                                //(USnd echo), 1 & 2 used for PWM
       ADCON1 = 0b00000110; //Set Port A for digital i/o
//Switch all outputs off
       PORTA = PORTB = PORTC = 0;
```

 Line 137

```
/*Initialise Timer 0: interrupt enabled,16-bit operation, internal clock,
prescaler divide by 16, hence (with 4MHz clock) input cycle period of 16us*/
       OpenTimer0 (TIMER_INT_ON & T0_SOURCE_INT & T0_16BIT & T0_PS_1_16);
       WriteTimer0 (64918);              //and initialise
//Initialise PWM
       OpenTimer2 (TIMER_INT_OFF & T2_PS_1_1 & T2_POST_1_1);
       OpenPWM1 (0xFF); //Enable PWM1 and set period
       OpenPWM2 (0xFF); //Enable PWM2 and set period
//Set Port B Interrupt on change to be low priority
       RCONbits.IPEN = 1; //Enable low priority interrupts
       INTCON2bits.RBIP = 0; //Set port change bit to be low priority
       INTCONbits.RBIE = 1;//Enable Port B change interrupt
}
```

 Line 151

```
/******************************************************************************

Main

******************************************************************************/

void main( void )
{
//Initialise Microcontroller
       Micro_Init();
       Delay10KTCYx (250);          //pause 2.5secs with conventional delay
//Initialise RTOS
       OSInit();
//Create Tasks and Message
       OSCreateTask(Motor_Task, OSTCBP(1), 4);
       OSCreateTask(USnd_Task, OSTCBP(2), 2);
       OSCreateMsg(Msg_to_Motor, (OStypeMsgP)0);
//Enable Global interrupts
       OSEi();
```

Program Example 19.5 Continued

```
//Scheduling Loop
       for (;;)
       OSSched();
}
```

```
/************************************************************************
Motor Drive Functions
************************************************************************/
...
```

(Same motor drive functions as Program Example 15.3.

Program Example 19.5 Continued

19.7.1   Selecting the library and configuration

As far as configuration is concerned, this program is no different from the previous one. They both have two tasks and one event. In this case the event is a message. Therefore, the **sfc18sfa.lib** library remains the right one to use and the **salvocfg.h** file from **rtos_ex2** can be copied in to this project.

19.7.2   The task: *USnd_ Task*

This task periodically pulses the ultrasound sensor, to detect whether there is an overhang above the AGV. If an overhang is detected, it sends a message, which is received by the **Motor_Task** task. As the action of the motors depends on a measurement made in this task, it was accordingly given the higher priority.

The structure of the **USnd_Task** is drawn from Program Example 9.7, using software delays to first generate the pulse and then to time the response. It starts from line 96 in the listing. The task uses the **OS_Delay( )** function, placed at the start of its 'infinite' loop, to make the function occur every 20 clock ticks. When a delay period is up, it outputs a pulse by setting Port C bit 5 high, calling a delay of 20 µs and then setting it low again.

The measurement timing loop, from line 110, needs be approximate only, and is only used for short-range measurement. Therefore, an effective time-out of 50 cycles is incorporated in the **while** loop. If the echo pulse (on Port C, bit 6) is seen to fall low during this timing loop, a message is generated. In this case the task then enters a significant delay. This uses two iterations of the **OS_Delay( )** function, as it can only apply 8-bit delay values. This inhibits the ultrasound action while the Derbot turns (under the control of the other task) and moves away from the overhang that it has detected. If, however, the echo pulse does not fall low, then the timing loop simply times out and the task yields through a call to **OS_Yield( )**.

19.7.3   The task: *Motor_ Task*

The **Motor_Task** task undertakes all motor settings, doing this based on the messages it receives from the Port B interrupt and the **USnd_Task** task. It starts from line 58 in the listing. The task opens by declaring the variable **msge**, where the incoming message will be stored. It then defines the message pointer **msgP** using a special Salvo data type **OSTypeMsgP** (Table 19.5).

Having started the two motors, the task waits for a message to arrive, through a call to the **OS_WaitMsg( )** function. Depending on the message received, execution moves to different states. As new motor states are set up, delays are forced using the **OS_Delay( )** function. During waiting for both message and delay, the task is completely inactive, as it is the scheduler which initially receives the message and determines whether the task should be enabled.

19.7.4   The use of messages

Part of the power of this program lies in its use of messages. Significant lines are replicated below. In every case they make use of the services summarised in Table 19.5. Only one Salvo message, **Msg_to_Motor**, is created, but it is used to carry different messages from different places in the program. All are received by the **Motor_Task** function.

```
...
//Carries messages from microswitch and ultrasound
#define Msg_to_Motor OSECBP(1)
...
From main
OSCreateMsg(Msg_to_Motor, (OStypeMsgP)0);
...
From Motor_Task Function
static char msge;    //hold message once recd
OStypeMsgP msgP;     //Declare msgP as special Salvo pointer type
...
//Wait for message
OS_WaitMsg(Msg_to_Motor,&msgP,OSNO_TIMEOUT,Motor_Task1);
//Proceed when message arrives
msge = *(char*)msgP;
...
```

Early in the program listing (line 50) the name of the message, **Msg_to_Motor**, is defined. As with the semaphore, it is actually the pointer to the ECB (event control block) which is being named, as this is used to identify the message in all three of the Salvo functions that are used.

The message is created in **main** using **OSCreateMsg( )**. Very early in the first call to **Motor_Task( )**, program execution reaches an **OS_WaitMsg( )** call. This causes a context switch and forces the task to wait until the message identified, **Msg_to_Motor**, is signalled. Time-out is explicitly not applied, through use of the **OSNO_TIMEOUT**. The pointer for the incoming message signal is specified as **msgP**, which is declared at the beginning of the function.

The message is signalled from within the **uswitch_isr** (one example shown below) and from the **USnd_Task** function. The format is as shown. The predefined name, **Msg_to_Motor**, is again used to supply the message pointer. The value of the message signal, named **Rt_usw**, has been defined earlier. The pointer to the message signal is then supplied, using the special Salvo **OStypeMsgP** data type.

```
From uswitch_isr Function
...
```

```
char Rt_usw = 0x01;
...
if (PORTBbits.RB4 == 0)    //Test right uswitch
OSSignalMsg(Msg_to_Motor,(OStypeMsgP)&Rt_usw); //Send message
...
```

19.7.5 The use of interrupts, and the ISRs

This is the first (and only) example in the book where both a high and a low priority 18 Series interrupt are applied, so it is worth checking the settings. A Timer 0 interrupt is used, exactly as in Program Example 19.3, to establish a 10 ms clock tick. To this is added a Port B interrupt on change, to detect microswitch presses.

The interrupts are configured mainly at the end of the **Micro_Init( )** function. The registers applied are seen in Figures 12.8–12.10 and the role of the **IPEN** bit is seen in Figure 12.7. This bit is first set high (line 147), enabling the low priority interrupt path. In the next two lines the Port Change Interrupt is enabled and set to low priority. The Global Interrupt is enabled through the call to **OSEi( )** in **main**.

The two ISRs are shown in Program Example 19.6. The Microswitch ISR, **uswitch_isr( )**, starts with a delay call of 8 ms, to allow microswitch bounce to settle. Both switches are then tested. If one is found to be low, the corresponding message is generated. It is quite possible that neither is found to be low, as the ISR is called on switch release, as well as switch activation. The ISR ends with the interrupt flag being cleared.

```
/*****************************************************************
ISR for rtos_ex3
There are two interrupt sources:
High Priority: Timer 0 for "clock tick",
Low Priority: microswitch (Port B change)for collision

TJW 3.1.06                                      Tested 5.1.06
*****************************************************************/
#include <salvo.h>
#include <p18F242.h>
#include <timers.h>        //header file for timers
#include <delays.h>        //header file for delays

//function prototypes
void timer0_isr (void);
void uswitch_isr (void);

static unsigned int tick_counter = 0;

//These are values for messages
      char Rt_usw = 0x01;
      char Left_usw = 0x80;

//Carries messages from microswitch and ultrasound
#define Msg_to_Motor OSECBP(1)
```

Program Example 19.6 ISRs for Program Example 19.5

```
/*****************************************************************************
Timer Interrupt (High Priority)
*****************************************************************************/
//Define the high priority interrupt vector to be at 0008h
#pragma code high_vector=0x08
void interrupt_at_high (void)
{
  _asm GOTO timer0_isr _endasm //jump to ISR
}


#pragma code //Return to default code section

//Function timer0_isr specified as high-priority ISR
#pragma interrupt timer0_isr

//timer0_isr function.
void timer0_isr (void)
{
      WriteTimer0 (64918);   //Timer reload value gives 625 cycles to
            //overflow, less compensation for interrupt latency
      OSTimer();
      tick_counter++;        //increment tick counter, (for simulation)
      INTCONbits.TMR0IF = 0; //Clear TMR0 interrupt flag
}
/*****************************************************************************
Microswitch Interrupt (Low Priority)
*****************************************************************************/
//Define the low priority interrupt vector to be at 0018h
#pragma code low_vector=0x18
void interrupt_at_low (void)
{
  _asm GOTO uswitch_isr _endasm  //jump to ISR
}
#pragma code                     //Return to default code section

//Function uswitch_isr specified as low-priority ISR
#pragma interruptlow uswitch_isr

//uswitch_isr function.
void uswitch_isr (void)
{
      Delay1KTCYx(8);            //8ms delay to ensure debounce
      if (PORTBbits.RB4 == 0)    //Test right uswitch
      OSSignalMsg(Msg_to_Motor,(OStypeMsgP)&Rt_usw); //Send message
      if (PORTBbits.RB5 == 0)    //Test left uswitch
      OSSignalMsg(Msg_to_Motor,(OStypeMsgP)&Left_usw); //Send message
//quite possible to land here with neither switch low any more, as interrupt
//will sense switch release
      INTCONbits.RBIF = 0;       //Clear Port B interrupt flag
}
```

Program Example 19.6 Continued

19.7.6 Simulating or running the program

This is a very pleasing program to run if you have a Derbot AGV. With its dependence on the ultrasound sensor and its extensive use of delays, it is less easy to simulate it in a satisfactory way.

19.8 The RTOS overhead

This chapter has aimed to give an introduction to the power of working with an RTOS. It is important to recognise that in making use of Salvo Lite we have only used a limited subset of what Salvo has to offer, or what is available in the wider world of the RTOS.

While appreciating the power of this type of programming, it is important also to recognise the costs. These fall into three categories:

(1) *Financial cost*. Once we move beyond using a free RTOS like Salvo Lite or equivalent, it will be necessary to buy a commercially available RTOS or else to spend time (and hence money) in designing an RTOS from scratch.
(2) *Program size cost*. A program written with an RTOS inevitably occupies more memory space.
(3) *Execution time cost*. With significantly more code to execute, due to the RTOS overheads, the program will run slower.

These costs are similar to those experienced in moving from Assembler to C programming. The programming technique is more powerful, but it comes with a cost. The last two of the above list, of course, determine the actual performance of the program. It is desirable, but not easy, to quantify these. The difficulty lies in the fact that both code size and execution time are dependent on the precise implementation of the program in question. One cannot, for example, simply say that an RTOS-based program occupies twice as much memory as a convention sequential program or runs at half the speed. A practice that is applied sometimes is to use 'benchmark' programs to allow comparisons to be drawn between different programming implementations of the same functional outcome.

The Salvo user manual [Ref. 19.1] has a useful chapter on 'Performance' (Chapter 9). This gives, in some detail, performance characteristics of the Salvo RTOS and its component parts.

As far as this book is concerned, we have a small number of programs that could be used as informal benchmarks. The Derbot 'blind navigation' program appeared in Assembler version as Program Example 8.4 and in C as Program Example 15.3, albeit with a different microcontroller. An interrupt-based 'flashing LEDs' program appeared first in a conventional C version as Program Example 17.3 and then in an RTOS-based version as Program Example 19.3. One simple comparison can be made between each of these pairs by looking at their memory usage, as seen in their **.map** files, described in Section 17.10.3 of Chapter 17. If you do this, you may then like to go on to explore optimising either the C or the RTOS-based versions. There is scope to do this in both cases, with some options being described in both the C compiler manual and the Salvo manual.

Summary

- Salvo is an effective Real Time Operating System for the small embedded environment; it illustrates in a practical way all key RTOS features and is very well suited to the PIC environment.
- Working with an RTOS leads to a new approach to programming, where tasks, priorities and events become the key features.
- There is a cost associated with using the RTOS, including financial, memory space and execution time. These need to be understood and evaluated when deciding whether to use an RTOS.

References

19.1.   Salvo^{TM} – The RTOS that runs in tiny places^{TM}, User Manual, Version 3.2.2. Pumpkin Inc.; http://www.pumpkininc.com

19.2.   Salvo Compiler Reference Manual – Microchip MPLAB-C18 (2005). Code RM-MCC18. Pumpkin Inc.; http://www.pumpkininc.com

19.3.   Building a Salvo Application with Microchip's MPLAB-C18 C Compiler and MPLAB IDE v. 6 (2004). Pumpkin Inc., Application Note AN-25.2004; http://www.pumpkininc.com

Section 5
Techniques of Connectivity and Networking

This closing chapter surveys techniques of connectivity and networking, an essential field of activity in the current world of embedded systems.

20
Connectivity and networks

Despite all that has been covered in the past 19 chapters, there are still important areas in embedded systems which have been little mentioned. One area is so important that this final chapter is dedicated to it.

If you implemented the hand controller board on the Derbot AGV, and maybe another I^2C device, then you created a little network. This approach is being replicated in every sort of situation, where different systems or subsystems are communicating with each other. In the home, workplace, motor vehicle, factory and across the world, thinking things are organising themselves into networks. The means of communication in these are becoming increasingly diverse: not just electrical, but also optical fibre, infrared or radio.

This chapter deals with issues of connectivity and networking – the medium of connection used to create data links, and the means by which data is actually formatted, moved and interpreted over those links. Essentially, the chapter is a survey of certain communication techniques and technologies. Like the diverse life-forms which inhabit the earth, we will find that network mechanisms are incredibly varied, each adapting itself to the needs of its very particular environment.

In this chapter, you will learn about:

- Some underlying concepts of setting up a network
- Alternatives for connectivity, including the wireless options of infrared and radio
- A range of network protocols
- How PIC® microcontrollers can be applied in these areas.

As each topic is a major field in itself, the chapter does not aim to provide complete solutions. It just gives overviews and ideas for further exploration. There are no design examples.

20.1 The main idea – networking and connectivity

There are many situations where we need to provide connection between different systems or subsystems. In the domestic environment, the automated household is on the verge of becoming a reality. Here different household appliances and gadgets may all be connected together, for example through the Internet. Elsewhere, there are other needs for networking. The modern motor vehicle may contain dozens of embedded systems, all engaged in very specific activity, but all interconnected. In the home or car situation, connections may be long term and stable. However, other networks or connections are transitory, for example when data is downloaded from a personal organiser to a laptop over a wireless link. All of these are of interest to the embedded designer and they pose very diverse challenges.

The traditional means of providing network connectivity has been through cabling, allowing electrical signals to flow from one subsystem to another. The computer I am currently writing at, no longer one of the newest, is festooned with cables, linking mouse, printer, scanner, Internet connection, speakers and all my PIC development tools! This need not be the only way. It is, of course, possible to make a data connection without any physical link. The most common alternatives to a cable connection are light or radio. There are many variations on each. We have long had TV remote controls, communicating by infrared. We have also had radio communication for many years; this has been adapted most effectively for data links within the computer environment.

Providing a network is about much more than just providing connectivity, important though this is. In a complex system it is also essential to deal in depth with how data is formatted and interpreted, how addressing is achieved, and how error correction can be implemented. All of this is pretty much independent of the physical interconnection itself. In order for different nodes to communicate on a network, there must therefore be very clear rules about how they create and interpret messages. We have already seen aspects of this with definitions of standards like I^2C. This set of rules is called a *protocol*, taking the word from its diplomatic and legal origins. Let us explore the concept of the protocol.

20.1.1 A word on protocols

With large networked systems, protocols can become incredibly complex, defining every aspect of the communication link. Some of these aspects are obvious and others less so. To aid in the complex process of defining a protocol, the International Organisation for Standardisation (ISO) devised a 'protocol for protocols', called the *Open Systems Interconnect* (OSI) model. This is shown in Figure 20.1. The OSI model sweeps up from the mundane and physical (defining what type of connectors we use or what voltages are recognised), to the more abstract (defining, for example, how data is encrypted and how error correction can be achieved).

Each layer of the OSI model provides a defined set of services to the layer above, and each therefore depends on the services of the layer below. The lowest three layers depend on the network itself and are sometimes called the *media layers*. The physical layer defines the physical and electrical link, specifying, for example, what sort of connector is used and how the data is represented electrically. The link layer is meant to provide

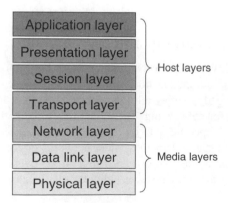

Figure 20.1 The ISO Open Systems Interconnect model

reliable data flow, and includes activities such as error checking and correcting. The network layer places the data within the context of the network and includes activities such as node addressing.

The upper layers of the OSI model are all implemented in software. This takes place on the host computer, and the layers are sometimes called the *host layers*. The software implementation is often called a *protocol stack*. For a given protocol and hardware environment, it can be supplied as a standard software package. A designer adopting a protocol stack may need to interface with it at the bottom end, providing physical interconnect, and at the top end, providing a software interface with the application.

This model forms a framework against which new protocols can be defined and a useful point of reference when studying the various protocols already available. Further information on the ISO OSI model may be found in Ref. 20.1. In practice, any one protocol is unlikely to prescribe for every layer of the OSI model, or it may only follow it in an approximate way.

A number of network protocols and means of connection are outlined in the sections that follow. To implement any of these, a physical and a software system will, of course, be required. We will look at a number of options for these, focusing on available hardware subsystems and available software drivers.

20.2   Infrared connectivity

Infrared (IR) data communication has been with us for many years, becoming widely seen first in applications like TV remote controls. Infrared signals are easy to generate and detect, using low-cost semiconductor devices. Being out of the visible spectrum, it is comparatively easy to apply optical filters which exclude visible light and hence to avoid interference.

The characteristics of all IR links are that data is communicated by a modulated beam of light. The link must therefore be line of sight. It is generally short range and communication is on a one-to-one basis. In this simple characterisation lie a number of interesting advantages and disadvantages. Because it is directional and local, the risk of interference is little and security is good. Because it must be line of sight, however, the ability to engage in wider networks is restricted. Infrared communication can be very low cost and enjoys the advantage of not being regulated by law.

While the early applications were mainly in control, like the TV remote, it was equally evident that IR was good for data communication. This has become a huge growth area for the technology, particularly in situations where a single cable can be replaced, for example in transfers of data from a hand-held device (like a personal organiser or digital camera) to a computer, or between computer and printer.

The Infrared Data Association (IrDA) [Ref. 20.2] is a group of manufacturers who have defined a series of standards for IR links, ranging from simple control to intensive data transfer.

20.2.1   The IrDA and the PIC microcontroller

Infrared communication is a natural area of activity for the small embedded system. Microchip offers several IR encoder/decoder ICs. An example is the MCP2122 [Ref. 20.3], whose pin connections are shown

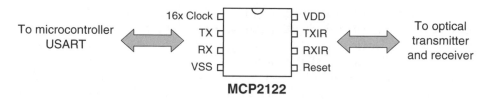

Figure 20.2 The Microchip MCP2122 infrared encoder/decoder

in Figure 20.2. This is intended to interface between a microcontroller on one side, and infrared source and receiver on the other. Thus, the **TXIR** pin can directly drive an IR LED, while the **RXIR** can connect to a sensor. Four interconnections are required with the host microcontroller: **TX**, **RX**, **Reset** and **16x Clock**. The **TX** and **RX** lines connect to the USART of the host microcontroller, just as described in Section 10.10 of Chapter 10. A clock source, running at 16 times the intended baud rate, is also required. This is connected to the **16x Clock** line. It can be generated through the microcontroller CCP module, as described in Ref. 20.4. A **Reset** input allows the host microcontroller to return the IC to its reset condition.

The previous paragraph demonstrates that the physical connection of an infrared port to a PIC microcontroller is simple, and from this stage informal links can be explored between two such nodes. The usual application is to implement data communication under an IrDA standard. The detail of this is beyond the scope of this book, but can be found in a number of Microchip application notes, for example Ref. 20.5.

20.3   Radio connectivity

While IR communication has some clear advantages, its need for line-of-sight communication is in many cases a significant disadvantage. Therefore, radio links are of very great interest. A low-power radio system can have local connectivity and can communicate through walls or other (non-conductive) obstructions. Yet now it is not line of sight, there is a major risk of interference between networks trying to occupy the same space. Imagine a place, say a hotel lobby, full of people with radio-enabled data communication devices. How do we avoid the risk of massive interference between them all? This section surveys a couple of approaches used.

20.3.1   Bluetooth

A major new player in the field of radio data communication is Bluetooth, which faces the challenges of data communication by radio in an interesting way. The development of Bluetooth is controlled by a group of electronics manufacturers, the Bluetooth Special Interest Group [Ref. 20.6]. It operates between 2.402 and 2.480 GHz, a band originally reserved by international agreement for industrial, scientific and medical (ISM) applications, but now also widely used for local wireless data networks.

Bluetooth provides data links between such devices as cell phones, computers, digital cameras or headphones. It has these characteristics:

- A low-power radio link – power is around 1 mW, compared to that of a mobile phone of 3 W
- A typical range of 10 m

- A data rate originally of 1 Mbps and currently (Bluetooth 2.0) of 3 Mbps
- Up to eight devices can be linked simultaneously
- Spread-spectrum frequency hopping is applied, with the transmitter changing frequency in pseudo-random manner 1600 times per second.

When Bluetooth devices detect one another, they determine automatically whether they need to interact with each other, for example through data exchange. This is without any user interaction. Each device has an address, and it is by the address that it determines whether another device that it has detected is of interest to it. Bluetooth systems in contact with each other in this way then form a piconet. Once communication is established, members of the piconet synchronise their frequency hopping, so they remain in contact. A single room could contain several piconets, each containing devices which relate to each other. Each piconet is switching together. For the occasions of momentary clash, there is software that can detect and reject the corrupted data.

There is considerable cleverness in Bluetooth, in the way it can autonomously configure a network and maintain high data rates. This does, however, make it costly and complex for the small or simple system. Therefore, we turn to look at an alternative, *Zigbee*. This carries some of the Bluetooth attributes, but is far simpler.

20.3.2 Zigbee

Zigbee is a very recent standard, managed by members of the Zigbee Alliance [Ref. 20.7]. It gains its inspiration from Bluetooth, but aims to be simpler and cheaper, with a smaller software overhead requirement. It applies the IEEE 802.15.4 Low-Rate Wireless Personal Area Network standard. Like Bluetooth it operates in the ISM bands of the radio spectrum.

Zigbee is particularly appropriate for home automation, and other measurement and control systems, with the ability to use small, cheap microcontrollers. Data rates are low and power consumption minimal

There are three types of Zigbee node:

- Zigbee Coordinator – each network has only one coordinator, which is the most powerful Zigbee type. It can communicate with other networks and store information about its own network.
- Full Function Device (FFD) – this can pass data from other devices, so can take on a routing role.
- Reduced Function Device (RFD) – this is the simplest device. It can communicate with the network, but do nothing else.

As the capability of each device reduces, so does its needs for memory and processing power, and hence its cost.

The minimal power requirement is possible because slave nodes can spend most of their time in Sleep mode. A slave wakes up briefly, just to confirm it is still part of the network.

20.3.3 Zigbee and the PIC microcontroller

Zigbee is an emerging standard and an interesting one to engage with. It is also a natural one to apply with PIC microcontrollers. A possible physical implementation of a Zigbee node is illustrated in Figure 20.3.

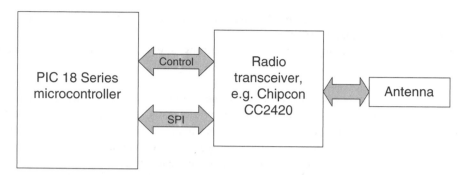

Figure 20.3 Possible PIC-based Zigbee implementation

The link is through a single-chip radio transceiver, such as the Chipcon CC2420 [Ref. 20.8]. A microntroller interfaces to this through an SPI link and certain control lines. Microchip has made available a Stack which can be adapted to apply the protocol. This is described in Ref. 20.9 and allows a Zigbee Coordinator or Reduced Function Device to be implemented.

20.4   Controller Area Network (CAN) and Local Interconnect Network (LIN)

We met in Chapter 10 the three main 'work horses' of serial communication in the embedded environment – SPI, I^{2}C and asynchronous. While these are good standards, they each have their own limitations. In particular, none are fault tolerant. This section looks at two serial standards which are developed for very specific applications, where high reliability is a key requirement.

20.4.1   Controller Area Network (CAN)

The concept of CAN was developed as the demand was growing for data communication in the motor vehicle environment. With its high level of electromagnetic interference, wide temperature and humidity range, this is a hostile environment for any signal and indeed for any electronic device. Moreover, very high reliability is essential. The serial standards developed for the benign environment of home or office were completely inappropriate and a new standard was therefore needed. Initially, CAN was developed by the German company Bosch. They published Version 2.0 of the standard in 1991, and in 1993 it was adopted by the ISO as an international standard, ISO 11898. At the time of writing Version 2.0 can be downloaded from the Bosch website [Ref. 20.10].

The CAN standard addresses only the lower two levels of the ISO/OSI model of Figure 20.1, but takes some fairly revolutionary approaches in so doing. With its very high level of data security, it is inevitably complex and we do not go into every detail here. The main features are listed below:

- Communication is asynchronous, half duplex, with (for a given system) a fixed bit rate. The maximum for this is 1 Mbit/s.

- The configuration is 'peer to peer', i.e. all nodes are viewed as equals. There is, however, a mechanism for prioritisation. Master and slave designation is not used.
- Logic values on the bus are defined as 'dominant' or 'recessive', where dominant overrides recessive. Physical interconnect is not otherwise defined.
- The bus access is flexible. With all nodes being peers, any can start a message. An ingenious arbitration process is applied in the case of simultaneous access, which does not lead to loss of time or data. The arbitration process recognises prioritisation.
- There are an unlimited number of nodes.
- Bus nodes do not have addresses, but apply 'message filtering' to determine whether data on the bus is relevant to them.
- Data is transferred in frames, which have a complex format. This starts with identifier bits, during which arbitration can take place. Eight data bytes are allowed per frame.
- There is an exceptionally high level of data security, with exhaustive error checking. A node that recognises that it is faulty can disconnect itself from the bus.

CAN is now very widely applied in the motor vehicle environment. Figure 20.4 shows the block diagram of just part of a hypothetical car network. Each of the blocks represents a small embedded system – the radio, door, seat and so on. One of these, the door, we met right at the very beginning of the book, in Figure 1.2.

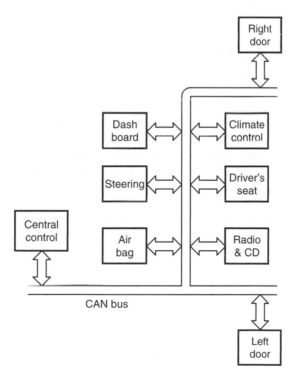

Figure 20.4 Part of car body control network

All subsystems are networked together, by the CAN bus. Another network, not shown, also connects to the Central Control module. It controls the vehicle locomotion – for example, engine, brakes and transmission.

20.4.2   CAN and the PIC microcontroller

Just as we have seen PIC microcontrollers with I^2C or SPI ports, so there are others with on-chip CAN modules. The current version of the Microchip CAN module is called the ECAN^{TM} – the Enhanced CAN module, which distinguishes it from earlier Microchip CAN modules. An example is the PIC 18F2480. Its data sheet [Ref. 20.11] contains details of the ECAN module.

The ECAN module is complex, containing features to buffer data, to format it in the required way and to check for errors. It has numerous control registers. It would be extremely time-consuming to write code for it from scratch. Therefore, Microchip supplies a set of C routines, described in Ref. 20.12, which can immediately be used to build up a program.

Whichever microcontroller is used, it must still be interfaced to the physical bus, meeting the electrical needs of the connection. This is usually done with a special interface IC, of which the Microchip MCP2551 is an example. Its pin connection diagram is shown in Figure 20.5 and data in Ref. 20.13. The CAN bus implementation used is differential, and connects to the **CANH** and **CANL** pins. An external resistor connected at **Rs** controls the slew rate of the data signal, with a slower rate minimising electromagnetic interference. The microcontroller connects from its ECAN module to the **RXD** and **TXD** pins.

20.4.3   Local Interconnect Network (LIN)

While CAN has proved itself as the provider of very high reliability data communication, it is therefore complex and costly. Not all links in the motor vehicle environment actually require the full capability of CAN. Therefore, LIN was developed to work alongside CAN. The standard is managed by the LIN consortium [Ref. 20.14]. First released in 1999, the current LIN version is 2.0.

The LIN bus is intended to be small and slow, communicating mainly with intelligent sensors and actuators. The network topology is fixed. There is a single master and all other nodes are slaves. The master consequently has greater processing power. The slaves need only have very limited processing power or can just be dedicated hardware. This makes them potentially very low cost. An interesting way that cost is minimised is that slaves can use simple RC oscillators, continuously resynchronising themselves as data is exchanged. The maximum data rate is a very modest 20 Kbit/s. The master initiates all data transfers and

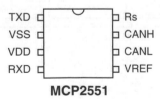

Figure 20.5 The MCP2551 CAN transceiver

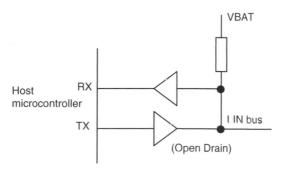

Figure 20.6 The LIN physical interface

only one slave may respond. There is no mechanism to cope with multiple access to the bus. The data link is single wire, with the concept shown in Figure 20.6. Like the CAN bus, logic states are recessive (logic high) and dominant (logic low).

A LIN data frame consists of *header* and *response*. The header is always from the master and data response from a single slave. The speed can be chosen from 1 to 20 Kbaud. The header consists of:

- A *Break* field, which alerts the slaves to a coming message
- A *Sync* byte at the intended data rate, containing the value 55_H, against which slaves calibrate their own clock period
- an *Identifier*, which identifies a slave or slaves and specifies an action to be undertaken.

The response is a message of up to 8 bytes, placed on the bus by just one slave. This is followed by a *Checksum*. In earlier versions of the bus only data bytes were checked; in the most recent the Identifier value is included in the checking process.

There are two bus states, *Sleep* and *Active*. Nodes enter sleep after a time-out period and can be reactivated by the *Wakeup* frame.

Interestingly, the LIN standard also includes software elements. The LIN API (Applications Programmers Interface) provides a set of standard C function calls, which between them implement all LIN functionality. This makes it easy to develop the software and then to test it.

In many environments, a LIN bus system may be designed to interface with a CAN bus system. A possible LIN/CAN system is shown in Figure 20.7. Central to this is a microcontroller, such as the 18F2480, which has both CAN and USART capability. The microcontroller shown acts as the LIN master and as a peer node on the CAN bus.

20.4.4   LIN and the PIC microcontroller

To implement a LIN master or slave in hardware, all that is needed is a microcontroller with USART capability and a bus interface IC. An example is the Microchip MCP201, shown in Figure 20.8.

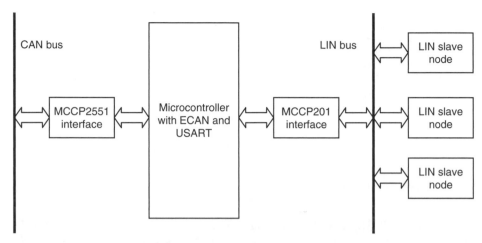

Figure 20.7 A communication system using CAN and LIN buses

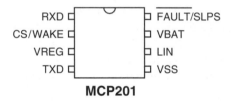

MCP201

Figure 20.8 Example interface chip, the MCP201

This implements the buffering seen in Figure 20.6. Connection is made between the host microcontroller's USART, and the **RXD** and **TXD** pins of the MCP201. Power is supplied to the 201 via the **VBAT** pin. There is an internal voltage regulator on board and the interface chip is able to supply power from its **VREG** pin to the host microcontroller. The **Fault/SLPS** pin is used both to flag a fault, in which case it acts as output, and to set the data slope. The **LIN** line connects to the LIN bus. Full data on this device can be found in Ref. 20.15.

As with most other standards discussed, Microchip has published example firmware as a starting point for developing code. Reference 20.16 provides code for both master and slave, in the (currently) latest 2.0 version of the bus.

20.5 Embedded systems and the Internet

Needless to say, the Internet has transformed communication worldwide over the past decade. The interest of this book is very much with the small embedded system, and the internet is viewed commonly as being something linked to the latest of desktop computers. Do the two have anything in common? The answer, maybe surprisingly, is yes! It *is* possible to link the small embedded system to the Internet and thence to network devices which are under embedded control. Once the link is there, it can be used for

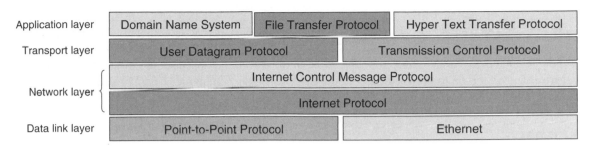

Figure 20.9 Internet protocol stack

many things, to monitor status or to exert control, or even to cause program or data downloads. Examples include the washing machine that can alert the repair man to an impending fault, the vending machine that can tell the Head Office it is empty, the manufacturer who can download a new version of firmware to an installed burglar alarm, or the home owner who can switch on the oven from the office or check that the garage door is closed.

Internet communication makes use of a suite of protocols, usually called collectively after the two most important: TCP/IP (Transmission Control Protocol/Internet Protocol). An example Internet protocol stack is shown in Figure 20.9. The relationship of this to the ISO/OSI model is also shown.

20.5.1 Connecting to the Internet with the PIC microcontroller

The Internet is the most complex of the protocols reviewed in this chapter. Microchip, however, provides good support, in terms of Application Notes, demonstration hardware (the PICDEM.net^{TM} board) and, importantly, a modular TCP/IP stack. This is implemented in C and intended for 18 Series microcontrollers. It can be downloaded from the Microchip website and is described in Ref. 20.17. The great benefit of this is that the user does not need to get involved in the fine detail of the protocol. The stack forms a set of utilities that service the needs of the TCP/IP protocol, forming an equivalent to Figure 20.9. It is responsive to both the user's program, as well as to events on the external connection. The stack occupies around 20 Kbytes of code.

To gain familiarity with networking PIC microcontrollers to the Internet, use of the PICDEM.net board and/or its accompanying documentation is recommended.

20.6 Conclusion

We have covered a remarkable range of ground over the past 19 chapters. Starting from almost nowhere, we have gradually developed a sophisticated picture of the structure of a microcontroller. We have programmed it in both Assembler and C, have interfaced it to a range of sensors and actuators, and linked it with a second microcontroller, itself linked to another microcontroller, thus creating a tiny network. We have gone on to place our programs under the discipline of a real time operating system. We have successfully powered all of this from a modest battery supply. All this represents a tremendous achievement, and if you have followed it all, you have done well. This final chapter has given some possible directions for future activity.

I hope you have enjoyed this voyage of exploration in the world of embedded systems, and wish you much enjoyment as you go on to design, build and program many more 'thinking things' of your own.

Summary

- Modern systems place very great demands on the need to communicate and network. The characteristic of the chosen network will be driven by the application. Considerations of flexibility, reliability, ease of use, data speed, power consumption and cost are all very important.
- Embedded systems often, but not exclusively, have an interest in low-speed, low-data-volume systems. High reliability is essential in a number of situations.
- Interconnection traditionally is electrical. Other techniques are possible and frequently advantageous. These include infrared and radio.
- In some situations embedded systems need to access small and local networks, in other cases major networks, including the Internet.
- Small networks may be dedicated and fixed, like an LIN system, or they may be flexible and possibly transitory, like a Bluetooth or Zigbee system.
- Very high reliability, accompanied by some complexity, is available in the CAN network.
- Microchip supply valuable support to implement a range of networks and interconnections, in the form of microcontrollers with dedicated communication modules, recommended circuits and free published firmware.

References

20.1. Green, D. C. (1995). *Data Communication*, 2nd edn. Longman. ISBN 0-582-24520-6.
20.2. Infrared Data Association (IrDA) website: http://www.irda.org/
20.3. MCP2122 Infrared Encoder/Decoder Data Sheet. (2004). Microchip Technology, Document No. DS21894B.
20.4. Interfacing the MCP2122 to the Host Controller (2004). Microchip Technology, Application Note AN946, Document No. DS00946A.
20.5. Programming the Pocket PC OS for Embedded IR Applications (2004). Microchip Technology, Document No. DS00926A.
20.6. The official Bluetooth website: http://www.bluetooth.com/
20.7. The Zigbee Alliance website: http://www.zigbee.org/
20.8. CC2420 2.4 GHz IEEE 802.15.4/ZigBee-ready RF Transceiver Data Sheet (undated). Chipcon, Document No. SWRS041; http://www.chipcon.com/
20.9. Microchip Stack for the ZigBee[TM] Protocol (2004). Microchip Technology, Application Note AN965, Document No. DS00965A.
20.10. The CAN section of the Bosch website: www.can.bosch.com/
20.11. PIC18F2480/2580/4480/4580 Data Sheet (2003). Microchip Technology, Document No. DS21667D.

20.12. PIC18C ECAN 'C' Routines (2003). Microchip Technology, Application Note AN878, Document No. DS00878A.

20.13. MCP2551 High-Speed CAN Transceiver Data Sheet (2003). Microchip Technology, Document No. DS21667D.

20.14. The LIN Consortium website: http://www.lin-subbus.org/

20.15. MCP201 LIN Transceiver with Voltage Regulator (2003). Microchip Technology, Document No. DS21730E.

20.16. LIN 2.0 Compliant Driver Using the PIC18XXXX Family Microcontrollers (2003). Microchip Technology, AN1009, Document No. DS01009A.

20.17. The Microchip TCP/IP Stack (2002). Microchip Technology, AN833, Document No. DS00833B.

Appendix 1
The PIC® 16 Series instruction set

Table A1.1 PIC 16 Series Instruction Set Summary

Mnemonic, Operands		Description	Cycles	14-Bit Opcode MSb			LSb	Status Affected	Notes
BYTE-ORIENTED FILE REGISTER OPERATIONS									
ADDWF	f, d	Add W and f	1	00	0111	dfff	ffff	C,DC,Z	1,2
ANDWF	f, d	AND W with f	1	00	0101	dfff	ffff	Z	1,2
CLRF	f	Clear f	1	00	0001	1fff	ffff	Z	2
CLRW	-	Clear W	1	00	0001	0xxx	xxxx	Z	
COMF	f, d	Complement f	1	00	1001	dfff	ffff	Z	1,2
DECF	f, d	Decrement f	1	00	0011	dfff	ffff	Z	1,2
DECFSZ	f, d	Decrement f, Skip if 0	1 (2)	00	1011	dfff	ffff		1,2,3
INCF	f, d	Increment f	1	00	1010	dfff	ffff	Z	1,2
INCFSZ	f, d	Increment f, Skip if 0	1 (2)	00	1111	dfff	ffff		1,2,3
IORWF	f, d	Inclusive OR W with f	1	00	0100	dfff	ffff	Z	1,2
MOVF	f, d	Move f	1	00	1000	dfff	ffff	Z	1,2
MOVWF	f	Move W to f	1	00	0000	1fff	ffff		
NOP	-	No Operation	1	00	0000	0xx0	0000		
RLF	f, d	Rotate Left f through Carry	1	00	1101	dfff	ffff	C	1,2
RRF	f, d	Rotate Right f through Carry	1	00	1100	dfff	ffff	C	1,2
SUBWF	f, d	Subtract W from f	1	00	0010	dfff	ffff	C,DC,Z	1,2
SWAPF	f, d	Swap nibbles in f	1	00	1110	dfff	ffff		1,2
XORWF	f, d	Exclusive OR W with f	1	00	0110	dfff	ffff	Z	1,2
BIT-ORIENTED FILE REGISTER OPERATIONS									
BCF	f, b	Bit Clear f	1	01	00bb	bfff	ffff		1,2
BSF	f, b	Bit Set f	1	01	01bb	bfff	ffff		1,2
BTFSC	f, b	Bit Test f, Skip if Clear	1 (2)	01	10bb	bfff	ffff		3
BTFSS	f, b	Bit Test f, Skip if Set	1 (2)	01	11bb	bfff	ffff		3
LITERAL AND CONTROL OPERATIONS									
ADDLW	k	Add literal and W	1	11	111x	kkkk	kkkk	C,DC,Z	
ANDLW	k	AND literal with W	1	11	1001	kkkk	kkkk	Z	
CALL	k	Call subroutine	2	10	0kkk	kkkk	kkkk		
CLRWDT	-	Clear Watchdog Timer	1	00	0000	0110	0100	TO,PD	
GOTO	k	Go to address	2	10	1kkk	kkkk	kkkk		
IORLW	k	Inclusive OR literal with W	1	11	1000	kkkk	kkkk	Z	
MOVLW	k	Move literal to W	1	11	00xx	kkkk	kkkk		
RETFIE	-	Return from interrupt	2	00	0000	0000	1001		
RETLW	k	Return with literal in W	2	11	01xx	kkkk	kkkk		
RETURN	-	Return from Subroutine	2	00	0000	0000	1000		
SLEEP	-	Go into standby mode	1	00	0000	0110	0011	TO,PD	
SUBLW	k	Subtract W from literal	1	11	110x	kkkk	kkkk	C,DC,Z	
XORLW	k	Exclusive OR literal with W	1	11	1010	kkkk	kkkk	Z	

Note 1: When an I/O register is modified as a function of itself (e.g., MOVF PORTB, · 1), the value used will be that value present on the pins themselves. For example, if the data latch is '1' for a pin configured as input and is driven low by an external device, the data will be written back with a '0'.

2: If this instruction is executed on the TMR0 register (and, where applicable, d = 1), the prescaler will be cleared if assigned to the Timer0 Module.

3: If Program Counter (PC) is modified or a conditional test is true, the instruction requires two cycles. The second cycle is executed as a NOP.

Appendix 2
The electronic ping-pong

This small embedded system is a game for two players, each of whom has a push-button 'paddle'. Either player can start the game by pressing his/her paddle. The ball, represented by the row of eight LEDs, then flies through the air to the opposing player, who must press his/her paddle only when the ball is at the end LED and at none other. The ball continues in play until either player violates this rule of play. Once this happens, the non-violating player scores and the associated LED is briefly lit up. When the ball is out of play, an 'out-of-play' LED is lit.

A picture of the ping-pong is shown in Figure 1.3. Its circuit diagram appears in Figure A2.1, with its program listing forming the remainder of this appendix.

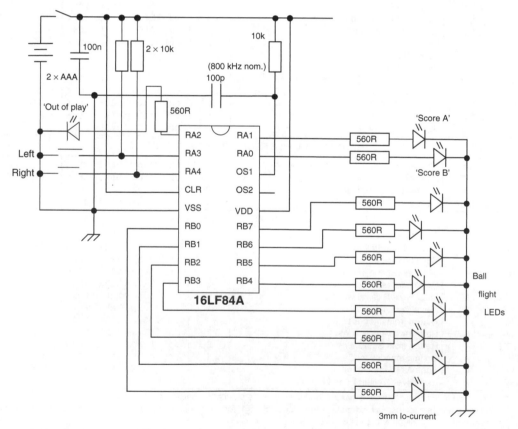

Figure A2.1 The electronic ping-pong circuit diagram

```
;****************************************************************
;ELECTRONIC PING-PONG!
;This program drives the electronic ping-pong game,
;fixed speed, single mode of play.
;TJW 21.6.01
;****************************************************************
;
;Clock freq 800kHz approx (RC osc.)
;Port A 4    right paddle (ip)
;       3    left paddle (ip)
;       2    "out of play" led (op)
;       1    "Score Left" led (op)
;       0    "Score Right" led (op)
;Port B 7-0  "play" leds (all op)
;
;Config Word:      RC oscillator, WDT off,
;            power-up timer on, code protect off
;No Interrupts used
;
        list   p=16F84A
;specify SFRs
status   equ          03
porta    equ          05
trisa    equ          05
portb    equ          06
trisb    equ          06
;
;specify a constant
led_durn equ          20    ;no. of time inner loop is iterated, hence
                                  ;time duration each led is lit.
;
;specify RAM locations
delcntr1 equ  10    ;used in 5ms delay SR
delcntr2 equ  11    ;used in 500ms delay SR
led_posn equ  12    ;holds current ball led posn.
loop_cntr equ 13    ;preloaded with value led_durn for every
                           ;led illumination, and counts down to 0 before
                           ;ball moves on
;
        org    00
        goto   start
;
; **"Initialise" State**
; Initialise
        org    0010
start bsf    status,5 ;select memory bank 1
      movlw  B'00011000'
      movwf  trisa  ;port A according to above pattern
      movlw  00
      movwf  trisb  ;all port B bits op
      bcf    status,5 ;select bank 0
;
```

Program Example A2.1 The electronic ping-pong program

```
;**"Wait"** State
;set up initial led patterns
wait  movlw  04
      movwf  porta   ;switch on "out of play" led
      movlw  00
      movwf  portb   ;all play leds off
;
;check that both paddles are clear before allowing play to commence
      btfss  porta,4 ;right paddle pressed?
      goto   wait    ;yes, so wait
      btfss  porta,3 ;left paddle pressed?
      goto   wait    ;yes, so wait
;
;now ready for action, now wait until paddle pressed
wait1 btfss  porta,4 ;right paddle pressed?
      goto   r_to_l  ;yes, so start play
      btfss  porta,3 ;left paddle pressed?
      goto   l_to_r  ;yes, so start play
      goto   wait1
;
;**"Right-to-Left" State**
;play has started
r_to_l movlw 00      ;switch off "out of play"
      movwf  porta
      movlw  80      ;define ball start posn.
      movwf  led_posn
;loop to here every time led is to change
rtl_0 movlw  led_durn
      movwf  loop_cntr   ;preset length of led illumination
      movf   led_posn,w  ;output new ball posn
      movwf  portb
;loop to here n times for every led, where n = led_durn.
;Check for rule violations. Special conditions apply if
;ball is at start or end.
rtl_1 btfss  led_posn,7  ;is ball at start (ie posn 7)?
      goto   rtl_2  ;no, so move on
      ;yes, it's OK if right paddle still pressed, so don't test
      btfss  porta,3     ;left paddle pressed?
      goto   rt_myscore  ;yes, so score
      goto   rtlend
rtl_2 btfss  led_posn,0  ;is ball at end (ie posn 0)?
      goto   rtl_3  ;no, so move on
;here if ball at end, left paddle can force direction change
      btfss  porta,3     ;left paddle pressed?
      goto   l_to_r ;yes, so change direction - **state exit**
      btfss  porta,4     ;right paddle pressed?
      goto   rt_yrscore  ;yes, so left scores
      goto   rtlend
;here if neither start nor end posn.
rtl_3 btfss  porta,4     ;right paddle pressed?
      goto   score_left  ; yes, so score
      btfss  porta,3     ;left paddle pressed?
      goto   rt_myscore  ;yes, so score
```

Program Example A2.1 Continued

```
;at then end of each loop call a delay
rtlend call   delay5
      decfsz loop_cntr   ;decrement loop counter, check if led is to move
      goto   rtl1
;here if ball moving on
      bcf    status,0
      rrf    led_posn,1
      btfsc  status,0     ;ball off end?
      goto   rt_myscore   ;yes, right's point
      goto   rt10
;**state exit**
rt_myscore    goto score_right
rt_yrscore    goto score_left
;
;**"Left-to-Right" State**
l_to_r movlw  00           ;switch off "out of play"
      movwf  porta
      movlw  01           ;define ball start posn.
      movwf  led_posn
ltr_0 movlw  led_durn
      movwf  loop_cntr    ;determine length of led illumination
;go round this loop "duration" times, for every ball position
ltr_1 movf   led_posn,w ;output new ball posn
      movwf  portb
      btfss  led_posn,0 ;is ball at start (ie posn 0)?
      goto   ltr_2        ;no, so move on
      ;yes, OK if left paddle still pressed (so only test rt paddle)
      btfss  porta,4      ;right paddle pressed?
      goto   lft_myscore  ;yes, so score
      goto   ltrend
ltr_2 btfss  led_posn,7 ;is ball at end (ie posn 7)?
      goto   ltr_3        ;no, so move on
;here if ball at end, right paddle will change dirn, score right if left paddle
      btfss  porta,4      ;right paddle pressed?
      goto   r_to_l       ;yes, so change direction
      btfss  porta,3      ;left paddle pressed?
      goto   lft_yrscore ;yes, so right score
      goto   ltrend
;here if neither start nor end posn.
ltr_3 btfss porta,4       ;right paddle pressed?
      goto   lft_myscore ;yes, so score
      btfss  porta,3      ;left paddle pressed?
      goto   lft_yrscore ;yes, so score
ltrend call   delay5
      decfsz loop_cntr   ;decrement loop counter, check if led is to move
      goto   ltr_1
;here if ball moving on
      bcf    status,0     ;Clear Carry, as rlf rotates through it
      rlf    led_posn,1
      btfsc  status,0   ;ball off end?
      goto   lft_myscore ;yes, left's point
      goto   ltr_0
```

Program Example A2.1 Continued

```
;**state exit**
lft_myscore goto score_left
lft_yrscore goto score_right
;
;**"Score" State**
;here if Left has scored
score_left
      movlw  00
      movwf  portb  ;all play leds off
      bsf    porta,1
      call   delay500
      bcf    porta,1
      goto   wait
;here if Right has scored
score_right
      movlw  00
      movwf  portb  ;all play leds off
      bsf    porta,0
      call   delay500
      bcf    porta,0
      goto   wait
;
;************************************************
;SUBROUTINES
;************************************************
;Delay of 5ms approx. Instruction cycle time is 5us.
delay5  movlw D'200' ;200 cycles called,
                               ;each taking 5x5=25us
      movwf  delcntr1
del1 nop                    ;5 inst cycles in this loop
      nop
      decfsz delcntr1,1
      goto   del1
      return
;
; Delay of 500ms (approx) - 100 calls to delay5
delay500 movlw D'100'
      movwf  delcntr2
del2 call    delay5
      decfsz delcntr2,1
      goto   del2
      return
;
      end
```

Program Example A2.1 Continued

Appendix 3
The Derbot AGV – hardware design details

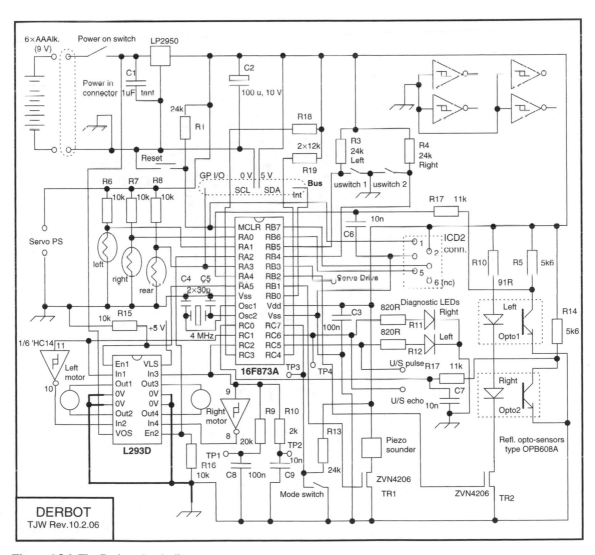

Figure A3.1 The Derbot circuit diagram

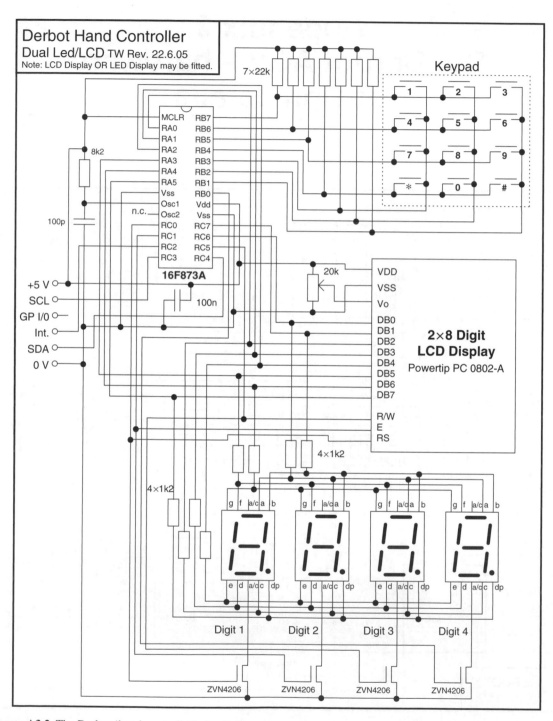

Figure A3.2 The Derbot 'hand controller' circuit diagram

Figure A3.3 A Derbot with hand controller fitted (hand controller connector is disconnected to show mating connector detail)

The Derbot incremental shaft encoder

Chapter 8 describes how a reflective opto-sensor is used to create a simple incremental shaft encoder. Three patterns are shown, with eight, 16 and 32 black/white cycles. All can be used, although the spacing between wheel pattern and sensor must be very carefully controlled for the 32-cycle pattern. The wheel diameter used in all prototypes is 56.0 mm and circumference hence 176.0 mm. The wheel patterns therefore provide forward resolutions of $176.0/8 = 22.0$ mm, $176.0/16 = 11.0$ mm and $176.0/32 = 5.5$ mm respectively. Furthermore, if the wheel is rotating at n r.p.m. and the 16-cycle encoder disc is used, then the shaft encoder frequency is $16n/60$.

Reference data [Ref. 8.8] for the motor used in the Derbot prototype, reproduced in Tables A3.1 and A3.2, indicates that the motors when supplied with 9 V run at 210 r.p.m. This is therefore the expected maximum motor speed for the Derbot. The resulting maximum shaft encoder frequency, using the 16-cycle disc, would

Figure A3.4 Patterns for incremental shaft encoder: eight, 16 and 32 cycles

Table A3.1 Published data for MFA/Como motor/gearbox type 918D30112 – RE280/1 motor characteristics

Operating voltage	No load		At maximum efficiency					Stall torque (g-cm)
	Speed (r.p.m.)	Current (A)	Speed (r.p.m.)	Current (A)	Torque (g-cm)	Output (W)	Efficiency (%)	
12 V	8400	0.1	6300	0.3	25.0	1.62	45	100

Table A3.2 Published data for MFA/Como motor/gearbox type 918D30112 – with 30:1 reduction gearbox: speed variation with supply voltage

6 V	12 V	18 V	24 V
140	280	420	560

be $16 \times 210/60 = 56$ Hz. In practice, the free-running in-circuit motor speed was measured to be 154 r.p.m., when supplied from 9 V. This translates to a maximum shaft encoder frequency of 41 Hz. The fact that the speed is less than the value predicted for a 9 V supply is primarily due to the voltage drops in the L293D drive IC.

Appendix 4
Some basics of Autonomous Guided Vehicles

This appendix covers certain introductory aspects of the design of Autonomous Guided Vehicles, such as are essential for developing the Derbot project.

Locomotion and wheel layout

It is interesting to take note of wheeled vehicles around us before speculating on the optimum arrangement for a small AGV. We note, of course, in passing that wheels are not our only option for locomotion – there are tracked and walking vehicles of different sorts, but these are either too complex or inefficient for our purposes. The most common wheel layout is certainly the motor car, generally with two fixed driven wheels at the back and two steerable wheels at the front. This gives very high stability, attractive for a vehicle that carries humans. It also gives moderate, but not outstanding, manoeuvrability – remember all those difficulties with parking a car in a narrow space.

Of the many wheel configurations possible (see Ref. A4.1), a very popular one has been chosen for the Derbot. The vehicle is supported at three points. Two are independently driven wheels and the third a low-friction roller-ball or slider. The latter could be replaced by a caster, but this tends to influence steering, so is worth avoiding if possible. The centre of gravity is placed just behind the wheels. To the spacing between the centre-line of the wheels we give the variable A.

The advantage of this configuration is its simplicity, combined with its extreme manoeuvrability. A disadvantage is its limited stability; with only three points of support, it is not too difficult to tip it over. A second slider can be placed at the opposite end of the vehicle, in front of the wheels. In this case, if the vehicle tilts forward, its front simply rests on this other slider.

Motor, gearbox and wheel

The Derbot uses small DC motors for its drive. Stepper motors, the alternative, are more power hungry and were therefore not selected. DC motors generally rotate at a speed that is too high for the purposes of the application. A step-down gearbox is therefore used. This is depicted in Figure A4.2.

An AGV having motors running at a steady speed of ω_m rad s^{-1}, with gearbox ratio of N and wheel radius r, will move forward at a steady speed v, given by:

$$v = (\omega_m/N)r \, \text{m s}^{-1}.$$

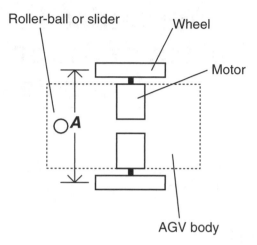

Figure A4.1 The Derbot wheel layout

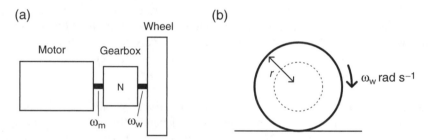

Figure A4.2 Interaction between motor, gearbox and wheel

Turning geometries

The generalised diagram of a vehicle such as the Derbot making a turn, by driving the wheels at different speeds, is shown in Figure A4.3. The wheels are each driven a certain different amount, with the left wheel travelling distance d_L and the right wheel distance d_R. The result is that the vehicle turns through the arc of a circle, centred on point O, and with radius R. It turns an angle θ. As d_L and d_R are the variables that can be controlled as the wheels are driven, it is useful to relate other variables back to them.

As the angle subtended by arcs d_L and d_R are the same, we can say:

$$\frac{d_L}{(R + A/2)} = \frac{d_R}{(R - A/2)} = \theta$$

leading to:

$$d_L \times (R - A/2) = d_R \times (R + A/2)$$

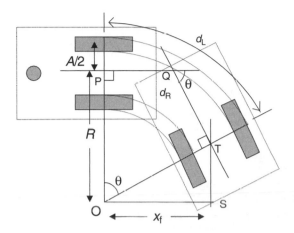

Figure A4.3 Derbot turning through $\theta°$ (final position of slider not shown)

and finally:

$$R = \frac{A}{2} \times \frac{(d_\mathrm{L} + d_\mathrm{R})}{d_\mathrm{L} - d_\mathrm{R}}.$$

Moreover, knowing

$$\theta = \frac{d_\mathrm{L}}{(R + A/2)},$$

we substitute for R into this, to get:

$$\theta = \frac{(d_\mathrm{L} - d_\mathrm{R})}{A}.$$

Also, by inspecting triangle OTS, the distance moved 'forward' from the initial frame of reference, x_f, can be seen to be $R \sin \theta$.

There are two important special cases of this turn, illustrated in Figure A4.4. One is when only one wheel is driven and the AGV rotates approximately round the other wheel. For a turn of 90°, d_L is $\pi A/2$ and d_R is zero. The approximation is due to the asymmetry of the AGV mass about the point of rotation and the drag of the slider, which may cause some slippage. A 90° turn is shown, though any angle can easily be implemented.

The other special case of a turn, shown in Figure A4.4(b), is when both wheels rotate equally in opposite directions. Again, some slippage may occur. The AGV rotates theoretically about a point midway between the wheels, and $d_\mathrm{L} = -d_\mathrm{R}$.

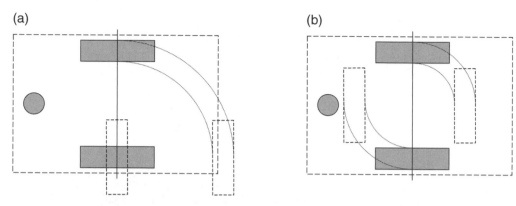

Figure A4.4 Derbot turning through 90°. (a) Pivoting on one wheel. (b) Turning on the spot – wheels turning equally (final positions of chassis outline and slider not shown in either case)

In the current implementation of the Derbot, distance A is measured as 16 cm. For a 180° turn on the spot, $R = 0$ and $d_L = -d_R = \pi A/2 = 25.1$ cm (for the Derbot).

Odometry

Odometry is the technique of measuring distance travelled by a vehicle, generally by measuring wheel revolutions or parts thereof. A shaft encoder, as described in Appendix 3, is able to measure angular displacement of a wheel. By knowing the wheel dimensions it is not difficult to calculate the actual distance moved. It is a simple technique and reasonably accurate. However, it can take no account of wheel slippage, or inaccuracies in wheel dimension or other parts of the measurement chain. Odometry can be used both for measuring forward motion and for implementing controlled turns, such as are described above.

In Appendix 3 the circumference of the Derbot wheel was calculated to be 176.0 mm. With a 16-cycle pattern adopted on the 'home-made' shaft encoder, each resultant pulse represents a forward move of 11 mm. Applying this, forward displacement or controlled turns can be implemented, the latter using the formulae derived above.

Reference

A4.1. Siegwart, R. and Nourbaksh, I. (2004). *Introduction to Autonomous Mobile Robots*. MIT Press, Cambridge, MA. ISBN 0-262-19502-X.

Appendix 5
PIC® 18 Series instruction set (non-extended)

Table A5.1 PIC 18 Series Instruction Set Summary

Mnemonic, Operands		Description	Cycles	16-Bit Instruction Word				Status Affected	Notes
				MSb			LSb		
BYTE-ORIENTED FILE REGISTER OPERATIONS									
ADDWF	f, d, a	Add WREG and f	1	0010	01da0	ffff	ffff	C, DC, Z, OV, N	1, 2
ADDWFC	f, d, a	Add WREG and Carry bit to f	1	0010	0da	ffff	ffff	C, DC, Z, OV, N	1, 2
ANDWF	f, d, a	AND WREG with f	1	0001	01da	ffff	ffff	Z, N	1,2
CLRF	f, a	Clear f	1	0110	101a	ffff	ffff	Z	2
COMF	f, d, a	Complement f	1	0001	11da	ffff	ffff	Z, N	1, 2
CPFSEQ	f, a	Compare f with WREG, skip =	1 (2 or 3)	0110	001a	ffff	ffff	None	4
CPFSGT	f, a	Compare f with WREG, skip >	1 (2 or 3)	0110	010a	ffff	ffff	None	4
CPFSLT	f, a	Compare f with WREG, skip <	1 (2 or 3)	0110	000a	ffff	ffff	None	1, 2
DECF	f, d, a	Decrement f	1	0000	01da	ffff	ffff	C, DC, Z, OV, N	1, 2, 3, 4
DECFSZ	f, d, a	Decrement f, Skip if 0	1 (2 or 3)	0010	11da	ffff	ffff	None	1, 2, 3, 4
DCFSNZ	f, d, a	Decrement f, Skip if Not 0	1 (2 or 3)	0100	11da	ffff	ffff	None	1, 2
INCF	f, d, a	Increment f	1	0010	10da0	ffff	ffff	C, DC, Z, OV, N	1, 2, 3, 4
INCFSZ	f, d, a	Increment f, Skip if 0	1 (2 or 3)	0011	11da	ffff	ffff	None	4
INFSNZ	f, d, a	Increment f, Skip if Not 0	1 (2 or 3)	0100	10da	ffff	ffff	None	1, 2
IORWF	f, d, a	Inclusive OR WREG with f	1	0001	00da	ffff	ffff	Z, N	1, 2
MOVF	f, d, a	Move f	1	0101	00da	ffff	ffff	Z, N	1
MOVFF	f$_s$, f$_d$	Move f$_s$ (source) to 1st word	2	1100	ffff	ffff	ffff	None	
		f$_d$ (destination) 2nd word		1111	ffff	ffff	ffff		
MOVWF	f, a	Move WREG to f	1	0110	111a	ffff	ffff	None	
MULWF	f, a	Multiply WREG with f	1	0000	001a	ffff	ffff	None	
NEGF	f, a	Negate f	1	0110	110a	ffff	ffff	C, DC, Z, OV, N	1, 2
RLCF	f, d, a	Rotate Left f through Carry	1	0011	01da	ffff	ffff	C, Z, N	
RLNCF	f, d, a	Rotate Left f (No Carry)	1	0100	01da	ffff	ffff	Z, N	1, 2
RRCF	f, d, a	Rotate Right f through Carry	1	0011	00da	ffff	ffff	C, Z, N	
RRNCF	f, d, a	Rotate Right f (No Carry)	1	0100	00da	ffff	ffff	Z, N	
SETF	f, a	Set f	1	0110	100a	ffff	ffff	None	
SUBFWB	f, d, a	Subtract f from WREG with borrow	1	0101	01da	ffff	ffff	C, DC, Z, OV, N	1, 2
SUBWF	f, d, a	Subtract WREG from f	1	0101	11da	ffff	ffff	C, DC, Z, OV, N	
SUBWFB	f, d, a	Subtract WREG from f with borrow	1	0101	10da	ffff	ffff	C, DC, Z, OV, N	1, 2
SWAPF	f, d, a	Swap nibbles in f	1	0011	10da	ffff	ffff	None	4
TSTFSZ	f, a	Test f, skip if 0	1 (2 or 3)	0110	011a	ffff	ffff	None	1, 2
XORWF	f, d, a	Exclusive OR WREG with f	1	0001	10da	ffff	ffff	Z, N	
BIT-ORIENTED FILE REGISTER OPERATIONS									
BCF	f, b, a	Bit Clear f	1	1001	bbba	ffff	ffff	None	1, 2
BSF	f, b, a	Bit Set f	1	1000	bbba	ffff	ffff	None	1, 2
BTFSC	f, b, a	Bit Test f, Skip if Clear	1 (2 or 3)	1011	bbba	ffff	ffff	None	3, 4
BTFSS	f, b, a	Bit Test f, Skip if Set	1 (2 or 3)	1010	bbba	ffff	ffff	None	3, 4
BTG	f, d, a	Bit Toggle f	1	0111	bbba	ffff	ffff	None	1, 2

continued

Mnemonic, Operands		Description	Cycles	16-Bit Instruction Word				Status Affected	Notes
				MSb			LSb		
CONTROL OPERATIONS									
BC	n	Branch if Carry	1 (2)	1110	0010	nnnn	nnnn	None	
BN	n	Branch if Negative	1 (2)	1110	0110	nnnn	nnnn	None	
BNC	n	Branch if Not Carry	1 (2)	1110	0011	nnnn	nnnn	None	
BNN	n	Branch if Not Negative	1 (2)	1110	0111	nnnn	nnnn	None	
BNOV	n	Branch if Not Overflow	1 (2)	1110	0101	nnnn	nnnn	None	
BNZ	n	Branch if Not Zero	2	1110	0001	nnnn	nnnn	None	
BOV	n	Branch if Overflow	1 (2)	1110	0100	nnnn	nnnn	None	
BRA	n	Branch Unconditionally	1 (2)	1101	0nnn	nnnn	nnnn	None	
BZ	n	Branch if Zero	1 (2)	1110	0000	nnnn	nnnn	None	
CALL	n, s	Call subroutine1st word	2	1110	110s	kkkk	kkkk		
		2nd word		1111	kkkk	kkkk	kkkk		
CLRWDT	—	Clear Watchdog Timer	1	0000	0000	0000	0100	$\overline{TO}, \overline{PD}$	
DAW	—	Decimal Adjust WREG	1	0000	0000	0000	0111	C	
GOTO	n	Go to address1st word	2	1110	1111	kkkk	kkkk	None	
		2nd word		1111	kkkk	kkkk	kkkk		
NOP	—	No Operation	1	0000	0000	0000	0000	None	
NOP	—	No Operation	1	1111	xxxx	xxxx	xxxx	None	4
POP	—	Pop top of return stack (TOS)	1	0000	0000	0000	0110	None	
PUSH	—	Push top of return stack (TOS)	1	0000	0000	0000	0101	None	
RCALL	n	Relative Call	2	1101	1nnn	nnnn	nnnn	None	
RESET		Software device RESET	1	0000	0000	1111	1111	All	
RETFIE	s	Return from interrupt enable	2	0000	0000	0001	000s	GIE/GIEH, PEIE/GIEL	
RETLW	k	Return with literal in WREG	2	0000	1100	kkkk	kkkk	None	
RETURN	s	Return from Subroutine	2	0000	0000	0001	001s	None	
SLEEP	—	Go into Standby mode	1	0000	0000	0000	0011	$\overline{TO}, \overline{PD}$	

Mnemonic, Operands		Description	Cycles	16-Bit Instruction Word				Status Affected	Notes
				MSb			LSb		
LITERAL OPERATIONS									
ADDLW	k	Add literal and WREG	1	0000	1111	kkkk	kkkk	C, DC, Z, OV, N	
ANDLW	k	AND literal with WREG	1	0000	1011	kkkk	kkkk	Z, N	
IORLW	k	Inclusive OR literal with WREG	1	0000	1001	kkkk	kkkk	Z, N	
LFSR	f, k	Move literal (12-bit) 2nd word	2	1110	1110	00ff	kkkk	None	
		to FSRx 1st word		1111	0000	kkkk	kkkk		
MOVLB	k	Move literal to BSR<3:0>	1	0000	0001	0000	kkkk	None	
MOVLW	k	Move literal to WREG	1	0000	1110	kkkk	kkkk	None	
MULLW	k	Multiply literal with WREG	1	0000	1101	kkkk	kkkk	None	
RETLW	k	Return with literal in WREG	2	0000	1100	kkkk	kkkk	None	
SUBLW	k	Subtract WREG from literal	1	0000	1000	kkkk	kkkk	C, DC, Z, OV, N	
XORLW	k	Exclusive OR literal with WREG	1	0000	1010	kkkk	kkkk	Z, N	
DATA MEMORY ↔ PROGRAM MEMORY OPERATIONS									
TBLRD*		Table Read	2	0000	0000	0000	1000	None	
TBLRD*+		Table Read with post-increment		0000	0000	0000	1001	None	
TBLRD*-		Table Read with post-decrement		0000	0000	0000	1010	None	
TBLRD+*		Table Read with pre-increment		0000	0000	0000	1011	None	
TBLWT*		Table Write	2 (5)	0000	0000	0000	1100	None	
TBLWT*+		Table Write with post-increment		0000	0000	0000	1101	None	
TBLWT*-		Table Write with post-decrement		0000	0000	0000	1110	None	
TBLWT+*		Table Write with pre-increment		0000	0000	0000	1111	None	

Note 1: When a PORT register is modified as a function of itself (e.g., MOVF PORTB, 1, 0), the value used will be that value present on the pins themselves. For example, if the data latch is '1' for a pin configured as input and is driven low by an external device, the data will be written back with a '0'.

2: If this instruction is executed on the TMR0 register (and, where applicable, d = 1), the prescaler will be cleared if assigned.

3: If Program Counter (PC) is modified or a conditional test is true, the instruction requires two cycles. The second cycle is executed as a NOP.

4: Some instructions are 2-word instructions. The second word of these instructions will be executed as a NOP, unless the first word of the instruction retrieves the information embedded in these 16-bits. This ensures that all program memory locations have a valid instruction.

5: If the Table Write starts the write cycle to internal memory, the write will continue until terminated.

Table A5.2 Summary of Opcode Operand Symbols

Symbol	Description (MPLAB Assembler default value underlined)
a	RAM Access bit. **a** = 0: Memory location is in Access RAM **a** = 1: RAM Bank specified by Bank Select Register (BSR)
b	bit number in byte
d	Destination select bit. **d** = 0: result stored in W Register **d** = 1: result stored in file register f (i.e. data memory location)
f	8-bit data memory address
fd	12-bit data memory address, destination address in a data move
fs	12-bit data memory address, source address in a data move
k	Literal value, constant data or label (8, 12 or 20-bit)
n	relative address (2's complement) for relative branch instructions, OR direct address for Call and Return instructions
s	Fast Call/Return mode select bit. **s** = 0: do not update into or from shadow registers **s** = 1: update W, Status and BSR registers into or from shadow registers

Appendix 6
Essentials of C

This appendix provides summary information on key aspects of the C programming language, as a set of tables. Example usage and further explanation of most of these features appears in Chapters 14–19 of this book.

Table A6.1 C keywords associated with data type and structure definition

Word	Summary meaning	Word	Summary meaning
char	A single character, usually 8-bit	**signed**	A qualifier applied to **char** or **int** (default for **char** and **int** is signed)
const	Data that will not be modified	**sizeof**	Returns the size in bytes of a specified item, which may be variable, expression or array
double	A 'double precision' floating-point number	**struct**	Allows definition of a data structure
enum	Defines variables that can only take certain integer values	**typedef**	Creates new name for existing data type
float	A 'single precision' floating-point number	**union**	A memory block shared by two or more variables, of any data type
int	An integer value	**unsigned**	A qualifier applied to **char** or **int** (default for **char** and **int** is signed)
long	An extended integer value; if used alone, integer is implied	**void**	No value or type
short	A short integer value; if used alone, integer is implied	**volatile**	A variable which can be changed by factors other than the program code

Table A6.2 C keywords associated with program flow

Word	Summary meaning	Word	Summary meaning
break	Causes exit from a loop	**for**	Defines a repeated loop – loop is executed as long as condition associated with **for** remains true
case	Identifies options for selection within a **switch** expression	**goto**	Program execution moves to labelled statement
continue	Allows a program to skip to the end of a **for**, **while** or **do** statement	**if**	Starts conditional statement; if condition is true, associated statement is executed
default	Identifies default option in a **switch** expression, if no matches found	**return**	Returns program execution to calling routine, causing also return of any data value specified by function
do	Used with **while** to create loop, in which statement following **do** is repeated as long as **while** condition is true	**switch**	Used with **case** to allow selection of a number of alternatives; **switch** has an associated expression which is tested against a number of **case** options
else	Used with **if**, and precedes alternative statement used if **if** condition is not true	**while**	Defines a repeated loop – loop is executed as long as condition associated with **while** remains true

Table A6.3 C keywords associated with data storage class

Word	Summary meaning	Word	Summary meaning
auto	Variable exists only within block within which it is defined. This is the default class	**register**	Variable to be stored in a CPU register; thus, address operator (&) has no effect
extern	Declares data defined elsewhere	**static**	Declares variable which exists throughout program execution; the location of its declaration affects in what part of the program it can be referenced

Table A6.4 C data types, as implemented by the MPLAB C18 C compiler

Data type	Description	Length (bytes)	Range
char	Character	1	-128 to $+127$
signed char	Character	1	-128 to $+127$
unsigned char	Character	1	0 to 255
int	Integer	2	$-32\,768$ to $+32\,767$
unsigned int	Integer	2	0 to 65 535
short	Integer	2	$-32\,768$ to $+32\,767$
unsigned short	Integer	2	0 to 65 535
short long	Integer	3	$-8\,388\,608$ to $8\,388\,607$
unsigned short long	Integer	3	0 to 16 777 215
long	Integer	4	$-2\,147\,483\,648$ to $2\,147\,483\,647$
unsigned long	Integer	4	0 to 4 294 967 295
float	Floating point	4	From 1.17549×10^{-38} to $6.80565 \times 10^{+38}$
double	Floating point, double precision	4	From 1.17549×10^{-38} to $6.80565 \times 10^{+38}$

Table A6.5 C operators

Prec. and order	Operation	Symbol	Prec. and order	Operation	Symbol
Parentheses and array access operators					
1, L to R	Function calls	( )	1, L to R	Point at member	X–>Y
1, L to R	Subscript	[ ]	1, L to R	Select member	X.Y
Arithmetic operators					
4, L to R	Add	X+Y	3, L to R	Multiply	X*Y
4, L to R	Subtract	X–Y	3, L to R	Divide	X/Y
2, R to L	Unary plus	+X	3, L to R	Modulus	%
2, R to L	Unary minus	–X			
Relational operators					
6, L to R	Greater than	X>Y	6, L to R	Less than or equal to	X<=Y
6, L to R	Greater than or equal to	X>=Y	7, L to R	Equal to	X==Y
6, L to R	Less than	X<Y	7, L to R	Not equal to	X!=Y
Logical operators					
11, L to R	AND (1 if both X and Y are not 0)	X&&Y	2, R to L	NOT (1 if X=0)	!X
12, L to R	OR (1 if either X or Y are not 0)	X||Y			
Bitwise operators					
8, L to R	Bitwise AND	X&Y	2, L to R	Ones complement (bitwise NOT)	~X
10, L to R	Bitwise OR	X|Y	5, L to R	Right shift. X is shifted right Y times	X≫Y
9, L to R	Bitwise XOR	X^Y	5, L to R	Left shift. X is shifted left Y times	X≪Y
Assignment operators					
14, R to L	Assignment	X=Y	14, R to L	Bitwise AND assign	X&=Y
14, R to L	Add assign	X+=Y	14, R to L	Bitwise inclusive OR assign	X|=Y
14, R to L	Subtract assign	X–=Y	14, R to L	Bitwise exclusive OR assign	X^=Y
14, R to L	Multiply assign	X*=Y	14, R to L	Right shift assign	X≫=Y
14, R to L	Divide assign	X/=Y	14, R to L	Left shift assign	X≪=Y
14, R to L	Remainder assign	X%=Y			
Increment and decrement operators					
2, R to L	Preincrement	++X	2, R to L	Postincrement	X++
2, R to L	Predecrement	– –X	2, R to L	Postdecrement	X– –
Conditional operators					
13, R to L	Evaluate *either* X (if Z≠0) *or* Y (if Z=0)	Z?X:Y	15, L to R	Evaluate X first, followed by Y	X,Y
'Data interpretation' operators					
2, R to L	The object or function pointed to by X	*X	2, R to L	The address of X	&X
2, R to L	Cast – the value of X, with (scalar) type specified	*(type)* X	2, R to L	The size of X, in bytes	Sizeof X

Key: Prec. = precedence; L = left; R = right.

Table A6.6 Example preprocessor directives

Directive	Summary description	Directive	Summary description
#if	Used for conditionally compiling code, based on evaluation of associated expression. Must be terminated by **#endif**	**#define**	Defines string constants which are used in source code and are substituted before code line is compiled
#ifdef	Similar to **#if**, but tests if specified symbol has been defined. Terminated by **#endif**	**#error**	Generates user-defined error message
#ifndef	Identical to **#ifdcf**, but tests if specified symbol has not been defined	**#include**	Include at this point the full text from file specified, which may contain unlimited C code, and is then compiled with the source program
#else	Used with **#if** to provide alternative section for compilation	**#line**	Allows user to specify line number within code
#elif	Used within an **#if** section to test a new expression	**#pragma**	Allows further directive-like information to be sent to the preprocessor, generally compiler specific
#endif	Terminates an **#if** section.	**#undef**	Reverses the action of **#define**, on string constant specified

Table A6.7 Some applications of punctuation marks

Symbol	Application	Example
:	Terminates a label	loop:
:	Used in bit-field format (seen also in Table A6.5)	`unsigned RB0:1;`
;	Terminates a statement or declaration	`unsigned RB0:1;`
.	Used to denote member of a structure (seen also in Table A6.5)	`PORTAbits.RA2 = 1;`
\	Next line is continuation of this one	–
{ }	Defines block of code	–

Table A6.8 MPLAB C18 additional storage qualifiers

	rom	ram
far	Variable can be anywhere in program memory	Variable can be anywhere in data memory; bank switching instruction needed to access, **far** and **ram** is the default combination
near	Variable is in program memory, with address <64K	Variable is in Access RAM

Table A6.9 MPLAB C18 memory models

Memory model	Pointer size	Default ROM qualifier	Memory size
small (default)	16 bit	**near**	Program memory $\leq$ 64K
large	24 bit	**far**	Program memory > 64K

Index

12F508, introduction to, 18–20
16F84A, introduction to, 26
16F873A, introduction to, 26
18F242, introduction to, 336
18F2420, introduction to, 383

Access RAM, 343, 347
Acquisition time, *see* Sample and hold
ADC, *see* Analog-to-digital converter
ADC module:
    of 16F873A, 312–319
    of 18F242, 380
Addressing, indirect, 97, 104–106
Agilent, *see* Oscilloscope
AGV, *see* Autonomous Guided Vehicle
ALU, *see* Arithmetic Logic Unit
Analog multiplexer, 308
Analog quantities, 304
Analog-to-digital converter (ADC):
    principles of, 306–308
    resolution, 307, 324
    signal conditioning for, 308
    see also Successive Approximation ADC, ADC
        module, *and* Voltage reference
Architecture:
    of 12F508, 18–20
    of 16F84A, 27
    of 16F873A, 146–149
    of 18F242, 337–340
Arithmetic Logic Unit (ALU), 9
    of PIC 16 Series, 69
Assembler programming:
    format, 71
    High Level Language comparison, 66
    principles, 66–69
    simplifying, 106
    see also MPASM
Asynchronous, *see* serial communication

Atmel:
    AT89C2051, 40
        interrupts, 140
Autonomous Guided Vehicle (AGV), 537–540
    see also Derbot AGV

Background Debug Mode (BDM), 170
Banked addressing, 33
BCD, *see* Binary Coded Decimal
BDM, *see* Background Debug Mode
Binary:
    binary to BCD conversion, 323
    fractional numbers, 322
    see also Fixed Point
Binary Coded Decimal (BCD), 323
    see also Binary
Bit banging, *see* Serial communcation
Bluetooth, 516
Breakpoints, 109
Brown-out, *see* Power supply

C programming language:
    arrays, 431–434
    bit-fields, 459
    bit testing and setting, 411
    break keyword, 415
    compiler, 394
    declarations, 388
    decrement operator, 439
    do keyword, 403
    duration, 454
    else keyword, 411, 429
    essential elements, 544–548
    for keyword, 416
    function prototype, 413
    functions, 391, 413–415
    goto keyword, 403
    header files, 394, 460–462
    if keyword, 411, 429
    increment operator, 439

C programming language (*continued*)
  introduction to, 387
  keywords, 390
  libraries, 394
  linkage, 454
  linker, 395
  main keyword, 391
  object file, 395
  operators, 392
  pointers, 431–434
  preprocessor, 394
  radix specification, 396
  scope, 454
  statements, 388
  storage classes, 453–456
  strings, 431, 433
  structures, 459
  translation unit, 394
  unions, 459
  variables in, 390
  while keyword, 393, 403, 434
C18 C compiler:
  ADC module, use of, 423, 428
  Assembler inserts, 445
  configuration word setting, 447
  data formatting with, 440–442
  delays, implementing, 415
  I^{2}C, use of, 437–439
  interrupts, use of, 448–452
  introduction to, 395
  libraries, 403–406
  linker, 396, 462–465
  memory allocation, 446
  memory models, 455
  pragmas, 447
  PWM, use of, 417–422
  simulating with, 400, 412, 429–431, 435–437, 452, 457–459
  start-up files, 444, 456
  Timers, use of, 417–421, 450
  tutorial, 396–400
CAN, *see* Controller Area Network
Capture/Compare/PWM (CCP) module:
  of 16F873A, 235–237
  of 18F242, 376–378

Carry flag, 102
CCP, *see* Capture/Compare/PWM module
Central Processing Unit (CPU), 9–11
  of 12F508, 18
  of 16F84A, 29
  of 16F873A, 146
  of 18F242, 337
Ceramic, *see* Clock
CISC, *see* Complex Instruction Set Computer
Clock:
  ceramic oscillator, 60
  impact on timing, 37
  of 16F84A, 60
  of 16F873A, 161
  of 18F242, 360–364
  quartz oscillator, 59, 168
  R-C oscillator, 59, 168
Clock tick, 229, 468
CMOS, *see* Complementary Metal Oxide Semiconductor
Commissioning, 165–167
Comparator, 327
  of 16F873A, 329
Compare, *see* Capture/Compare/PWM module
Complementary Metal Oxide Semiconductor (CMOS), 16
Complex Instruction Set Computer (CISC), 9, 86
Computed goto:
  with PIC 16 series, 99
  with PIC 18 series, 349
Computer, 8–11
Configuration word/register:
  of 16F84A, 35
  of 16F873A, 154, 179
  of 18F242, 350
Connectivity, *see* Networks
Controller Area Network (CAN), 518–520, 522
  PIC applications, 520
  see also Local Interconnect Network
Counting:
  Counter, 131–134
  object or event counting, 136
  see also Timer

DAC, *see* Digital to analog conversion
Data acquisition system, 305
   in microcontroller, 311
   see also Analog-to-digital converter
Data memory, *see* Memory, data
DC Motor, 212
Debugger, 170
   in-circuit debugger (ICD), 170
Delay, *see* Timing
Derbot AGV:
   ADC application, 319–321
   blind navigation program:
     in Assembler, 222–224
     in C, 417–420
   circuit diagram (main), 533
   hand controller, 186–188, 194, 201
     circuit diagram, 534
   initial builds, 172–175
   I^2C application, 286–292
   intermediate builds, 220, 261, 303
   introduction to, 7
   light-seeking program:
     in Assembler, 326, 332
     in C, 424–428
   load switching on, 217
   motor switching on, 220
   odometry example, 228–231
   program, 176–178
   PWM application, 241–244
   shaft encoder, 535
   speed control, 255–257
   speed measurement, 252–255
   with 18 Series microcontroller, 382
Diagnostics, *see* Commissioning
Digital quantities, 304
Digital to analog conversion, *see* Pulse Width
     Modulation
Displays:
   liquid crystal (LCD), 199–202
     design example, 200–202
   seven segment LED, 193
     design example, 194–199
Downloading program, 83
   see also In-Circuit Serial Programming

EEPROM, *see* Electrically Erasable Programmable
     Read Only Memory
Electrically Erasable Programmable Read Only
     Memory (EEPROM), 31
   of 16F84A, 35–37
   of 16F873A, 155
Electronic ping-pong, *see* Ping-pong
Embedded System, 3–8
   definition, 3
   examples of, 4–8
EPROM, *see* Erasable Programmable Read Only
     Memory
Erasable Programmable Read Only Memory
     (EPROM), 31
Extended instruction set, *see* Instruction set

Fault finding, *see* Commissioning
Fibonacci series, 103, 401
File Select Register (FSR):
   of 16F84A, 97
   of 18F242, 347
Filtering (analog), 306
Fixed point, 322
   see also Binary
Flash (Memory), 31
Floating point, 322
Flow diagram, 89
Foreground/background structure, *see* Multi-tasking
Freescale, 20
   interrupt strategy, 140
Frequency measurement, 252
FSR, *see* File Select Register

H Bridge, *see* Switching
Hand controller, *see* Derbot
Harvard Architecture, 11
High Level Language (HLL), 67
Hitachi, 199
Human Interface, 184–187

ICD, *see* Debugger
ICD 2, 171
   applying, 178–180
ICE, *see* In-Circuit Emulator
ICSP, *see* In-Circuit Serial Programming

IDE, *see* Integrated Development Environment
I²C bus, *see* Inter-Integrated Circuit bus
In-Circuit Emulator (ICE), 170
In-Circuit Serial Programming (ICSP):
    of 16F873A, 156–158
Include files, 106
Indirect addressing, *see* File Select Register
Inductive Loads, switching of, *see* Switching
Infrared connectivity, 515
Infrared Data Association (IrDA), 515
    PIC applications, 515
Input conditioning, digital, *see* Interface, digital
Instruction set:
    of PIC 16 series, 70, 74, 87, 527
        arithmetic instructions, 102–104
        branching instructions, 92–94
        logical instructions, 101
        subroutines, use of, 94
    of PIC 18 series, 340–345, 365, 541–543
        extended, 384
Instruction cycle, 37
Integrated Development Environment (IDE), 69
    see also MPLAB
Interface, digital, 208–212
Inter-Integrated Circuit bus:
    principles of, 275–277
    see also Master Synchronous Serial Port
Internet, 522
    PIC applications, 523
Interrupts:
    context saving, 127–130, 161
    critical regions, 130
    introduction to, 121, 124
    latency, 141
    masking, 122, 130
    of 16F84A, 122
        INTCON register, 123
        programming with, 125–131
    of 16F873A, 158–161
        INTCON register, 159
    of 18F242, 353–358
        INTCON register, 357
    on change, 55
    prioritisation in hardware, 353–354
    repetitive, 231

vector, 124
    see also Salvo Real Time Operating System
Interrupt Service Routine (ISR), 124
    see also Interrupts
IrDA, *see* Infrared Data Association
ISO Open System Interconnect, 514
ISR, *see* Interrupt Service Routine

Keypad, 187
    design example, 188–192
Kingbright, 53, 193

Latency, *see* Interrupt
LCD, *see* Displays
LED, *see* Light Emitting Diode
LDR, *see* Light Dependent Resistor
Light Dependent Resistor (LDR), 204
Light Emitting Diode (LED), 53
    see also Displays
Light meter (Derbot configured as), 331
LIN, *see* Local Interconnect Network
Liquid Crystal Display, *see* Displays
Local Interconnect Network (LIN), 520–522
    PIC applications, 521
Logic analyser, 167–169
Look-up table:
    with 16 Series, 98–101
    with 18 Series, 349
Low voltage detect
    of 18F242, 380–382

Machine cycle, 37
Macro, *see* MPASM
Master Synchronous Serial Port (MSSP)
    of 16F873A:
        in I²C mode, 277–281
        in I²C master mode, 283–285
        in I²C slave mode, 281
        in SPI mode, 267–274
    of 18F242, 378–380
    see also Serial communication
Memory:
    non-volatile, 10
    of 16F84A, 32–35
    of 16F873A, 147–150

of 18F242, 345–353
technologies, 29
volatile, 10
Memory, data, 9–11
of 16F84A, 33, 34
of 16F873A, 152
of 18F242, 345–347
Memory, program, 9–11
of 16F84A, 32
of 16F873A, 150–152, 155
of 18F242, 347–349
Microchip Technology Inc.:
PIC 12 Series family, 17
PIC 16 Series family, 25–27
PIC 18 Series family, 336
PIC Microcontrollers, 15–17
Microcontroller:
general features, 12–15
packaging, 14
Microprocessor, 11
Microswitch, 204
Microwire, 266–275
Motor (DC), *see* DC Motor
MPASM:
directives, 72, 248
introduction to, 71
introductory programming with, 73 76
macros, 107
number representation, 73
see also Assembler
MPLAB:
file structure, 77
introduction to, 76
special instructions, 108
tutorial with, 77–81
with PIC 18 series, 364
see also C18 C compiler
MPSIM, introduction to, 81–83
Multiplexer, analog, *see* Analog
multiplexer
Multiplication, 324
Multi-tasking:
deadlines, 465
foreground/background structure, 467
introduction to, 464–466

priorities, 465–468
sequential programming implementation,
467–470
tasks, 465–468
see also Real time, *and* Real Time Operating
System
MSSP, *see* Master Synchronous Serial Port

Nanowatt technology, 383
Networks, introduction to, 513

Odometry, *see* Derbot
One Time Programmable (OTP), 31
Open Drain, 50–52
Operating System, general purpose, 471
see also Real Time Operating System
OPTION register, 134
Opto-isolator, 211
Opto-sensor, reflective, 205
Oscillator (clock), *see* Clock
Oscilloscope, 167–169
Agilent, 168
OTP, *see* One Time Programmable

Parallel input/output port:
electrical characteristics, 49–52
introduction to, 46–49
of 16F84A, 55–58
of 16F873A, 161–165
input characteristics, 207
of 16F874A/877A (Ports D, E),
180–182
of 18F242, 369–371
Parallel slave port, *see* Parallel
input/output port, Ports D, E
Philips Semiconductors, 275
PIC, *see* Microchip Technology Inc.
PICSTART Plus, 84–86
Ping-pong (electronic), 6
circuit diagram, 528
hardware design, 63
program, 112–116, 529–531
simulating, 116–118
Pipelining, 38
Port, *see* Parallel Input/Output Port

Power supply, 61
  brown-out, 161
  of 16F84A, 62
  of 16F873A, 161
  of 18F242, 358
  power-up, 38
Priority/prioritization, *see* Interrupts *and/or*
      Multi-tasking
Program Counter:
  of 16F84A, 32
  of 16F873A, 147, 150
  of 18F242, 349
Program memory, *see* Memory, program
Programming:
  principles, 66
  structure, 89–92
Protocols, 514
Pulse Width Modulation:
  digital to analog converter application,
      249–252
  generating with 16F873A, 239–241
  low pass filtering of, 249
  principles of, 237–239
  software generation of, 244–248
PWM, *see* Pulse Width Modulation

Quartz Crystal *see* Clock

Radio connectivity, 516
Random Access Memory (RAM), 10
  see also Memory, data
Read Only Memory (ROM), 10
  see also Memory, program
Real time:
  introduction to, 466
Real Time Operating System (RTOS):
  Introduction to, 472
  Overhead, 509
  Scheduler/Scheduling, 473–477
  Semaphore, 478
  Tasks, developing, 477
    Task states, 474
  see also Multi-tasking
Reduced Instruction Set Computer
      (RISC), 9, 86

Reference (Voltage), *see* Voltage Reference
Reset, 38
  of 16F84A, 39, 41–43
  of 16F873A, 161
  of 18F242, 358, 360
Reset Vector:
  of 16F84A, 32
  of 16F873A, 150
  of 18F242, 349
RISC, *see* Reduced Instruction Set Computer
ROM, *see* Read Only Memory
RS-232, 293
RTOS, *see* Real Time Operating System

Salvo Real Time Operating System:
  clock tick with, 491–496
  configuration file, 482, 489, 494
  delays with, 496
  interrupts with, 491–496, 507–508
  introduction to, 480–485
    example program, 485–490
  libraries, 482
  messages, 499, 506
  scheduling with, 484
  semaphores, use of, 496
  simulating, 490, 497–499
  tasks with, 485, 488
  see also Real Time Operating System
Sample and hold, 309
  acquisition time, 309
Samsung, 199
Schmitt Trigger, 50
Semaphore, *see* Real Time Operating
      System
Serial communication:
  asynchronous, 264, 293–295
  bit banging, 303
  principles of, 263–266
  protocols, 264
  synchronous, 264, 266, 293
  see also Master Synchronous Serial Port, *and*
      Universal Synchronous Asynchronous
      Receiver Transmitter

Serial Peripheral Interface, 266–275
    see also Master Synchronous
            Serial Port
Servo, 214
SFR, *see* Special Function Registers
Shaft Encoder, 205
Simulation, 81
    graphical, 118
    see also MPSIM, *and* C18 C compiler
Sleep mode:
    of 16F84A, 139
    of 16F873A, 260
    of 18F2420, 383
Special Function Registers:
    of 16F84A, 33, 34
    of 16F873A, 152
    of 18F242, 345–347
SPI, *see* Serial Peripheral Interface
SRAM, *see* Static RAM
Stack:
    action of, 94
    of 16F84A, 32
    of 16F873A, 150
    of 18F242, 352
State diagram, 91
Static RAM (SRAM), 30
Status Register:
    of 16F84A, 29
    of 16F873A, 146
    of 18F242, 340
Stepping (Stepper) Motor, 212
Stopwatch (simulator), 110
Strings (data), *see* C programming
            language
Subroutines, 94
Successive approximation ADC, 306
Superloop, 90, 469
    see also Multi-tasking
Switch (electro-mechanical), 52
    debouncing, 212
    interfacing, 52
    see also Keypad
Switching:
    inductive loads, 217
    reversible (H bridge), 218

simple DC, 215–217
transistors as switches, 215
Synchronous, *see* serial communication

Tables (data), *see* Look-up table
Tape measure (Derbot configured as), 329
Tasks, *see* Multi-tasking
Test, *see* Commissioning
Timer, 131–134
Timer 0 module:
    of 16F84A, 134–138
    of 16F873A, 226
    of 18F242, 371–373
Timer 1 module:
    of 16F873A, 226–228
    of 18F242, 373
Timer 2 module:
    of 16F873A, 232–234
    of 18F242, 373
Timer 3 module:
    of 18F242, 373–375
Timing:
    delays, hardware-generated, 137
    delays, software-generated, 95–97,
            99–101
    time measurement in software,
            258–260
Trace (simulator), 110
Track and hold, *see* Sample and hold
Transistor transistor logic (TTL), 61
Trouble-shooting, *see* Commissioning

Universal Synchronous Asynchronous Receiver
            Transmitter (USART):
    of 16F873A, 295–302
        example program, 300
    of 18F242, 378, 380
TTL, *see* Transistor transistor logic

Ultrasonic sensor, 207
USART, *see* Universal Synchronous
            Asynchronous Receiver Transmitter

Voltage reference:
   of ADC, 308
   of ADC module, 312
   of 16F873A, 329
Voltmeter (Derbot configured as), 331
Von Neumann, 10

Watchdog Timer (WDT):
   of 16F84A, 138
   of 18F242, 376
Weak Pull-up, 55

Zigbee, 517
   PIC applications, 517